AF410822

National Parks and Protected Areas

National Parks and Protected Areas

Editors

Rui Yang
Yue Cao
Steve Carver
Le Yu

MDPI • Basel • Beijing • Wuhan • Barcelona • Belgrade • Manchester • Tokyo • Cluj • Tianjin

Editors
Rui Yang
Tsinghua University
Beijing, China

Yue Cao
Tsinghua University
Beijing, China

Steve Carver
University of Leeds
Leeds, UK

Le Yu
Tsinghua University
Beijing, China

Editorial Office
MDPI
St. Alban-Anlage 66
4052 Basel, Switzerland

This is a reprint of articles from the Special Issue published online in the open access journal *Land* (ISSN 2073-445X) (available at: https://www.mdpi.com/journal/land/special_issues/national_parks_protected_areas).

For citation purposes, cite each article independently as indicated on the article page online and as indicated below:

LastName, A.A.; LastName, B.B.; LastName, C.C. Article Title. *Journal Name* **Year**, *Volume Number*, Page Range.

ISBN 978-3-0365-8312-9 (Hbk)
ISBN 978-3-0365-8313-6 (PDF)

Cover image courtesy of Pei Wang

Contents

Article

New Approaches to Modelling Wilderness Quality in Iceland

Steve Carver [1,*], Sif Konráðsdóttir [2], Snæbjörn Guðmundsson [3], Ben Carver [4] and Oliver Kenyon [4]

1 School of Geography, University of Leeds, Leeds LS2 9JT, UK
2 Attorney-at-Law, ÓFEIG Nature Conservation Society, 107 Reykjavík, Iceland
3 Icelandic Museum of Natural History, 108 Reykjavík, Iceland
4 Wildland Research Limited, Northallerton DL7 8FF, UK
* Correspondence: s.j.carver@leeds.ac.uk; Tel.: +44-(0)113-3433318

Abstract: Much of Europe's remaining wilderness areas are found in Iceland, yet few are formally protected despite ongoing threats from renewable energy exploitation and 4 × 4 usage. Robust and repeatable approaches are required to map wilderness landscape qualities in support of developing policy on designations that meet international standards. We present an approach to mapping wilderness that is based on internationally recognised methods and customised to suit the unique nature of Icelandic landscapes. We use spatially explicit models of wilderness attributes that measure human impact from vehicular access, land use and visible human features rather than relying on proxy measures such as buffer zones. Seventeen wilderness areas are identified across the Central Highlands and surrounding areas, totalling some 28,470 km². These are compared to existing mapping projects. The character of these areas is described using additional spatial data models on openness, ruggedness and accessibility from settlements, together with information on mobile phone coverage and grazing patterns. This is the most detailed mapping of wilderness in Iceland to date and an important step towards the formal definition of boundaries of wilderness areas meeting IUCN Category 1b and Wild Europe Working Definition in Iceland.

Keywords: wilderness quality; wilderness character; Iceland; Central Highlands

Citation: Carver, S.; Konráðsdóttir, S.; Guðmundsson, S.; Carver, B.; Kenyon, O. New Approaches to Modelling Wilderness Quality in Iceland. *Land* **2023**, *12*, 446. https://doi.org/10.3390/land12020446

Academic Editors: Alexander Khoroshev and Kenneth R. Young

Received: 29 November 2022
Revised: 6 January 2023
Accepted: 7 February 2023
Published: 9 February 2023

1. Introduction

Wilderness is an increasingly rare landscape resource characterised by the IUCN as "protected areas that are usually large, unmodified or slightly modified areas, retaining their natural character and influence, without permanent or significant human habitation, which are protected and managed so as to preserve their natural condition" [1] (p. ii). Recent research using global datasets has highlighted alarming rates of loss with estimates ranging from a nearly 10% loss between 1993 and 2009 [2] to 175 km² of wilderness lost per day [3], most of it due to land-take for agriculture and urban expansion [4]. These rapid rates of attrition comprise a principal threat to biodiversity conservation and UN Sustainable Development Goals [5] such that the post-2020 Global Biodiversity Framework of the Convention on Biological Diversity has placed "retaining wilderness areas" as the first of 21 action-oriented targets for 2030 [6].

The European Parliament recognised the importance of protecting Europe's wilderness areas in February 2009 with a subsequent policy paper calling for wilderness to be defined, mapped, and protected at all levels [7]. The resulting guideline document on wilderness within the Natura 2000 protected area network refines the definition of wilderness in Europe as "an area governed by natural processes … composed of native habitats and species, and large enough for the effective ecological functioning of natural processes. It is unmodified or only slightly modified and without intrusive or extractive human activity, settlements, infrastructure, or visual disturbance" [8], p. 10. An EU wilderness register and map published in 2013 highlighted disparities in wilderness protection across Europe. This revealed interesting patterns in remaining wilderness within EU states and partner countries based on the mapping of potential naturalness of vegetation, remoteness from settlements and other human infrastructure and remoteness from roads [9]. This work shows clear altitudinal and latitudinal trends in these data with most of Europe's wildest landscapes being found in high-latitude (Arctic and near-Arctic) and high-altitude (mountainous) areas. Other interesting trends are seen in the level of protection afforded to the mapped wilderness with many large areas, particularly in northern Scandinavia and Iceland, remaining unprotected despite possessing all the attributes of wilderness [10].

Retaining wilderness is one of the stated objectives of the Icelandic Nature Conservation Act No 60/2013 (NCA), Article 3. Article 5 (19) of the Act defines wilderness as "An uninhabited area that is in principle at least 25 km^2 in size or in such a way that one can enjoy solitude and nature there without disturbance from man-made structures or the traffic of motorized vehicles and in principle at least 5 km away from structures and other technical traces, such as power lines, power plants, reservoirs and built roads" [11]. While different from those definitions provided by the EU and IUCN, this is closely linked to the conditions for designating lands as wilderness protected areas given in Article 46, which states to retain the wilderness as "Large areas, in principle untouched by human activities, where nature can evolve independently, may be legally designated as wilderness protected areas" and that "The designation shall aim at protecting the characteristics of the areas, for example to maintain diverse and unique landscape, openness and/or protecting large ecosystems; and to ensure that present and future generations can enjoy solitude and the nature without disturbance from man-made structures or the traffic of motorized vehicles" [11].

These provisions were a novelty when the nature conservation law reform entered into force late 2015, with the preparatory legislative work referring explicitly to IUCN Category 1b for wilderness designation. However, to date, no area within the Central Highlands has been designated as a wilderness protected area despite provision for doing so within the NCA. A more recent legal novelty, entering into force early 2021 and adding Article 73a, together with a temporary provision to the NCA, provides for the mapping of the wilderness areas across Iceland "in line with internationally recognized methodology" [11]. The work presented here was initiated locally and developed by the paper's authors against this legal background.

Iceland is a unique and important case as regards wilderness in Europe and as such is worth careful attention. The work of Kuiters et al. [9] shows that as much as 43% of Europe's top 1% wildest areas fall within Iceland, and as such, Iceland represents a significant resource for nature protection as well as tourism and recreation [12]. While much of this presents as the extensive icecaps of the Vatnajökull, Hofsjökull, etc., large areas of the Central Highlands comprising ice-free hills, mountains, rivers, lakes and expansive gravel plains are also included in the 43% figure. The fact that many of these areas are currently unprotected highlights the need for appropriate and locally specific methods to assist the authorities in identifying variations in wildness across Iceland building on the IUCN, European and Icelandic definitions of wilderness as stated in the text of the 2013 NCA and subsequent amendments. An Iceland-specific approach to modelling wilderness quality that builds on existing recognised methods is therefore required to identify boundaries of wilderness areas for designation and ensure future protection. Such methods are needed to support the planning process through strategic and responsive "what if?" modelling of proposed developments (e.g., renewable energy projects) to reliably predict and illustrate the likely impacts should they go ahead [13].

Different countries and their local cultures often project different understandings of what is meant by "wilderness" and what it means for landscapes and protected areas. In Iceland, óbyggð víðerni (usually shortened to víðerni) is used as a legal term, which literally translated means "uninhabited wilderness". This corresponds broadly to IUCN Category 1b areas. However, in local vernacular, it is usual to use words such as óbyggðir (literally meaning "uninhabited area") and miðhálendi (as a place term referring specifically to the uninhabited areas of the Central Highlands) [14]. Words aside, much of Iceland's interior landscapes may reliably and reasonably be classified as wilderness once away from roads and influences from other human infrastructure and land use.

The landscape of Iceland's interior is unique within Europe, and perhaps the rest of the world. It is characterised by a spectacular mix of glaciers and icecaps, wide flat gravel plains (or 'sandurs'), rolling hills and rugged mountains interspersed with glacier-fed rivers, hot springs, and deep valleys [15]. The overall impression is of a primeval, almost moon-like landscape shaped entirely by the forces of nature. Geologically speaking, Iceland is young (the oldest exposed rocks are approximately 15–16 million years old) with volcanic landforms of lava flows, cinder cones, geothermal areas and active volcanoes as key characteristics along the volcanic rift zone of the Central Highlands [16]. Water, either in the form of snow and ice or huge glacial rivers, lakes, ponds and springs, is also a key element that provides interest and often forms a barrier to movement, thus increasing remoteness. Vegetation is often sparse or non-existent with Arctic/Alpine plant communities and moss carpets dominating, with its low stature creating an open landscape feel across much of the interior. Example landscapes of the Central Highlands are shown in Figure 1.

In this paper, we develop an Iceland-specific approach to modelling wilderness quality as a basis for robust mapping wilderness boundaries; the principal aim being to support the Icelandic government in their designation process in meeting both the objectives of the NCA (2013) and UN Sustainable Development goals. The specific objectives of the paper are to: (a) modify existing and recognised wilderness quality models to create a custom approach suitable for the Icelandic landscape; (b) apply IUCN and European wilderness definitions and criteria to define existing wilderness areas and map their boundaries; and (c) describe the wilderness character of the resulting areas based on additional spatial attributes. We propose a 4Rs approach utilising:

1. Rigorous, spatially explicit models of attributes influencing wilderness quality;
2. Robust measurement of wilderness attributes describing human landscape impacts such as remoteness (time taken to walk from nearest point of mechanised access), visual impact (proportion of the landscape occupied by human features), and land use (affecting perceived naturalness of ecosystems);
3. Repeatable analyses that can achieve the same results each time the model is run enabling accurate predictions of impacts from proposed developments and associated changes in wilderness quality; and
4. Reliable interpretations of wilderness definitions using best available data at high enough resolutions enabling comparability of work at both local and national scales.

Previous mapping work has tended to focus primarily on the size and distance thresholds outlined in the NCA and previous versions of the wilderness definition. While some attempts have been made at visual impact analysis, the resulting maps interpret the more objective part of the definition of wilderness from the NCA using simple buffers to identify areas at least 3 or 5 km away from roads, buildings, and other human infrastructure, and then reselecting those resulting areas that are at least 25 km^2 in size [17]. One exception has been the innovative use of Participatory GIS (PGIS) by Ólafsdóttir and Sæþórsdóttir [18] to compare these areas with crowd-sourced perceptions of wilderness among local people and tourists. Here, an online map is used together with a spray can tool (Map-Me) to allow users to define their own wilderness areas by spraying directly on the map [19].

We suggest that buffer zones and reselections based on the distance and area thresholds alone, as taken from the objective part of the NCA definition, are proxy measures and do not measure actual impacts associated with human infrastructure within the Central Highlands. As such, these fail to capture the core of the wilderness definition as intended by the legislator. The application of such proxy measures needs to be carried out with care, as the results can be misleading. For example, a rough, single-track gravel road can have the same effect as a paved and elevated dual carriageway road, whereas its true impact is dependent on its type (and traffic volume), how visible it is and how long it takes to walk from it into the surrounding landscape. Weighted buffer zones using different buffer widths to account for road type and traffic volume can go some way towards estimating variations in the degree of impact [20] but cannot accurately measure impacts in terms of naturalness, visibility and remoteness. Other uncertainties and differences can be further introduced in deciding which roads to include in the mapping exercise. Ostman et al. [20] exclude all unpaved gravel roads from their maps with the result that the size and extent of wilderness areas within the Central Highlands are greatly over-estimated despite these roads having a similar impact to paved roads, at least in terms of remoteness from motorized access. This is inconsistent with the legal text and interpretation of the NCA definition itself, and furthermore, such a categorisation of roads is not supported by the NCA's reference to IUCN Category 1b criteria.

Rather than rely on proxy measures, we develop an Iceland-specific approach to modelling impacts from human infrastructure and land use on wilderness quality that is based on the actual measurement of these impacts using spatial interpretations of the EU and IUCN wilderness definitions as suggested in the preparatory work of the NCA. Our approach is based on the legal interpretation of reformed Icelandic law in the field of nature conservation and wilderness as described above. Our research builds on existing, internationally recognised methods, as suggested in the latest amendments to the NCA, and applies these to Iceland with regard to the characteristics of the Central Highlands landscape. Existing examples include mapping wildness in Scottish National Parks [21,22] and wild land areas (WLAs) across Scotland by Scottish Natural Heritage [23]; mapping Haute Naturalité, or high naturalness, across France for IUCN France [24]; mapping variations in wilderness characteristics in designated wilderness areas for the US National Park Service [25]; and modelling variations in wilderness quality across China [26]. Adapting and enhancing these approaches enables us to model impacts on wilderness quality with reference to the 4Rs and then apply EU and IUCN wilderness definitions to draw wilderness protected area boundaries and describe their character. The resulting models represent a more rigorous, robust, and reliable representation of actual patterns of wilderness than those achievable using proxy measures and a tool with which the impacts of proposed future developments and planning decisions can be accurately predicted through repeat mapping.

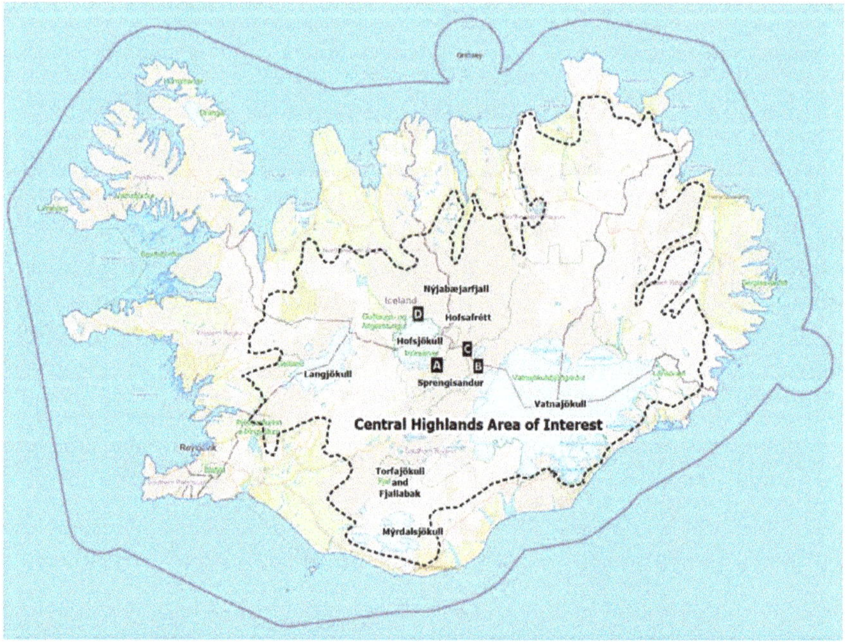

Figure 1. *Cont.*

(**a**) Hofsjökull

(**b**) Vonarskarð

(**c**) Jökuldalur

(**d**) Ingólfsskáli

Figure 1. Map and landscapes of the Central Highlands: (**a**) Hofsjökull, (**b**) Vonarskarð, (**c**) Jökuldalur, (**d**) Ingólfsskáli.

2. Materials and Methods

A simple approach to modelling wilderness quality in Iceland would be to just apply one of the existing methods such as that employed in Scotland [22]. However, the variety seen in surface form and geographical context within the Central Highlands of Iceland creates the need for a two-part model that can firstly model variations in wilderness quality and secondly categorise individual areas depending on their landscape character and those human features affecting public perceptions of wilderness.

The first part of the method is a more traditional "Wilderness Quality Index" (WQI), based on a multi-criteria evaluation (MCE) of three principal attributes: (1) remoteness from mechanised access (or time taken to walk from a motorised vehicle); (2) lack of visual intrusion from modern human artefacts; and (3) perceived naturalness of land cover. When used together, these key attributes can model the spatial variation in wilderness quality, which can then help define wilderness core, buffer and transition zones by careful application of appropriate size and areas thresholds derived from EU and IUCN wilderness definitions. The second part of this model focuses on wilderness character using additional spatial datasets to describe, map and tabulate the unique characteristics of the areas defined in part 1 of the method. This includes further detail from spatial models of openness, ruggedness and accessibility (time taken to drive from human settlements) and additional information provided from maps of mobile phone coverage, livestock grazing and broader landscape character assessments. This two-part method provides detail and nuance in the mapping of key attributes and overall wilderness quality while providing further information about the character of each of the resulting core wilderness areas, thus meeting the need for a reliable, rigorous, robust and repeatable method that can be confidently used to inform decisions about policy on protected areas. This is summarised in Figure 2.

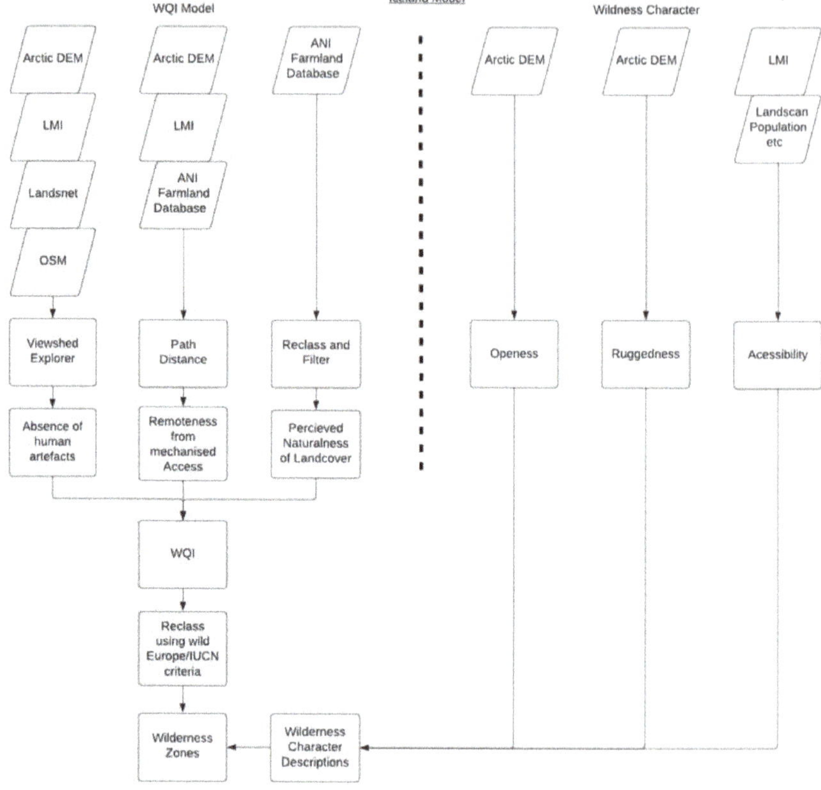

Figure 2. Model flow chart.

2.1. Method Development

Earlier work on wilderness quality mapping by Lesslie and Maslen [27] for the Australian National Wilderness Inventory (ANWI) and adapted by Carver et al. [22] for Scotland's national parks uses four wilderness attributes to create a combined map of wilderness or a WQI. Many wild areas are often characterised by their rugged nature (thus limiting their utility), but this is not always the case, leading to bias in mapped wildness towards mountainous areas or rugged coastlines. For example, in the Scottish wild land mapping, areas such as the low-lying Flow Country in the far northeast of the Scottish mainland are under-represented due to the flat nature of the terrain, despite this landscape being extremely challenging and difficult to cross due to its boggy nature. This is true also for Iceland's Central Highlands, where wilderness areas span a range of landscape types from the many wide open gravel plains such as Sprengisandur and Hofsafrétt, and icecaps including the Vatnajökull and Hofsjökull, while enclosed and rugged valleys are found locally in other areas such as Nýjabæjarfjall in the north and Torfajökull/Fjallabak area in the south (see Figure 1). Variations in topography thus have a marked influence on sense of space and openness as well as impacting on patterns of visual impact and remoteness.

To control for this, the attributes used to map wilderness quality are restricted to remoteness from mechanised access, absence of modern human artefacts, and perceived naturalness of land cover, thereby avoiding possible bias by inclusion of a ruggedness layer in this part of the method. These attributes, together with the data sources and approaches used to map them, are described in Section 2.2 below.

Potential wilderness areas are defined by classifying the resulting WQI into interior core, core, buffer, transition, and non-wild zones using statistical methods. Here, a Jenks

Natural Breaks model is applied as per the Scottish WLA mapping [23]. The size and area thresholds from the Wild Europe Working Definition for wilderness [28] are then applied to these zones to produce a set of wilderness area boundaries meeting the criteria from IUCN Category 1b guidelines and the NCA definition.

These areas are then described using additional information (including openness, ruggedness, accessibility to centres of population, etc.) to create individual maps and tabulate wilderness character descriptions building on the work and experience of the US National Park Service 'Keeping It Wild' wilderness character mapping [25].

2.2. Wilderness Attributes

Three attributes are used to model spatial patterns of wilderness and create a WQI for the Central Highlands area. Justification for their inclusion, data sources, models used, and outputs are described for each attribute below.

2.2.1. Remoteness from Mechanised Access

Remoteness is a key element determining wilderness quality since it affects how a human subject feels being separated from the modern world and our mechanical modes of transportation and also reflects both the effort required to obtain a location by non-mechanical means and personal risk/safety should something go wrong (e.g., injury or bad weather). Remoteness is modelled here using Naismith's Rule [29] as described by Carver et al. [22] for mapping wildness in Scotland's National Parks. Given the varied and challenging nature of the terrain found within the Central Highlands, it is essential to include terrain as a principal variable governing remoteness across the area. A GIS implementation of Naismith's Rule used here incorporates detailed terrain and land cover information to estimate the time in seconds required to walk from the nearest point of mechanised access, be that a paved road or gravel track, taking the effects of distance, relative slope, ground cover and barrier features such as open water, large rivers, crevassed areas of icecaps and very steep ground into account. This assumes remoteness to be directly proportional to the time taken to walk from the nearest road across varied terrain and land cover types. This is performed in ArcGIS Pro 3.0 using the Distance Accumulation tools. The implementation of this model of remoteness requires a detailed terrain model and ancillary data layers that are used to modify walking speeds according to ground cover. The model incorporates barrier features as null values which force a detour to find a safe and suitable crossing point. Datasets used are listed in Table 1.

Table 1. Data sources.

Name	Data	Type	Source	Use
Arctic DEM	10 m digital terrain model	Raster	https://www.pgc.umn.edu/data/arcticdem/ (accessed on 1 February 2021)	Remoteness, viewsheds, openness, ruggedness
LMÍ Landmælingar Íslands	Roads, coastline, buildings, etc.	Vector	https://www.lmi.is/is (accessed on 4 March 2021)	Remoteness, viewsheds, accessibility
Landsnet	Power line routes	Vector	https://www.landsnet.is/ (accessed on 19 June 2021)	Viewsheds
Open Street Map	Roads	Vector	https://www.openstreetmap.org/ (accessed on 1 February 2021) https://download.geofabrik.de/europe/iceland.html (accessed on 1 February 2021)	Remoteness, viewsheds, accessibility
AUI Farmland Database	Land cover	Raster	https://www.moldin.net/nytjaland---aui-farmland-database.html (accessed on 1 February 2021)	Naturalness, remoteness
Landscan	Population	Raster	https://landscan.ornl.gov/ (accessed on 22 August 2021)	Accessibility

2.2.2. Absence of Modern Human Artefacts

This attribute refers to the lack of obvious human constructions within the visible landscape, including roads, vehicle tracks, pylons, dams, reservoirs, buildings and other built structures. A subject's feeling of both naturalness and remoteness is significantly affected by the number of human features that are visible at any location within the area of interest and their distance from them. The choice of which human features to include here is driven largely by what is understood to act as a wilderness detractor [30]. Early work on the effects of human artefacts on wilderness quality has tended to focus on simple distance measures [31], with more recent work using measures of visibility of human artefacts derived from viewshed analyses and digital terrain models [22] to calculate the area from which a given artefact can be seen using line-of-sight from one point of a terrain surface to another [32]. A similar approach to that used by Carver et al. [22,25] is adopted here using artefacts that are deemed to have an impact on wilderness, together with a detailed digital surface model (DSM) and a rapid viewshed assessment method developed for the earlier Cairngorm wildness mapping project [33].

It has been shown that the reliability of viewsheds produced in GIS is strongly dependent on the accuracy of the terrain model used and the inclusion of intervening features (buildings, woodland, etc.) or terrain clutter in the analysis [34]. Modern human artefacts are extracted from appropriate datasets (see Table 1) and assigned appropriate height values reflecting how tall they are and, therefore, how prominent they appear in the landscape. Roads are modelled with a 3 m height value used to represent an average vehicle height. Cumulative viewsheds, weighted according to artefact type and distance, are produced using the Viewshed Explorer tool [32] to show the relative effects associated with the presence and absence of human artefacts, and the results processed in ArcGIS Pro 3.0. Bishop's work [35] on the determination of thresholds of visual impact were used to help define the limits of viewsheds and the distance decay function used.

An inverse square distance function is used in calculating the significance of visible cells in the GIS database. This function gives the relative area in the viewer's field of view that a cell or feature occupies in comparison to the background terrain surface taking distance decay effects and the intervening terrain into account. The output is a unitless grid, the numbers in which are dependent on the area of terrain and input features visible from any point on the terrain surface.

2.2.3. Perceived Naturalness of Land Cover

Perceived naturalness is described here as the extent to which land management, or lack of it, creates a pattern of vegetation and land cover which appears natural to the casual observer. Perceptions of wilderness are in part related to evidence of land management activities such as fencing, improved pasture and stocking rates, as well as presence of natural or near-natural vegetation patterns. Here, the AUI Farmland [36] data were used to describe perceived naturalness in the Central Highlands. Aspects of land management are identifiable from national land cover datasets and enables their reclassification using additional input from local experts (including mountain guides and park rangers) into the naturalness classes shown in Table 2.

To account for the influence that patterns of land cover within the area immediately around the observer location has upon perceived naturalness, the mean naturalness class is calculated for each location within a 250 m radius neighbourhood using the Focal Statistics tool in ArcGIS Pro 3.0. This unitless value is then assigned to the target cell to represent the overall naturalness score for that location.

2.3. WQI and Zone Definition

A simple weighted linear summation MCE model is used to combine all three wilderness attributes into a final WQI. All input attribute layers are normalised onto a common unitless scale that enables cross comparison. This is accomplished by rescaling values onto a 1–256 scale (256 values) using the equal intervals option in ArcGIS Pro 3.0 Slice tool, where low values are indicative of lower wildness. These normalised values are then applied using an equally weighted MCE analysis within the ArcGIS Pro 3.0 Raster Calculator. This allows the effects of each value to be accounted for and a final value for wildness calculated. Weighting of individual attribute layers may then be altered to account for different perceptions on priorities attached to each attribute but are maintained as equal in this exercise assuming each input layer to the model is of equal importance.

Table 2. Naturalness classifications applied to AUI Farmland Data.

Naturalness Class	Land Cover Class (from AUI Farmland Database)
0	No Data
1	Built
2	Cultivated Land/Shrubland
3	Grassland/Unknown (Lowland Vegetated)
4	Rich Heathland/Poor Heathland
5	Mossland/Damp Wetland/Wetland/Poorly Vegetated/Barren/Lakes/Glacier/Unknown

This is a continuous model that ranges from least to most wild, and while useful as an indication of these internal patterns, it needs to be reclassified into zones for it to be useful in a planning and policy context for supporting decisions about protected area boundaries. The WQI is therefore reclassified into Interior Core, Core, Buffer and Transition zones based on a Jenks "Natural Breaks" Classification model. This follows the approached used by SNH in their 2014 Phase 2 map of Wild Land Areas in Scotland [23]. The method examines the distribution of the WQI values across the mapped area and divides these into a specified number of classes such that the difference from the mean within each class is minimised. The classification used here uses 5 classes as per the SNH 2014 methodology, with class 5 being labelled 'Interior Core', class 4 as 'Core', class 3 as 'Buffer', class 2 as 'Transition' and class 1 being 'Not Wild'. The Wild Europe Working Definition for wilderness areas is used to identify 'Core' and 'Core plus Contiguous Buffer' areas larger than 3000 ha (30 km^2) and >10,000 ha (100 km^2), respectively [28]. Jenks class 3 areas not contiguous with 'Core' areas > 3000 ha (together with any class 4 areas < 3000 ha) are classified as 'Buffer' and all class 2 areas as 'Transition' zones. All class 1 areas are classified as 'Not wild'.

2.4. Wilderness Character

The wilderness zones derived using the above classification are further classified according to a range of variables describing their geographical nature and wilderness character. This includes area, elevation range, openness, ruggedness, accessibility, mobile phone coverage, livestock grazing and landscape character classes. Further spatial models are needed to map openness, ruggedness and accessibility to centres of population.

2.4.1. Openness

Openness follows the method developed by Yokoyama et al. [37] as a measure to display surface features on a terrain model using a method independent of a light source and as an alternative to other methods such as hillshading. The method allows for the enclosure of each cell to be represented graphically, thus differentiating between wide open spaces and closely enclosed valleys, assisting in defining the openness characteristics of each identified wildland area. Topographic Openness is calculated from the terrain model using the Skyview tool within the QGIS SAGA toolbox. This generates values representing the proportion (percentage) of visible sky for each cell within the dataset.

2.4.2. Ruggedness

Ruggedness is taken to refer to the physical characteristics of the landscape including effects of steep and rough terrain that is frequently found across the Central Highlands. A terrain model is used to derive indices of terrain complexity based on total slope curvature (rate of change of slope in both plan and profile). Areas where curvature changes frequently are identified because they are deemed to represent rapidly changing terrain and hence ruggedness. A simple index defined as the standard deviation (SD) of total terrain curvature within a 250 m radius of the target location is used to map variations in terrain ruggedness utilizing the Curvature and Focal Statistics tools in ArcGIS Pro 3.0.

2.4.3. Accessibility

While there is a relatively well-developed network of gravel roads across parts of the Central Highlands, with corresponding effects on remoteness from mechanised access as described in Section 3.2, much of Iceland's interior has a remote feel due in part to the time it takes to get there from the main centres of population. This is an essential aspect of the Central Highlands' wilderness character and is modelled here using a population-weighted accessibility surface taking the road network, road type and average speed of driving into account. A combination of a Cost Distance surface calculated using the Distance Accumulation tool in ArcGIS Pro 3.0 and a simple weighted linear summation model in the Raster Calculator is used with centres of population extracted from LandScan global population data. Here we use population density thresholds ($n = 10$) to identify a range of population centres from farmsteads and villages to major towns and the city of Reykjavik. These are used as journey source locations (origins) for the Cost Distance calculations based on average estimated driving speeds according to road type and a background offroad walking speed of 5 km/h. This enables the calculation of isochrone surfaces providing a 'time taken to travel' surface for each of the population density thresholds which are then combined using the Raster Calculator in a linear weighted summation model using the relative population thresholds as weights.

Maps from other existing sources are used to derive wilderness character information pertaining to mobile phone coverage, livestock grazing and landscape character assessments. Mobile phone coverage is remarkably good across much of Iceland, including the Central Highlands. This is an important additional factor influencing wilderness character since it affects the sense of remoteness. The ability to make an emergency call to summon help should it be needed (e.g., in case of personal injury, vehicle breakdown, navigational error, etc.) along with access to digital maps and GPS location has a significant impact on wilderness character, self-reliance, solitude and risk. Livestock grazing is carried out over the summer in parts of the Central Highlands. This includes both sheep and horses, the latter being used principally for recreation. Associated with this grazing activity is fencing, 4x4 tracks and small huts/shelters. As a human economic land use, grazing of animals and associated infrastructure has an influence on wilderness character in the areas where it takes place. Finally, landscape character has been mapped across Iceland and the 27 different landscape type units across 7 categories described in a recent report prepared by EFLA and Land Use Consultants, Scotland [38]. The boundaries of these landscape units and the information contained in the report are used here to supplement the information wilderness character.

3. Results

Results from the analysis and models applied are presented as a series of three normalized and unitless wilderness quality attribute maps. These are combined to create a WQI which is in turn classified into wilderness zones and a series of seventeen separate wilderness areas meeting the criteria for European wilderness areas. Three wilderness character maps are also presented to illustrate how further spatial data models can be used and combined with existing maps to describe the unique characteristics of each of the seventeen wilderness areas.

3.1. Remoteness from Mechanised Access

Remoteness from mechanised access is calculated here using the above methods described in Section 2.2.1 for both summer and winter conditions to account for differences that occur between the two main seasons. During the summer months, vehicles are restricted to established roads, and off-road driving is specifically prohibited. However, during the winter months these rules are relaxed, and except for some restricted areas, vehicles may travel anywhere in Iceland provided there is sufficient snow and ice cover. The difference in relative remoteness between walking (summer or winter) and off-road driving in 4 × 4 "super jeep" vehicles (winter) is very noticeable, with these vehicles being able to cover greater distances in shorter times. This has potentially far-reaching implications for the designation of areas of IUCN Category 1b wilderness, as described later in the paper. Both summer and winter remoteness surfaces are shown in Figure 3.

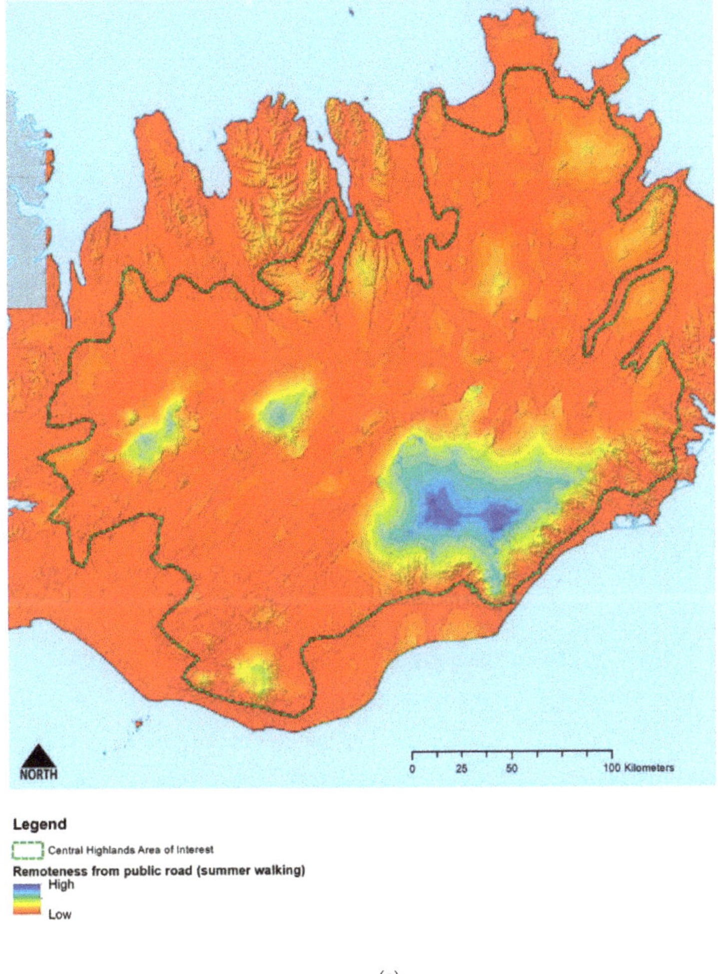

(a)

Figure 3. *Cont.*

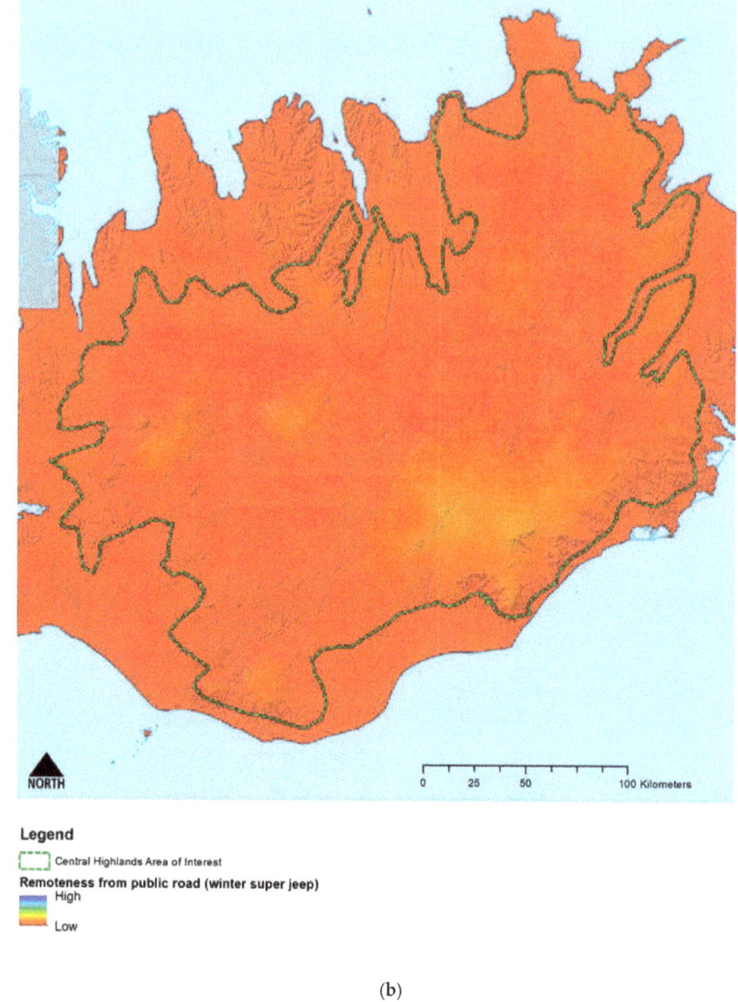

(**b**)

Figure 3. Summer (**a**) and winter (**b**) remoteness surface.

3.2. Absence of Modern Human Artefacts

Absence of modern human artefacts is used to represent the degree of visual intrusion from built structures in the landscape. The model additionally highlights areas which are in total shadow from all visual features owing to the shape of the local landscape. Such areas of zero visual intrusion from modern human artefacts currently comprise a significant portion of the core areas of the Central Highlands, many of which occupy the interior and valleys which are entirely shielded by their topography. While occurring less frequently in the proximity of modified areas, pockets entirely bereft of visual intrusion can be found everywhere, owing to the high relief and general ruggedness of the terrain. The output layer describing the absence of modern human artefacts, including buildings and other structures, roads, hydro-power schemes and power lines, is shown in Figure 4, with areas of zero visual intrusion highlighted in white.

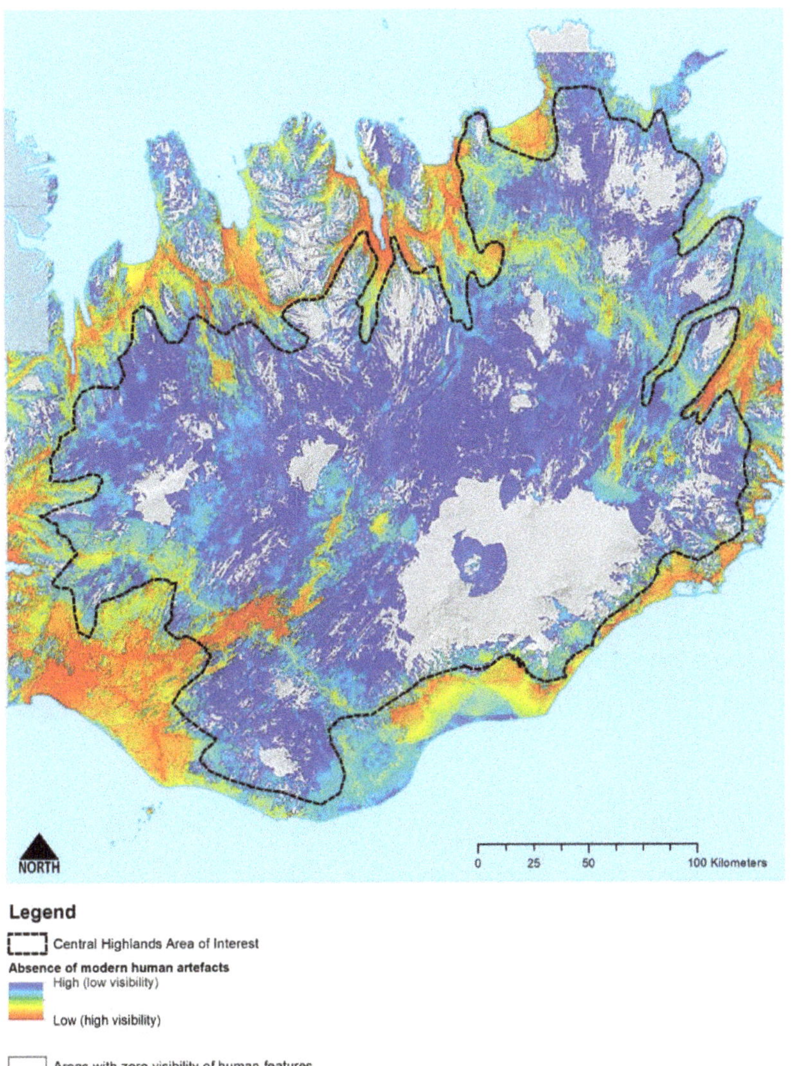

Figure 4. Absence of modern human artefacts.

3.3. Perceived Naturalness of Land Cover

Perceived naturalness of land cover is mapped from the AUI Farmland Database using the methods described in Section 2.2.3. The resulting attribute map is shown in Figure 5. Except for the areas immediately surrounding roads, huts, reservoirs and associated power infrastructure, the vast majority of the Central Highlands presents as the highest category on the naturalness scale. The effects of farming and urban areas around the coast fringe are clearly visible in the lower naturalness scores seen in these regions.

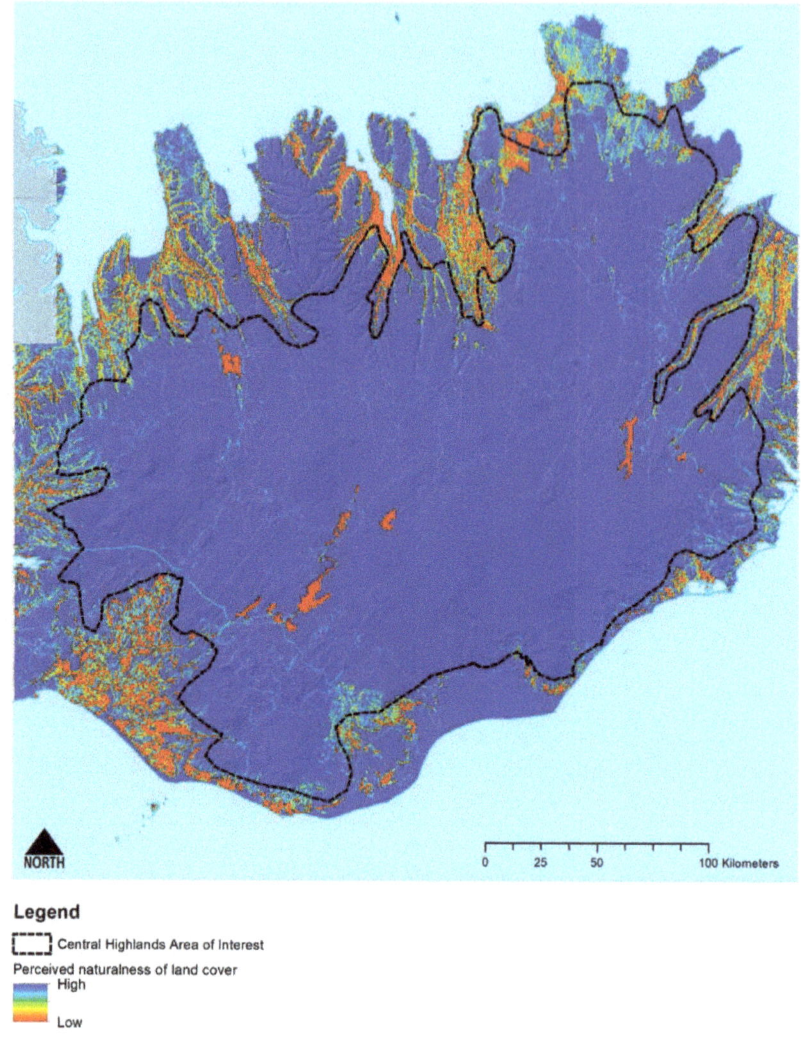

Figure 5. Perceived naturalness of land cover.

3.4. Wilderness Quality Index

The final WQI is shown in Figure 6. This shows the pattern in spatial variations in wilderness quality across the whole of the Central Highlands study area taking the three wilderness attributes of remoteness, visual impact from human features and naturalness of land cover into account. A series of five wilderness zones based on the reclassification of the WQI is shown in Figure 7. Strong spatial patterns influenced by the major icecaps of the Vatnajökull, Hofsjökull, Langjökull and Mýrdalsjökull can be seen as defining the Interior Core wilderness zones and the network of gravel roads, powerlines, hydro-power schemes and other human infrastructure playing a major role in defining the pattern of buffer and transition zones. Hydro-power reservoirs are large unnatural features and so stand out particularly strongly in Figure 6. Roads and power lines emanating from these complete the picture, dissecting the Central Highlands area into a series of large wilderness areas (Core and Interior Core zones) and their surrounding Buffer and Transition zones in Figure 7.

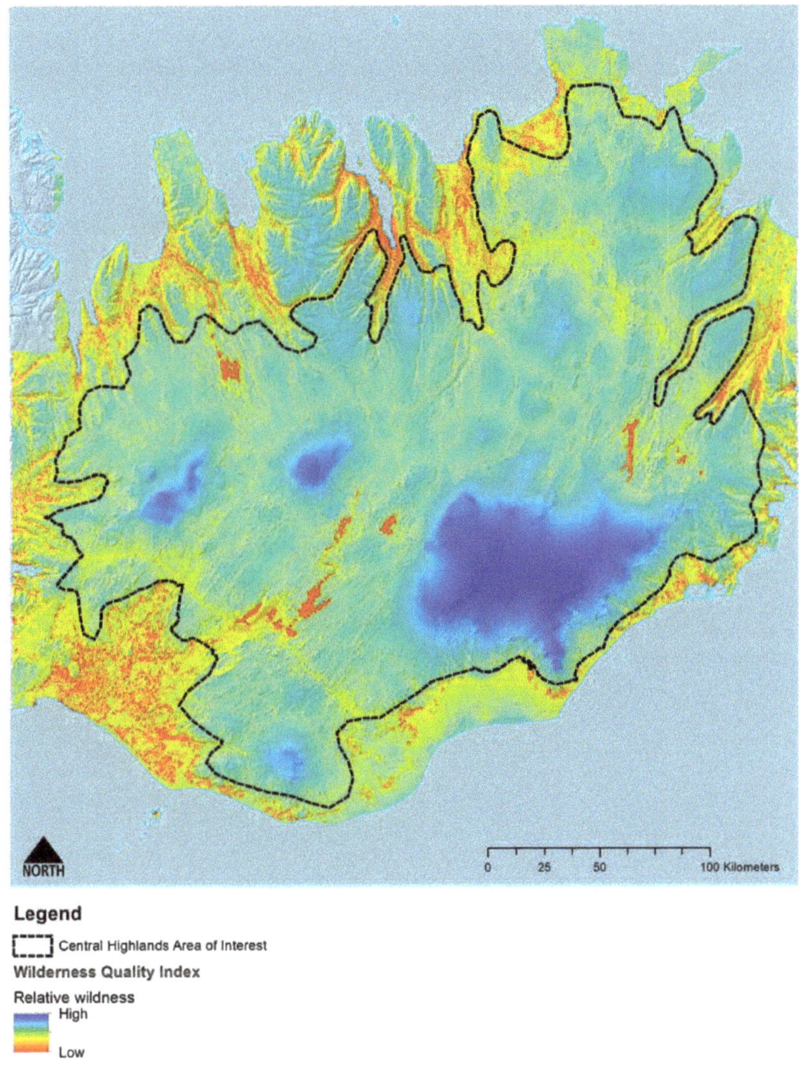

Figure 6. Wilderness Quality Index (WQI) for the Central Highlands.

3.5. Wilderness Area Definition

Applying the size/area constraints from the Wild Europe Working Definition identifies Core wilderness zones as Interior Core and Core areas (Jenks classes 4 and 5) larger than 3000 ha (30 km^2) together with contiguous buffer zones (Jenks class 3) larger than 10,000 ha (100 km^2) as wilderness. These are shown in Figure 8 together with core areas less than the required 3000 ha and transition zone (Jenks class 2) as possible IUCN Category 2 areas. This results in the delineation of seventeen wilderness areas across the Central Highlands and adjacent landscapes. Of these, fourteen lie inside the Central Highlands and three outside, totaling some 28,470 km^2, of which 26,404 km^2 is inside and 2066 km^2 is outside the area of interest. Together, these cover over 47 percent of the Central Highlands area of interest (55,400 km^2), plus three in adjacent areas, of which 19,500 km^2 is public land and 8970 km^2 privately owned. Also shown on this map are the existing protected areas. These include the internationally important Vatnajökull National Park, the Mývatn-Laxá and Þjósárver Ramsar Sites and the Þjórsáver and Fjallabak Nature Reserves, but crucially in respect to the work and results presented here, there are no extant designated wilderness areas. While these wilderness areas are geographically distinct, some are divided and fragmented by narrow corridors created by gravel roads, further illustrating the significance of mechanised access on remoteness and visual impact.

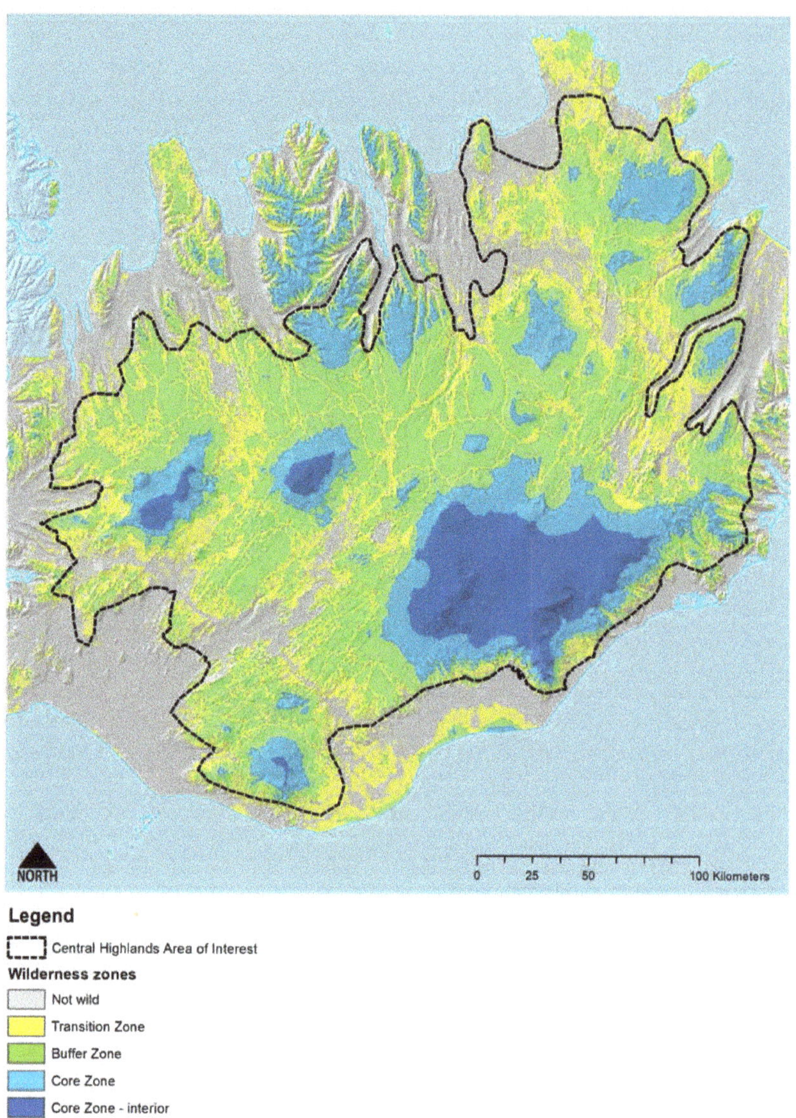

Figure 7. Wilderness zones in the Central Highlands.

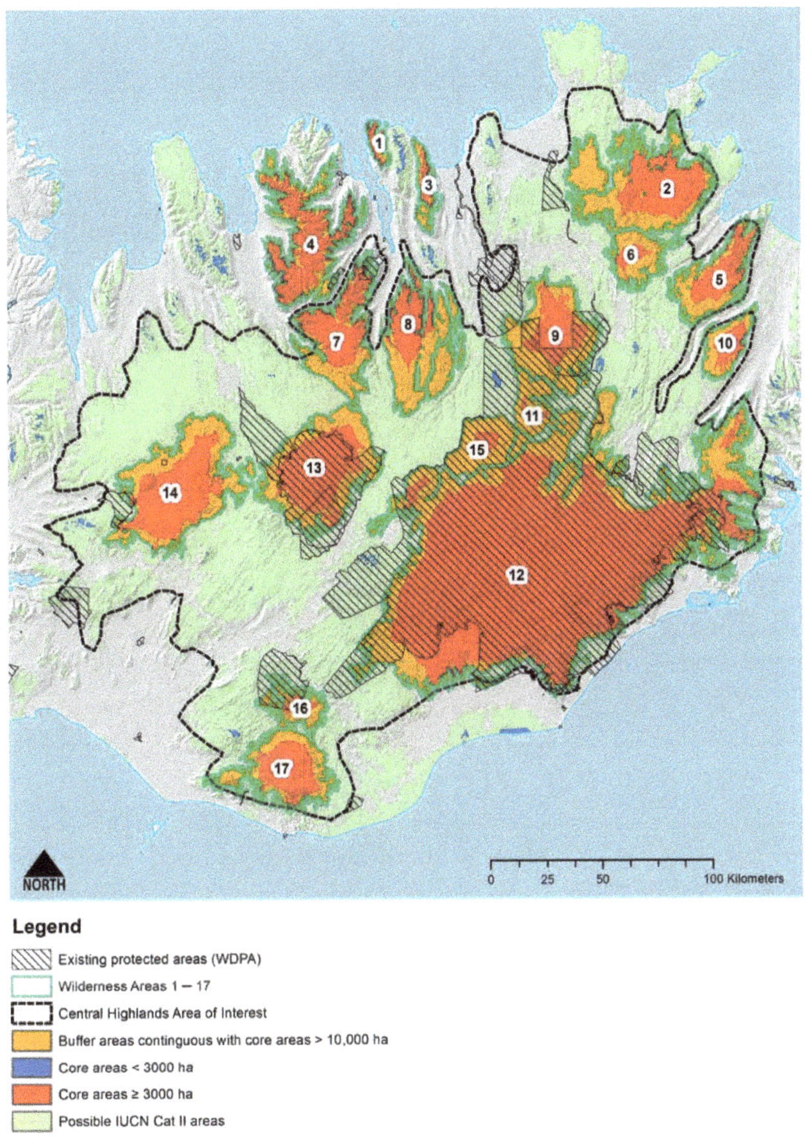

Figure 8. Wilderness areas 1–17 meeting Wild Europe Working Definition.

3.6. Wilderness Character

The wilderness areas shown in Figure 8 are further classified according to a range of variables describing their geographical nature and wilderness character, including the modelled and normalized variables for openness, ruggedness and accessibility as shown in Figures 9–11. Table 3 summarises each of the seventeen wilderness areas by their geographical characteristics. The character of each wilderness area is described in further detail. Area 12 Vatnajökulssvæðið is provided here as an example (see Figure 12).

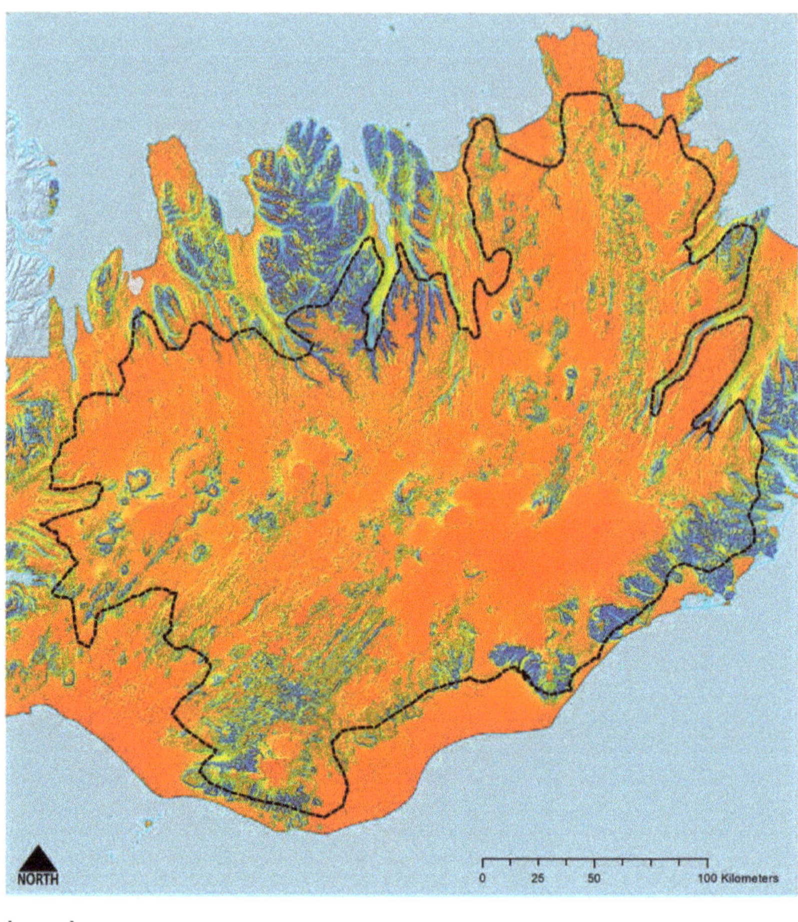

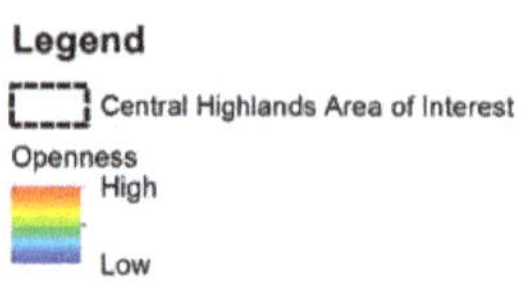

Legend

Central Highlands Area of Interest

Openness
High

Low

Figure 9. Obtained values for Openness for the Central Highlands.

Table 3. Wilderness character summary figures.

No [1]	Name	Area (km^2)	Altitude (m)	Openness (Mean, %)	Ruggedness [2] (Mean)	Accessibility [3] (Mean)
1	Keflavík og Látraströnd	124	17–1168	88	1.54	22,180
2	Heljardalsfjöll	2083	30–983	97	0.40	30,213
3	Náttfaravíkur og Kinnarfjöll	237	9–1214	91	1.11	20,507
4	Tröllaskagi	1478	34–1440	89	1.33	18,167
5	Smjörfjöll	870	109–1255	96	0.53	29,108
6	Dimmifjallgarður	511	351–1037	96	0.52	25,968
7	Nýjabæjarfjall	1198	189–1541	93	0.93	19,060
8	Bleiksmýrardalur	1402	130–1254	96	0.62	20,225
9	Ódáðahraun	1379	382–1678	98	0.44	29,226
10	Fljótsdalsheiði	413	297–710	99	0.25	29,548
11	Askja í Dyngjufjöllum	380	523–1517	96	0.60	29,530
12	Ríki Vatnajökuls	12,315	4–2108	97	0.53	30,002

Table 3. *Cont.*

No [1]	Name	Area (km^2)	Altitude (m)	Openness (Mean, %)	Ruggedness [2] (Mean)	Accessibility [3] (Mean)
13	Hofsjökull og Þjórsárver	1907	554–1789	98	0.35	18,796
14	Langjökull	2095	294–1670	97	0.45	14,472
15	Trölladyngja	546	750–1465	98	0.38	25,674
16	Fjallabak	408	67–1383	93	1.26	14,115
17	Mýrdalsjökull og Eyjafjallajökull	1124	56–1637	95	0.87	13,426

[1] Number code for each of the seventeen wilderness area corresponding to the numbers and locations shown in Figure 8. [2] Ruggedness is a unitless number calculated as standard deviation of slope curvatures (rate of change of slope) within a 250 m radius. Higher numbers indicate greater ruggedness. [3] Accessibility is a unitless number calculated as a population and distance weighted surface taking typical road class driving speeds into account. Lower numbers indicate an area closer to more populated areas, such as Reyjavik and Akureyri (with shorter driving times), and higher numbers indicate those further away (with longer driving times).

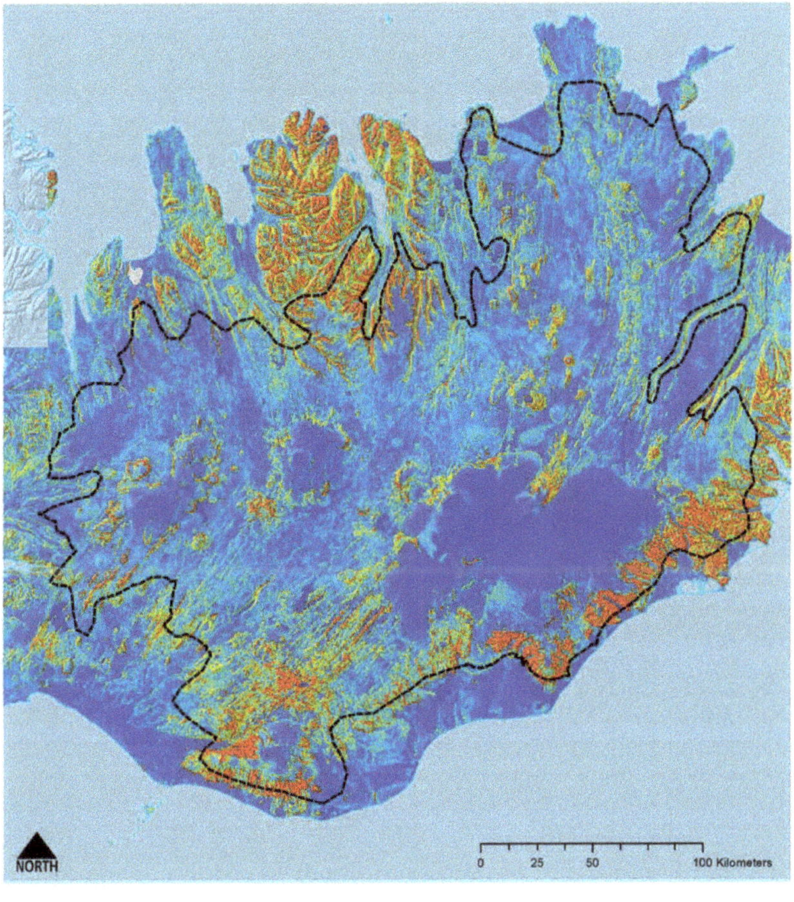

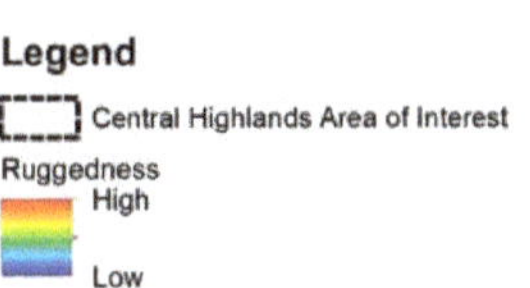

Figure 10. Obtained values for Ruggedness for the Central Highlands.

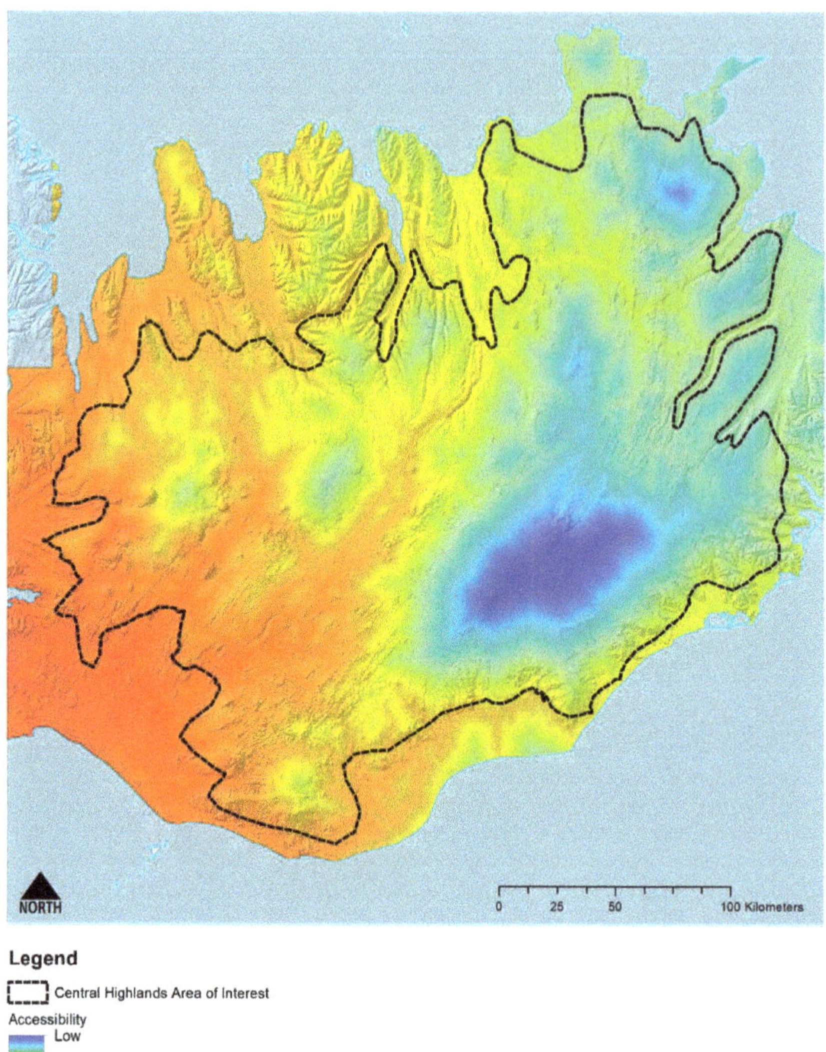

Figure 11. Obtained values for accessibility for the Central Highlands.

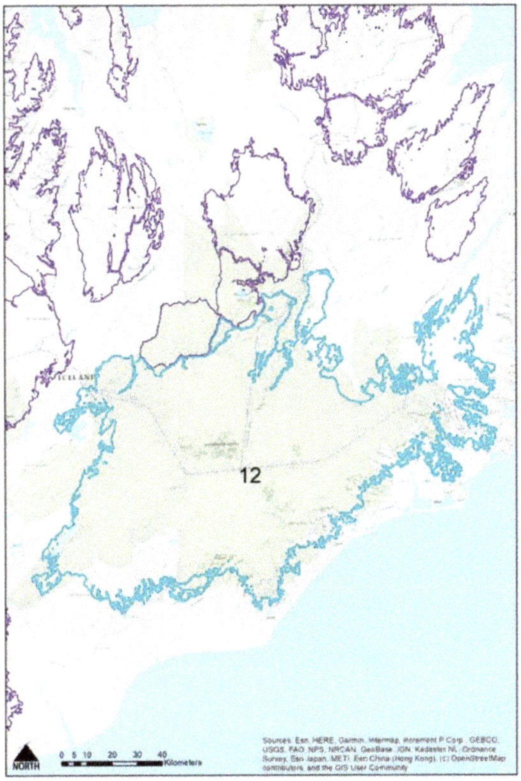

General setting and description: The Vatnajökull and its surrounding landscape is perhaps the most iconic of all Icelandic wilderness landscapes. The area is large and is the largest ice cap in Iceland.

Topography The topography of Vatnajökulssvæðið is dominated by the vast ice cap, sub-glacial volcanoes (Grímsvötn, Bárðarbunga, Tungnafellsjökull, Kverkfjöll and Öræfajökull), glacier flows from around its edges and rugged mountain ridges. The Vatnajökulssvæðið also includes an area of coastal mountains and uninhabited valleys in the east.

Landscape assessment: The landscape is varied with large expansive open areas on the ice cap and to the north and complex, enclosed, and rugged areas around the edges of the glacier to the south and east. Here, the ice flows have carved deep valleys with further open areas across the expansive lakes and rivers along its southern coastal margin.

Land use: Most of the landscape is snow- and ice-covered and there is no livestock grazing to the west and north of the glacier, but sheep graze in the southeast valleys where there is also reindeer hunting. The main land use is recreation and tourism. Mobile phone coverage is generally poor, with many areas without signal. Much of the area is extremely remote.

Figure 12. Vatnajökulssvæðið wilderness character and description.

4. Discussion

The use of proxy measures for wilderness area mapping has its origins in some of the earliest global scale mapping. McCloskey and Spalding [31] defined the world's remaining wilderness as those areas more than six kilometres from the nearest settlement, road, railway or navigable river using 1:2 million scale Jet Navigation Charts. Ibisch et al. [39] provide a more up-to-date estimate of the world's remaining roadless areas using a buffer distance of 1 km, finding that only 7% of the world's land surface is covered by roadless areas greater than 100 km^2. While such buffers are useful as global proxies, remoteness and visual impact are better modelled using more sophisticated methods at national or local scales. For example, is it safe to assume that all roads are equal? Does a paved highway exert a greater influence than a gravel track? Does a small cluster of farm buildings have the same impact as a large town or city? How does topography and associated barriers to movement and resistance to travel affect their impact? Does the fact that you can or cannot see the nearest road from where you stand alter how you think about remoteness? All these factors and their influence are too complex to map using simple buffer zones and thus require more nuanced models that measure their impact in terms of remoteness and visibility.

It is instructive to compare the wilderness areas in Figure 8 with previous wilderness maps drawn for Iceland. These include the EU Wilderness Index [9], the map provided by Ólafsdóttir and Runnström [40], and the most recent map by Ostman et al. [20]. Figure 13 shows these maps superimposed over the seventeen wilderness areas from Figure 8. A simple visual comparison of the wilderness areas developed here and those based on the EU-level WQI from Kuiters et al. [9] in Figure 13a demonstrates a reasonable degree of similarity. This is only to be expected since, despite differences in criteria, data and approach, these maps are dealing with the same landscape and the same underlying characteristics of wilderness, namely, remoteness and naturalness, measured along a continuum from least to most wild. Comparisons with those maps derived from simple buffer zones around selected human features show much larger levels of disagreement, with the maps from Ólafsdóttir and Runnström [40] and Ostman et al. [20] including substantially greater areas of wilderness when compared to the results of the current analysis.

The Ólafsdóttir and Runnström [40] map in Figure 13b is a straightforward spatial mapping of the criteria described in the previous text of the NCA No 44/1999, which maps those areas more than 5 km from a road or building as simple buffers and then selects those that are more than 25 km^2 in size. Here, all buildings and public roads are used regardless of road grade or building size, with the result that a small hut or shelter has the same effect as a large geothermal power station on the wilderness buffers. The scale of development and the influence or impact that this has on the landscape is therefore not considered. The work by Ólafsdóttir and Runnström [40] does expand the mapping further by including a binary viewshed analysis to show the zones of theoretical visibility (ZTVs) of human features, but this is not included in the final wilderness map.

The Ostman et al. [20] map shown in Figure 13c employs the same criteria but excludes gravel roads from consideration, despite their proven impact on remoteness and visibility. Previous work by Árnason et al. [41] applied the 5 km buffer to all roads in the national register of the Road Authority, producing a map that is much nearer to that by Ólafsdóttir and Runnström [40]. Ostman et al. [20] apply buffers of 3 km and 5 km around power lines depending on the voltage level. There is an attempt to take relative level of impact into account by varying the buffer distances applied based on a scoring system calculated from the use and number of buildings/structures present, their surface area, visibility and connection to the road network, while paved roads are buffered at a uniform 5 km. The resulting wilderness area boundaries are much more extensive than those presented by Ólafsdóttir and Runnström [40] or in the work presented here and conform more closely to the suggested IUCN Category 2 areas shown in Figure 8. This is largely due to the exclusion of gravel roads from consideration and the use of simple buffering, albeit modified with a scoring system.

The exact boundaries of the core areas and buffer/transition zones drawn here in Figure 8 are, in contrast, derived from detailed spatial data and models that measure the impact of human artefacts, remoteness and naturalness to create a WQI rather than relying on simple proxies such as distance buffers. The WQI is classified using statistical methods that take the full range of wilderness quality measures across the Central Highlands into account. As a result, the boundaries at this stage tend to be complex and quite fragmented as seen in Figure 8. It is suggested here that these will need to be simplified for planning and policy use (as with the Phase 3 WLA boundaries produced by SNH [23]), but that the maps provide a rigorous and robust approach to informing such policy decisions at a later stage in the designation process.

The reliability and repeatability of the methods developed here naturally lend themselves to "what if?" analyses of proposed future developments. This, again, can provide an invaluable source of information to support planning and policy decisions regarding development proposals for significant infrastructure within or adjacent to wilderness areas. Such repeat modelling of wilderness quality with and without the features in place can be used to gauge the impact of the proposed development and quantify the area of wilderness lost should the development be allowed to move forward.

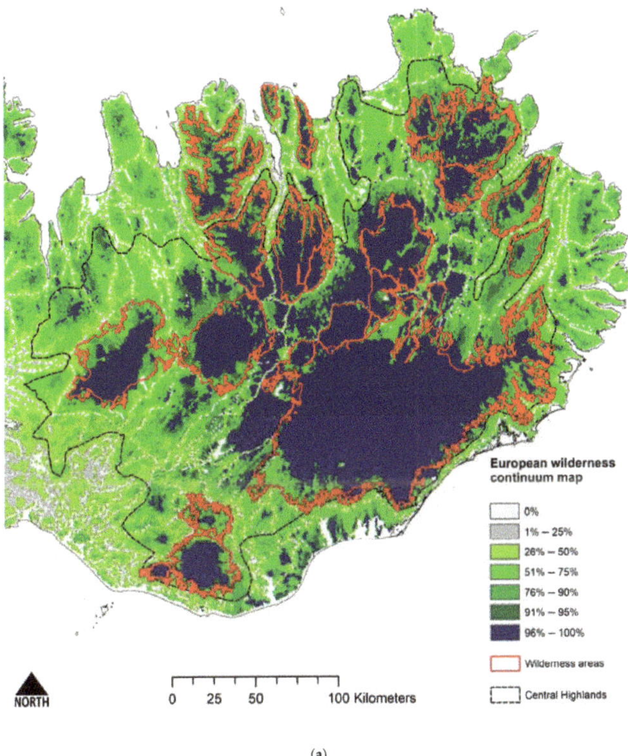

(a)

Figure 13. *Cont.*

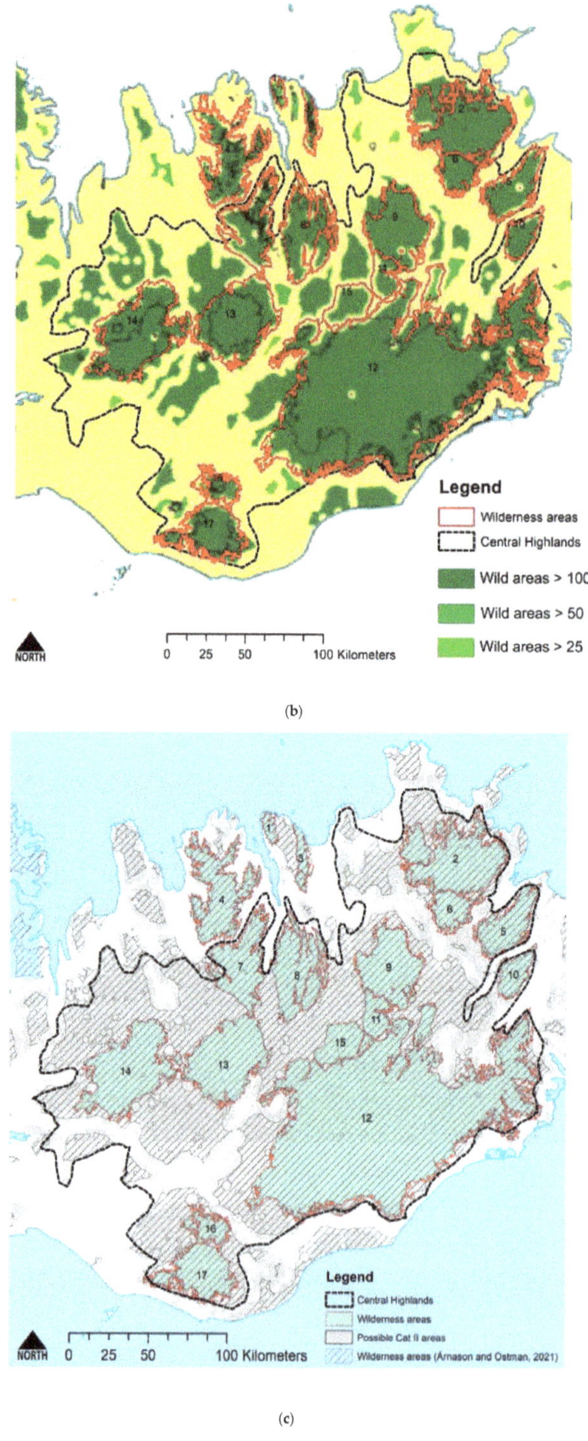

(b)

(c)

Figure 13. Comparison with wilderness maps from: (**a**) Kuiters et al. (2013), (**b**) Ólafsdóttir and Runnström (2011), and (**c**) Ostman et al. (2021).

Winter driving offroad over snow and ice remains an issue that requires further attention. While much of the mapping and analysis carried out here relates to summer conditions and rules (e.g., limiting vehicles to those roads usable by the public), the maps in Figure 3 demonstrate the potential effect of winter offroad driving in greatly reducing remoteness. This is an issue that could potentially limit opportunities for the Icelandic government to designate large areas of the Central Highlands under IUCN Category 1b due to the explicit exclusion of mechanical means of transportation in IUCN wilderness area guidelines. This requires careful engagement with the 4x4 community to explore options for limiting offroad winter driving to certain areas outside of mapped wilderness cores as mentioned in Article 46(2) of the NCA " … and to ensure that present and future generations can enjoy solitude and the nature without disturbance from man-made structures or the traffic of motorized vehicles".

5. Conclusions

The co-related aims of protecting pristine nature and facilitating tourism and recreational use is a key challenge facing the Icelandic government in the Central Highlands. This requires striking a careful balance between visitor use, resource exploitation and the preservation of nature [42]. Nowhere is this more important than in the potential conflicts between winter offroad driving, renewable energy developments and wilderness designation. Detailed and accurate mapping of landscape attributes and human impacts are key to sustainable decision making about wilderness landscapes in this regard. This paper presents a significant improvement on existing approaches to mapping wilderness areas in Iceland both in terms of detail and methods used and one that carefully considers and takes account of local nature conservation legislation.

The work described is the most detailed and accurate mapping of wilderness quality and wilderness character for the Central Highlands of Iceland that has been carried out to date. This has enabled the definition of seventeen separate and distinct wilderness areas along with surrounding buffer and transition zones. A key advantage over existing studies is the use and adaptation of internationally recognised methods and wilderness standards which use direct measurement and modelling of spatial factors determining wilderness quality. This is supplemented by wilderness character assessments based on additional mapping and descriptions of spatial factors affecting the individual wilderness landscapes and their unique character. The use of a 4Rs approach ensures rigour, robustness, repeatability, and reliability in the work carried out.

The work and the maps presented in this paper differ significantly from previous work in that rather than using simple distance/area proxies, the attributes mapped here represent the actual measurement of human impacts from land use, settlement, and infrastructure development on wilderness landscapes. The WQI and seventeen wilderness areas identified can be seen as an important step towards the formal definition of boundaries of wilderness areas meeting IUCN Category 1b and Wild Europe Working Definition in Iceland. Further work is recommended to complete the mapping for the whole of Iceland as mandated in the amendment to the 2013 NCA in Article 73a 2021 [11]. This could be supplemented where necessary by additional models to account for variations in remoteness around the coastal areas and islands, where different modes of travel/access will play an important role, and by comparison with ecological data on protected habitats and species distributions.

Finally, we suggest that the 4Rs approach developed here, along the methods and models applied, could be usefully applied across all countries of Europe taking the individual national datasets and conditions pertaining to wilderness and its relevance to social, political and cultural understanding into account. This could, with cross-border collaboration where necessary, help better map the patterns of Europe's remaining wilderness areas and inform decisions regarding their future protection in meeting the recommendations from the European Parliament resolution on wilderness [7] and joint agreements on nature protection and restoration of degraded ecosystems under the UN Sustainable Development Goals [5], the Global Biodiversity Framework Convention on Biological Diversity action oriented targets and the recent Kunming–Montreal agreement calling on signatories to protect 30% of land and sea for nature by 2030 [43]. If we are to meet these commitments, then rigorous, robust, reliable and repeatable methods of mapping wilderness boundaries will be required in supporting the decisions made.

Author Contributions: Conceptualization, S.C., S.K. and S.G.; methodology, S.C.; software, S.C.; validation, S.C., S.K. and S.G.; formal analysis, S.C., B.C. and O.K.; investigation, S.C., S.K., S.G., B.C. and O.K.; resources, S.K. and S.G.; data curation, S.C.; writing—original draft preparation, S.C.; writing—review and editing, S.K., S.G. and B.C.; visualization, S.C., B.C. and O.K.; supervision, S.C. and S.K.; project administration, S.C., S.K. and S.G.; funding acquisition, S.K. and S.G. All authors have read and agreed to the published version of the manuscript.

Funding: This research was funded by SKH Foundation inc., the Nell Newman Foundation and the Viljandi Foundation.

Data Availability Statement: All the model outputs referred to in this paper will be made available through the Icelandic government in due course.

Acknowledgments: We gratefully acknowledge the following organisations and individuals for their assistance in developing the work described in the paper: David Ostman, Þorvarður Árnason, Rannveig Ólafsdóttir and Anna Dóra Sæþórsdóttir for providing data on previous mapping work; Veðurstofa Íslands, Landmælingar Íslands, NGA-NSF, Landbúnaðarháskóli, Landsnet, Landsvirkjun, RARIK and OpenStreetMap for access to datasets and expertise; Fanney Ásgeirsdóttir and colleagues in Vatnajökull National Park for local knowledge and field assistance; Snorri Baldursson of Skrauti; and university students Ester Alda Hrafnhildar Bragadóttir, Helga Østerby Þórðardóttir and Finnur Ricart Andrason.

Conflicts of Interest: The authors declare no conflict of interest.

References

1. Casson, S.A.; Martin, V.G.; Watson, A.; Stringer, A.; Kormos, C.F.; Locke, H.; Sonali, G.; Carver, S.; McDonald, T.; Sloan, S.S.; et al. Wilderness Protected Areas: Management Guidelines for IUCN Category 1b Protected Areas. Best Practice Protected Area Guidelines Series. 2016. Available online: https://www.iucn.org/news/protected-areas/201612/wilderness-protected-areas-management-guidelines (accessed on 1 February 2021).
2. Watson, J.E.; Shanahan, D.F.; Di Marco, M.; Allan, J.; Laurance, W.F.; Sanderson, E.W.; Mackey, B.; Venter, O. Catastrophic declines in wilderness areas undermine global environment targets. *Curr. Biol.* **2016**, *26*, 2929–2934. [CrossRef] [PubMed]
3. Theobald, D.M.; Kennedy, C.; Chen, B.; Oakleaf, J.; Baruch-Mordo, S.; Kiesecker, J. Earth transformed: Detailed mapping of global human modification from 1990 to 2017. *Earth Syst. Sci. Data* **2020**, *12*, 1953–1972. [CrossRef]
4. Cao, Y.; Tseng, T.H.; Wang, F.; Jacobson, A.; Yu, L.; Zhao, J.; Carver, S.; Locke, H.; Zhao, Z.; Yang, R. Potential wilderness loss could undermine the post-2020 global biodiversity framework. *Biol. Conserv.* **2022**, *275*, 109753. [CrossRef]
5. Lu, Y.; Nakicenovic, N.; Visbeck, M.; Stevance, A.S. Policy: Five priorities for the UN sustainable development goals. *Nature* **2015**, *520*, 432–433. [CrossRef] [PubMed]
6. Convention on Biological Diversity. First Draft of the Post-2020 Global Biodiversity Framework. 2017. Available online: https://www.cbd.int/doc/c/abb5/591f/2e46096d3f0330b08ce87a45/wg2020-03-03-en.pdf (accessed on 17 June 2022).
7. European Parliament. Wilderness in Europe. Resolution of 3 February 2009. 2008/2210(INI). 2009. Available online: https://www.europarl.europa.eu/doceo/document/TA-6-2009-0034_EN.html (accessed on 2 February 2021).
8. European Commission. *Guidelines on Wilderness in Natura 2000*; Technical Report 2013-069; European Commission: Brussels, Belgium, 2013. [CrossRef]
9. Kuiters, A.T.; van Eupen, M.; Carver, S.; Fisher, M.; Kun, Z.; Vancura, V. *Wilderness Register and Indicator for Europe Final Report*; EEA Contract No 0703072011610387 SERB3; European Environment Agency: Copenhagen, Denmark, 2013.
10. Bastmeijer, K. (Ed.) *Wilderness Protection in Europe: The Role of International, European and National Law*; Cambridge University Press: Cambridge, England, 2016.
11. Nature Conservation Act. No 60/2013 as Later Amended. 2013. Available online: https://www.althingi.is/lagas/nuna/2013060.html (accessed on 17 June 2021).
12. Bishop, M.V.; Ólafsdóttir, R.; Árnason, Þ. Tourism, Recreation and Wilderness: Public Perceptions of Conservation and Access in the Central Highland of Iceland. *Land* **2022**, *11*, 242. [CrossRef]
13. Tverijonaite, E.; Sæþórsdóttir, A.D.; Ólafsdóttir, R.; Hall, C.M. Renewable Energy in Wilderness Landscapes: Visitors' Perspectives. *Sustainability* **2019**, *11*, 5812. [CrossRef]

14. Ólafsdóttir, R.; Sæþórsdóttir, A.D. Public perception of wilderness in Iceland. *Land* **2020**, *9*, 99. [CrossRef]

15. UNESCO. Decision 43 COM 8B.8, Vatnajökull National Park—Dynamic Nature of Fire and Ice (Iceland). 2019. Available online: https://whc.unesco.org/en/decisions/7364 (accessed on 10 March 2021).

16. Guðmundsson, S. Exploring Iceland's Geology. Mál og menning, Iceland. 2016. Available online: https://www.abebooks.co.uk/9789979336259/Exploring-Icelands-Geology-2016-Gudmundsson-9979336250/plp (accessed on 10 March 2021).

17. Ólafsdóttir, R.; Sæþórsdóttir, A.D.; Runnström, M. Purism scale approach for wilderness mapping in Iceland. In *Mapping Wilderness*; Springer: Dordrecht, The Netherlands, 2016; pp. 157–176.

18. Ólafsdóttir, R.; Sæþórsdóttir, A.D. *Hálendið í hugum Íslendinga. 1. hluti: Merking hugtakanna víðerni, óbyggðir og miðhálendi [The Highlands in the Minds of Icelanders. Part 1: Meaning of the Terms Wilderness, Wilderness and Central Highlands]*; Náttúrufræðingurinn: Garðabær, Iceland, 2020; pp. 202–208.

19. Ólafsdóttir, R.; Sæþórsdóttir, A.D.; Guðmundsson, H.; Huck, J.; Runnström, M. Viðhorf og Upplifun Íslendinga á Víðernum, Óbyggðum og Miðhálendi Íslands [Attitudes and Experiences of Icelanders in the Wilderness, Wilderness and Central Highlands of Iceland]. 2016. Available online: https://www.ramma.is/media/rannsoknir-f2-ra3/Vidhorf-og-upplifun-Islendinga-a-vidernum,-obyggdum-og-midhalendi-Islands.pdf (accessed on 11 October 2021).

20. Ostman, D.; Neumann, O.; Árnason, Þ. Óbyggð Víðerni á Íslandi—Greining og Kortlagning á Landsvísu [Wilderness in Iceland—National Analysis and Mapping]. 2021. Available online: https://www.ramma.is/media/rannsoknir/OstmanEtal2021_WildernessIceland.pdf (accessed on 1 December 2021).

21. Comber, A.; Carver, S.; Fritz, S.; McMorran, R.; Washtell, J.; Fisher, P. Different methods, different wilds: Evaluating alternative mappings of wildness using fuzzy MCE and Dempster-Shafer MCE. *Comput. Environ. Urban Syst.* **2010**, *34*, 142–152. [CrossRef]

22. Carver, S.; Comber, A.; McMorran, R.; Nutter, S. A GIS model for mapping spatial patterns and distribution of wild land in Scotland. *Landsc. Urban Plan.* **2012**, *104*, 395–409. [CrossRef]

23. Scottish Natural Heritage. Landscape Policy: Wild Land. 2014. Available online: https://www.nature.scot/professional-advice/landscape-change/landscape-policy-and-guidance/landscape-policy-wild-land (accessed on 24 July 2021).

24. Benest, F.; Carruthers-Jones, J.; Guetté, A. Travaux actuels d'inventaire des forêts à forte naturalité à l'échelle nationale et européenne. *Rev. For. Française* **2022**, *73*, 161–178. [CrossRef]

25. Carver, S.; Tricker, J.; Landres, P. Keeping it wild: Mapping wilderness character in the United States. *J. Environ. Manag.* **2013**, *131*, 239–255. [CrossRef] [PubMed]

26. Cao, Y.; Carver, S.; Yang, R. Mapping wilderness in China: Comparing and integrating Boolean and WLC approaches. *Landsc. Urban Plan.* **2019**, *192*, 103636. [CrossRef]

27. Lesslie, R.G.; Maslen, M. *National Wilderness Inventory Australia*; Australia Government Pub. Service: Canberra, Australia, 1995.

28. Wild Europe. A Working Definition of European Wilderness and Wild Areas. 2013. Available online: https://www.europarc.org/wp-content/uploads/2015/05/a-working-definition-of-european-wilderness-and-wild-areas.pdf (accessed on 23 June 2021).

29. Naismith, W.W. Cruach Ardran, Stobinian, and Ben More. *Scott. Mt. Club J.* **1892**, *2*, 136.

30. Scottish Natural Heritage. *Wildness in Scotland's Countryside*; Scottish Natural Heritage: Edinburgh, UK, 2002.

31. McCloskey, J.M.; Spalding, H. A reconnaissance-level inventory of the amount of wilderness remaining in the world. *Ambio* **1989**, 221–227. Available online: https://www.jstor.org/stable/4313570 (accessed on 23 December 2021).

32. Washtell, J.; Carver, S.; Arrell, K. A viewshed based classification of landscapes using geomorphometrics. In Proceedings of the Geomorphometry Conference, Zurich, Switzerland, 31 August–2 September 2009; pp. 44–49.

33. Carver, S.; Comber, L.; Fritz, S.; McMorran, R.; Taylor, S.; Washtell, J. Wildness Study in the Cairngorms National Park. University of Leeds. 2008. Available online: https://www.geog.leeds.ac.uk/groups/wildland/Cairngorm2008.pdf (accessed on 1 December 2008).

34. Fisher, P.F. Algorithm and implementation uncertainty in viewshed analysis. *Int. J. Geogr. Inf. Sci.* **1993**, *7*, 331–347. [CrossRef]

35. Bishop, I.D. Assessment of visual qualities, impacts, and behaviours, in the landscape, by using measures of visibility. *Environ. Plan. B Plan. Des.* **2003**, *30*, 677–688. [CrossRef]

36. Gísladóttir, F.Ó.; Brink, S.H.; Og Arnalds, O. Nytjaland [Land use]. Fjölrit LbhÍ nr. 49, Landbúnaðarháskóli Íslands. 2014. Available online: https://www.lbhi.is/images/pdf/utgefid%20efni/fjolrit%20rannsoknastofnunar%20landbunadarins/rit_lbhi_nr_49_nytjaland.pdf (accessed on 17 June 2022).

37. Yokoyama, R.; Shirasawa, M.; Pike, R.J. Visualizing topography by openness: A new application of image processing to digital elevation models. *Photogramm. Eng. Remote Sens.* **2002**, *68*, 257–266.

38. EFLA/LUC. Landslag á Íslandi. Flokkun og Kortlagning Landslagsgerða á Landsvísu. 2020. Available online: https://www.skipulag.is/media/landsskipulagsstefna-vidbaetur/Skyrslan_Lokaeintak-2-.pdf (accessed on 3 May 2021).

39. Ibisch, P.L.; Hoffmann, M.T.; Kreft, S.; Pe'er, G.; Kati, V.; Biber-Freudenberger, L.; DellaSala, D.A.; Vale, M.M.; Hobson, P.R.; Selva, N. A global map of roadless areas and their conservation status. *Science* **2016**, *354*, 1423–1427. [CrossRef]

40. Ólafsdóttir, R.; Runnström, M.C. How wild is Iceland? Wilderness quality with respect to nature-based tourism. *Tour. Geogr.* **2011**, *13*, 280–298. [CrossRef]

41. Árnason, Þ.; Ostman, D.; Hoffritz, A. Kortlagning Víðerna á Miðhálendi Íslands: Tillögur að Nýrri Aðferðafræði. 2017. Available online: http://www.skipulag.is/media/pdf-skjol/Kortlagning_Viderna_Web2.pdf (accessed on 24 May 2021).

42. Sæþórsdóttir, A.D.; Wendt, M.; Ólafsdóttir, R. Tourism Industry Attitudes towards National Parks and Wilderness: A Case Study from the Icelandic Central Highlands. *Land* **2022**, *11*, 2066. [CrossRef]

43. Conference of the Parties to the Convention on Biological Diversity, Fifteenth Meeting—Part II. Montreal, Canada, 7–19 December 2022 Agenda Item 9A Kunming-Montreal Global Biodiversity Framework. Available online: https://www.cbd.int/doc/c/e6d3/cd1d/daf663719a03902a9b116c34/cop-15-l-25-en.pdf (accessed on 20 December 2022).

 *land* 

Article

A Protected Area Connectivity Evaluation and Strategy Development Framework for Post-2020 Biodiversity Conservation

Zhicong Zhao [1,2,†], Pei Wang [1,2,†], Xiaoshan Wang [1,2], Fangyi Wang [1,2], Tz-Hsuan Tseng [1,2], Yue Cao [1,2], Shuyu Hou [1,2], Jiayuan Peng [1,2] and Rui Yang [1,2,*]

[1] Institute for National Parks, Tsinghua University, Beijing 100084, China
[2] Department of Landscape Architecture, School of Architecture, Tsinghua University, Beijing 100084, China
* Correspondence: yrui@tsinghua.edu.cn; Tel.: +86-10-62797027
† These authors contributed equally to this work.

Citation: Zhao, Z.; Wang, P.; Wang, X.; Wang, F.; Tseng, T.-H.; Cao, Y.; Hou, S.; Peng, J.; Yang, R. A Protected Area Connectivity Evaluation and Strategy Development Framework for Post-2020 Biodiversity Conservation. *Land* 2022, *11*, 1670. https://doi.org/10.3390/land11101670

Academic Editor: Alexander Khoroshev

Received: 3 September 2022
Accepted: 22 September 2022
Published: 27 September 2022

Abstract: Maintaining and improving the connectivity of protected areas (PAs) is essential for biodiversity conservation. The Post-2020 Global Biodiversity Framework (GBF) aims to expand the coverage of well-connected PAs and other effective area-based conservation measures to 30% by 2030. We proposed a framework to evaluate the connectivity of PAs and developed strategies to maintain and improve the connectivity of PAs based on PA connectivity indicators, and we applied this framework to China's terrestrial PAs. We considered that the concept of PA connectivity is at the level of both PA patches and PA networks, including four aspects: intra-patch connectivity, inter-patch connectivity, network connectivity, and PA–landscape connectivity. We found that among China's 2153 terrestrial PA patches, only 427 had good intra-patch connectivity, and their total area accounted for 11.28% of China's land area. If inter-patch connectivity, network connectivity, and PA–landscape connectivity were taken as the criteria to evaluate PA connectivity, respectively, then the coverage of well-connected terrestrial PAs in China was only 4.07%, 8.30%, and 5.92%, respectively. Only seven PA patches have good connectivity of all four aspects, covering only 2.69% of China's land. The intra-patch, inter-patch, network, and PA–landscape connectivity of China's terrestrial PA network reached 93.41%, 35.40%, 58.43%, and 8.58%, respectively. These conclusions indicated that there is still a big gap between China's PA connectivity and the Post-2020 GBF target, which urgently needs to be improved. We identified PA patches and PA networks of ecological zones that need to improve PA connectivity and identified improvement priorities for them. We also identified priority areas for connectivity restoration in existing PAs, potential ecological corridors between PAs, and priority areas for PA expansion to improve the connectivity of PAs in China. Application of our framework elsewhere should help governments and policymakers reach ambitious biodiversity conservation goals at national and global scales.

Keywords: biodiversity conservation; connectivity; protected areas; dispersal probability; least-cost distance; ecological corridor

1. Introduction

Biodiversity loss and climate change are urgent and critical crises to which humanity must respond [1–3]. Connectivity can facilitate a range shift and the climate resilience of species [4,5]. Maintaining and improving connectivity is essential for achieving long-term biodiversity outcomes in response to climate change [6–8]. Research has shown that habitat connectivity is sensitive to climate change and may be lost more rapidly than habitat area [9,10]. In summary, connectivity loss has a robust, lasting, and negative impact on biodiversity and is, therefore, a major threat to biodiversity maintenance [11,12].

The establishment of protected areas (PAs) is a vital initiative for biodiversity conservation [13–16], and connectivity is necessary, and even of central importance, for the effectiveness of PAs [17,18]. Both the Conservation for Biodiversity Aichi Targets [19] and the Post-2020 Global Biodiversity Framework (GBF), which is under discussion globally, emphasize the importance of PA connectivity and set global PA connectivity targets. Aichi Targets and the First Draft of the Post-2020 GBF call for 17% and 30%, respectively, of the global land area to be conserved through well-connected PAs and other effective area-based conservation measures (OECMs) [19,20].

Research on connectivity evaluation has led to the development of different connectivity indicators [21,22]. The probability of connectivity (PC) is a widely used indicator to evaluate the connectivity of PAs [23–25]. Based on PC, Saura et al. (2018) used the ProtConn indicator and found that only 7.5% of global terrestrial land is covered by well-connected PAs, whereas in case of China, the value is 8–12% [26]. Ward et al. (2020) used the ConnIntact indicator and found that intact land structurally connected only 10% of the terrestrial PAs globally [27]. Among the existing global PA connectivity assessment studies, some focused on the connectivity of the PA network including intra-patch connectivity and inter-patch connectivity [25,26,28], while others concentrated on the connectivity between PA patches (inter-patch connectivity) [27,29]. It is necessary to integrate the connectivity at different levels and aspects into a unified framework to comprehensively describe connectivity and propose systematic approaches to address it accordingly.

We considered that the concept of PA connectivity includes the intra-patch connectivity, inter-patch connectivity, network connectivity, and PA–landscape connectivity of both PA patches and PA networks (Figure 1) based on previous studies [23,27,30–32]. For a PA network that includes several PA patches located in a landscape, we distinguish the above concepts of connectivity according to the following definition. The intra-patch connectivity of a PA patch means the connectivity within the PA patch. The inter-patch connectivity of a PA patch means the connectivity between it and other PA patches within the PA network. The network connectivity of a PA patch means its connectivity with the PA network that includes its intra-patch connectivity and its inter-patch connectivity with other PA patches. The PA–landscape connectivity of a PA patch means the connectivity between this PA patch and the whole landscape. The intra-patch connectivity of the PA network includes the connectivity within every PA patch of the PA network. The inter-patch connectivity of the PA network includes the connectivity between every patch pairs within the PA network. The network connectivity of the PA network includes the intra-patch connectivity of every PA patch within the PA network and the inter-patch connectivity between every patch pair within the PA network. The PA–landscape connectivity of the PA network includes the connectivity between every PA patch and the whole landscape.

This study proposed a set of indicators to evaluate the PA connectivity of both PA patches and PA networks based on dispersal probability and the PC indicator [23], and all of these indicators range from 0 to 1. The probability of connectivity of intra PA patches (PCintra) indicator measures intra-patch connectivity, the probability of connectivity of inter PA patches (PCinter) indicator measures inter-patch connectivity, the probability of connectivity with the PA network (PCnet) indicator measures network connectivity, and the probability of connectivity with the whole landscape (PCland) indicator measures PA–landscape connectivity. We established a PA connectivity evaluation and strategy development framework based on these PA connectivity indicators (Figure 1).

The aim of this study is to provide a framework on PA connectivity evaluation and improvement for post-2020 biodiversity conservation and illustrate how this framework can be applied and guide the management of PAs, using China as an example. In the methods section, we explain the calculation methods of different connectivity indictors and how to determine the connectivity maintenance or improvement strategies according to the connectivity evaluation results. In the results section, we show the calculation results of the connectivity indicators of the PA networks and PA patches in China, the connectivity strategy classification results of PAs based on connectivity indicators, and the spatial priority area to improve PA connectivity in China.

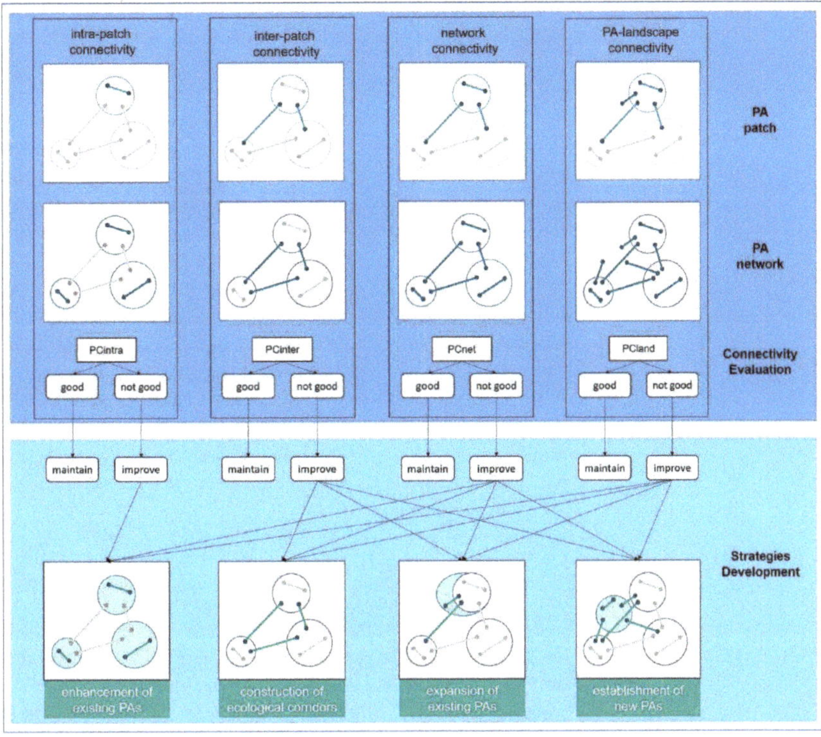

Figure 1. A framework for protected area (PA) connectivity evaluation and strategy development based on connectivity indicators and conducted from four perspectives: intra-patch, inter-patch, network and PA–landscape connectivity.

2. Materials and Methods

2.1. Protected Areas and Ecological Zones

The natural conservation geographical regionalization scheme of China [33], which aims to guide China's biodiversity conservation and establishment of the PA system, was adopted in this study. This biogeographic regionalization scheme divides China's land into 38 terrestrial ecological zones. The South China Sea island tropical humid zone (VIII2), which has no terrestrial PAs, was not included in the analysis. This study assumed that PAs need to connect with PAs within the same ecological zone, and we evaluated PA connectivity separately at the ecological zone scale.

We used data collected for various types of terrestrial PAs in China, including 819 polygons and 3163 points. The polygon data included data for 10 national parks, which were mapped according to the national park pilot area plans released by the Chinese government. Data for 252 national nature reserves and 377 local nature reserves were extracted from information published by the Chinese government. The data were merged with data on 180 PAs in China provided by the World Database on Protected Areas (WDPA) for September 2020 (https://www.protectedplanet.net (accessed on 5 January 2021)). The point data included scenic areas, forest parks, and geoparks, which we collected according to information released by the Chinese government. Areas of high ecological integrity within 2 km of the point data were used instead of the point data, as many studies have shown that it is reasonable to use areas of high ecological integrity for connectivity analysis [30,34–36]. In this study, global-scale, very low human impact areas [37] and China-scale wilderness areas [38] were selected to form high ecological integrity areas. The polygon data were merged with the high ecological integrity areas that replaced the point data. We intersected the PA patches with ecological zones and obtained 2153 PA patches covering 14.68% of China's land surface with a total area of 1,409,761 km^2.

2.2. Resistance Surface

The resistance surface measures how difficult it is for an organism or ecological flow to move successfully [36] or measures the relative cost of passing through a gridded mapped surface [39]. Many studies create resistance surfaces based on the degree of human modification, naturalness, or other similar indicators [30,39–41]. In this study, we created a resistance surface based on the global land-scale human modification indicator, HMc, which estimates the cumulative human modification of the land using 13 global human stressor datasets with 2016 as the median year; the value is between 0 and 1 and has a spatial resolution of 1 km [42]. The stressor datasets included human settlement (population density, built-up areas), agriculture (cropland, livestock), transportation (major roads, minor roads, two tracks, railroads), mining and energy production (mining, oil wells, wind turbines), and electrical infrastructure (powerlines, nighttime lights) [42]. Despite the uncertainties that global data might bring, this was the best available data on human modification of China's land. We performed an exponential transformation of HMc, similar to Cao et al. (2020) [30], and we formed a resistance surface R between 1 and 1000 using the following equation:

$$R = 1 + 999 * \frac{e^{HM_c} - 1}{e - 1} \tag{1}$$

Finally, we removed areas covered by water bodies and glaciers extracted from land use data of China from the resistance surface (Figure S1), assuming that terrestrial animals do not pass over glaciers or through water bodies during dispersal. The land use data were obtained from the Resources and Environmental Science Data Center, Chinese Academy of Sciences (Beijing, China; http://www.resdc.cn/ (accessed on 16 June 2021)).

2.3. PA Connectivity Evaluation

For a PA network in a landscape that includes n PA patches, the area of PA patch i was noted as $a_i (i = 1, 2 \ldots., n)$, the total area of the PA network was $A_N = \sum_{i=1}^{n} a_i$, and the total area of the landscape was A_L. We evaluated the connectivity of PAs by dispersal probability, which can be estimated as a negative-exponential function of distance [32,43].

2.3.1. Intra-Patch Connectivity

As the distance an animal can disperse within a certain time duration is limited, the intra-patch connectivity of a patch can be simplified as the probability of a successful dispersal of a fixed distance from every point in a patch. We created a dispersal probability surface (with a value of P) (Figure S2) from the resistance surface. When the resistance surface is raster data with a cell side length D and a value R, for any cell on the raster surface, the cost distance is R when animal dispersal in the cell moves a distance D, and the dispersal probability P is as follows:

$$P = e^{-h*R} \tag{2}$$

In the present study, R was between 1 and 1000, so we defined h as 1/1000, considering that the dispersal probability is 1/e (0.3679) when R takes the maximum possible value of 1000, and $e^{-1/1000}$ (0.9990) when the resistance is the minimum value of 1, which is very close to 1.

The PCintra of PA patch i is defined as the probability of a successful dispersal of a fixed distance from any point within this patch and can be calculated as the average value of the dispersal probability surface within this patch:

$$PCintra_i = average(P)(patchi) \tag{3}$$

The PCintra of the PA network is defined as the probability of a successful dispersal of a fixed distance from any point within patches can be calculated as the average value of the dispersal probability surface within the PA network:

$$\text{PCintra} = \sum_{i=1}^{n} \frac{a_i}{A_N} * \text{PCintra}_i \tag{4}$$

After creating the dispersal probability surface, we used the partition statistics tool of ArcGIS 10.2 to calculate the PCintra of the PAs.

2.3.2. Inter-Patch Connectivity

Dispersal probability p_{ij} characterizes the feasibility of a step between patches i and j, where a step is defined as a direct movement of a disperser between two habitat patches without passing by any other intermediate habitat patches [23]. We considered that an animal that moves from one patch i to another patch j first needs to move from some point A inside patch i to some point B on the edge of patch i; then, it moves successfully from point B through the matrix, to some point C on the edge of patch j, and from C to some point D inside patch j. The probability of successful dispersal from points A to B is PCintra_i, and the probability of successful dispersal from points B to C can be estimated as a negative-exponential function of the inter-patch distance d_{ij} [32,43]. The probability of successful dispersal from points C to D is PCintra_j. Then, the probability of direct dispersal between patches i and j is calculated as follows (k is a constant):

$$p_{ij} = \text{PCintra}_i * e^{-kd_{ij}} * \text{PCintra}_j \tag{5}$$

The value of p_{ij}^* is the maximum product probability of all possible paths between patches i and j (including single-step paths) [23]. For the case of indirect dispersal from patch i through patch k to patch j, the probability is equal to the product of the probability of success of each step of the animal's movement:

$$p_{ij}' = p_i * e^{-kd_{ik}} * p_k * e^{-kd_{kj}} p_j \tag{6}$$

The inter-patch distance d_{ij} can be estimated by the Euclidean distance or least-cost distance [23,44]. Measuring the connectivity between patches based on Euclidean distance does not reflect spatial heterogeneity, and this approach is considered unreasonable by some researchers [45]. Therefore, the least-cost distance was used as the inter-patch distance in this study. The Linkage Pathways Tool of Linkage Mapper Toolbox 2.0 (available at http://www.circuitscape.org/linkagemapper (accessed on 4 March 2021)) was used to calculate the least-cost distance between patches and obtain the least-cost paths (LCPs). The median distance refers to the distance corresponding to a dispersal probability of 0.5 and can be used to define the factor *k* in the equation for calculating the dispersal probability [25]. In the latest global PA network connectivity evaluation study, 10 km was used as the median distance [28]. Thus, we multiplied 10 km by the average value of the resistance surface of China (219.34) as the median cost distance, and then, we set k = 0.000316.

The PCinter_I of PA patch i is defined as the probability that an animal randomly departs from any point within this patch and successfully disperses to any point in other patches, and it can be calculated as follows:

$$\text{PCinter}_i = \frac{\sum_{j \neq i}^{n} a_j p_{ij}^*}{\sum_{j \neq i}^{n} a_j} = \frac{\sum_{j \neq i}^{n} a_j p_{ij}^*}{A_N - a_i} \tag{7}$$

The PCinter of the PA network is defined as the probability that an animal randomly departs from any point within the network and successfully disperses to any point located in different patches from the departure point. The probability that the departure point falls in patch i is $a_i / A_N (i = 1, 2 \ldots, n)$; thus, the probability of successful dispersal is as follows:

$$PCinter = \sum_{i=1}^{n} \frac{a_i}{A_N} * PCinter_i \tag{8}$$

After calculating PCintra for each PA patch and d_{ij} between patches, we calculated the PCinter of each PA patch using the Conefor 2.6 software [46].

2.3.3. Network Connectivity

The $PCnet_i$ of patch i is defined as the probability that an animal randomly departs from any point within this patch and successfully disperses to any point in the network. The probability that the destination point falls in patch j is $a_j / A_N (j = 1, 2 \ldots, n)$; thus, the probability of successful dispersal of an animal from patch i can be calculated as follows:

$$PCnet_i = \frac{\sum_{j=1}^{n} a_j P_{ij}^*}{A_N} = PCintra_i * \frac{a_i}{A_N} + PCinter_i * \frac{A_N - a_i}{A_N} \tag{9}$$

The PCnet of the PA network is defined as the probability that an animal randomly departs from any point in the network and successfully disperses to any point in the network. The probability that the departure point falls in patch i is $a_i / A_N (i = 1, 2 \ldots, n)$; thus, the probability of successful dispersal was calculated as follows:

$$PCnet = \sum_{i=1}^{n} \frac{a_i}{A_N} * PCnet_i \tag{10}$$

The proportion of connectivity of intra PA patches (PROCintra) indicator and the proportion of connectivity of inter PA patches (PROCinter) indicator describe the proportion of network connectivity provided by intra-patch connectivity and inter-patch connectivity, respectively.

The $PROCintra_i$ and $PROCinter_i$ of patch i can be calculated as follows:

$$PROCintra_i = \frac{a_i P_{ii}^*}{\sum_{j=1}^{n} a_j P_{ij}^*} \tag{11}$$

$$PROCinter_i = \frac{\sum_{j=1, j \neq i}^{n} a_j P_{ij}^*}{\sum_{j=1}^{n} a_j P_{ij}^*} = 1 - PROCintra_i \tag{12}$$

The PROCintra and PROCinter of the PA network can be calculated as follows:

$$PROCintra = \frac{\sum_{i=1}^{n} a_i * PCnet_i * PROCintra_i}{\sum_{i=1}^{n} a_i * PCnet_i} \tag{13}$$

$$PROCinter = \frac{\sum_{i=1}^{n} a_i * PCnet_i * PROCinter_i}{\sum_{i=1}^{n} a_i * PCnet_i} = 1 - PROCintra \tag{14}$$

2.3.4. PA–Landscape Connectivity

The $PCland_i$ of patch i is defined as the probability that an animal randomly departs from any point within this patch and successfully disperses to any point in the landscape. In this study, we assumed that when the destination point is out of the PA patches, the animal could not disperse successfully. The probability that the destination point falls in patch j is $a_j / A_L (j = 1, 2 \ldots, n)$; thus, the probability of successful dispersal of an animal from patch *i* can be calculated as follows:

$$PCland_i = \frac{\sum_{j=1}^{n} a_j P_{ij}^*}{A_L} = PCnet_i * \frac{A_N}{A_L} \tag{15}$$

The PCland of the PA network is defined as the probability that an animal randomly departs from any point in the PA network and successfully disperses to any point in the landscape. The probability that the destination point falls in patch i is a_i/A_L $(i = 1, 2 \ldots, n)$; thus, the probability of successful dispersal was calculated as follows:

$$PCland = \sum_{i=1}^{n} \frac{a_i}{A_N} * PCland_i = PCnet * \frac{A_N}{A_L} \tag{16}$$

2.3.5. PAs with Good Connectivity

According to the Post-2020 GBF objectives for PAs, it is necessary to define good connectivity. For PA patches and PA networks, when the PCland indicator reaches 30%, its PA–landscape connectivity is considered to be well; otherwise, its PA–landscape connectivity is not well based on the Post-2020 GBF. Similarly, we considered whether the PCintra, PCinter, and PCnet reach 90%, 50%, and 60% as the standards to judge whether the intra-patch connectivity, inter-patch connectivity, and network connectivity are good. There is a relative lack of research on the standards of good connectivity. There are two main reasons why we decided on these standards. First, these indicators have a relative size relationship; that is, for a PA patch, the value of PCintra is greater than the value of PCnet, and the value of PCnet is greater than the value of PCinter, so they should be given different standards. Second, 90%, 50%, and 60% are values that are easier for managers of PAs to understand. We discussed the impact of standards on the coverage of well-connected PAs in the discussion section.

2.4. Strategy Development for PA Connectivity

2.4.1. Strategy Classification of PA Connectivity Based on Indicators

We classified PA patches and PA networks into 16 categories based on whether the four aspects of connectivity were good or not, and each category corresponded to a four-letter string, although some may not actually exist. When a PA patch's intra connectivity was good, it was marked as category A; otherwise, it was marked as category B. We classified inter-patch connectivity, network connectivity and PA–landscape connectivity in the same way. We combined those letters in order of intra, inter, network and PA–landscape to obtain a four-letter string. For example, PAs classified as AAAA had good intra-patch, inter-patch, network and PA–landscape connectivity, and class ABBB only had good intra-patch connectivity.

When the intra-patch, inter-patch, network or PA–landscape connectivity reaches good, it should be maintained, and when it is not good, it should be improved. For example, PAs classified as AAAA needed to maintain the four aspects of connectivity, and class AAAB needed to maintain intra-patch, inter-patch and network connectivity and improve PA–landscape connectivity.

There are four strategies to improve PA connectivity (Figure 1). The enhancement of existing PAs through habitat restoration, construction of wildlife crossings, and other methods is a strategy to improve the intra-patch connectivity, which then can improve inter-patch, network and PA–landscape connectivity. The construction of ecological corridors is a widely used effective measure to improve inter-patch connectivity [7,47,48], which then can improve network and PA–landscape connectivity. Similar to ecological corridors, the expansion of existing PAs and the establishment of new PAs can reduce the cost distance between existing PA patches and thus improve inter-patch, network and PA–landscape connectivity. These two methods can also improve PA–landscape connectivity by increasing PA coverage.

2.4.2. Spatial Priority Area for PA Connectivity Improvement

Within existing PA patches requiring improved intra-patch connectivity, areas with a dispersal probability of less than 90% (corresponding to the good intra-patch connectivity standard) were identified as priority areas for enhancing existing PAs. We identified the

LCPs between two PA patches that both needed to improve inter-patch connectivity as priority ecological corridors. We considered high ecological integrity areas along these LCPs as priority areas for the expansion of existing PAs and the establishment of new PAs because of both high integrity and high connectivity contribution.

3. Results

3.1. Connectivity of PAs in China

Our result showed that the PCintra of China's PA network was 93.41%, which indicated that the connectivity within China's PA network is good. However, the PCintra of the 2153 PA patches varied greatly from 99.90% to 43.17% (Figure 2a). A total of 427 patches had good intra-patch connectivity, accounting for 11.28% of China's land area. The intra-patch connectivity of the PA network was not good in 22 of the 37 ecological zones (Table 1).

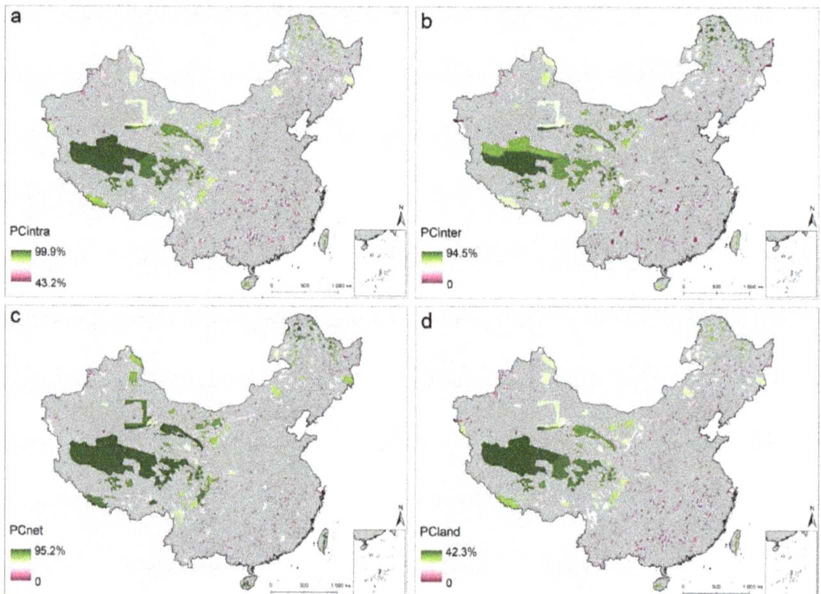

Figure 2. (**a**) Intra-patch connectivity of protected area (PA) patches in China based on PCintra indicator. (**b**) Inter-patch connectivity of PA patches in China based on PCinter indicator. (**c**) Network connectivity of PA patches in China based on PCnet indicator. (**d**) PA-landscape connectivity of PA patches in China based on PCland indicator.

Table 1. Connectivity indicators of ecological zones' PA network.

No.	Ecological Zone	PCintra of PAs (%)	PCinter of PAs (%)	PCnet of PAs (%)	PCland of PAs (%)
I1	Northern Daxing'anling cold-temperate semi-humid zone	96.33	65.06	67.17	10.18
I2	Southern Daxing'anling temperate semi-humid zone	92.11	20.59	23.16	3.32
I3	Xiaoxing'anling temperate semi-humid zone	91.28	49.57	50.87	8.63
I4	Northeast Plain temperate semi-humid zone	74.44	1.52	7.62	0.38
I5	Changbai Mountain temperate humid semi-humid zone	84.97	3.98	17.01	2.57
I6	Liaodong Peninsula warm-temperate semi-humid zone	77.20	2.21	6.20	0.31
II1	Yanshan Mountain warm-temperate semi-humid zone	77.64	1.26	7.78	0.36
II2	Haihe Plain warm-temperate semi-humid zone	59.36	0.09	12.87	0.33
II3	Shanxi Plateau warm-temperate semi-humid zone	82.92	2.51	8.48	0.24

Table 1. *Cont.*

No.	Ecological Zone	PCintra of PAs (%)	PCinter of PAs (%)	PCnet of PAs (%)	PCland of PAs (%)
II4	Northern Shaanxi and Longzhong Plateau warm-temperate semi-arid zone	81.11	4.07	14.31	0.41
II5	Southern Taihang and northern Qinling warm-temperate semi-humid zone	78.71	5.64	11.75	0.48
II6	Yellow and Huai River Plain warm-temperate semi-humid zone	55.20	0.01	10.60	0.29
II7	Shandong Peninsula warm-temperate semi-humid zone	64.65	0.10	9.22	0.16
III1	Middle and lower reaches of Yangtze River northern subtropical humid zone	69.97	0.49	8.84	0.38
III2	Middle and lower reaches of Yangtze River central subtropical humid zone	79.52	0.84	4.28	0.19
III3	Southeast China humid south subtropical zone	80.57	0.78	5.18	0.20
III4	Taiwan Island tropical subtropical humid zone	92.02	23.89	82.76	16.27
III5	Southeast China tropical humid zone	82.20	1.90	37.81	1.01
III6	Hainan Island tropical humid zone	84.57	21.78	73.35	12.19
IV1	Qinba Mountains northern subtropical humid zone	88.27	7.23	27.21	3.92
IV2	Sichuan basin and marginal mountains subtropical humid zone	88.25	30.58	48.33	5.68
IV3	Guizhou plateau and marginal mountains subtropical humid zone	79.79	2.57	6.56	0.24
IV4	Northern Transverse Mountains subtropical humid semi-humid zone	93.23	12.59	21.61	3.36
IV5	Southern Transverse Mountains central subtropical humid zone	81.58	9.18	15.38	1.60
IV6	Southwest China tropical subtropical humid zone	81.61	2.00	6.00	0.36
IV7	Eastern edge of the Himalayas tropical humid zone	93.56	33.41	79.35	6.59
V1	Xiliaohe River temperate semi-arid zone	81.07	4.50	16.99	1.48
V2	Eastern Inner Mongolia Plateau temperate semi-arid zone	90.28	6.20	17.50	2.78
V3	Ordos Plateau and surrounding mountains temperate semi-arid zone	83.97	14.15	21.96	2.42
V1	Western Inner Mongolia Plateau temperate arid zone	95.22	31.23	37.19	5.25
VI2	Northern Xinjiang temperate arid semi-arid zone	92.79	12.29	29.03	3.79
VI3	Southern Xinjiang temperate warm temperate arid zone	97.65	10.85	52.16	4.11
VII1	Kunlun Mountains alpine arid zone	99.68	32.30	87.34	39.65
VII2	Qaidam and Qilian Mountains alpine arid semi-arid zone	94.84	51.87	72.50	17.02
VII3	Qiangtang Plateau alpine arid zone	99.02	76.71	91.59	40.50
VII4	East Tibet and south Qinghai alpine semi-humid zone	95.20	48.74	81.37	27.88
VII5	Southern Tibetan alpine semi-humid semi-arid zone	93.29	13.45	55.92	9.17
VIII1	South China Sea islands tropical humid zone	—	—	—	—

The PCinter of China's PA network was 35.40%, which was not good. The PCinter of the PA patches varied from 94.50% to 0 (Figure 2b). A total of 116 patches had good inter-patch connectivity, accounting for 4.07% of China's land area. Only three ecological zones' PA network had good inter-patch connectivity (Table 1), including the Northern Daxing'anling cold-temperate semi-humid zone, the Qaidam and Qilian Mountains alpine arid semi-arid zone and the Qiangtang Plateau alpine arid zone (Ecological Zones I1, VII2 and VII3).

The PCnet of China's PA network was 58.43% and very close to good. The PCnet of the PA patches varied from 95.21% to 0 (Figure 2c). Only 90 PA patches had good network connectivity, accounting for 8.30% of China's land area. Eight ecological zones had good network connectivity (Table 1).

The PCland of China's PA network was 8.58%, which was not good. The PCland of the PA patches varied from 42.28% to 0 (Figure 2d). Only nine PA patches had good inter-patch connectivity, accounting for 5.92% of China's land area. Two ecological zones on the Qinghai-Tibet Plateau have good PA–landscape connectivity (Table 1), including the

Qiangtang Plateau alpine arid zone and the Kunlun Mountains alpine arid zone (Ecological Zones VII1 and VII3).

3.2. PA Connectivity Strategy Classification

Only seven PA patches located in Ecological Zones VII1 and VII4 were classified as AAAA, accounting for 2.67% of China's land area (Figure 3a). Two PA patches were classified as ABAA and also located in Ecological Zones VII1 and VII4, accounting for 3.26% of China's land area. A total of 72 PA patches were classified as AAAB, accounting for 1.16% of China's land area. These PA patches had good network connectivity and need to be extended or have new PAs established around them to improve their PA–landscape connectivity. Only 3, 5, 26, and 1 PA patches are classified as BAAB, ABAB, AABB, and BBAB, respectively. A total of 315 PA patches were classified as ABBB, accounting for 2.89% of China's land area. A total of 1714 PA patches were classified as BBBB, accounting for 2.89% of China's land area. These PA patches urgently needed to be improved in all aspects of connectivity. The connectivity of large PA patches was not necessarily good, and in fact, many large PA patches were classified as ABBB or BBBB (Figure 3b).

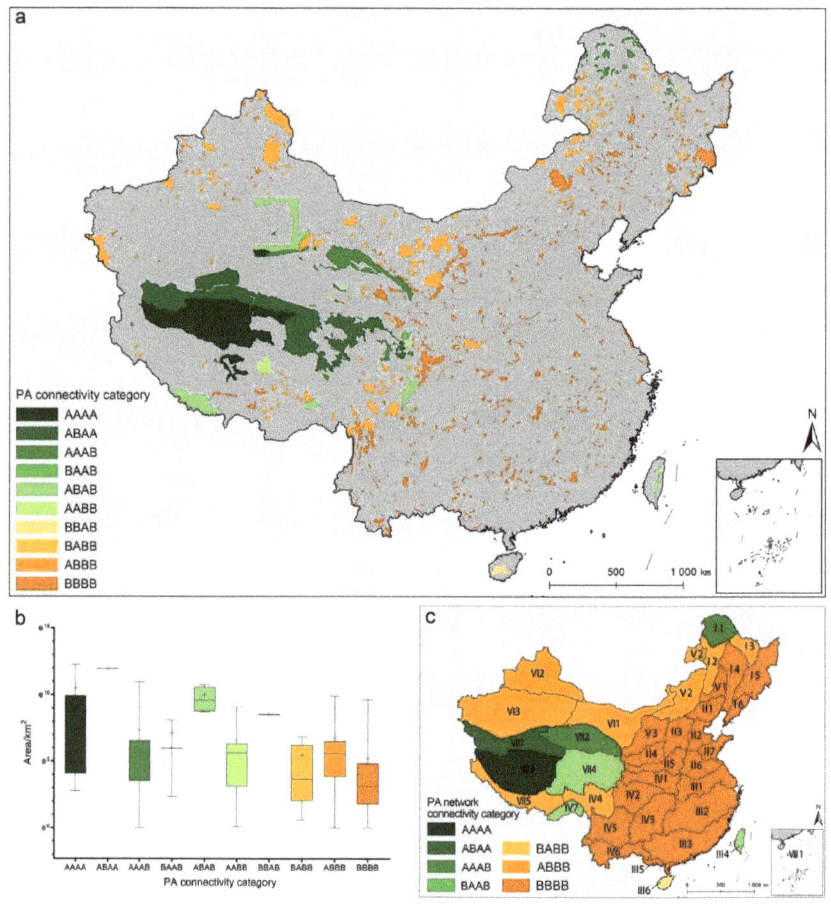

Figure 3. (**a**) Connectivity strategy classification of PA patches in China. (**b**) Distribution of area of PA patches under different categories of connectivity strategies in China. (**c**) Connectivity strategy classification of ecological zones' PA network in China.

Among the ecological zones, only the PA network of the Qiangtang Plateau alpine arid zone (Ecological Zone VII3) was classified as AAAA. Ecological Zone VII1 was classified as ABAA and should focus on improving inter-patch connectivity. Ecological Zones I1 and VII2 were classified as AAAB and should focus on improving PA–landscape connectivity by increasing PA coverage. Ecological Zones III4, IV7 and VII4 were classified as ABAB; this suggested that they should improve both inter-patch connectivity and PA–landscape connectivity. Ecological Zone III6 was classified as BBAB. Ecological Zones I2, I3, IV4, V2, VI1, VI2, VI3, and VII5 were classified as ABBB. The other 21 ecological zones were classified as BBBB and should urgently improve PA connectivity in multiple ways.

3.3. Spatial Priority Area to Improve PA Connectivity in China

A total of 17.24% of the area of existing PAs (243,060 km^2) were priority areas for connectivity enhancement to improve intra-patch connectivity (Figure 4). We identified 4344 potential priority ecological corridors between PAs (Figure 4). The priority area for expanding existing PAs included 1253 patches with a total area of 1,123,240 km^2, covering 11.70% of China's land area (Figure 4). The priority area for establishing new PAs included 9284 patches with a total area of 712,087 km^2, covering 7.41% of China's land area (Figure 4).

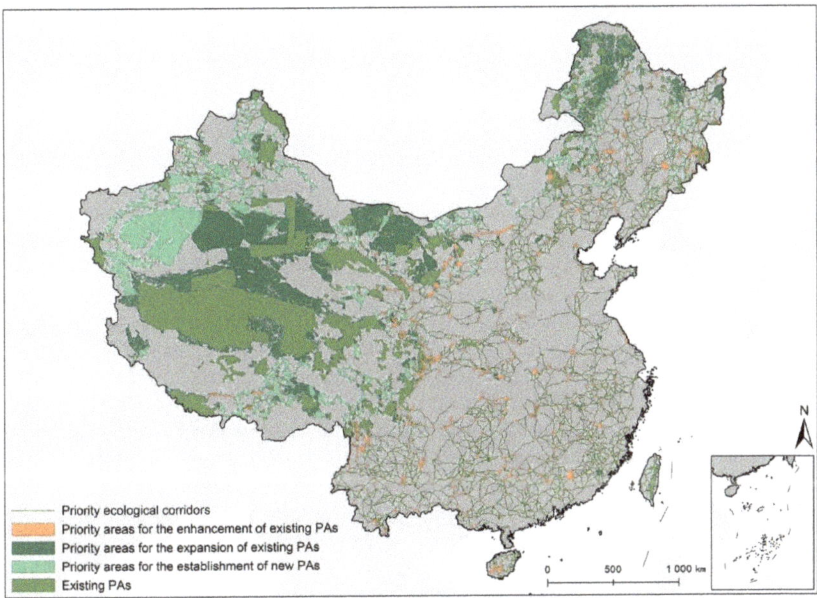

Figure 4. Spatial priority area to improve PA connectivity in China, including priority ecological corridor, priority areas for the enhancement of existing PAs, the expansion of existing PAs and the establishment of new PAs.

4. Discussion

4.1. Importance of Intra-Patch Connectivity

We suggested that the intra-patch connectivity should be regarded as important in both the evaluation and the improvement of PA connectivity. Some connectivity evaluation studies consider only inter-patch connectivity, ignoring the contribution of intra-patch connectivity to the overall connectivity, which can lead to erroneous conclusions in connectivity evaluations [31]. We calculated the PROCintra of China's PA network as 74.69%, which indicated that intra-patch connectivity contributed much more to network connectivity than inter-patch connectivity in China. The PROCintra values of 467 PA patches were higher than 75% and the PROCintra values of 213 patches were between 75% and 50% (Figure 5a). The PROCintra values of the PA network of 30 ecological zones were higher than the

PROCinter values (Figure 5b). We also found that there was no significant correlation between the value of PCintra indicator and PCnet for both PA patches and PA networks (Figure 5c,d). This indicated that the relationships between intra-patch, inter-patch, and network connectivity are complex.

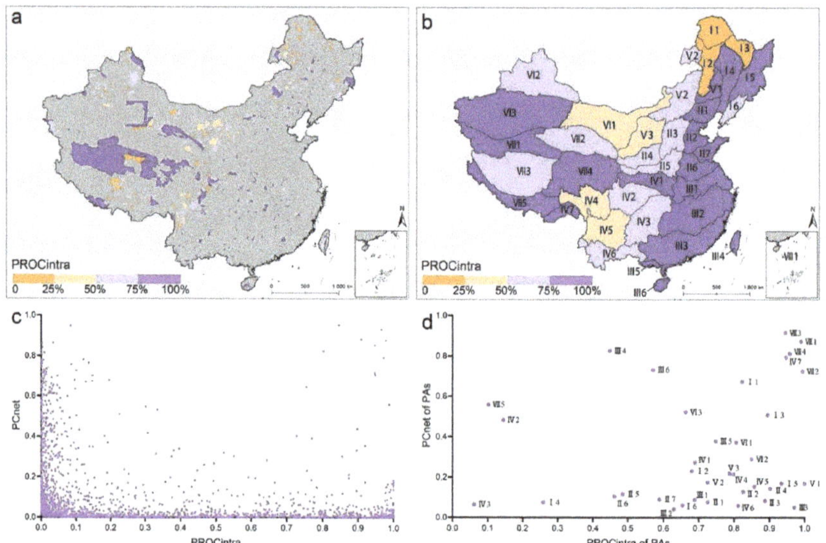

Figure 5. (**a**) The proportion of network connectivity of protected area (PA) patches provided by intra-patch connectivity in China based on PROCintra indicator. (**b**) The proportion of network connectivity of ecological zones' PA network provided by intra-patch connectivity in China based on PROCintra indicator. (**c**) Relationship between PCnet and PROCintra of PA patches in China. (**d**) Relationship between PCnet and PROCintra of ecological zones' PA network in China.

Research on the connectivity performance of PA management is lacking, leading to the assumption that PAs are effectively managed for connectivity in many studies [26]. Previous studies have generally assumed an excellent intra-patch connectivity (as a value of 1) [28,44]. We found that such assumptions may significantly overestimate the network connectivity of PAs. We calculated the PCnet and PCland indicator of each ecological zone's PA network assuming a PCintra of 1 for all PA patches (Table S1). Under this assumption, the PCnet of China would increase from 58.43% to 62.11%, and the Pcnet of Xiaoxing'anling temperate semi-humid zone (Ecological Zone I3) would increase from 50.87% to 71.07%. Clearly, overvalued network connectivity is not conducive to developing targeted enhancement strategies.

Improving intra-patch connectivity may effectively improve the connectivity of the PA network. For example, our findings showed that the Yellow and Huai River Plain warm-temperate semi-humid zone (Ecological Zone II6) had the poorest intra-patch connectivity of PAs in the ecological zones of China. If the PCintra of the PAs of this ecological zone is improved from 55.20% to 1, then the PCnet would improve from 10.60% to 19.44%. This result was consistent with previous studies suggesting that the connectivity within core areas is important [31]. This suggested that decision makers of PAs with similar circumstances should first begin to improve connectivity within PAs to ensure a high-quality PA system.

4.2. Evaluation of Connectivity at the Patch Scale

In the previous network connectivity analysis of PAs, some studies have discussed the contribution of patches to the connectivity of a PA network [44,49,50]. In addition, others have focused on mapping potential inter-patch dispersal routes [51]. The mapping studies have identified areas that are important as potential dispersal routes by applying concepts such as current density and betweenness centrality [40,52,53]. These studies have evaluated how well the PA network formed by the patches is connected, but they have not directly answered the question of how well connected the patches are. Therefore, the results might not directly guide managers in making decisions for PA patches.

Based on the dispersal probability between patches [43], we tried to extend the concept of PA connectivity from PA networks to PA patches. Our results showed that the connectivity strategy category of a PA network may be inconsistent with the connectivity strategy categories of PA patches within the network (Figure 3a,c). This indicated the need for connectivity evaluation at the patch scale.

Our PA connectivity evaluation framework for both PA patches and PA networks can support comparison and management decisions for the PA connectivity of countries, ecological zones, and administrative regions. Using our framework, the manager responsible for a PA can accurately assess the connectivity of the PA, apply a targeted approach to secure external funding and coordinate with managers of other PAs and external local governments. The manager of a region can clearly understand the connectivity of each PA in the region and how to enhance the connectivity of the regional PAs through coordination among the PAs.

4.3. Connectivity Indicators for Well-Connected PAs

It is important to identify connectivity's own target with accompanying indicators to guide global conservation efforts [54]. The four indicators we propose can be used as a basis to evaluate whether the PAs are well-connected. The coverage of well-connected PAs in a region or country can then be calculated to compare with the post-2020 biodiversity conservation targets. In fact, the coverage of well-connected PAs depends on the coverage of PAs and the indicator standard of good connectivity (Figure 6a). Future research can further discuss which indicators to choose and how to determine the standard of good connectivity. We believed that the combined use of these indicators would contribute to a comprehensive understanding of PA connectivity.

The Post-2020 GBF requires 30% global land area coverage of well-connected PAs, and according to this requirement, among the 37 ecological zones, only Ecological Zones VII1, VII3 and VII4 had more than 30% PA coverage (Figure 6b). No matter which indicator was chosen as a criterion for good connectivity, only these three ecological zones may have over 30% coverage of well-connected PAs. Our results showed that 11 ecological zones did not have PA patches with good intra-patch connectivity, 27 ecological zones did not have PA patches with good inter-patch connectivity, 25 ecological zones did not have PA patches with good network connectivity and 34 ecological zones did not have PA patches with good PA–landscape connectivity (Figure 6c–f). Compared with the results of connectivity indicators of the PA network, the coverage of well-connected PAs more strongly indicated that the PA connectivity of these ecological zones urgently needs to be improved. We suggest that specifying which indicator or series of indicators to use in the Post 2020 GBF objectives is necessary to facilitate global awareness and begin initiatives to improve connectivity. At the same time, we recommend that countries consider using the series of indicators in our framework to describe PA connectivity to drive comprehensive conservation and enhancement measures at all levels.

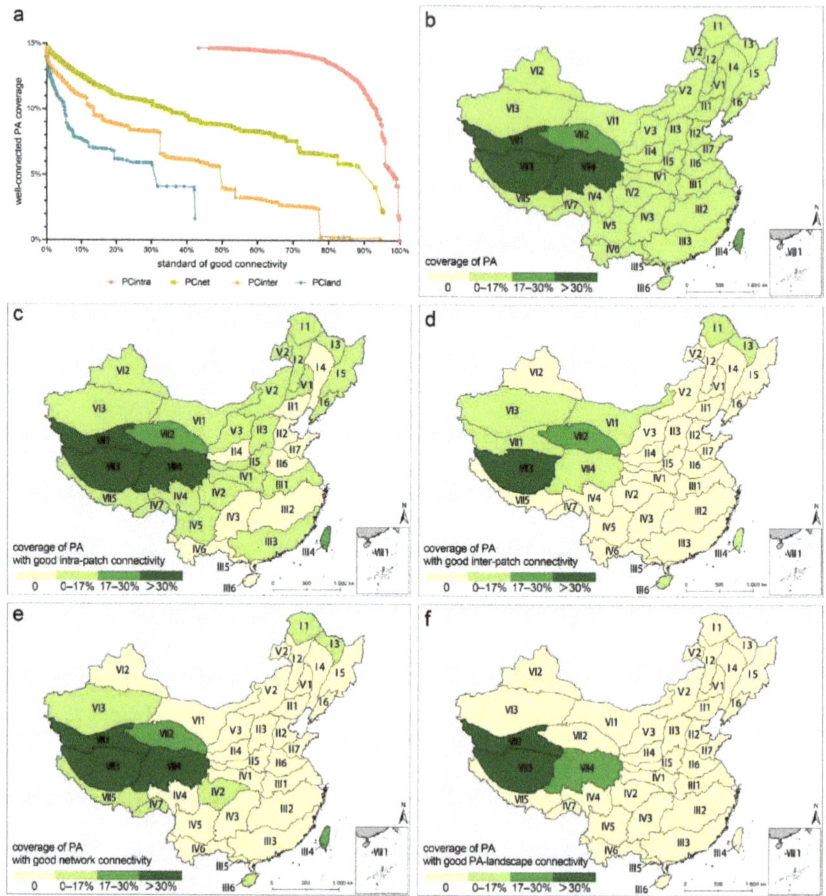

Figure 6. (**a**) Relationship between the standards of connectivity indicators to determine good PA connectivity and the well-connected PA coverage in China, including PCintra, PCinter, PCnet and PCland. (**b**) The PA coverage of ecological zones in China. (**c**) The well-connected PA coverage of ecological zones in China based on intra-patch connectivity. (**d**) The well-connected PA coverage of ecological zones in China based on inter-patch connectivity. (**e**) The well-connected PA coverage of ecological zones in China based on network connectivity. (**f**) The well-connected PA coverage of ecological zones in China based on PA–landscape connectivity.

4.4. Limitations and Future Research

First, uncertainties exist in the creation of the resistance surface. Both the selection of human modification data and the calculation method of transforming human modification data into resistance surface would bring uncertainty to the resistance surface. This has implications for the creation of dispersal probability surface and cost distances between PA patches based on resistance surfaces and thus creates uncertainties in the PA connectivity evaluation results. Many studies have discussed how to create resistance surfaces in connectivity research [30,39,53], but there is not a high degree of consensus among researchers on this question. Future research could focus on how to create resistance surfaces to evaluate PA connectivity.

Second, we did not consider the effect of PAs' shape and area on intra-patch connectivity and led to uncertainty in the evaluation result of PAs' intra-patch connectivity. A more reasonable evaluation method of intra-patch connectivity, such as the use of least-cost distance model or circuit model, is necessary in the future.

Third, the selection of median cost distance will bring uncertainty to the evaluation of connectivity between PAs. Some studies have analyzed the effect of median distance on inter-patch connectivity when using Euclidean distance to evaluate inter-patch connectivity [25,26], but scholars have not reached a high level of consensus on this issue. Future research should discuss how to determine the median cost distance when using least-cost distance to evaluate inter-patch connectivity.

5. Conclusions

In this study, we have proposed a unified framework to evaluate and develop strategies for PA connectivity, and the results can directly guide management decisions. This study proposed a conceptual framework for the connectivity of PAs that includes intra-patch, inter-patch, network and PA–landscape connectivity for both PA patches and PA networks, which can be evaluated logically and consistently in this framework. This framework provides a set of indicators for the post-2020 biodiversity conservation targets on well-connected PAs. The proposed framework considers the differences in the intra-patch connectivity of PAs and thus might provide a better evaluation of PAs' inter-patch connectivity and network connectivity. The framework also includes how to develop strategies and identify priority areas to improve PA connectivity based on the evaluation results of PAs' connectivity indicators. This study shows that the connectivity of China's PAs is not good and needs to be improved. At the same time, the PA connectivity of the Qinghai-Tibet Plateau is relatively good, and attention should be paid to maintaining the connectivity of existing PAs in this region. The method proposed in this study can be used for the evaluation, improvement, and spatial planning of the connectivity of PAs at regional, national, and global scales. Our conceptual framework, indicators, and evaluation methods for connectivity can also be widely used in landscape connectivity research.

Supplementary Materials: The following supporting information can be downloaded at: https://www.mdpi.com/article/10.3390/land11101670/s1, Figure S1: Resistance Surface of China; Figure S2: Dispersal Probability Surface of PAs in China; Table S1: Connectivity indicators of ecoregions' PA network in China under the assumption that the intra-patch connectivity of PA patches is very good, that is, the PCintra index of all the PA patches is 1.

Author Contributions: Conceptualization, Z.Z., P.W. and R.Y.; methodology, Z.Z. and P.W.; software, Z.Z. and P.W.; validation, Z.Z.; formal analysis, Z.Z. and P.W.; resources, R.Y.; data curation, P.W., X.W., F.W., T.-H.T., Y.C. and Z.Z.; writing—original draft preparation, Z.Z. and P.W.; writing—review and editing, X.W., F.W., Y.C., S.H., J.P. and R.Y.; visualization, P.W.; supervision, R.Y. and Z.Z.; project administration, R.Y.; funding acquisition, R.Y. and Z.Z. All authors have read and agreed to the published version of the manuscript.

Funding: This research was funded by the National Natural Science Foundation of China (Grant No. 51708323 and 51978365) and the Tsinghua University Initiative Scientific Research Program (Grant No. 20223080018).

Data Availability Statement: Not applicable.

Acknowledgments: We sincerely thank the researchers helping us to collect the data used in this study. We also sincerely thank four anonymous reviewers and the journal editors.

Conflicts of Interest: The authors declare no conflict of interest.

References

1. Díaz, S.; Fargione, J.; Chapin, F.S.; Tilman, D. Biodiversity Loss Threatens Human Well-Being. *PLoS Biol.* **2006**, *4*, e277. [CrossRef]
2. Patz, J.A.; Campbell-Lendrum, D.; Holloway, T.; Foley, J.A. Impact of Regional Climate Change on Human Health. *Nature* **2005**, *438*, 310–317. [CrossRef]
3. Wheeler, T.; von Braun, J. Climate Change Impacts on Global Food Security. *Science* **2013**, *341*, 508–513. [CrossRef]
4. Keeley, A.T.H.; Ackerly, D.D.; Cameron, D.R.; Heller, N.E.; Huber, P.R.; Schloss, C.A.; Thorne, J.H.; Merenlender, A.M. New Concepts, Models, and Assessments of Climate-Wise Connectivity. *Environ. Res. Lett.* **2018**, *13*, 073002. [CrossRef]
5. Littlefield, C.E.; McRae, B.H.; Michalak, J.L.; Lawler, J.J.; Carroll, C. Connecting Today's Climates to Future Climate Analogs to Facilitate Movement of Species under Climate Change. *Conserv. Biol.* **2017**, *31*, 1397–1408. [CrossRef] [PubMed]

6.   Gross, J.E.; Woodley, S.; Welling, L.A.; Watson, J.E.M. (Eds.) *Adapting to Climate Change: Guidance for Protected Area Managers and Planners*; Best Practice Protected Area Guidelines Series No. 24; IUCN: Gland, Switzerland, 2017.

7.   Hilty, J.; Worboys, G.L.; Keeley, A.; Woodley, S.; Lausche, B.J.; Locke, H.; Carr, M.; Pulsford, I.; Pittock, J.; White, J.W.; et al. *Guidelines for Conserving Connectivity through Ecological Networks and Corridors*; Best Practice Protected Area Guidelines Series No. 30; IUCN: Gland, Switzerland, 2020.

8.   Foden, W.B.; Young, B.E. (Eds.) *IUCN SSC Guidelines for Assessing Species' Vulnerability to Climate Change*; Occasional Paper of the IUCN Species Survival Commission No. 59; IUCN Species Survival Commission: Cambridge, UK; Gland, Switzerland, 2016.

9.   Liang, J.; Ding, Z.; Jiang, Z.; Yang, X.; Xiao, R.; Singh, P.B.; Hu, Y.; Guo, K.; Zhang, Z.; Hu, H. Climate Change, Habitat Connectivity, and Conservation Gaps: A Case Study of Four Ungulate Species Endemic to the Tibetan Plateau. *Landsc. Ecol.* **2021**, *36*, 1071–1087. [CrossRef]

10.   Grande, T.O.; Aguiar, L.M.S.; Machado, R.B. Heating a Biodiversity Hotspot: Connectivity Is More Important than Remaining Habitat. *Landsc. Ecol.* **2020**, *35*, 639–657. [CrossRef]

11.   Pascual-Hortal, L.; Saura, S. Comparison and Development of New Graph-Based Landscape Connectivity Indices: Towards the Priorization of Habitat Patches and Corridors for Conservation. *Landsc. Ecol.* **2006**, *21*, 959–967. [CrossRef]

12.   Haddad, N.M.; Brudvig, L.A.; Clobert, J.; Davies, K.F.; Gonzalez, A.; Holt, R.D.; Lovejoy, T.E.; Sexton, J.O.; Austin, M.P.; Collins, C.D.; et al. Habitat Fragmentation and Its Lasting Impact on Earth's Ecosystems. *Sci. Adv.* **2015**, *1*, e1500052. [CrossRef]

13.   Pimm, S.L.; Jenkins, C.N.; Li, B.V. How to Protect Half of Earth to Ensure It Protects Sufficient Biodiversity. *Sci. Adv.* **2018**, *4*, eaat2616. [CrossRef] [PubMed]

14.   Chape, S.; Harrison, J.; Spalding, M.; Lysenko, I. Measuring the Extent and Effectiveness of Protected Areas as an Indicator for Meeting Global Biodiversity Targets. *Philos. Trans. R. Soc. B Biol. Sci.* **2005**, *360*, 443–455. [CrossRef] [PubMed]

15.   Thomas, C.D.; Gillingham, P.K. The Performance of Protected Areas for Biodiversity under Climate Change. *Biol. J. Linn. Society* **2015**, *115*, 718–730. [CrossRef]

16.   Yang, R.; Cao, Y.; Hou, S.; Peng, Q.; Wang, X.; Wang, F.; Tseng, T.-H.; Yu, L.; Carver, S.; Convery, I.; et al. Cost-Effective Priorities for the Expansion of Global Terrestrial Protected Areas: Setting Post-2020 Global and National Targets. *Sci. Adv.* **2020**, *6*, eabc3436. [CrossRef] [PubMed]

17.   Carr, M.H.; Robinson, S.P.; Wahle, C.; Davis, G.; Kroll, S.; Murray, S.; Schumacker, E.J.; Williams, M. The Central Importance of Ecological Spatial Connectivity to Effective Coastal Marine Protected Areas and to Meeting the Challenges of Climate Change in the Marine Environment. *Aquat. Conserv.* **2017**, *27*, 6–29. [CrossRef]

18.   Bauduin, S.; Cumming, S.G.; St-Laurent, M.H.; McIntire, E.J.B. Integrating Functional Connectivity in Designing Networks of Protected Areas under Climate Change: A Caribou Case-Study. *PLoS ONE* **2020**, *15*, e0238821. [CrossRef] [PubMed]

19.   Convention on Biological Diversity. Aichi Biodiversity Targets. 2010. Available online: https://www.cbd.int/sp/targets/ (accessed on 30 November 2021).

20.   Convention on Biological Diversity. First Draft of the Post-2020 Global Biodiversity Framework. 2021. Available online: https://www.cbd.int/doc/c/abb5/591f/2e46096d3f0330b08ce87a45/wg2020-03-03-en.pdf (accessed on 30 November 2021).

21.   Hashemi, R.; Darabi, H. The Review of Ecological Network Indicators in Graph Theory Context: 2014–2021. *Int. J. Environ. Res.* **2022**, *16*, 1–26. [CrossRef]

22.   Keeley, A.T.H.; Beier, P.; Jenness, J.S. Connectivity Metrics for Conservation Planning and Monitoring. *Biol. Conserv.* **2021**, *255*, 109008. [CrossRef]

23.   Saura, S.; Pascual-Hortal, L. A New Habitat Availability Index to Integrate Connectivity in Landscape Conservation Planning: Comparison with Existing Indices and Application to a Case Study. *Landsc. Urban Plan.* **2007**, *83*, 91–103. [CrossRef]

24.   Saura, S.; Estreguil, C.; Mouton, C.; Rodríguez-Freire, M. Network Analysis to Assess Landscape Connectivity Trends: Application to European Forests (1990–2000). *Ecol. Indic.* **2011**, *11*, 407–416. [CrossRef]

25.   Saura, S.; Bastin, L.; Battistella, L.; Mandrici, A.; Dubois, G. Protected Areas in the World's Ecoregions: How Well Connected Are They? *Ecol. Indic.* **2017**, *76*, 144–158. [CrossRef]

26.   Saura, S.; Bertzky, B.; Bastin, L.; Battistella, L.; Mandrici, A.; Dubois, G. Protected Area Connectivity: Shortfalls in Global Targets and Country-Level Priorities. *Biol. Conserv.* **2018**, *219*, 53–67. [CrossRef] [PubMed]

27.   Ward, M.; Saura, S.; Williams, B.; Ramírez-Delgado, J.P.; Arafeh-Dalmau, N.; Allan, J.R.; Venter, O.; Dubois, G.; Watson, J.E.M. Just Ten Percent of the Global Terrestrial Protected Area Network Is Structurally Connected via Intact Land. *Nat. Commun.* **2020**, *11*, 4563. [CrossRef] [PubMed]

28.   Saura, S.; Bertzky, B.; Bastin, L.; Battistella, L.; Mandrici, A.; Dubois, G. Global Trends in Protected Area Connectivity from 2010 to 2018. *Biol. Conserv.* **2019**, *238*, 108183. [CrossRef] [PubMed]

29.   Brennan, A.; Naidoo, R.; Greenstreet, L.; Mehrabi, Z.; Ramankutty, N.; Kremen, C. Functional Connectivity of the World's Protected Areas. *Science* **2022**, *376*, 1101–1104. [CrossRef]

30.   Cao, Y.; Yang, R.; Carver, S. Linking Wilderness Mapping and Connectivity Modelling: A Methodological Framework for Wildland Network Planning. *Biol. Conserv.* **2020**, *251*, 108679. [CrossRef]

31.   Spanowicz, A.G.; Jaeger, J.A.G. Measuring Landscape Connectivity: On the Importance of within-Patch Connectivity. *Landsc. Ecol.* **2019**, *34*, 2261–2278. [CrossRef]

32.   Urban, D.; Keitt, T. Landscape Connectivity: A Graph-Theoretic Perspective. *Ecology* **2001**, *82*, 1205–1218. [CrossRef]

33. Guo, Z.; Cui, G. The Comprehensive Geographical Regionalization of China Supporting Natural Conservation. *Shengtai Xuebao Acta Ecol. Sin.* **2014**, *34*, 1284–1294. [CrossRef]

34. Hoctor, T.S.; Carr, M.H.; Zwick, P.D. Identifying a Linked Reserve System Using a Regional Landscape Approach: The Florida Ecological Network. *Conserv. Biol.* **2000**, *14*, 984–1000. [CrossRef]

35. Marulli, J.; Mallarach, J.M. A GIS Methodology for Assessing Ecological Connectivity: Application to the Barcelona Metropolitan Area. *Landsc. Urban Plan.* **2005**, *71*, 243–262. [CrossRef]

36. Beier, P.; Spencer, W.; Baldwin, R.F.; McRae, B.H. Toward Best Practices for Developing Regional Connectivity Maps. *Conserv. Biol.* **2011**, *25*, 879–892. [CrossRef] [PubMed]

37. Jacobson, A.P.; Riggio, J.; Tait, A.M.; Baillie, J.E.M. Global Areas of Low Human Impact ('Low Impact Areas') and Fragmentation of the Natural World. *Sci. Rep.* **2019**, *9*, 14179. [CrossRef] [PubMed]

38. Cao, Y.; Carver, S.; Yang, R. Mapping Wilderness in China: Comparing and Integrating Boolean and WLC Approaches. *Landsc. Urban Plan.* **2019**, *192*, 103636. [CrossRef]

39. Belote, R.T.; Dietz, M.S.; McRae, B.H.; Theobald, D.M.; McClure, M.L.; Irwin, G.H.; McKinley, P.S.; Gage, J.A.; Aplet, G.H. Identifying Corridors among Large Protected Areas in the United States. *PLoS ONE* **2016**, *11*, e0154223. [CrossRef]

40. Dickson, B.G.; Albano, C.M.; McRae, B.H.; Anderson, J.J.; Theobald, D.M.; Zachmann, L.J.; Sisk, T.D.; Dombeck, M.P. Informing Strategic Efforts to Expand and Connect Protected Areas Using a Model of Ecological Flow, with Application to the Western United States. *Conserv. Lett.* **2017**, *10*, 564–571. [CrossRef]

41. Correa Ayram, C.A.; Mendoza, M.E.; Etter, A.; Pérez Salicrup, D.R. Anthropogenic Impact on Habitat Connectivity: A Multidimensional Human Footprint Index Evaluated in a Highly Biodiverse Landscape of Mexico. *Ecol. Indic.* **2017**, *72*, 895–909. [CrossRef]

42. Kennedy, C.M.; Oakleaf, J.R.; Theobald, D.M.; Baruch-Mordo, S.; Kiesecker, J. Managing the Middle: A Shift in Conservation Priorities Based on the Global Human Modification Gradient. *Glob. Chang. Biol.* **2019**, *25*, 811–826. [CrossRef]

43. Bunn, A.G.; Urban, D.L.; Keitt, T.H. Landscape Connectivity: A Conservation Application of Graph Theory. *J. Environ. Manag.* **2000**, *59*, 265–278. [CrossRef]

44. Gurrutxaga, M.; Rubio, L.; Saura, S. Key Connectors in Protected Forest Area Networks and the Impact of Highways: A Transnational Case Study from the Cantabrian Range to the Western Alps (SW Europe). *Landsc. Urban Plan.* **2011**, *101*, 310–320. [CrossRef]

45. Naidoo, R.; Brennan, A. Connectivity of Protected Areas Must Consider Landscape Heterogeneity: A Response to Saura et al. *Biol. Conserv.* **2019**, *239*, 108316. [CrossRef]

46. Saura, S.; Torné, J. Conefor Sensinode 2.2: A Software Package for Quantifying the Importance of Habitat Patches for Landscape Connectivity. *Environ. Model. Softw.* **2009**, *24*, 135–139. [CrossRef]

47. Chetkiewicz, C.-L.B.; Clair, C.C.S.; Boyce, M.S. Corridors for Conservation: Integrating Pattern and Process. *Annu. Rev. Ecol. Evol. Syst.* **2006**, *37*, 317–342. [CrossRef]

48. Gilbert-Norton, L.; Wilson, R.; Stevens, J.R.; Beard, K.H. A Meta-analytic Review of Corridor Effectiveness. *Conserv. Biol.* **2010**, *24*, 660–668. [CrossRef]

49. Bodin, Ö.; Saura, S. Ranking Individual Habitat Patches as Connectivity Providers: Integrating Network Analysis and Patch Removal Experiments. *Ecol. Model.* **2010**, *221*, 2393–2405. [CrossRef]

50. Zhang, J.; Jiang, F.; Cai, Z.; Dai, Y.; Liu, D.; Song, P.; Hou, Y.; Gao, H.; Zhang, T. Resistance-Based Connectivity Model to Construct Corridors of the Przewalski's Gazelle (Procapra Przewalskii) in Fragmented Landscape. *Sustainability* **2021**, *13*, 1656. [CrossRef]

51. Diniz, M.F.; Cushman, S.A.; Machado, R.B.; de Marco Júnior, P. Landscape Connectivity Modeling from the Perspective of Animal Dispersal. *Landsc. Ecol.* **2020**, *35*, 41–58. [CrossRef]

52. Carroll, K.A.; Hansen, A.J.; Inman, R.M.; Lawrence, R.L.; Hoegh, A.B. Testing Landscape Resistance Layers and Modeling Connectivity for Wolverines in the Western United States. *Glob. Ecol. Conserv.* **2020**, *23*, e01125. [CrossRef]

53. Barnett, K.; Belote, R.T. Modeling an Aspirational Connected Network of Protected Areas across North America. *Ecol. Appl.* **2021**, *31*, e02387. [CrossRef] [PubMed]

54. Belote, R.T.; Beier, P.; Creech, T.; Wurtzebach, Z.; Tabor, G. A Framework for Developing Connectivity Targets and Indicators to Guide Global Conservation Efforts. *Bioscience* **2020**, *70*, 122. [CrossRef] [PubMed]

Article

Ecological Network Construction of a National Park Based on MSPA and MCR Models: An Example of the Proposed National Parks of "Ailaoshan-Wuliangshan" in China

Caihong Yang [1], Huijun Guo [2,*], Xiaoyuan Huang [1,*], Yanxia Wang [1], Xiaona Li [1] and Xinyuan Cui [1]

[1] School of Geography and Ecotourism, Southwest Forestry University (SWFU), Kunming 650224, China
[2] National Plateau Wetlands Research Center, Southwest Forestry University (SWFU), Kunming 650233, China
* Correspondence: hjguo@swfu.edu.cn (H.G.); hxy21cn@swfu.edu.cn (X.H.); Tel.: +86-138-8824-1697 (X.H.)

Abstract: The establishment of ecological networks facilitates genetic exchange among species in national parks and is an effective means of avoiding habitat fragmentation. Using the proposed "Ailaoshan-Wuliangshan" in Yunnan Province, China, as the study area, the identification of ecological source sites using the morphological spatial pattern analysis (MSPA) method, extraction of potential ecological corridors using the minimum resistance model (MCR) and construction of the ecological network of national parks were performed. Based on the gravity model, important ecological corridors were selected, and corresponding ecological network optimization strategies were presented. The results showed that (1) the core area identified by MSPA was 4440.08 km^2, with a low degree of fragmentation, and is distributed in strips within the woodland land classes in the study area; (2) the establishment of an ecological network model of least cost resistance based on 10 indicators in four dimensions of land tenure, geographic factors, vegetation characteristics, and human meddling; (3) the ecological network included 13 ecological source sites, 77 potential ecological corridors, 48 important ecological corridors and 25 pedestrian pathways and extracts an optimal ecological corridor connecting with the natural reserve; and (4) the network closure degree of the constructed ecological network was (1.18), line point rate (3.08), network connectivity (1.12), and cost ratio (0.98). By using the proposed ecological network construction method, ecological patches and potential corridors can be accurately identified to ensure the integrity and connectivity of the national park while minimizing the land demand pressure of the surrounding communities, which provides some reference for the construction of other national parks' ecological networks in China.

Keywords: national park; MSPA; MCR; ecological corridor

Citation: Yang, C.; Guo, H.; Huang, X.; Wang, Y.; Li, X.; Cui, X. Ecological Network Construction of a National Park Based on MSPA and MCR Models: An Example of the Proposed National Parks of "Ailaoshan-Wuliangshan" in China. *Land* **2022**, *11*, 1913. https://doi.org /10.3390/land11111913

Academic Editors: Rui Yang, Yue Cao, Steve Carver and Le Yu

Received: 23 September 2022
Accepted: 25 October 2022
Published: 27 October 2022

Publisher's Note: MDPI stays neutral with regard to jurisdictional claims in published maps and institutional affiliations.

1. Introduction

China is one of the hotspots of biological habitats and biodiversity [1], and several critical biodiversity areas, including the Hengduan Mountains, the Tibetan Plateau mountains, Xishuangbanna in southern Yunnan, the Qinling Mountains, the Changbai Mountains, and the Tianshan Mountains, have been proposed based on species richness and the number of endemic species [2]. With the rate of urbanization accelerating, the natural environment is being destroyed to varying degrees, and landscape connectivity is diminishing. Habitat fragmentation can lead to the isolation of biological populations, which greatly increases the likelihood of extinction and poses a serious threat to biodiversity [3]. The national park system is designed to effectively protect the originality and integrity of the most nationally representative natural ecosystems with a broad scope of protection and comprehensive ecological processes. In June 2019, the General Office of the Central Committee of the Communist Party of China and the General Office of the State Council issued the Guiding Opinions on the Establishment of a Nature Reserve System with National Parks as the Mainstay, emphasizing the main position of the conservation value and ecological functions of national parks in the national nature reserve system. No other types of nature reserves will be maintained or established in the same areas after national parks are established [4]. Official approved national parks due to the late start of national park research in China are Giant Panda National Park, SanJiangYuan National Park and Northeast Tiger National Park as well as Leopard National Park HaiNan Tropical Rainforest and WuYiShan National Park [5], which have a large gap with the construction of the proposed national park-based nature reserve system in China [6]. Despite the large number of nature reserves of various types and functions, China has played an important role in biodiversity protection and national ecological security maintenance. However, due to the fragmented distribution and fragmentation of nature reserves [7], their variable size and small protection areas, and the distribution of a large number of remaining forests, villages, towns, and agricultural lands around them, ecosystem integrity is blocked, and integrity and connectivity are not robust [8]. Both the Conservation for Biodiversity Aichi Targets and the Post-2020 Global Biodiversity Framework (GBF), which is under discussion globally, emphasize the importance of PA connectivity and set global PA connectivity targets [9]. Aichi Targets and the First Draft of the Post-2020 GBF call for 17% and 30%, respectively, of the global land area to be conserved through well-connected PAs and other effective area-based conservation measures (OECMs) [9,10]. Therefore, how to build an ecological network system and realize ecosystem integrity by relying on existing nature reserves is the key issue facing national park building.

Ecological networks are made up of patches and their connections to achieve effective conservation of species diversity through the establishment of ecological corridors through fragmented natural systems [11]. In terms of construction methods, "identifying ecological sources—constructing resistance surfaces—extracting ecological corridors" has become the basic framework for constructing ecological networks [12,13]. There are usually two methods for determining ecological source sites: one is to directly select nature reserves, attractions, and forest parks as ecological source sites based on the empirical judgement of professionals [14], which is subject to more subjective interference and ignores the connecting role of patches in the landscape [15]. Second, morphological spatial pattern analysis (MSPA), proposed by Vogt et al. to achieve the measurement and identification of spatial patterns of forest landscapes by correlating morphological features with specific shapes in raster images [16], is widely used in forest fragmentation and urban green space system research [17,18]. This method is different from the traditional method of selecting only nature reserves, forest parks, etc., as ecological source sites can classify the spatial pattern of raster images more precisely in terms of functional-type structures and then identify landscape types with different ecological meanings and increase the scientificity of ecological source sites [19]. Species migration and exchange between different ecological source sites can only be achieved if resistance is overcome, and the resistance surface is the total cost of overcoming multiple resistance factors formed between patches during species migration. The ease of species migration between different landscape units varies. The higher the suitability of the patch is, the lower the resistance of species migration, and the resistance is mainly influenced by factors such as topography, land use type and the intensity of human interference. Combining the basis of existing resistance surface-related studies at home and abroad [20–22], this study constructs an ecological network resistance surface based on 10 indicators in four dimensions: land tenure, geographic factors, vegetation characteristics and human interference. Ecological corridors can improve landscape connectivity and contribute to species dispersal and maintain gene flow between populations. Many methods have been used today to identify ecological corridors, such as individual-based movement models, connectivity probability (PC), and circuit theory [23–25]. Currently, the minimal cumulative resistance model (MCR) has become the mainstream method for identifying ecological corridors. The method was first proposed in 1992 by Knaapen et al. by calculating the minimum consumption path between the source and target and the optimal path for the outward spread, migration and dispersal of species [26], which can effectively avoid interference from the external environment and well reflect the possibilities and trends of movement of living species between habitat areas, thus protecting biodiversity [27–29]. There are an increasing number of studies combining MSPA and MCR models to construct ecological networks, but they are mostly used for urban ecological network construction by identifying the central nodes of urban ecological source sites and establishing urban ecological networks in combination with road and water networks and mainly focusing on the construction of ecological networks within the city (a whole) [30,31]. Currently, there are few studies on the construction of ecological networks in national parks [32,33]. For example, the ecological networks of Giant Panda National Park and Shuangzi Mountain National Forest Park were constructed by using 3S technology and the theory of landscape ecology to identify ecological source sites. The least cumulative resistance method was used to simulate important corridors and potential corridors, and an ecological network optimization strategy was proposed. The ecological network of urban parks based on birds is constructed by using several factors in the InVEST model to determine the suitable ecological source sites for birds, and the corridor is extracted by constructing a resistance surface with three indicators: land use type, road and water system. This study will address the problem of insufficient spatial connectivity in national parks composed of multiple nature reserves and provide a scientific basis for achieving national park connectivity and integrity by constructing potential ecological corridors and ecological networks using MSPA and MCR models. The objectives of this study were as follows: (1) Build an ecological network of national parks with multiple protected areas to improve the integrity and connectivity of national park ecosystems. (2) The identification of ecological corridors and ecological networks in national parks using a combination of MSPA and MCR models. (3) The determination of the resistance surface involves several factors, and the weight of the human interference factor was set to a higher value to form a more reasonable ecological resistance surface.

2. Materials and Methods

2.1. Study Area

The proposed "Ailaoshan-Wuliangshan" National Park, consisting of the Ailao Mountains National Nature Reserve, WuLiang Mountains National Nature Reserve and Dinosaur River State Nature Reserve, is located in the central part of Yunnan Province within Jingdong County's territory, Zhenyuan County, Xinping County, Chuxiong City, Shuangbai County, Nanhua County, and Nanjian County, which are linked by four states (cities), namely, Pu'er City, Yuxi City, Chien Yi Autonomous Prefecture, and Dali Bai Autonomous Prefecture, with geographic coordinates. The geographical coordinates are 23°46′50.75″~24°56′06.35″ north latitude and 100°19′07.95″~101°37′54.19″ east longitude (Figure 1).

With a length of 180 km from north to south, a width of 130 km from east to west, an elevation of 452~3348 m and a total area of 1652.82 km². The "Ailaoshan-Wuliangshan" Mountains Conservation Area is a major conservation area and boundary zone in China, located at the intersection of two geographic units, the Hengduan Mountains in western Yunnan and the plateau in eastern Yunnan, and is a significant corridor for tropical to temperate transition, species migration, and gene exchange in the Asian continent, as well as one of eight routes for global migration of birds. Ecological security, with the complex biodiversity composition and obvious transitional characteristics of the flora and fauna, directly affects Vietnam, Laos, Myanmar and even many countries in the southern and southeastern subregion bordering Yunnan [34]. In parallel, the region is also the water catchment area and ecological conservation area of two major cross-border rivers, namely, the Lancang River (Mekong) and Yuanjiang River (Red), which has great significance to maintain the international ecological security of the area. The Mount Ailao Mountains–Wuliang Mountains are a typical representative of the subtropical forest ecosystem and belong to the priority land protection ecosystem of China, among which the Zhongshan broadleaf wet evergreen forest area is the largest and most comprehensive broadleaf leaf wet evergreens in China, preserving the largest area of broadleaf mountain evergreen forests in the subtropics of China, which is a completely primitive state, has stable natural breeding resources, and has extremely rich wildlife resources [35,36]. In both mountainous areas, the diversity of animals is extremely rich, with over 90% of the western black-crowned gibbon (*Nomascus concolour*) populations inhabiting both mountainous terrains [37]. There are also nationally protected keystone I animals such as the grey langur (*Trachypithecus phayrei*), nationally protected keystone I birds (*Syrmaticus humiae*), and green peafowl (*Pavo muticus*) in the region, and the biodiversity composition is complex [38]. Lamentation "Ailaoshan-Wuliangshan" National Park can more effectively integrate conservation efforts, establish perfect protection mechanisms and maintain the originality and integrity of ecosystems. In this study, the three nature reserves identified by the proposed national park and the area between them are taken as a whole, with a 1.3 km buffer outwards as a study region.

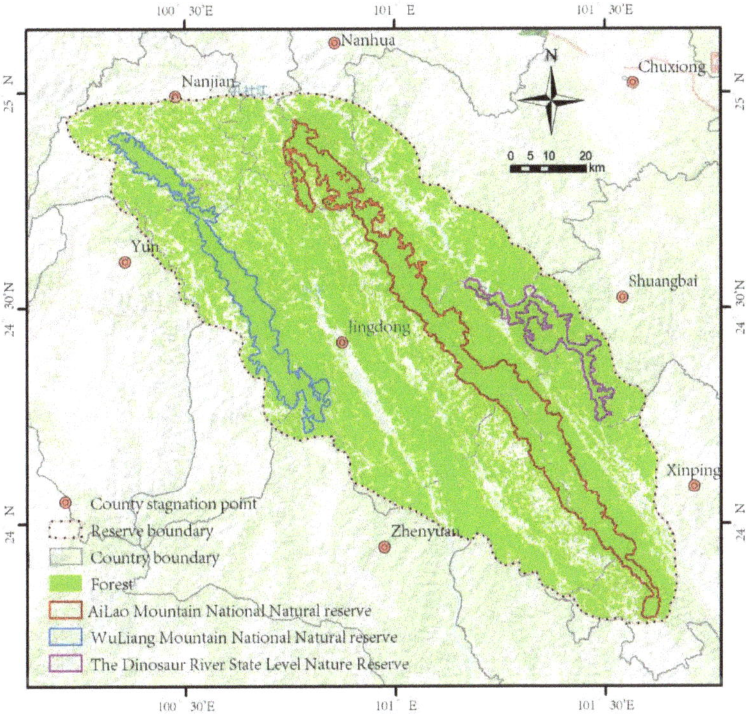

Figure 1. Geographic location map of national parks.

2.2. Data Sources and Preprocessing

In the study area, vector data were the land use type data of 2020 obtained from the Pu 'er Third and Fourth Forestry Resources Type II Survey database sponsored by the Pu 'er Forestry and Grassland Bureau. This database was created using SPOT-5 satellite imagery at 2.5 m resolution combined with field surveys; vector data are grouped by multiple attributes, such as land use type, land ownership, vegetation origin and forest type. Forest-land categories were used in this study to determine forest distribution, with farmland and built-up areas reflecting anthropogenic disturbance. We obtained township boundaries using BIGEMAP, a Google package that facilitates the editing of satellite maps for upload. Elevation and slope data were generated from a 30 m digital elevation grid (DEM) provided by the Geospatial Data Cloud Platform of the Computer Network Information Center of the Chinese Academy of Sciences "http://www.gscloud.cn (20 September 2021)". The study area of 9914.24 km² was obtained by buffering outward from the three protected areas as a whole. The land use types in the study area were classified into eight categories, including forest, shrubland, unstocked forest, agricultural land, water bodies, cropland, buildings, and other non-forest land. The habitat characteristics of the western black-crowned gibbon, the flagship species of the proposed Mourning Mountains–Wuilangshan National Park, have been identified as an important basis for the construction of the resistance surface of the ecological system [39].

2.3. Methods

2.3.1. Identifying Ecological Sources Based on MSPA

Morphological spatial pattern analysis (MSPA) is a method for the quantitative identification of ecological source lands and is mainly used to identify and classify ecological source sites by image methods to derive a more scientific distribution of ecological source sites [40]. This study, based on the accurately calibrated land use data of the national park, reclassifies the already classified primary land classes, extracts forestland as the foreground, uses remaining arable land, water, other land, and building land as the background, and converts it into binary tiff maps. The data were analyzed for morphological spatial patterns using Guidos' eight-neighborhood software method to obtain seven landscape element categories that are independent of each other and have different landscape functions, namely, Core, Islet, Bridge, Edge, Perforation, Loop and Branch [41]. Of these, core areas are the largest habitat patches of the seven landscape types, have higher connectivity, are more conducive to species survival and spread and are important for maintaining ecosystem integrity and biodiversity [42]. Lastly, the classification results were tallied, and considering the balanced distribution of ecological patches, 30 ecological source sites were selected for landscape importance analysis based on the area and spatial distribution area of the MSPA core patches.

2.3.2. Evaluating the Importance of Ecologically Sourced Landscapes

Higher importance indices of patches represent more stable ecosystems within the source site. Referring to the relevant literature [43], two landscape indices, *pc* and *dpc*, were selected in this study and calculated using Conefor2.6 software.

This is example 1 of an equation:

$$PC = \sum_{i=1}^{n} \sum_{j=1}^{n} \alpha_i \times \alpha_j \times q_{ij}^* / A, dpc = 100\% \times (pc - pc_{remove}) / pc, \tag{1}$$

where *pc* is the probability connectivity index, *dpc* is the patch importance index, and *A* represents the total area of the landscape. N is the total number of patches, α_i and α_j are the areas of patches *i* and *j*, respectively, q_{ij} is the maximum distance for organisms to spread in different patches, and pcremove is the overall connectivity index of the landscape after removing a patch. The larger the *dpc* value is, the more important the interelement ratio is. Considering that the distance threshold was set too large, which will result in splitting some large patches and vanishing small patches, the patch connectivity distance threshold value was set to 500 m and 0.5 as the connectivity probability between patches [44].

2.3.3. Ecological Network Construction Method

1.   Construction of Integrated Resistance Surface

In the proposed Ailao and WuLiang Mountains National Park, species are inevitably hampered by different factors and degrees in the migration process of each source location. Currently, most studies select three resistance factors, namely, land use type, topography, and slope, to construct resistance surfaces [15]. In this study, the resistance surface was constructed by combining land ownership, human disturbance (town center, village, road, land type), vegetation (vegetation type and origin of tree species) and geographic factors (elevation and slope), and each resistance factor was assigned different weight values, with higher weight values indicating a greater influence of the resistance factor on the migration of biological species; in contrast, lower weight values indicated less influence.

Land tenure factor weight, human disturbance factor weight value, vegetation factor weight and geographical factor weight were determined according to related studies [45,46]

(Table 1). The land tenure factor was divided into two types, national and collective, each with a weight value of 0.5 each; the vegetation factor was set to 0.6 and 0.4 strength values for the vegetation type and the origin of tree species according to habits and activity characteristics of the species; the influence of anthropogenic disturbance was set at 0.3, 0.2, and 0.2, respectively, according to the distance of species from town centers, villages, and roads during the migration process. The land use type was set at 0.3 according to the influence of the land type on species; the geographical factors included elevation and slope, with the elevation set at 0.5 according to the characteristics of species' activities; and the slope was set at 0.5 according to the standard grading table of woodland slopes. The weighted overlay operation by the ArcGis matrix calculator was used to build the comprehensive Two Mountains National Park resistance surface as the cost data of the MCR model [46]. The equation is as follows:

Table 1. Assignment of resistance factors.

Resistance Factors	Weight	Classification Indications	Resistance Value
Land ownership	0.2	State owned	0.5
		Community owned	0.5
Artificial	0.4	Town center	0.3
		Village	0.2
		Road	0.2
		Land use type	0.3
Vegetation	0.3	Type	0.6
		Origin	0.4
Landform	0.1	Altitude	0.5
		Slope	0.5

This is example 2 of an equation:

$$F_i = \sum_{j=1}^{n} W_j \times A_{ij},\tag{2}$$

where i represents the grid, j represents the resistance factor, F_i represents the integrated resistance value of grid i, n represents the number of resistance factors, W_j represents the proportion of j and A_{ij} represents the strength value of j in grid i.

2. Potential ecological corridor construction based on the MCR model

The minimum cumulative resistance (MCR, minimal cumulative resistance model) model was first introduced into China by Yu Kongjian [47]. It can determine pathways by calculating the minimum cumulative resistance distance between the source and target to better reflect the physical energy of the landscape and the likelihood of biological species moving between habitat patches and trends [27]. The cumulative surface area of minimum resistance for the expansion of ecological source sites in all directions can be obtained by using the model of minimum strength [26].

This is example 3 of an equation:

$$MCR = f_{min} \sum_{j=n}^{i=m} (D_{ij} \times R_i),\tag{3}$$

where MCR refers to the minimum cumulative resistance value of the ecological source to one another point; fmin is the minimum cumulative resistance value (MCR), representing the positive correlation function; D_{ij} indicates the spatial distance to be crossed for a point j to reach another point i; and R_i is the resistance value to be overcome across space i.

The cost distance tool in ArcGis distance analysis was used to generate the minimum cumulative resistance surface using ecological source sites and integrated resistance surfaces [41], and the cost path tool was used to calculate the minimum cost path from the source site to the target to generate potential ecological corridors.

3. Determination of the ecological nodes

Ecological nodes are the point of intersection of the pathways and the shortest routes needed by species during migration and are the weakest ecological functions with a "stepping stone" role [48]. For organisms that migrate long distances, increasing the number of "stepping stones" and decreasing the distance between "stepped stones" may effectively improve species survival rates during migration [44]. In combination with the study area environment, the collection points of the minimum cost path and the ecological patches of the bridging area are used as the ecological nodes.

4. Identify important ecological corridors

The gravity model can scientifically and quantitatively evaluate the strength of interactions between patches, and the larger the value of interaction force is, the more important the position of the corridor between them in the ecosystem of the study area [49–51].

This is example 4 of an equation:

$$G_{ab} = \frac{N_a N_b}{D_{ab}^2} = \frac{L_{max}^2 \ln s_a \ln s_b}{L_{ab}^2 P_a P_b} \tag{4}$$

where G_{ab} is the interaction strength between ecological source sites a and b; N_a and N_b represent the corresponding weight values of source sites a and b; D_{ab} is the standard value of corridor resistance between source sites; P_a and P_b represent the average resistance values of source sites a and b; S_a and S_b are the areas of source sites a and b; L_{ab} is the value of corridor resistance between source sites a and b; and L_{max} is the minimum cumulative resistance in the area of the maximum value.

According to the construction of potential ecological corridors, the interaction matrix between ecological source sites was computed using a gravity model to quantitatively analyze the strength of interactions among patches. Higher values of interaction force between source patches indicate less resistance and closer contact at ecological sources, the more frequent the material–energy transfer, information transfer, species migration and the more important the corridors connected between them [52]. According to the calculation results of the gravity model and the actual situation in the research area, the interaction strength of the potential ecological corridors greater than 700 is regarded as important corridors and other corridors as general corridors. Finally, the ecological network map of the proposed Ailao–Wuliang Mountains National Park is obtained.

2.3.4. Ecological Network Connectivity Evaluation

The graph theory and network analysis method were used to assess the ecological network connectivity of Ailao–Wuliang Mountains National Park and explore the effectiveness of its internal structure. The four factors of network closure (α), line point rate (β), network connectivity (γ) and cost ratio (c) were used to determine the connectivity of the ecological network in the study area [53].

This is example 5 of an equation:

$$\alpha = (L - v + 1)/(2v - 5)$$
$$\beta = \frac{L}{v}$$
$$\gamma = \frac{L}{L_{max}} = \frac{L}{3(v-2)} \tag{5}$$
$$c = 1 - \frac{L}{d}$$

where L denotes the number of corridors, v denotes the number of ecological nodes, and d is the total length of all corridors in the ecolgical corridor. A higher α index indicates a greater number of circuits in the ecological network and greater material circulation and energy mobility [54]. β is the number of corridors corresponding to the ecological nodes, $\beta < 1$ is a tree-like ecological corridor, $\beta = 1$ is a single-loop ecological network, and $\beta > 1$ is a complex ecological network structure. γ in [0,1], characterizing the degree of interconnection of ecological nodes in the network, and a larger value of γ indicates a higher degree of interconnection of ecological nodes. c indicates the input/output relationship, and a lower value is more favorable for building ecological networks.

3. Results and Analysis

3.1. Subsection Analysis of Ecological Source Results Based on MSPA

The MSPA was performed with forested land in the study area as the foreground (Figure 2), and the area and proportion of each type of landscape were also counted (Table 2). Among them, the total foreground area is 7820.08 km^2, accounting for 79% of the total survey area, which mainly consists of the core and bridge areas. Among the various types of foreground landscapes, the core area is the largest, accounting for 56.78% of the total area, with more large patches in the core region and distributed in bands within the national park, which are not far apart and conducive to the overall connectivity of the region under study. Bridging zones with a larger area of 22.08% indicates that the connectivity between core areas of prospective patches is high, which is conducive to the circulation of organisms between core zones; the edge zone and the pore space both have edge effects and can maintain the stability of the core areas, with proportions of 9.45% and 2.72%, respectively, which indicates that core areas in this study area are relatively stable. In addition, island plaques accounted for the lowest proportion, at percent, indicating that there were few isolated, fragmented and disconnected patches within the study; the proportions of ring roads and spurs were 4.88% and 3.09%, respectively. In general, the large ecological patches in the study area are more concentrated, the landscape connectivity is better and the edges are more stable, which are conducive to the construction and optimization of ecological networks.

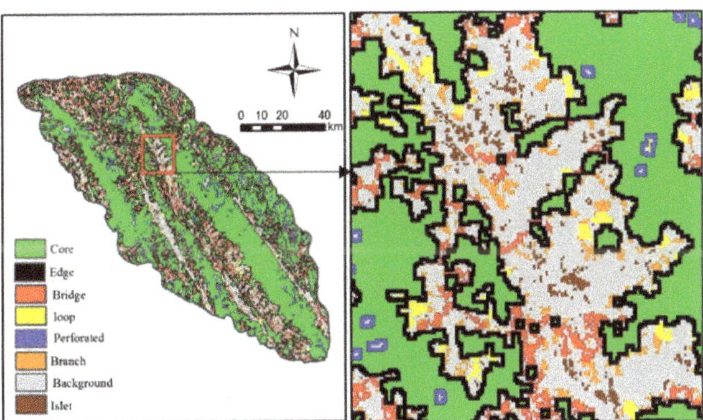

Figure 2. Landscape classification map based on MSPA.

Table 2. Area of each landscape type based on MSPA.

Landscape Type	Area (km^2)	Proportion of Forest and Areas (%)	Proportion of Total Area (%)
Core	4440.08	56.78	44.78
Islet	77.91	1	0.79
Perforation	212.54	2.72	2.14
Edge	739.37	9.45	7.46
Loop	381.35	4.88	3.85
Bridge	1727.05	22.08	17.42
Branch	241.79	3.09	2.44
Total	7820.08	100	78.88

3.2. Analysis of the Importance of Ecological Source Landscapes

Thirteen ecological patches with large values of the importance of patches were selected as habitats for migration and development and reproduction of biological species based on the calculation results of the software Conefor2.6 (Table 3). From Table 3 and Figure 3f, the importance index of patch 9 is 70.9, with an area of 1231.16 km^2, and it is located in the protection zone of Ailao Mountain, which indicated that the landscape connectivity within this region is good, which is conducive to species migration activity among patches. Figure 4 is followed by patch 3, with an importance index of 17.73 and an area of 750.89 km^2, which is located within the boundaries of the Wuliang Mountains Nature Reserve. Ecological patches 4 and 11, with a larger remaining area and higher importance index, were also distributed in the nature reserve. We can see from (f) that these patches serve both as habitats for species and corridor connectivity across the landscape. Additionally, other large patches were densely distributed around the reserve, such as 5, 6, 8 and 10. These patches facilitate species migration between reserves and promote connectivity of the overall landscape of the study area.

Table 3. Evaluation table of the landscape importance of the ecological source.

Code	Area/km^2	dPC	Code	Area/km^2	dPC
1	31.94	0.03	8	447.37	7.00
2	9.39	0.00	9	1231.16	70.90
3	750.89	17.73	10	57.09	1.41
4	128.18	14.98	11	191.73	17.26
5	184.95	8.60	12	20.33	0.94
6	33.19	1.38	13	17.64	0.23
7	74.59	1.78			

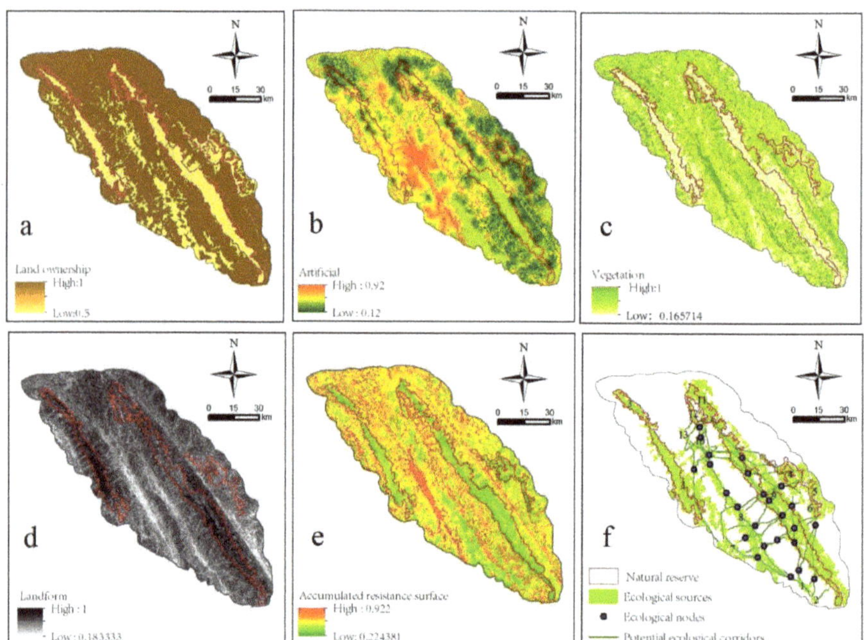

Figure 3. Composite resistance surface of species migration in national park. (**a**) Land ownership; (**b**) artificial; (**c**) vegetation; (**d**) landform; (**e**) accumulated resistance surface; and (**f**) potential ecological corridors.

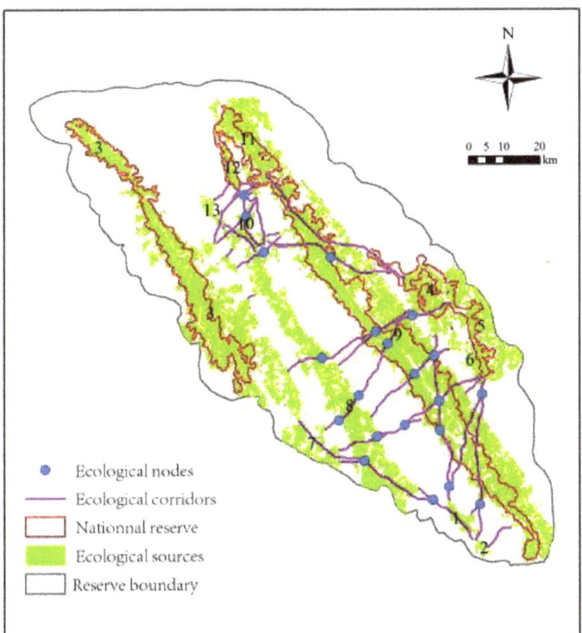

Figure 4. Ecological network of the national park.

3.3. Ecological Network Construction

3.3.1. MCR Based on the Analysis of Potential Ecological Corridor Extraction

The integrated resistance surface (e) of the Ailao–Wuliang Mountains National Park was constructed based on the vector data of land tenure factor (a), geographic factor (b), vegetation factor (c) and anthropogenic factor (d) in the study area of Figure 3. It can be observed from Figure 3e that the high resistance values in the integrated resistance surface are concentrated in the city centers and villages, which are located outside the scope of the national park and are disturbed by human activities, and their corresponding resistance values are correspondingly higher, while the low resistance values are mostly distributed within the scope of the national park and in the ecological source areas mostly in woodlands. Based on the landscape resistance surface, the MCR model was used to calculate the minimum cumulative resistance value between each ecological source site, and a total of 77 potential ecological corridors were generated, with a total length of 3589 km, to construct the potential ecological corridor of the Ailao–Wuliang Mountains National Park (Figure 3f). As we can see from the figure, the potential ecological corridors in the national park show a denser network with a more uniform spatial distribution, connecting large ecological patches in the park, with more complex corridors among source locations and higher connectivity.

3.3.2. Important Analysis of Ecological Corridors

We numbered the 13 ecological source sites and calculated the interaction strength between different ecological source locations in the study area by the gravity model (Table 4), and the stronger the interaction force between ecological source locations was, the more meaningful the construction of intersource corridors. Based on the study, the gravity threshold was set to 700, and 48 corridors were selected as important corridors with a total length of 865 km (Figure 4). According to Table 4, the interaction strength between source site 9 and source site 11 was the largest at 647,208,185, indicating the strongest spatial association between the two, and the less resistance species encounter when migrating and spreading between the two patches, the more beneficial for regional ecological conservation.

We should therefore strengthen ecological corridor protection between source sites 9 and 11, maintain connectivity of both patches and avoid destruction due to expansion of regional landmasses. Sources 4 and 5 have stronger interaction strengths with sources 5 and 6, indicating that the connectivity between source 4 and source 5 is stronger. The species need to overcome less resistance when propagating movement through the corridor, and the possibility of material and energy exchange is higher, so the ecological corridor between sources 4, 5, and 6 can be established to increase the possibility for species migration between sources 3, 4, 5 and 6 and expand the species' range of activities. Additionally, ecological corridors built between source sites 8 and 9 may link species exchange between source locations 3, 4, 5, and 5. On this basis, the migration and dispersal channels of species between sources 3, 4, 5 and 6 were established, compensating for the high resistance and habitat fragmentation of migration among the sources. For example, the G values between patches 2 and 12 and between patches 2 and 13 were split into 33 and 34, which were distant and poorly connected. The possibility of western black-crowned gibbon dispersion between them was small, and the cost of building ecological corridors was steep if necessary. Accordingly, to improve the possibility of species migration, 25 footstones were established at the convergence point (ecological corridor intersection) and bridge zone where the least expensive paths were selected, and the presence of footstones may compensate for the lack of connectivity of the corridors.

Table 4. Level of interaction of ecological corridors.

Code	1	2	3	4	5	6	7	8	9	10	11	12	13
1	0	313,473	742	788	1244	861	1122	22,956	88,786	151	168	77	79
2		0	267	263	413	279	334	2419	12,682	64	68	33	34
3			0	1507	1002	678	2,176,830	204,027	11,100	209,177	5797	3804	34,679
4				0	27,378,454	5,488,676	894	3684	5,649,277	769	765	319	307
5					0	18,938,739	738	2071	60,051	545	555	242	237
6						0	528	1648	137,802	279	289	129	126
7							0	127,908	3154	341	259	140	174
8								0	255,194	10,341,338	9047	3752	7646
9									0	122,793	647,208,185	367,09	7410
10										0	246,249	32,476	2,019,026
11											0	1,145,712	7539
12												0	11,611
13													

3.3.3. Analysis of Ecological Corridor Construction in Nature Reserves

The ecological network constructed by important ecological corridors is more suitable for areas with fewer villages and farmland. Considering that, in reality, the proposed ecological network will occupy a large amount of land, causing land pressure and aggravating the human–land conflict affecting socioeconomic development. The ecological corridors were constructed by using the patches in the three protected areas as ecological sources, and a total of three ecological corridors, corridors 3-11, 3-12 and 3-9, were generated in the protected areas of the Ailao–Wuliang Mountains. The buffer area resistance cost accumulation values were calculated for each 200 m buffer on each side of the three ecological corridors (Table 5). The table shows that the lowest resistance value for corridors 3-12 is 90.88, which indicates that building this ecological corridor is the least expensive and easiest to achieve. The corridors generated by the Wailing Mountains and Shuangbai Reserve overlap by 11-4, 12-4 and 9-4, respectively, and finally, 3-12-4 can be identified as the optimal ecological corridor for the three reserves (Figure 5), connecting the three nature reserves of the Ailao Mountains, the Wuliang Mountains and the Dinosaur River and included in the scope of the national park.

Table 5. Ecological corridor cost table.

Code	MIN	MAX	MEAN	SUM	
3-12	0.308381	0.750667	0.441144	90.875575	√
3-11	0.300714	0.785	0.481216	125.116101	
3-9	0.309476	0.814667	0.501656	129.42734	

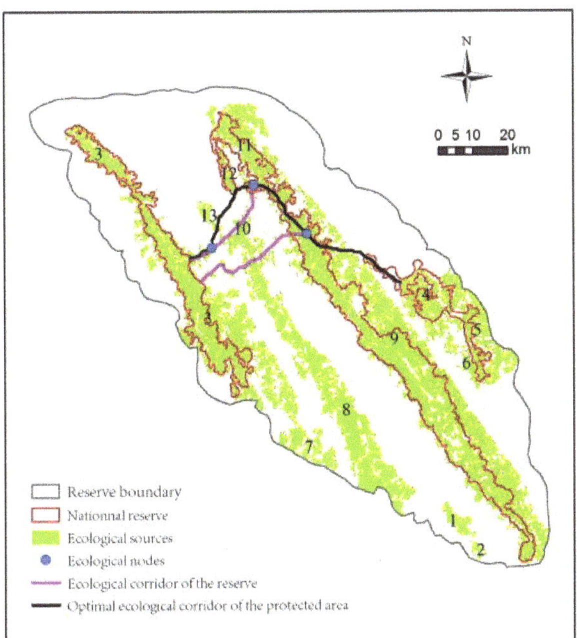

Figure 5. Reserve ecological corridors.

3.3.4. Ecological Network Connectivity Evaluation

The structural rationality of potential ecological corridors, important ecological corridors and optimal ecological corridors in protected areas was assessed based on graph theory and network analysis methods (Table 6). The results of the study are summarized in Table 6. Table 6 shows that the α values of potential ecological corridors, important ecological corridors and optimal ecological corridors of protected areas are 1.18, 0.76 and 2, indicating that the optimal ecological corridor of protected areas has the best structural connectivity and better routes for the migration and dispersal of species. The β values were 3.08, 2.28, and 1.33, all with $\beta > 1$, which indicated that all were complex structures of ecological networks with high connectivity of ecological corridors. The γ values were 1.12, 0.84, and 1.33, where potential ecological corridors and optimal ecological corridors of protected areas had larger γ, which indicated that their ecological nodes were well connected. The c values were 0.98, 0.94, and 0.97, indicating that the cost values of building both potential ecological corridors, important ecological corridors and optimal ecological corridors in protected areas were higher, and the reason for their higher cost might be interference from anthropogenic activities such as farmland, construction land, and cities in plots between protected areas. Together with the complex geomorphology and fragmented protected areas of the Ailao Mountains and Wuliang Mountains National Park, which lead to the complex structure of ecological corridors, if ecological corridors are constructed in reality, comprehensive consideration is given to the priority of constructing optimal ecological corridors within protected areas.

Table 6. Ecological corridor connectivity evaluation table.

Connectivity Index	Potential Ecological Corridors	important Ecological Corridors	Optimal Ecological Corridor of the Protected Area
α	1.18	0.76	2
β	3.08	2.28	1.33
γ	1.12	0.84	1.33
c	0.98	0.94	0.97

4. Discussion

4.1. Advantages and Challenges of Research Methods Based on MSPA and MCR Models

In this study, an integrated construction method for ecological networks based on the MSPA and MCR was proposed. Compared to the ecological source site identification and ecological network construction of Giant Panda National Park and Shuangzi Mountain National Forest Park using 3S technology and landscape ecology theory [30,31]. MSPA was a widely used method for ecological source site identification, which is simpler and more scientific in distinguishing spatial patterns in the landscape and identifying patches with more suitable conditions as ecological source sites. The combined approach with the MCR model to construct ecological networks has become mature and is commonly used in cities with good results. However, the ecological network constructed by combining these two models has rarely been studied for connected isolated Chinese nature reserves. The ecological source sites in the study area analyzed by the MSPA method in this study were more concentrated and have better landscape connectivity, which is very favorable for the construction and optimization of the ecological network [20]. The ecological corridors and ecological networks constructed by the MCR model are more reasonable and improve the connectivity and integrity of the proposed Ailaoshan–Wuliang Mountain National Park. However, the selection of ecological source sites was the key to improving landscape connectivity and building ecological networks, and in the process of ecological source site selection, source sites can be identified from a multi-indicator integrated evaluation method of ecosystem functional importance [55], biodiversity [56], and species distribution, which can consider the functions, processes, and patterns of ecological source sites in an integrated manner and may lead to one-sided results if ecological source sites were identified from a single level [55]. However, for most areas, species movement and distribution data are often difficult to obtain [57]. Therefore, how to introduce species distribution into the construction of ecological networks using other models remains to be investigated.

4.2. Proposed Construction of the Ecological Network of Ailaoshan-Wuliang Mountain National Park

Due to increasing human activities, nature reserves are becoming "islands", which are mostly unable to protect species populations and natural ecological processes in the long term [58]; therefore, there is a need to integrate the reserves into a larger spatial scale to enhance the ecological connectivity among the reserves [59]. China first proposed the establishment of a national park system in 2013 [60], which is comparable to the national parks established internationally, such as Yellowstone National Park in the United States [61], Canadian national parks [62] and the national parks now established in China, which belong to a concentrated contiguous area and were relatively large, but most of the nature reserves in China are insular [63], varying in size and fragmented in distribution, with little connectivity and integrity [64]. By constructing ecological corridors and ecological networks, connectivity among nature reserves can be enhanced and constitute large national parks. Ailaoshan–Wuliangshan National Park has intact wet evergreen broad-leaved forest ecosystems, and both are home to a large number of flagship species of western black-crowned gibbons [65] surrounded by a large amount of remnant forest. History suggests that the Ailaoshan–Wuliang Mountains may have strong connectivity, providing the possibility for national parks to construct potential ecological networks and ecological corridors.

Influenced by human activities, the construction of a reasonable ecological network requires highlighting the role of anthropogenic disturbance factors in the resistance surface. As shown in Figure 3f, the proposed ecological corridor of Ailaoshan–Wuliangshan Mountain National Park is more evenly distributed. Corridors with high resistance values are located mainly in town centers near villages and roads. In contrast, ecological corridors within protected areas and far from human activities have lower resistance values and can better connect ecological source sites. The resistance surface is usually constructed using unidimensional indicators such as slope, elevation, land use type, roads and human activities [66,67], and the weight values are set by the expert scoring method [22]. In this study, the resistance surface was constructed from multiple dimensions of land tenure, vegetation type, topography and human interference, and the integrated resistance surface model established by using the expert scoring method to set higher weight values for human interference factors, including town center, village, road and land use type, achieves better results in this empirical evidence.

4.3. The Impact of Building Ecological Networks on Surrounding Land

The ecological network formed by the proposed 77 ecological corridors is an ideal ecological network, and the types of land they pass through include agricultural land, natural forests, and construction land. If all of them are to be realized, they will occupy a large amount of land and aggravate the conflict between people and the land around them. Although protected areas in Yunnan, China, are located in remote mountainous areas, the surrounding population is large and dependent on land resources [68]. The three protected areas will be set up as ecological source sites, and an optimal ecological corridor will be screened out by calculating the cumulative value of resistance costs in the buffer zone and incorporated into the land area of the national park to implement strict protection management and maximize coordination between the national park and local community residents for conservation and development. In response to the problem of constructing ecological corridors that occupy the surrounding residents' farmland, local special resources can be developed through community participation in co-management and the establishment of ecological compensation mechanisms [69,70]. By encouraging community participation, the conservation and development of national parks coexist. In most cases, large corridors do not preclude reasonable human use of their resources [58]. Combining conservation with the benefits of social, economic, and peripheral development allows residents to share the benefits of natural resource conservation [71], which can weaken the negative effects on the economic development of local communities caused by the occupation of land resources due to the establishment of ecological corridors.

5. Conclusions

This study attempted to build an ecological network of the proposed Ailao–Wuliang Mountains National Park based on the MSPA and MCR models. First, based on the MSPA method, we can directly identify and quantify the ecological source sites in the study area and provide important data for the building of ecological networks in national parks. The landscape importance among ecological source locations was further analyzed scientifically using the more scientific Conefor 2.6 software. The integrated resistance surface is generated by the four dimensions of geography, human activities, vegetation, and land tenure, where the pattern of human activities plays a key role in generating the integrated resistance area, and vegetation and land permanence play a major role in the integrated resistant surface. Potential ecological corridors of the national park are generated using the MCR model, and important ecological corridors in the study area were judged based on the assessment of the gravity model. Form an ecological network of the study area, install footstones to optimize the ecological network, and finally screen an ecological corridor to communicate and link protected areas to form a comprehensive ecological system. The construction of an ecological network of national parks solves the problem of insularity of nature reserves, improves the connectivity and integrity of reserves, solves the real problem of socioeconomic development of the surrounding area caused by the construction of ecological networks by screening the optimal ecological corridors, relieves the pressure on humans and land, and provides the possibility of species migration in the reality of protected areas. The methods based on the MSPA and MCR models are generally applicable to the construction of ecological networks of national parks with multiple nature reserves in isolation, and the results of the study can maximize the conservation of species habitats and biodiversity for the proposed Ailaoshan–Wuliangshan Mountain National Park.

Author Contributions: Conceptualization, C.Y. and X.H.; methodology, C.Y. and X.H.; software, X.H.; validation, C.Y.; formal analysis, C.Y. and X.H.; investigation, C.Y., X.L. and X.H.; resources, H.G.; data curation, C.Y. and X.H.; writing—original draft preparation, C.Y. and X.C.; writing—review and editing, Y.W. and X.H.; visualization, C.Y.; supervision, Y.W. and X.H.; project administration, H.G. and Y.W.; funding acquisition, Y.W. All authors have read and agreed to the published version of the manuscript.

Funding: This research was funded by the National Natural Science Foundation of China [No. 42061004].

Data Availability Statement: Not applicable.

Acknowledgments: The authors gratefully acknowledge the support of the funding.

Conflicts of Interest: The authors declare no conflict of interest.

References

1. Cai, J.; Yu, W.B.; Zhang, T.; Wang, H.; Li, D.Z. China's biodiversity hotspots revisited: A treasure chest for plants. *Phyto Keys* **2019**, *130*, 1–24. [CrossRef] [PubMed]
2. Di-Qiang, L.; Yan-Ling, S. Review on hot spot and GAP analysis. *Biodivers. Sci.* **2000**, *8*, 208–214. [CrossRef]
3. Cook, E.A. Urban landscape networks: An ecological planning framework. *Landsc. Res.* **1991**, *16*, 7–15. [CrossRef]
4. The General Office of the CPC Central Committee The General Office of the State Council issued the Guiding Opinions on Establishing a Protected Natural Areas System with National Parks as the main body. *Bull. State Counc. People's Repub. China* **2019**, *19*, 16–21.
5. Yang, Y.; Zhu, Y. China has officially established the first batch of national parks. *Eco-Econ.* **2021**, *37*, 9–12.
6. Zhang, D.; Xia, E.; Liu, C.; Yang, W.; Ma, Y.; Fan, S. The origin and innovation of China's National Park concept. *World For. Res.* **2022**, *35*, 1–7. [CrossRef]
7. Tang, X.; Luan, X. Build a protected natural area system with national parks as the main body. *For. Resour. Manag.* **2017**, *6*, 1–8. [CrossRef]
8. Wang, W.; Li, J. Development course of biodiversity conservation Policy in China. *Environ. Sustain. Dev.* **2021**, *4 6*, 26–33. [CrossRef]
9. Convention on Biological Diversity. Aichi Biodiversity Targets. 2010. Available online: https://www.cbd.int/sp/targets/ (accessed on 1 October 2022).
10. Secretariat, C.B.D. First Draft of the Post-2020 Global Biodiversity Framework. 2021. Available online: https://www.cbd.int/doc/c/abb5/591f/2e46096d3f0330b08ce87a45/wg2020-03-03-en.pdf (accessed on 1 October 2022).
11. Luo, Y.; Wu, J.; Wang, X.; Wang, Z.; Zhao, Y. Can policy maintain habitat connectivity under landscape fragmentation? A case study of Shenzhen, China. *Sci. Total Environ.* **2020**, *715*, 136829. [CrossRef]
12. Yang, L.; Suo, M.; Gao, S.; Jiao, H. Construction of an Ecological Network Based on an Integrated Approach and Circuit Theory: A Case Study of Panzhou in Guizhou Province. *Sustainability* **2022**, *14*, 9136. [CrossRef]
13. Liu, H.; Niu, T.; Yu, Q.; Yang, L.; Ma, J.; Qiu, S. Evaluation of the Spatiotemporal Evolution of China's Ecological Spatial Network Function–Structure and Its Pattern Optimization. *Remote Sens.* **2022**, *14*, 4593. [CrossRef]
14. Weber, T.; Sloan, A.; Wolf, J. Maryland's Green Infrastructure Assessment: Development of a comprehensive approach to land conservation. *Landsc. Urban Plan.* **2006**, *77*, 94–110. [CrossRef]
15. Chen, Z.; Kuang, D.; Wei, X.; Zhang, L. Construction of Yujiang County Ecological network based on MSPA and MCR model. *Yangtze River Basin Resour. Environ.* **2017**, *26*, 1199–1207.
16. Soille, P. *Morphological Image Analysis: Principles and Applications*; Springer: Berlin/Heidelberg, Germany, 1999. [CrossRef]
17. Sudhakar Reddy, C.; Vazeed Pasha, S.; Satish, K.V.; Saranya, K.R.L.; Jha, C.S.; Krishna Murthy, Y.V.N. Quantifying nationwide land cover and historical changes in forests of Nepal (1930–2014): Implications on forest fragmentation. *Biodivers. Conserv.* **2018**, *27*, 91–107. [CrossRef]

18. Huang, X.Y.; Ye, Y.H.; Zhang, Z.Y.; Ye, J.X.; Gao, J.; Bogonovich, M.; Zhang, X. A township-level assessment of forest fragmentation using morphological spatial pattern analysis in Qujing, Yunnan Province, China. *J. Mt. Sci.* **2021**, *18*, 3125–3137. [CrossRef]

19. Shi, Y. Research on Ecological Network Construction in Zixing City Based on MSPA and MCR Models. Master's Thesis, Central South University of Forestry and Technology, Changsha, China, 2019.

20. Chen, N.; Kang, S.; Zhao, Y.; Zhou, Y.; Yan, J.; Lu, Y. Construction of the Qinling Mountains (Shaanxi section) mountain ecological network based on MSPA and MCR models. *J. Appl. Ecol.* **2021**, *32*, 1545–1553. [CrossRef]

21. Zhou, D.; Song, W. Identifying ecological corridors and networks in mountainous areas. *Int. J. Environ. Res. Public Health* **2021**, *18*, 4797. [CrossRef]

22. Chen, C.D.; Wu, S.J.; Meurk, C.D.; Lu, M.Q.; Wen, Z.F.; Jiang, Y.; Chen, J.L. Impact of resistance assignment on landscape connectivity simulation. *J. Ecol.* **2015**, *35*, 7367–7376.

23. Allen, C.H.; Parrott, L.; Kyle, C. An individual-based modelling approach to estimate landscape connectivity for bighorn sheep (*Ovis canadensis*). *Peer J* **2016**, *4*, e2001. [CrossRef]

24. Saura, S.; Pascual-Hortal, L. A new habitat availability index to integrate connectivity in landscape conservation planning: Comparison with existing indices and application to a case study. *Landsc. Urban Plan.* **2007**, *83*, 91–103. [CrossRef]

25. Yin, Y.; Liu, S.; Sun, Y.; Zhao, S.; An, Y.; Dong, S.; Coxixo, A. Identifying multispecies dispersal corridor priorities based on circuit theory: A case study in Xishuangbanna, Southwest China. *J. Geogr. Sci.* **2019**, *29*, 1228–1245. [CrossRef]

26. Knaapen, J.P.; Scheffer, M.; Harms, B. Estimating habitat isolation in landscape planning. *Landsc. Urban Plan.* **1992**, *23*, 1–16. [CrossRef]

27. Huang, L.; Chen, J. Evaluation of urban construction land in Huadu District, Guangzhou—Based on MCR surface feature extraction. *Resour. Sci.* **2014**, *36*, 1347–1355.

28. Li, F.; Ye, Y.; Song, B.; Wang, R. Evaluation of urban suitable ecological land based on the minimum cumulative resistance model: A case study from Changzhou, China. *Ecol. Model.* **2015**, *318*, 194–203. [CrossRef]

29. Ye, H.; Yang, Z.; Xu, X. Ecological corridors analysis based on MSPA and MCR model—A case study of the Tomur World Natural Heritage Region. *Sustainability* **2020**, *12*, 959. [CrossRef]

30. Luo, Y.; Tan, X.; He, L.; Li, C. Construction and optimization of ecological network in Qionglai Mountain-Daxiangling area of Giant Panda National Park. *Landsc. Archit.* **2022**, *29*, 93–101. [CrossRef]

31. Zhao, S.M.; Ma, Y.F.; Wang, J.L.; You, X.Y. Landscape pattern analysis and ecological network planning of Tianjin City. *Urban For. Urban Green.* **2019**, *46*, 126479. [CrossRef]

32. Wang, H.; Wang, T.; Zhao, Z. Construction and optimization of ecological network in Shuangzi Mountain National Forest Park. *Heilongjiang Agric. Sci.* **2018**, *10*, 116–120. [CrossRef]

33. Yang, Y.; Zhou, Y.; Feng, Z.; Wu, K. Making the case for parks: Construction of an ecological network of urban parks based on birds. *Land* **2022**, *11*, 1144. [CrossRef]

34. Yuan, J. Study on biodiversity and conservation measures in Wuliang Mountain National Nature Reserve. *Fujian For. Sci. Technol.* **2010**, *37*, 131–135.

35. Wen, H.; Lin, L.; Yang, J.; Hu, Y.; Cao, M.; Liu, Y.; Lu, Z.; Xie, Y. Species composition and community structure of a 20 hm^2 plot of mid-mountain moist evergreen broad-leaved forest on the Mts. Ailaoshan, Yunnan Province, China. *J. Plant Ecol.* **2018**, *42*, 419–429. [CrossRef]

36. Peng, H.; Wu, Z. A preliminary study on the humid evergreen broad-leaved forest and its flora in the middle mountains of Wuliang Mountain. *Yunnan Plant Res.* **1998**, *1*, 12–22.

37. Hua, Z.; Yang, D.; Bi, Y.; Yan, L.; Song, J.; Zheng, J. Current status and conservation measures of western black-crowned gibbons in Yunnan Province. *For. Surv. Plan.* **2013**, *38*, 55–60+66. [CrossRef]

38. Kong, G. The pearl of Jingdong is in splendor–Introduction to the nature reserve of Wailun Mountain and Wuliang Mountain. *Yunnan For.* **1994**, *4*, 21.

39. Luo, Z. Population number and distribution of black-crowned gibbon view East Asian species in western Yunnan Wuliangshan National Nature Reserve. *Sichuan Anim.* **2011**, *30*, 283–287. [CrossRef]

40. Xiao, L.; Cui, L.; Jiang, Q.O.; Wang, M.; Xu, L.; Yan, H. Spatial structure of a potential ecological network in Nanping, China, based on ecosystem service functions. *Land* **2020**, *9*, 376. [CrossRef]

41. Dai, J.; Zhu, K.; Zhou, T.; Peng, J. Construction of Urban Ecological Space Network in Tengchong City Center. *For. Resour. Manag.* **2021**, *5*, 131–138. [CrossRef]

42. Liu, Y.; He, Z.; Chen, J.; Xie, C.; Xie, T.; Mu, C. Research on Ecological Network Construction Method Based on MSPA and MCR Model—Takes Nanchong City as an example. *Southwest J. Agric.* **2021**, *34*, 354–363. [CrossRef]

43. An, Y.; Liu, S.; Sun, Y.; Shi, F.; Beazley, R. Construction and optimization of an ecological network based on morphological spatial pattern analysis and circuit theory. *Landsc. Ecol.* **2021**, *36*, 2059–2076. [CrossRef]

44. Xu, F.; Yin, H.; Kong, H.; Xu, J. Construction of Ecological Network of Midwest New Town Based on MSPA and MinPath Method. *Ecol. J.* **2015**, *35*, 6425–6434.

45. Zhang, Y.; Yin, W.; Hu, X.; Li, X.; Yang, X. The—construction of resource-saving ecological network based on MSPA and MCR model takes Dongshan Island of Fujian Province as an example. *J. Northwest For. Coll.* **2021**, *36*, 254–261.

46. Zheng, Q.; Hu, J.; Shen, M. Construction of Hunan ecosystem network based on MSPA and MCR models. *J. Nat. Sci. Hunan Norm. Univ.* **2021**, *44*, 1–10.

47. Yu, K. Landscape ecological security pattern of biological protection. *J. Ecol.* **1999**, *1*, 10–17.
48. Chen, X.; Chen, W. Construction and Evaluation of Ecological Network in Poyang Lake Ecological Economic Zone. *J. Appl. Ecol.* **2016**, *27*, 1611–1618. [CrossRef]
49. Wang, H.; Zhang, B.; Liu, Y.; Liu, Y.; Xu, S.; Zhao, Y.; Chen, Y.; Hong, S. Urban expansion patterns and their driving forces based on the center of gravity-GTWR model: A case study of the Beijing-Tianjin-Hebei urban agglomeration. *J. Geogr. Sci.* **2020**, *30*, 297–318. [CrossRef]
50. Li, S.; Wang, Z.; Zhong, Z. Gravitational model of tourist space interactions and its applications. *J. Geogr.* **2012**, *67*, 526–544.
51. Yu, Y.; Yin, H.; Kong, F.; Wang, J.; Xu, W. Time and Space-time Change Analysis of Green Infrastructure Network pattern in Nanjing Based on MSPA. *J. Ecol.* **2016**, *35*, 1608–1616. [CrossRef]
52. Guo, J.; Hu, Z.; Li, H.; Liu, J.; Zhang, X.; Lai, X. Construction of Urban Ecological-space Network Based on MCR Model. *J. Agric. Mach.* **2021**, *52*, 275–284.
53. Cook, E.A. Landscape structure indices for assessing urban ecological networks. *Landsc. Urban Plan.* **2002**, *58*, 269–280. [CrossRef]
54. Kong, F.; Yin, H. Construction of Urban Green Space Ecological Network in Jinan. *J. Ecol.* **2008**, *4*, 1711–1719.
55. Li, S.; Li, T.; Peng, C.; Huang, Z.; Qi, Z. Construction of ecological network of green areas in Dongting Lake area based on comprehensive evaluation method. *J. Appl. Ecol.* **2020**, *31*, 2687–2698. [CrossRef]
56. Sun, J.; Huang, J.; Wang, Q.; Zhou, H. A method of delineating ecological red lines based on gray relational analysis and the minimum cumulative resistance model: A case study of Shawan District, China. *Environ. Res. Commun.* **2022**, *4*, 045009. [CrossRef]
57. 57. Fagan, W. F. Quantifying connectivity: Balancing metric performance with data requirements. *Connect. Conserv.* **2006**, *12*, 297–317.
58. Yu, F.; Zhang, B.; Wang, J.; Zhang, X.; Yao, Y. Progress of large-scale ecological corridor conservation abroad and the construction of Qinling National Park. *J. Nat. Resour.* **2021**, *36*, 2478–2490. [CrossRef]
59. Robillard, C.M.; Coristine, L.E.; Soares, R.N.; Kerr, J.T. Facilitating climate-change-induced range shifts across continental land-use barriers. *Conserv. Biol.* **2015**, *29*, 1586–1595. [CrossRef]
60. Wang, Q. Research on national park management and conservation. Master's Thesis, Chengdu University of Technology, Chengdu, China, 2018; p. 62.
61. Wang, L.; Holenhst, S. Creating a Unified Chinese National Park System—American Historical Experience. *Geogr. Res.* **2014**, *33*, 2407–2417.
62. Liu, H. The Construction and Management of Canadian National Parks and Their Implications for China. *J. Ecol.* **2001**, *06*, 50–55.
63. Deng, X.; Qiao, H. Thoughts on China's National Park Construction. *Garden* **2018**, *1*, 34–37.
64. Wang, W.; Xin, L.; Du, J.; Chen, B.; Liu, F.; Zhang, L.; Li, J. Assessment of Nature Conservation: Progress and Outlook. *Biodivers.* **2016**, *24*, 1177–1188.
65. Li, G.; Yan, X.; Zhang, H.; Li, W. Distribution and group number of black-crowned gibbons in West Ailao Mountain, Xinping, Yunnan province. *Zool. Stud.* **2011**, *32*, 675–683. [CrossRef]
66. Yang, Q.; Ye, Y. Measurement and evaluation of suitable ecological land based on the minimum cumulative resistance model: A case study in Shanghai, China. *IOP Conf. Ser. Earth Environ. Sci.* **2019**, *227*, 062041. [CrossRef]
67. Yu, F.; Zhang, B.; Yao, Y.; Wang, J.; Zhang, X.; Liu, J.; Li, J. Identifying Connectivity Conservation Priorities among Protected Areas in Qinling-Daba Mountains, China. *Sustainability* **2022**, *14*, 4377. [CrossRef]
68. Wang, X.; Cui, G. *Construction and Management of Nature Reserves*; Chemical Industry Press: Beijing, China, 2003; pp. 178–200.
69. Hiwasaki, L. Toward sustainable management of national parks in Japan: Securing local community and stakeholder participation. *Environ. Manag.* **2005**, *35*, 753–764. [CrossRef] [PubMed]
70. Jing, L.; Hong, M.; Zhiyun, O.; Weihua, X.; Hua, Z. Analyzing the effectiveness of community management in Chinese nature reserves. *Biodivers. Sci.* **2008**, *16*, 389. [CrossRef]
71. Wangchuk, S. Maintaining ecological resilience by linking protected areas through biological corridors in Bhutan. *Trop. Ecol.* **2007**, *48*, 177.

land

Article

Knowledge Mapping on Nepal's Protected Areas Using CiteSpace and VOSviewer

Liang Chang [1], Teiji Watanabe [1,*], Hanlin Xu [2] and Jiho Han [2]

[1] Faculty of Environmental Earth Science, Hokkaido University, Sapporo 060-0810, Japan; liang_ch@frontier.hokudai.ac.jp
[2] College of Tourism, Rikkyo University, Niiza Campus, 1-2-26 Kitano, Niiza-shi 352-8558, Japan; hanlinxu@rikkyo.ac.jp (H.X.); hanjiho@rikkyo.ac.jp (J.H.)
* Correspondence: twata@ees.hokudai.ac.jp

Abstract: Protected areas (PAs) play a vital role in environmental conservation, particularly in Asian countries. Numerous studies were conducted on PAs in Nepal. We analyzed 864 papers from the Web of Science database using two visualization tools: VOSviewer and CiteSpace. This study identified the most influential journals, institutions, countries, and regions. In addition, we investigated the changing trend of research hotspots on PAs in Nepal. Keyword mapping was conducted for each type of PA and their differences were compared. We found that the research hotspots are changing with the shifting of conservation policies in Nepal. We suggest conducting more predictive studies on the future development of PAs. Currently, PA research is mainly conducted in traditional disciplines, but with the impact of climate change and the consequent increase in its negative impacts, academic contributions from other disciplines are expected to increase much more. We found that there was a shift in research power in countries and regions. We also detected an imbalanced distribution in which "protected areas" and "national parks" have been studied the most. Only 12 publications were about the hunting reserve, despite its importance to snow leopard conservation and economic significance to the buffer zone communities.

Keywords: knowledge mapping; bibliometrics; VOSviewer; CiteSpace; protected areas; Nepal

Citation: Chang, L.; Watanabe, T.; Xu, H.; Han, J. Knowledge Mapping on Nepal's Protected Areas Using CiteSpace and VOSviewer. *Land* **2022**, *11*, 1109. https://doi.org/10.3390/land11071109

Academic Editors: Rui Yang, Le Yu, Yue Cao and Steve Carver

Received: 30 May 2022
Accepted: 14 July 2022
Published: 19 July 2022

Publisher's Note: MDPI stays neutral with regard to jurisdictional claims in published maps and institutional affiliations.

1. Introduction

Protected areas (PAs) play a vital role in conservation around the world [1–4]. According to the International Union for Conservation of Nature (IUCN), a protected area is "a clearly defined geographical space, recognized, dedicated, and managed, through legal or other effective means, to achieve the long-term conservation of nature with associated ecosystem services and cultural values" [5,6]. These include national parks, national forests, natural reserves, conservation areas, wilderness areas, marine protected areas, wildlife refuges, and sanctuaries. PAs have significantly increased in number and coverage over the last century [7]. There were 248,754 designated PAs as of November 2021, encompassing approximately 15.72% of the Earth's land surface area and 7.91% of the Earth's ocean surface area [8]. The rapid increase in the number and area of PAs combined with wide support from different social groups has increased the worldwide expectations from the performance of PAs [9]. PAs also play an important role in biodiversity conservation and environmental stability [10,11]. Furthermore, as part of the Millennium Development Goals, PAs are projected to play a direct role in national development and poverty alleviation [9].

Although PAs serve as powerful tools to ensure conservation and sustainable development, they face major challenges arising from various aspects that undermine their efficiencies. Therefore, site selection is of great significance. However, some PAs have been designated merely because of the low cost of management rather than conservation priorities [12–14]. Various other issues, either inside or outside the PAs, also hamper management efficacy. Due to inadequate management staff and budgets [15] and lack of management schemes [16], many problems can occur within PAs, including competition over natural resources with PAs and conflicts between humans and wildlife [17–20], illegal poaching [21–24], illegal logging [25–27], and invasion of alien plant species [28–30]. However, challenges have also been identified outside the PAs. These challenges are largely related to the pressures and impacts of anthropogenic factors from adjacent areas [31–33]—notably, human encroachment [34–38]. Changes in land use and activities occurring in the surrounding regions can evoke a majority of negative impacts on PAs [39–41]. Therefore, extensive studies have been conducted to examine the relationship between PAs and their surrounding areas [42–44] because these impacts can destroy the conservation-development balance within and around PAs [45]. Much effort has been made to overcome these problems with mixed success. These efforts include conservation of PAs, sustainable development, and community-based management, which have been well documented by Du et al. [46].

PAs are usually densely populated by rural communities and bordered by agricultural land and are largely established in areas of the world where poverty is common and 92% of the world's poor rely on natural resources for their survival [47]. Thus, PAs are expected to contribute to community livelihoods and well-being [48], which is an important aspect of advancing sustainable development. Asia's PAs have great ecological value [9] while maintaining large concentrations of people, supporting local livelihoods and development, yet suffering from commercial pressures such as tourism and the construction of roads, mines, and dams [49]. Given widespread poverty [9], rapid population increase [50], and political instability [51], managing protected areas in developing countries poses significant challenges. Furthermore, PAs have attracted significant investment at the cost of opening remote areas for logging, oil exploitation, and mining [52]. Hence, to understand the various aspects of PAs in developing nations, much research has been conducted, resulting in a substantial and expanding corpus of literature.

Pritchard proposed bibliometric analysis, which is a mathematical and statistical strategy for analyzing relevant literature and understanding worldwide research patterns in a particular field [53,54]. Bibliometric analysis approaches have been employed in environmental engineering and science, soil science, ecology, food safety, new energy use, and other domains to provide quantitative evaluations of the academic literature [55]. A bibliometric study aids in identifying research gaps and directions in a certain field [56].

Bibliometric studies have been successfully applied in several fields to review and detect research trends and hot topics. For instance, Pratikshya et al. filled the research gap in the limited data on ecosystem science [57]. They revealed temporal trends, geographical distribution, and patterns of authors, institutions, and topics. Yang et al. conducted a systemic and objective review of climate change and tourism [58], identifying the most urgent issues in this field. In the field of regenerative medicine, Chen et al. identified the most active topics and revealed emerging trends and new developments in the interplay between basic and applied research [59].

Nepal is one of the world's 46 least developed and lowest income countries [60], sandwiched between two economic heavyweights—India and China [61]. Nepal is an ecologically and culturally diversified country with a large area of PAs [61,62] and some globally important ecoregions [63] Moreover, Nepal is ambitious and enthusiastic about advocating for PA strategies. It has signed many international conventions and treaties to promote conservation courses, including the Convention on Biological Diversity, Ramsar Convention, the Convention on International Trade in Endangered Species of Wild Fauna and Flora, and the World Heritage Convention [64]. Nepal has also had various policy and plan transitions, from state control to community-based management [65]. Therefore, drawing a holistic picture of Nepal's PAs can provide insights into relevant studies on PAs.

The increasing number of academic, governmental, and (inter)national entities investigating, implementing, and managing PAs in Nepal has resulted in an increase in the literature, which includes a constantly rising body of research in academic journals, books, and conference proceedings. The volume of scientific literature available on PA research continues to grow, making it difficult for researchers and practitioners to obtain a thorough and structured overview of essential data. A large number of review studies have been conducted on management issues [66], environmental policy [64], community forestry and livelihood [67], ecotourism [68], conservation issues [64,69], human-wildlife conflict [70], biodiversity [71], and climate change impacts [72]. These perspectives are interdependent on one another and conducted separately, focusing on a certain perspective. Thus, it is difficult to grasp the whole picture using traditional literature review methods. However, scientific knowledge mapping analysis based on bibliometrics is a more practical method for extracting insightful information from large amounts of data [73].

Therefore, this study employed a combination of performance analysis, which reveals the number of articles, as well as the main journals and research areas, and science mapping analysis, which reveals the main research topics, their structure, evolution, and trends. It aimed to understand the performance, lineage of research, main aspects, and trends of research on PAs in Nepal from a vast amount of literature to provide a reference for other scholars in related research. In this study, we have used bibliometrics as a research method for the first time to conduct a study on PAs in Nepal. It provides a more comprehensive and systematic analysis compared to the common literature review to deal with large amounts of data. In addition, in terms of research methodology, we analyzed each type of PA in Nepal. This is because different types of PAs have different conservation objectives and priorities, and they face different problems. By doing so, this study not only provides a panoramic view but also allows comparison between different protected area types.

This study contributes to the literature in several ways. First, it reflects the status quo and content of the research more immediately, making it easier to trace the field's origins and trends. Second, it depicts the evolution of research, allowing scholars to better comprehend the field's evolution and identify new directions. Third, it displays the most prominent institutions and journals, allowing scholars to more precisely search for journals and articles.

2. Materials and Methods

2.1. Study Area

These are of five types: national parks, buffer zones, wildlife reserves, hunting reserves, and conservation areas (Figure 1). They are spread across Nepal's high mountains, mid-hill areas, and lowland areas, covering 23.63% of the country's total land area in 2021 [74] ranking eighth among Asian countries and regions as of 2021 [75]. Details of Nepal's PAs are presented in the Supplementary Materials (Table S1).

However, PAs in Nepal are facing increasing issues as the country's human and cattle populations grow [65]. Nepal counts on the tourism industry to alleviate poverty, and it has already been confirmed as a powerful tool for reducing the degree of poverty in Nepal [77]. However, tourism-related negative impacts have also received considerable attention. In several of Nepal's protected regions, issues of tourism pressure and waste control are evident [78]. Furthermore, Nepalese PAs are not fully representative of conservation priorities. It has been identified that although vulnerable animal species are effectively protected, the existing PA system does not cover a vast number of threatened plant species [79]. Given the fact that Nepal is located in the Himalayas, one of the world's top 20 biodiversity hotspots and is a biodiversity-rich country that contributes significantly to global biodiversity [79], the success of its PAs can have an impact beyond its own territory.

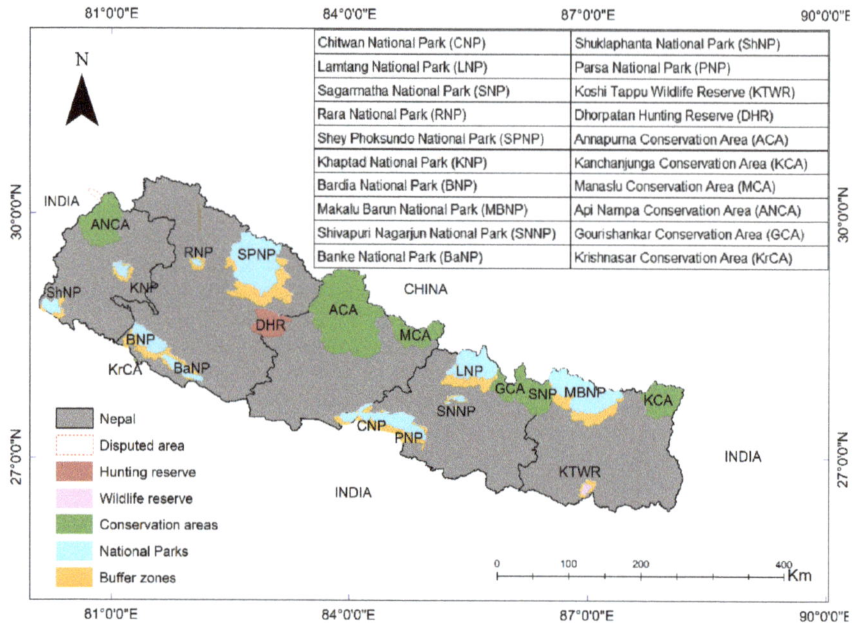

Figure 1. Nepal's PAs [76].

2.2. Methods

Bibliometric analysis is a quantitative tool for evaluating academic work on a certain topic by reviewing previous publications [80]. This is a quantitative analysis of scientific production, allowing us to track the growth of a scientific subject in detail. By examining secondary data obtained from a digital database from a quantitative and objective standpoint, bibliometric analysis can introduce a systematic, transparent, and repeatable review procedure, thereby improving the reliability and quality of the results [81].

2.2.1. Software

There is no consensus on which method is the best among existing bibliographic software [82]. Therefore, VOSviewer (1.6.18) and CiteSpace (5.8. R3) were used to create knowledge maps. They are both Java-based research tools that are widely used for visualizing and analyzing knowledge maps, as stated earlier. Both use scientometric theory to present the structure, patterns, distribution, and potential knowledge of scientific knowledge; they can produce collaboration networks of authors, countries, or regions, and co-occurrence of authors and keywords. The combination of the two can help achieve accurate visualization of the literature. According to Fu and Ding [83], CiteSpace was found to have specific advantages in revealing the dynamic development of disciplines and detecting citation bursts. VOSviewer can be used to create a knowledge map when there is a clear relationship between subjects or when the amount of data is substantial.

2.2.2. Indicators of Analysis

We employed descriptive and relational bibliometric indicators and methods. Countries and institutions contribute to a better understanding of the socio-demographic context. The publication year frequency aids in visualizing and establishing stages in the history of research. Keywords aid the comprehension of how concepts and research are classified and linked in this context. This clarifies which of these have not been thoroughly examined.

2.2.3. Data Sourcing and Analysis Method

In terms of the database selection, Google Scholar lacks the quality control needed for its use as a bibliometric tool; the larger coverage it provides consists in some cases of items not comparable with those provided by other similar databases [84]. We did not choose Scopus either as it has a more comprehensive list of contemporary sources. However, our study aims to cover a broader time range, that is, starting from the earliest documents. Based on the discussion above, Google Scholar and Scopus have been excluded from this study.

This study used datasets from Web of Science (WoS). WoS is a well-known and widely used digital database that provides researchers with high-quality publications of various types [73,85,86]. WoS has over 21,000 peer-reviewed journals in over 250 categories and covers a wide spectrum of publications from many fields [87]. Furthermore, WoS is an appropriate database because it contains a variety of data, including titles, authors, institutions, countries, abstracts, keywords, references, citation counts, impact factors, and other information [82,88]. As a result, the datasets can be used for bibliometric analysis and information visualization.

The data were retrieved from the WoS Core Collection (WoSCC) database on 10 January, 2022, and the time span was set from "1 January, 1900 to 31 December, 2021." There are five types of PAs in Nepal [76]. Therefore, the search formula used was "TS = Nepal protected area* OR Nepal national park* OR Nepal wildlife reserve* OR Nepal buffer zone* OR Nepal hunting reserve* OR Nepal conservation area*" and the document type was chosen as "ARTICLE" and in "English", yielding a total of 864 documents. We only selected journal articles because they are regarded as "certified knowledge" and because they are the outcome of an evaluation procedure, which gives the results credibility [89]. As a result, we excluded proceedings papers, news articles, or other documents (Table 1).

Table 1. A summary of searching criteria.

Data Source	Web of Science Core Collection
Citation indexes	SCI-EXPANDED; SSCI; AHCI; ESCI
Date range	1 January 1900–31 December 2021
Keywords	"Nepal protected areas OR Nepal national parks OR Nepal wildlife reserves OR Nepal buffer zones OR Nepal Hunting reserves"
Document types	"Articles"
Language	"English"
Sample size	864

We did not analyze "Hunting reserve" because the sample size was too small (only 12) to be used for knowledge mapping (Table 2) since the ideal sample size should be more than 50 documents [90]. After searching and screening, 864 articles covering 73 research areas were collected. These papers were by 2057 authors affiliated with 1026 institutions in 64 countries and regions. These were published in 315 journals and cited 13,014 references (Table 3).

Table 2. Counts of keywords.

Keywords	Counts	%
All	864	100.0
National Park	622	71.9
Protected area	327	37.8
Wildlife reserve	68	7.8
Conservation area	171	19.8
Buffer zone	118	13.7
Hunting reserve	12	1.39

Table 3. Descriptive results.

Criteria	Quantity
Publications	864
Research categories	73
Authors	2057
Journals	315
Institutions	1026
Countries and regions	64
Cited references	13,014

Using the WoS "Analyze the Results" function, descriptive statistics on year and count, research categories, countries, and regions were conducted; SPSS 26.0 was used to conduct statistical analysis on the stages of publication; CiteSpace V and VOSviewer were used to conduct the mapping process.

3. Results

3.1. Publication Performance Statistics

Figure 2 depicts the publication counts over the years and the cumulative publications. All data were imported into SPSS 26.0 for a correlation test. This shows that there is an exponential relationship between the volume of the literature and time (Table 4).

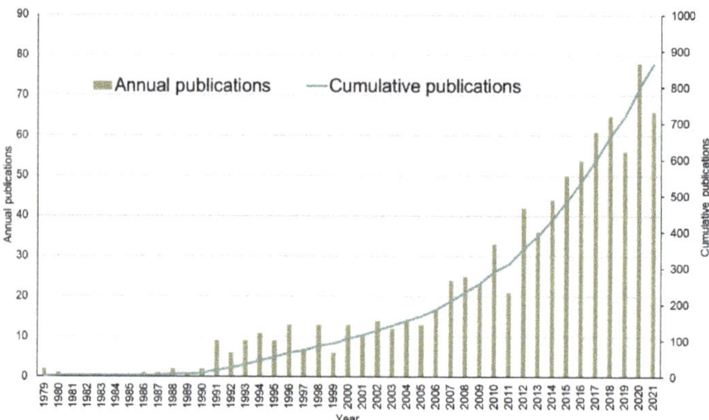

Figure 2. Publication counts and cumulative counts over years.

Table 4. Statistical analysis of counts and years.

		Year	Count
Year	Pearson Correlation	1	0.893 **
	Sig. (2-tailed)		0.000
	N	38	38
Count	Pearson Correlation	0.893 **	1
	Sig. (2-tailed)	0.000	
	N	38	38

** Correlation is significant at the 0.01 level (2-tailed).

From 1979 to 1990, there was a period in which only a few publications were produced, with a barren period between 1979 and 1990, the incipient period. The second phase (1991–2006) witnessed a nearly 10-fold increase in the number of publications on average. Although the third period (2007 to 2014) had some fluctuations, it still showed a significant increase in the number of articles, indicating that the study had progressed. After 2015, the number of articles increased sharply. The year 2020, with 78 articles, had the most publications. By the end of 2021, the cumulative number of publications reached 864. A further increase is expected for 2022.

A total of 73 research categories were included. The research domain was broad in scope, encompassing a wide range of topics and disciplines. Figure 3 shows the top 15 with more than 20 publications. Environmental sciences came first with 249 papers, followed by ecology with 235 papers. Biodiversity conservation contributed 194 publications, and zoology 118 publications. Publications can also be found in other disciplines.

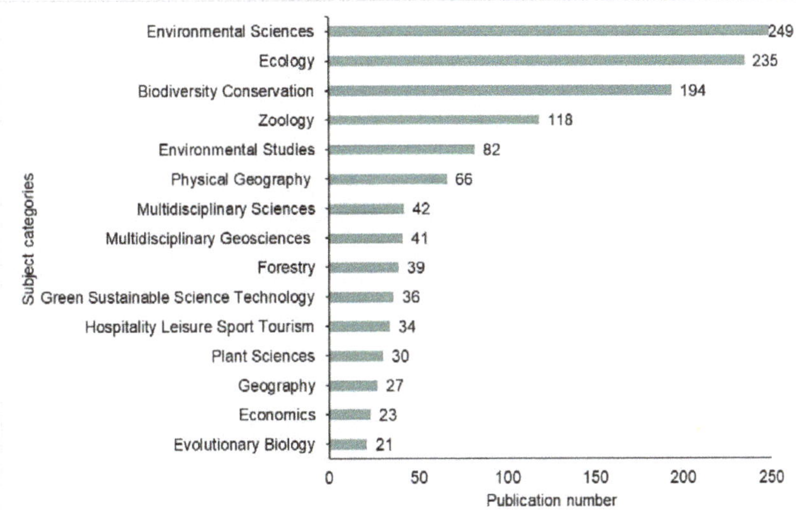

Figure 3. Research categories.

In total, there were 864 publications in 315 journals. Although many journals supported a wide range of research themes and multidisciplinary characteristics of studies on Nepal's PAs, 33% (n = 286) of the journals had published no more than five publications. Table A1 provides a list of journals with more than 10 publications; Table A2 shows the top 10 most cited articles.

The visualization map produced using VOSviewer provides a more direct impression of the journals' citation correlation (Figure 4). The threshold was set at five to study the connections and clusters of the most prolific journals. The map shows five clusters (five colors). The cluster shown on the right part of the map consists of five journals of geoscience and appears slightly distant from the other four clusters, which are closely connected to one another. The journals were extensively connected to each cluster. The node size denotes the number of journal publications, as illustrated in the map.

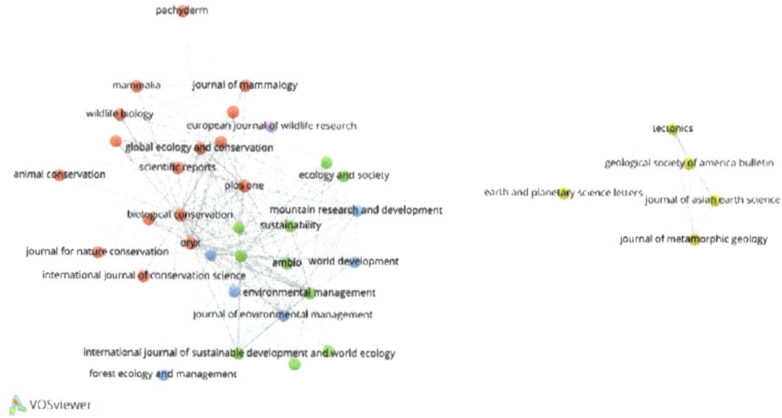

Figure 4. Journals citation correlation.

3.2. Analysis of Countries and Regions and Major Research Institutions

3.2.1. Countries and Regions

A countries/regions co-authorship network visualization map (Figure 5) was built to show their relationships. The minimum document criterion is set at 5. Of the 64 units, 33 were identified as visualization objects. The number of papers is represented by the size of the circles, with larger circles indicating more documents. Seven clusters can be recognized by their distinct colors. For example, Nepal and the United States collaborated extensively, and their contributions were obviously larger than others. Nepal contributed 360 publications, while the United States contributed 313. Other countries and regions have also contributed to this research field as well. However, many of them are far from each other on the map, showing weak cooperation.

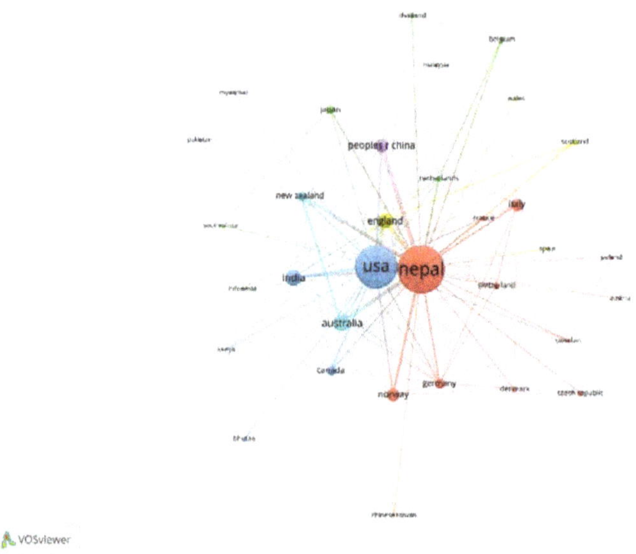

Figure 5. Countries and regions.

3.2.2. Major Research Institutions

VOSviewer was used to create an organization citation visualization map to investigate primary collaboration among the 976 organizations (Figure 6). There were 73 powerful organizations (7.5%) that remained when the threshold value was set to 6. The map shows that these organizations are grouped into four clusters (shown in 10 colors in Figure 6).

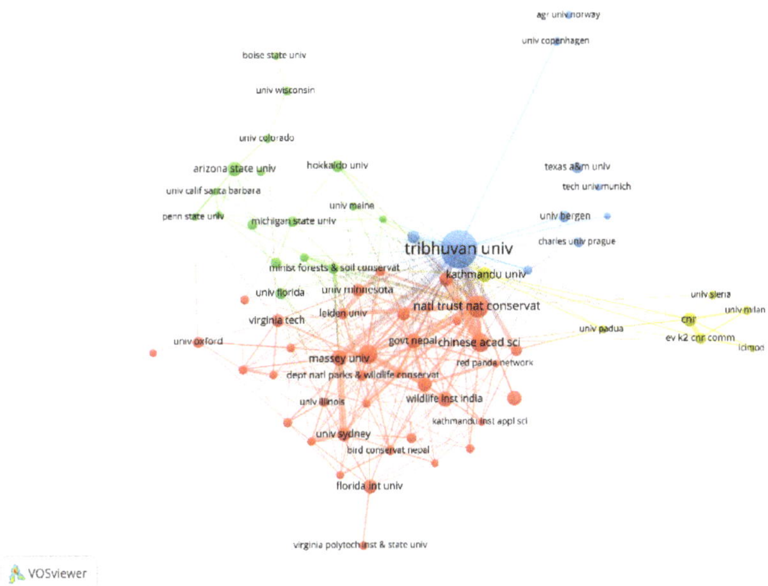

Figure 6. Map of "Research Institutions".

The size of the node symbolizes the number of publications, and the line connecting the two nodes indicates the academic connection between the two organizations. The stronger the connection, the shorter the line. All institutes are labeled with their abbreviations. As shown in the map, the red cluster has the most members (39). The National Trust for Nature Conservation (natl trust nat conservat) led the red cluster in terms of publication production (46), followed by the Chinese Academy of Sciences (chinese acad sci) with 35 documents. Tribhuvan University (tribhuvan univ) led to a blue cluster. Tribhuvan University contributed the most to both publications (138) and linkages (146). Arizona State University (arizona state univ) led its cluster (in green color) with 20 publications followed by Hokkaido University (hokkaido univ) (14). In the yellow cluster, Kathmandu University (kathmandu university) contributed 23 documents. All clusters showed a close internal connection, except for the blue cluster. On top of the map, two institutes, the Agricultural University of Norway (agr univ norway) and University of Copenhagen (univ copenhagen), are remotely related to Tribhuvan University and bear no connection to any other clusters (Figure 6).

3.3. Analysis of Research Lineage

Diverse Research Aspects

Keywords are nouns or phrases that express the important substance of an article [91]. The keywords used in the publications were analyzed to provide both the most important themes and significant research trends in the field [92]. VOSviewer was used to create a keyword co-occurrence map that visualizes variations in scientific production [93]. We set the threshold as the default value (10), and a binary counting method from both titles and abstracts, ignoring structured abstract labels and copyright statements, was adopted. A total of 339 (out of 20,916) items were discovered and sorted into three clusters (separated by color, as shown in Figure 7). The most frequently used keywords are shown in larger nodes. These nodes are connected to each other at various distances. The greater the association between the terms, the shorter the distance between the different nodes.

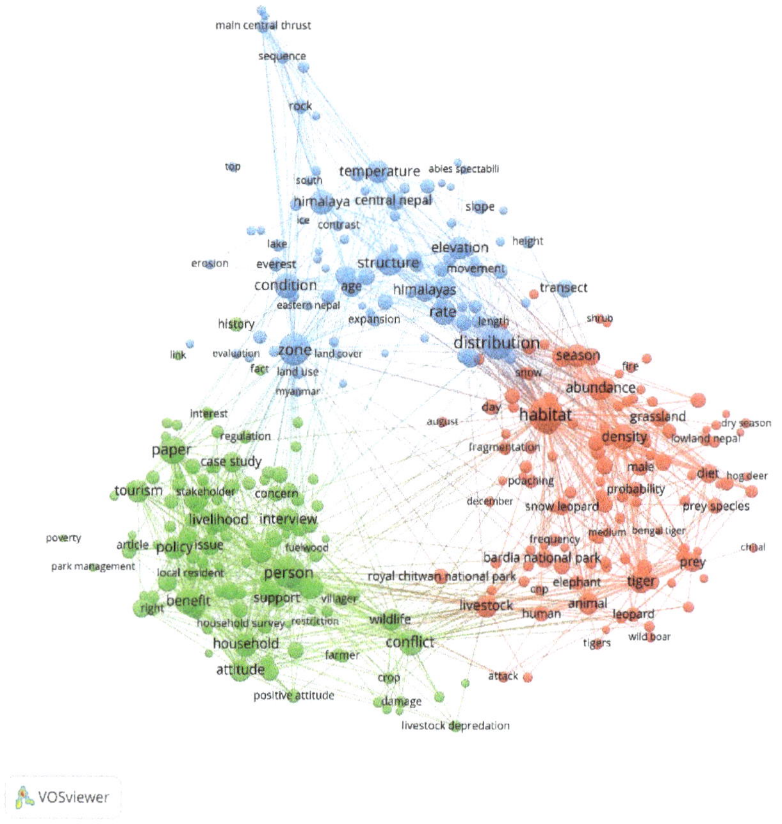

Figure 7. Mapping of "All Keywords".

The sizes of the nodes in Figure 7 show the frequency of the terms used. Larger nodes indicate more frequently used keywords. The term "person", "zone", "distribution", and "habitat" had the most power. The nodes in the same cluster indicated that these publications had a common theme. As illustrated in the red cluster (cluster 1, right, 130 nodes), the primary nodes like "habitat" and "density" were found. In this cluster, other keywords such as "animal", "livestock", "tiger", and "poaching" indicate a research interest in major animals and related topics. Keywords like "person", "wildlife", "conflict", "household", "livelihood", "income", "policy", and "tourism" formed core topics in the green cluster (cluster 2, bottom left, 115 items). We can determine that this cluster's main concern is related to people's lives and their interactions with wildlife. Other keywords such as "interview" and "case study" indicated the most adopted research methods in this cluster. Next, nodes such as "zone", "distribution", "elevation", "himalaya", "temperature", and "transect" focused on the aspect of geographical and geological studies in the blue cluster (cluster 3, top, 91 items).

Burst detection is a valuable analytic tool for identifying keywords that attract considerable attention from connected scientific communities over time. Keyword citation bursts occur when the number of citations for a certain keyword spike is dramatic. Here, 25 bursts discovered on the keywords were calculated using CiteSpace (parameter settings: years per slice: 1; node types: keyword) to investigate the PA-relevant studies and to explore the intensely explored directions (Figure 8). The top 25 keywords with bursts were mirrored by the discovered hotspot keywords displayed in Figure 8. The period during which the citation boom occurred is indicated in red.

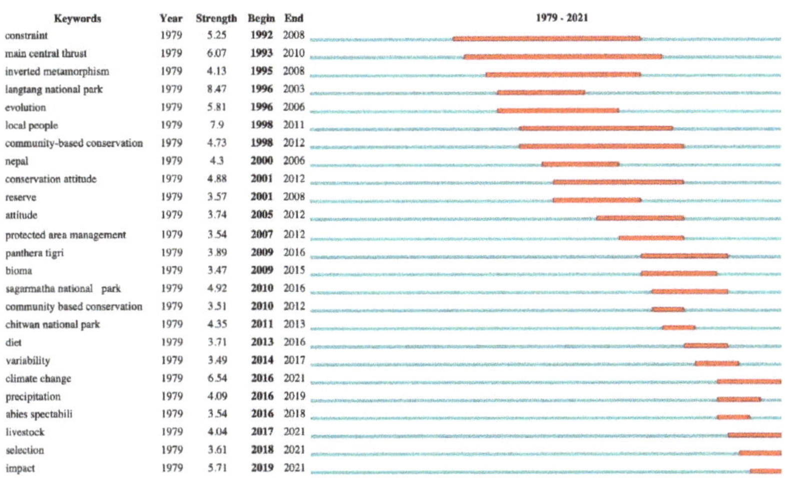

Figure 8. Citation burst detection.

The keywords of the early stage were "constraint", "main central thrust", and "inverted metamorphism", denoting a period of research interest in geological studies. Then, from 1998 to 2015, "local people", "community", "conservation", and "protected area management" and wildlife attracted intense research enthusiasm. Two of Nepal's famous national parks, Sagarmatha National Park and Chitwan National Park, have received intense attention from the scientific community. In recent years, topics related to climate change have enjoyed a boom. Keywords "climate change" and "precipitation" are now in the burst range, along with the two other burst leading keywords, "abies spectabili" and "impact", indicating the latest research hotspot related to climate change.

By assessing the burst keyword order, such as "main central thrust", "local people", "conservation", "attitude", "climate change", and "impact", the dynamic process can be found in Figure 8. The keyword bursts also revealed that the focus of the study shifted rapidly over time.

3.4. Comparison among Different Keywords

3.4.1. Protected Areas

PA subjects were divided into five colored clusters (120 items) (Figure 9). The red cluster with the most terms (60) is led by "park". Other keywords in this cluster include "local person", "perception", "policy", "attitude", "interview", and "tourism". Most of the keywords in this cluster were related to parks and interactions with local people/communities. "Species", "population", "distribution", "threat", "habitat", and "landscape" are the primarily associated terms in the green cluster (43 items), which is related to research on wildlife and their habitats as well as the impact of human. The blue cluster contains 14 items with "forest", "diversity", and "community forest" being bigger nodes, indicating a research interest in forests and interaction with the community. The other two clusters contained too few items to be analyzed.

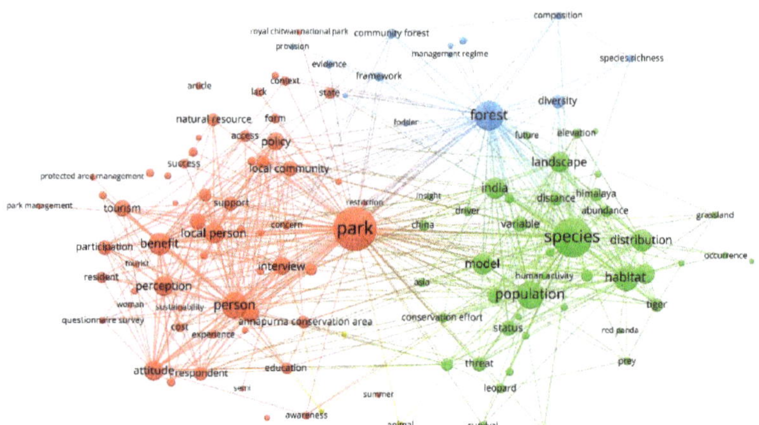

Figure 9. Mapping of "Protected Areas".

3.4.2. National Parks

The "National Parks" topic has three colored clusters with 228 items (Figure 10). The red cluster holds the most terms (87) with "person" at the center. Other keywords in this cluster include "livelihood", "attitude", "perception", "conflict", "wildlife", and "tourism", suggesting studies related to local people's relationship with national parks. The green cluster contains 82 items with bigger nodes of "range", "density", "animal", "abundance", "livestock", and "tiger", which demonstrates a research interest in wildlife and their living environment and interaction with the community. Most of the keywords in the blue cluster (59 items) were related to abiotic studies, such as geological and climate change. "Himalaya", "structure", "climate change", and "glacier" are important terms in this cluster.

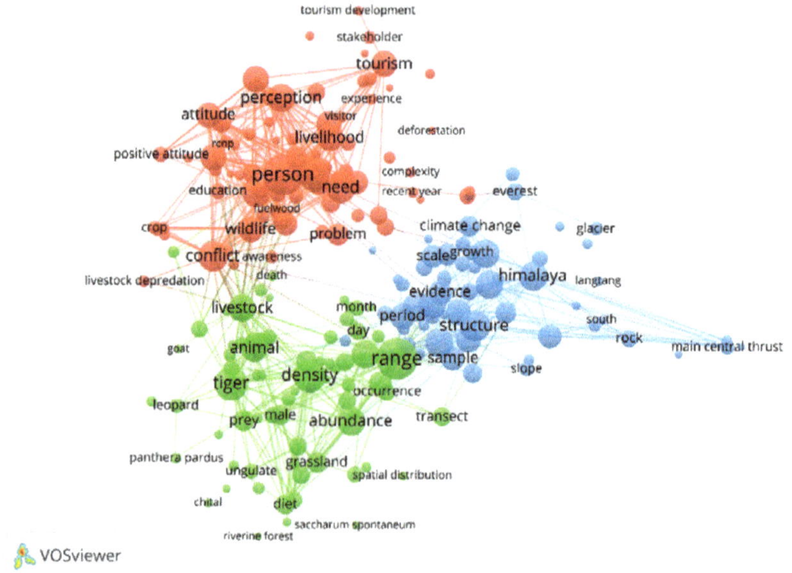

Figure 10. Mapping of "National Parks".

3.4.3. Buffer Zones

"Buffer zones" brought up 221 items and were grouped into three clusters (Figure 11). The keywords such as "region", "tourism", "development", "dynamic", and "sagarmatha", led the red cluster (92 items). In the green cluster, studies were more related to "royal chitwan national park", "resident", "cost", and "place" places (67 items). The blue cluster has 62 items with "conflict", "increase", "wildlife", "tiger", "attack", and "victim" being eye-catching. This cluster concerns wildlife attacks and their impact.

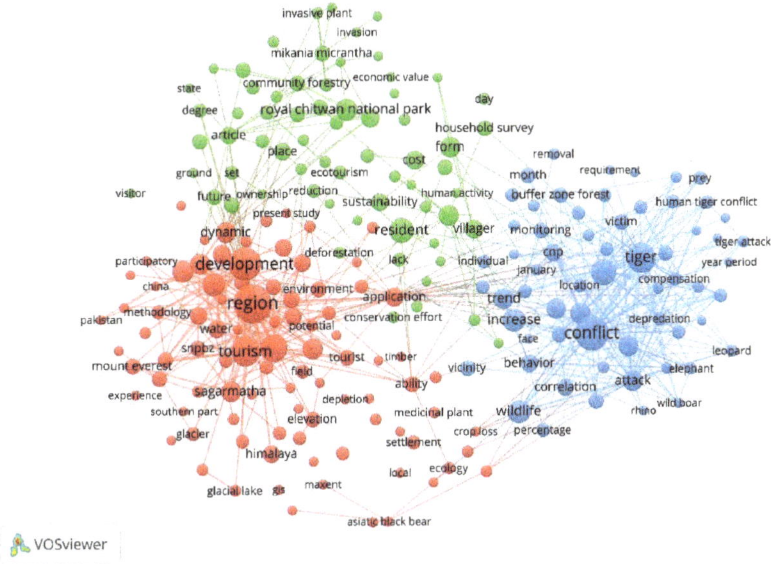

Figure 11. Mapping of "Buffer Zones".

3.4.4. Conservation Areas

This field of "conservation areas" generated three clusters as well, containing 144 items (Figure 12). The threshold for this map was set at five because the sample size was too small to produce a satisfactory map. It is worth noting that the keyword "park" is centered on the whole map, showing a close connection to the other two clusters.

The red cluster holds 65 items with important nodes of "development", "participation", "local community", "income", "tourism", and "aca (Annapurna Conservation Area)" pointing to studies on tourism-led development and local participation. The green cluster (49 items) is led by items of "habitat", "park", "distribution", "range", "species richness", "temperature", "treeline ecotone", and "musk deer" showing a mixed research focus on wildlife, their habitat and plants' correlation with temperature. In the blue cluster (30 items), "snow leopard", "conflict", "blue sheep", "density", "ecology", and "prey" are bigger nodes showing intense research interest in animal and ecological perspectives. The word "conflict" is very close to the red cluster, indicating a close relationship between wildlife and "development".

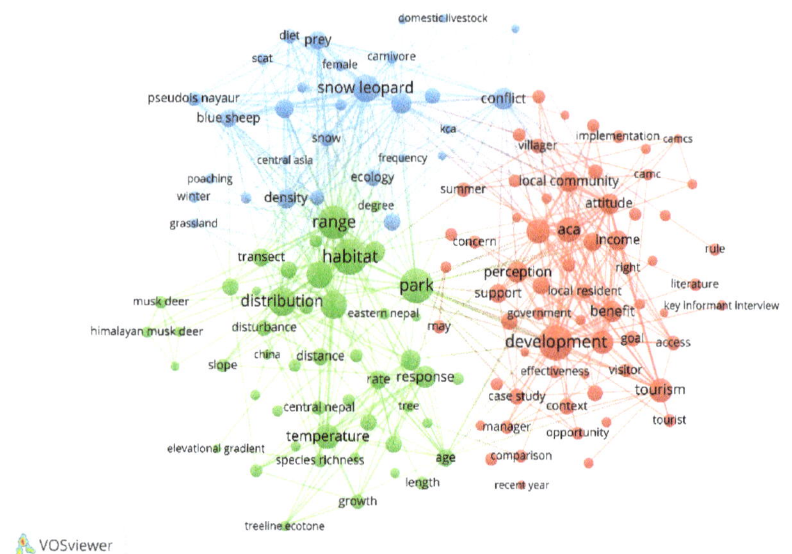

Figure 12. Mapping of "Conservation Areas".

3.4.5. Wildlife Reserves

Mapping of this topic produced 119 items that were organized into three clusters (Figure 13). "Park" and "reserve" are considerably larger than the other keywords with the former being center of the map. Apart from "reserve", the red cluster (46 items) concentrates on "threat", "individual", "water buffalo", and "poaching". It is also a field of research on animals and their living environments. Green (38 items) is mainly about animals, such as "ungulate", "axis porcinus", and "prey". However, this cluster's studies have been mostly low with regard to "lowland Nepal". The blue one is clearly related to studies on local people and conservation because this cluster is led by keywords such as "person", "household", "benefit", "local community", "conflict", "damage", and "compensation".

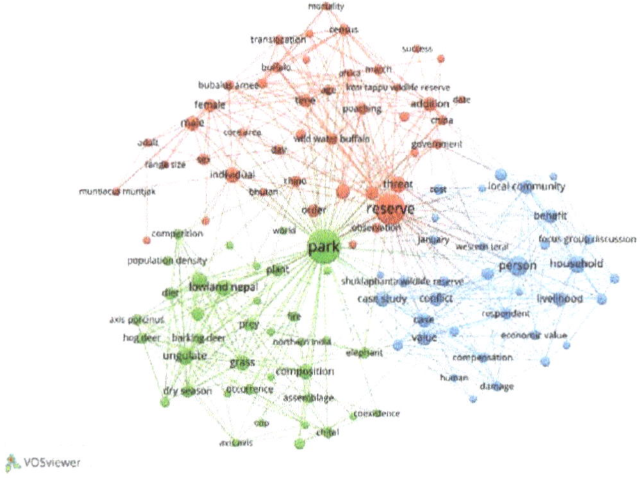

Figure 13. Mapping of "Wildlife Reserves".

4. Discussion

4.1. Research Trends Shifting

The multidisciplinary nature of research on PAs, as well as the numerous and complicated interrelationships between these fields, have made it difficult to identify its trends shifting pattern. We discovered buried information underpinning this major body of research.

The conservation strategy of protected areas is in a process of dynamic change. National policies and socio-economics are the drivers of these changes [65]. The emergence of new changes leads to changes in research hotspots accordingly. This trend of changes is also reflected in our study. In 1973, the Chitwan National Park (CNP) was established and the Nepalese conservation paradigm entered the "Yellowstone paradigm", with strict management and a ban on people living in the park [65]. This phase was dominated by research focused on geology.

From the 1980s onward, the Nepalese government recognized the importance of a participatory conservation and development model. The government legislated in 1989 to define the approach, that is, to recognize the indispensable role of local people in the conservation process [65]. Our keyword burst analysis also reflects this trend. The burst of the keyword "community-based conservation" (Figure 8) from 1998 to 2012 marks the emergence of a great number of relevant studies. The large number of studies also provides a strong theoretical basis for Nepal to be a successful model of biodiversity conservation [94]. As the new conservation approach no longer completely excludes people from PAs, it has also led to some new thinking. For example, studies on people's attitudes toward conservation, on the relationship between people and animals have been conducted.

The latest research trends are mainly related to climate change and its impacts, as Nepal is a country prone to climate change disasters [95], which is in accordance with the global concern regarding this topic.

It is important to note that we observed a lag in the changes in research hotspots relative to policy changes. This is because it takes time for policies to take effect and for research to progress. Based on this, we highly encourage research on future projections based on the previous research findings and changes in research hotspots. Research on PAs in Nepal is mostly conducted in traditional academic disciplines. However, with the impact of climate change and the resultant increase in natural disasters, studies from other research areas, such as remote sensing, meteorology, and atmospheric sciences, are thus expected to contribute much more [96].

4.2. Power Shifting and New Players

By using the WoS function "analyze results", we found that Kenya and Austria were among the pioneers of studying Nepal's Pas (Figure 14). Kenya contributed to animal research, with its first publication concerning the ivory trade in 1998 [97]. Austria started its studies on Nepal's PAs with a publication in 1994 in Germany about the impact of tourism [98]. As new players, the Netherlands and Belgium have contributed mostly to studies on biodiversity conservation and ecology. Meanwhile, Poland is interested in the management of national parks, waste management, plants, and tourism.

Something interesting can be found in the density map of countries and regions (Figure 15). Besides the US and Nepal, another contributor to the research on PAs in Nepal is China. This is not unusual because China is naturally interested because of its neighboring location in Nepal. To a certain degree, these two countries are connected to one another ecologically. Other close neighbors of Nepal, such as Bhutan and Bangladesh, seem to be less active in this party. However, bordering on each other means that they are bound to have mutual benefits or losses. Transboundary PAs exist in Nepal. For example, the Sacred Himalayan Landscape (SHL) connects Nepal, India, and Bhutan. These PAs also play the role of ecological corridors for some iconic animals [65] between countries and regions. Hence, here we highly suggest that these neighboring countries and regions conduct joint research, which will bring more benefit to a larger regional, even international scale.

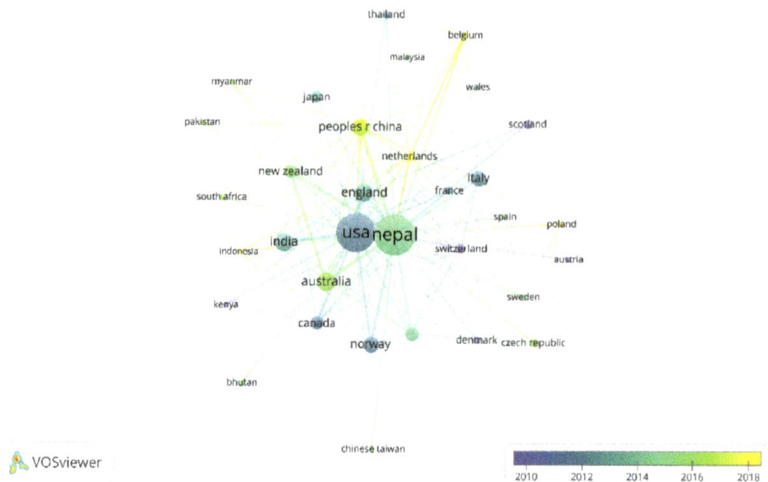

Figure 14. Overlay map of countries and regions.

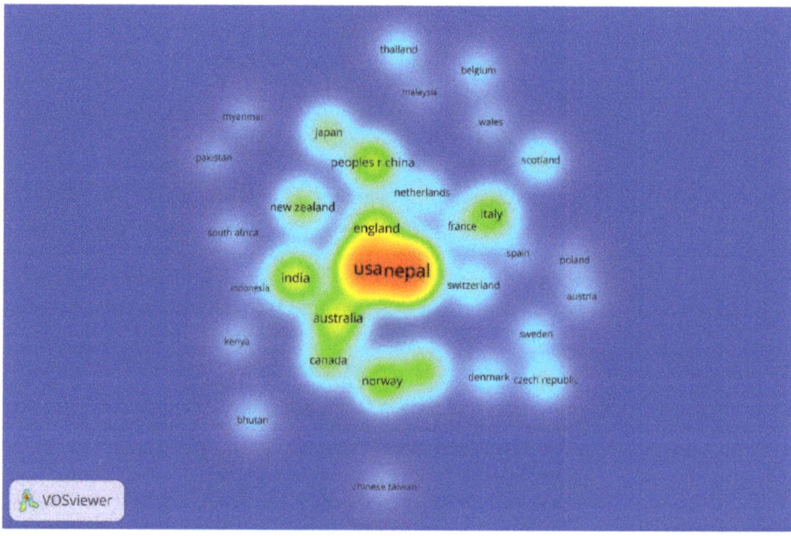

Figure 15. Density map of countries and regions.

4.3. Uneven Research Interest and Homogenization of Research Methods about Each PA Type

Among the mapping results, we found that the keywords about people, animals, development, and conflict were the most prevalent research topics. The management policies at the early stage did not allow people to live in PAs, and people's demand for natural resources created conflicts with the reserve [99]. The community-based conservation recognizes the indispensable role of local people in the conservation process and no longer completely prevents people from living and working in PAs. However, it increases the chances of encounters between people and wildlife, which can lead to conflicts. For example, in CNP alone, there were over 4000 wildlife-attack losses to humans, livestock, and property from 1998 to 2016 [100]. In other PAs around the world, human–animal conflict is also of widespread concern [101]. In addition, it is difficult to strike a perfect balance between conservation and development, for example, using tourism to promote the economy will inevitably bring some environmental pressure.

Figure 16 shows the top five keywords that appeared to fall into each category. The first two, "protected areas" and "national parks", have received much more attention from researchers. One of the reasons that national parks have been accumulating more publications would be their longest history in Nepal's conservation progress since the establishment of the first national park, Chitwan National Park [79], a sign of formal conservation in the country. Another reason may be the dominant number of parks (12 national parks), which is the most common type of PAs in the country (Figure 1). Being different from national parks, buffer zones, conservation areas, and wildlife reserves allow local people to use forest products in a sustainable way in Nepal [65]. Under such circumstances, the management goals and practical needs of local people often lead to "park–people conflicts" [102,103]. The buffer zone is thought to be a major conservative priority, but few studies have been conducted to test its effectiveness in Nepal [104]. Hence, more studies in these less-investigated areas should be conducted in the future.

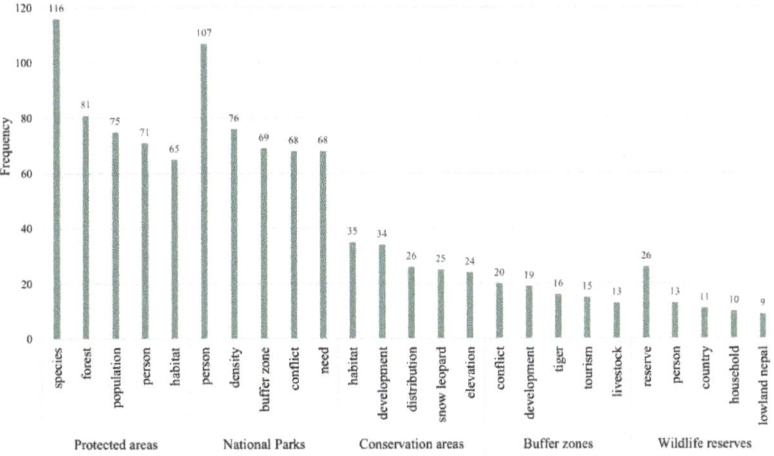

Figure 16. Comparison of top five keywords under each topic.

Many studies related to "people", such as perception and attitude studies, employ the research methods of questionnaire surveys or interviews. However, it is worth noting that even in different studies, repeated questions may evoke the "memory effect" leading to unreliable answers [105]. Therefore, more studies should be pursued, and new relevant topics and previously under-studied disciplines should be investigated.

4.4. Insufficient Attention to "Hunting Reserves"

The Dhorpatan Hunting Reserve (DHR; Figure 1) is the only hunting reserve in Nepal and is home to many mammalian species [101]. We did not analyze the keywords of "hunting reserves" because there were only 12 papers concerning this topic, which was not sufficient to generate a preferable map for reasonable analysis. However, this does not indicate that this hunting reserve deserves no academic attention. There are approximately 350–500 snow leopards (*Uncia uncia*) living in Nepal's northern frontier, and their presence has been suggested in many PAs of Nepal, with DHR being one of them [106]. However, their survival is threatened by conflict with humans [107]. As a controlled hunting area, the DHR has the potential to contribute to the conservation of snow leopards. Many scholars have studied this endangered species in conservation areas, as suggested by the large node (Figure 12). However, little research has been conducted on snow leopards in the DHR. We suggest that special attention be paid to this issue. However, buffer zone communities depend on the revenue generated by the DHR. This avenue usually comes from the government's sale of hunting permits, and DHR creates certain job opportunities [99]. Given the importance of DHR in terms of its social and ecological aspects, more relevant studies are needed.

4.5. Limitations of This Study

Proceedings were excluded from our study. However, papers in proceedings derived from international conferences usually contain hot topics. Book chapters were also excluded, but many important social science studies have been described. Here, we suggest that future studies consider an analysis that includes proceeding papers and book chapters. We did not perform thesaurus removal because of the large number of keywords analyzed in our study. This may cause some inaccuracies in the node size and links.

5. Conclusions

To draw a holistic and systematic picture of research on PAs in Nepal, we undertook an integrative study using bibliometric analysis. An increase in the number of papers indicates that the topic is growing and has attracted intense research interest. This research did not receive widespread attention in the early years. However, the exponential growth trend in the literature shows a high level of enthusiasm for research on this topic in Nepal. We identified the changing trend in this field from geological aspects in the early stage to the recent hotspots of climate change-related perspectives. There has been a shift of "research powers" in countries and regions. Kenya, Canada, Norway, Switzerland, and the US were among the earliest players. Nepal contributed the most in the middle stage. China also became interested in this period. Belgium and Poland contributed the latest publications.

We found that the research hotspots are changing with the shifting of conservation policies in Nepal. We suggest conducting more predictive studies on the future development of PAs. Currently, PAs research is mainly conducted in traditional disciplines, but with the impact of climate change and the consequent increase in its negative impacts, academic contributions from other study disciplines, such as remote sensing, meteorology, and atmospheric sciences, are expected to contribute much more. Research enthusiasm toward each keyword showed some imbalance with "protected areas" and "national parks", attracting much more attention than others. Although there is currently only one hunting reserve, we suggest that more relevant studies should be conducted.

Supplementary Materials: The following are available online at https://www.mdpi.com/article/10.3390/land11071109/s1, Table S1: Detailed information of Nepal's PAs.

Author Contributions: Conceptualization and methodology, L.C., T.W., H.X. and J.H.; validation, formal analysis, investigation, resources, data curation, and visualization, L.C. and T.W.; writing—original draft preparation, L.C.; writing—review and editing, L.C., T.W., H.X. and J.H.; supervision and project administration, T.W. All authors have read and agreed to the published version of the manuscript.

Funding: This work was supported by JSPS KAKENHI (Grant-in-Aid for Scientific Research), Grant Number JP21H04371 (T.W.).

Institutional Review Board Statement: Not applicable.

Informed Consent Statement: Not applicable.

Data Availability Statement: The data supporting the findings of this study are available from the corresponding author upon reasonable request.

Acknowledgments: We thank Sun Yujie, Atupelye Komba, Anushilan Acharya, and Mengyu for their kind assistance in collecting publications and suggestions.

Conflicts of Interest: The authors declare no conflict of interest.

Appendix A

Table A1. The journals with more than 10 publications.

Ranking	Journals	Publication Number
1	Mountain Research And Development	39
2	Oryx	31
3	Environmental Conservation	29
4	Global Ecology And Conservation	26
5	Biological Conservation	23
6	Environmental Management	22
7	Ecology And Evolution	19
8	Plos One	19
9	Biodiversity And Conservation	18
10	Journal Of Environmental Management	14
11	Sustainability	14
12	Journal Of Mountain Science	12
13	International Journal Of Sustainable Development And World Ecology	11

Table A2. The top 10 most cited articles.

Rank	Title of Publications	Journals	Publication Year	Citation Count
1	Tectonic evolution of the central Annapurna Range, Nepalese Himalayas	Tectonics	1996	423
2	Neogene foreland basin deposits, erosional unroofing, and the kinematic history of the Himalayan fold-thrust belt, western Nepal	Geological Society of America Bulletin	1998	361
3	Shisha Pangma leucogranite, south Tibetan Himalaya: Field relations, geochemistry, age, origin, and emplacement	Journal of Geology	1997	332
4	Isotopic constraints on the age and provenance of the Lesser and Greater Himalayan sequences, Nepalese Himalaya	Geological Society of America Bulletin	1996	323
5	Insights on linking forests, trees, and people from the air, on the ground, and in the laboratory	Proceedings of the National Academy of Sciences of the United States of America	2006	319
6	Decompression And Anatexis of Himalayan Metapelites	Tectonics	1994	301
7	Tectonometamorphic evolution of the Himalayan metamorphic core between the Annapurna and Dhaulagiri, central Nepal	Journal of Metamorphic Geology	1996	254
8	Exhumation, crustal deformation, and thermal structure of the Nepal Himalaya derived from the inversion of thermochronological and thermobarometric data and modeling of the topography	Journal of Geophysical Research-solid Earth	2010	217
9	P-T-t data from central Nepal support critical taper and repudiate large-scale channel flow of the Greater Himalayan Sequence	Geological Society of America Bulletin	2008	213
10	Local attitudes towards conservation and tourism around Komodo National Park, Indonesia	Environmental Conservation	2001	208

References

1. Rodrigues, A.S.L.; Akçakaya, H.R.; Andelman, S.J.; Bakarr, M.I.; Boitani, L.; Brooks, T.M.; Chanson, J.S.; Fishpool, L.D.C.; da Fonseca, G.A.B.; Gaston, K.J.; et al. Global Gap Analysis: Priority Regions for Expanding the Global Protected-Area Network. *BioScience* **2004**, *54*, 1092–1100. [CrossRef]
2. Chape, S.; Harrison, J.; Spalding, M.; Lysenko, I. Measuring the Extent and Effectiveness of Protected Areas as an Indicator for Meeting Global Biodiversity Targets. *Philos. Trans. R. Soc. Lond. B Biol. Sci.* **2005**, *360*, 443–455. [CrossRef]
3. Loucks, C.; Ricketts, T.H.; Naidoo, R.; Lamoreux, J.; Hoekstra, J. Explaining the Global Pattern of Protected Area Coverage: Relative Importance of Vertebrate Biodiversity, Human Activities and Agricultural Suitability. *J. Biogeogr.* **2008**, *35*, 1337–1348. [CrossRef]

4. Craigie, I.D.; Baillie, J.E.M.; Balmford, A.; Carbone, C.; Collen, B.; Green, R.E.; Hutton, J.M. Large Mammal Population Declines in Africa's Protected Areas. *Biol. Conserv.* **2010**, *143*, 2221–2228. [CrossRef]
5. UN. *UNEP-WCMC State of the World's Protected Areas: An Annual Review of Global Conservation Progress*; UN: Cambridge, UK, 2008.
6. Dudley, N. *Guidelines for Applying Protected Area Management Categories*; IUCN: Gland, Switzerland, 2008; ISBN 9782831710860.
7. Watson, J.E.M.; Dudley, N.; Segan, D.B.; Hockings, M. The Performance and Potential of Protected Areas. *Nature* **2014**, *515*, 67–73. [CrossRef] [PubMed]
8. WDPA February 2021 Update for the WDPA. Available online: https://livereport.protectedplanet.net/chapter-2 (accessed on 13 February 2021).
9. Naughton-Treves, L.; Holland, M.B.; Brandon, K. The Role of Protected Areas in Conserving Biodiversity and Sustaining Local Livelihoos. *Ann. Rev. Environ. Res.* **2005**, *17*, 219–252. [CrossRef]
10. Adhikari, S.; Southworth, J. Simulating Forest Cover Changes of Bannerghatta National Park Based on a CA-Markov Model: A Remote Sensing Approach. *Remote Sens.* **2012**, *4*, 3215–3243. [CrossRef]
11. Watson, J.E.M.; Venter, O.; Lee, J.; Jones, K.R.; Robinson, J.G.; Possingham, H.P.; Allan, J.R. Protect the Last of the Wild. *Nature* **2018**, *563*, 27–30. [CrossRef]
12. Venter, O.; Fuller, R.A.; Segan, D.B.; Carwardine, J.; Brooks, T.; Butchart, S.H.M.; Di Marco, M.; Iwamura, T.; Joseph, L.; O'Grady, D.; et al. Targeting Global Protected Area Expansion for Imperiled Biodiversity. *PLoS Biol.* **2014**, *12*, e1001891. [CrossRef]
13. Pouzols, F.M.; Toivonen, T.; di Minin, E.; Kukkala, A.S.; Kullberg, P.; Kuustera, J.; Lehtomaki, J.; Tenkanen, H.; Verburg, P.H.; Moilanen, A. Global Protected Area Expansion Is Compromised by Projected Land-Use and Parochialism. *Nature* **2014**, *516*, 383–386. [CrossRef]
14. Joppa, L.N.; Pfaff, A. High and Far: Biases in the Location of Protected Areas. *PLoS ONE* **2009**, *4*, e8273. [CrossRef] [PubMed]
15. Geldmann, J.; Coad, L.; Barnes, M.D.; Craigie, I.D.; Woodley, S.; Balmford, A.; Brooks, T.M.; Hockings, M.; Knights, K.; Mascia, M.B.; et al. A Global Analysis of Management Capacity and Ecological Outcomes in Terrestrial Protected Areas. *Conserv. Lett.* **2018**, *11*, 1–10. [CrossRef]
16. Hockings, M.; Stolton, S.; Leverington, F. *Evaluating Effectiveness: A Framework for Assessing Management Effectiveness of Protected Areas*, 2nd ed.; IUCN: Gland, Switzerland, 2006.
17. Soliku, O.; Schraml, U. Making Sense of Protected Area Conflicts and Management Approaches: A Review of Causes, Contexts and Conflict Management Strategies. *Biol. Conserv.* **2018**, *222*, 136–145. [CrossRef]
18. Redpath, S.M.; Young, J.; Evely, A.; Adams, W.M.; Sutherland, W.J.; Whitehouse, A.; Amar, A.; Lambert, R.A.; Linnell, J.D.C.; Watt, A.; et al. Understanding and Managing Conservation Conflicts. *Trends Ecol. Evolut.* **2013**, *28*, 100–109. [CrossRef] [PubMed]
19. Vedeld, P.; Jumane, A.; Wapalila, G.; Songorwa, A. Protected Areas, Poverty and Conflicts. A Livelihood Case Study of Mikumi National Park, Tanzania. *For. Policy Econ.* **2012**, *21*, 20–31. [CrossRef]
20. Abel, N.; Blaikie, P. Elephants, People, Parks and Development: The Case of the Luangwa Valley, Zambia. *Environ. Manag.* **1986**, *10*, 735–751. [CrossRef]
21. De Matos Dias, D.; Ferreguetti, Á.C.; Rodrigues, F.H.G. Using an Occupancy Approach to Identify Poaching Hotspots in Protected Areas in a Seasonally Dry Tropical Forest. *Biol. Conserv.* **2020**, *251*, 108796. [CrossRef]
22. Moore, J.F.; Mulindahabi, F.; Masozera, M.K.; Nichols, J.D.; Hines, J.E.; Turikunkiko, E.; Oli, M.K. Are Ranger Patrols Effective in Reducing Poaching-Related Threats within Protected Areas? *J. Appl. Ecol.* **2018**, *55*, 99–107. [CrossRef]
23. Ghoddousi, A.; Soofi, M.; Hamidi, A.K.; Ashayeri, S.; Egli, L.; Ghoddousi, S.; Speicher, J.; Khorozyan, I.; Kiabi, B.H.; Waltert, M. The Decline of Ungulate Populations in Iranian Protected Areas Calls for Urgent Action against Poaching. *Oryx* **2019**, *53*, 151–158. [CrossRef]
24. Cooney, R.; Roe, D.; Dublin, H.; Phelps, J.; Wilkie, D.; Keane, A.; Travers, H.; Skinner, D.; Challender, D.W.S.; Allan, J.R.; et al. From Poachers to Protectors: Engaging Local Communities in Solutions to Illegal Wildlife Trade. *Conserv. Lett.* **2017**, *10*, 367–374. [CrossRef]
25. Brancalion, P.H.S.; De Almeida, D.R.A.; Vidal, E.; Molin, P.G.; Sontag, V.E.; Souza, S.E.X.F.; Schulze, M.D. Fake Legal Logging in the Brazilian Amazon. *Sci. Adv.* **2018**, *4*, 8. [CrossRef] [PubMed]
26. Müller, J.; Noss, R.F.; Thorn, S.; Bässler, C.; Leverkus, A.B.; Lindenmayer, D. Increasing Disturbance Demands New Policies to Conserve Intact Forest. *Conserv. Lett.* **2019**, *12*, e12449. [CrossRef]
27. Abman, R. Rule of Law and Avoided Deforestation from Protected Areas. *Ecol. Econ.* **2018**, *146*, 282–289. [CrossRef]
28. Hulme, P.E.; Pyšek, P.; Pergl, J.; Jarošík, V.; Schaffner, U.; Vilà, M. Greater Focus Needed on Alien Plant Impacts in Protected Areas. *Conserv. Lett.* **2014**, *7*, 459–466. [CrossRef]
29. Allen, J.A.; Brown, C.S.; Stohlgren, T.J. Non-Native Plant Invasions of United States National Parks. *Biol. Invas.* **2009**, *11*, 2195–2207. [CrossRef]
30. Foxcroft, L.C.; Spear, D.; van Wilgen, N.J.; McGeoch, M.A. Assessing the Association between Pathways of Alien Plant Invaders and Their Impacts in Protected Areas. *NeoBiota* **2019**, *43*, 1–25. [CrossRef]
31. Meyer, C.G. The Impacts of Spear and Other Recreational Fishers on a Small Permanent Marine Protected Area and Adjacent Pulse Fished Area. *Fish. Res.* **2007**, *84*, 301–307. [CrossRef]

32. Environmental Impacts of Tourism on the Australian Alps Protected Areas. Available online: https://bioone.org/journals/mountain-research-and-development/volume-23/issue-3/0276-4741_2003_023_0247_EIOTOT_2.0.CO_2/Environmental-Impacts-of-Tourism-on-the-Australian-Alps-Protected-Areas/10.1659/0276-4741(2003)023[0247:EIOTOT]2.0.CO;2.full (accessed on 1 April 2022).

33. Guetté, A.; Godet, L.; Juigner, M.; Robin, M. Worldwide Increase in Artificial Light at Night around Protected Areas and within Biodiversity Hotspots. *Biol. Conserv.* **2018**, *223*, 97–103. [CrossRef]

34. Jones, K.R.; Venter, O.; Fuller, R.A.; Allan, J.R.; Maxwell, S.L.; Negret, P.J.; Watson, J.E.M. One-Third of Global Protected Land Is under Intense Human Pressure. *Science* **2018**, *360*, 788–791. [CrossRef]

35. Geldmann, J.; Joppa, L.N.; Burgess, N.D. Mapping Change in Human Pressure Globally on Land and within Protected Areas. *Conserv. Biol. J. Soc. Conserv. Biol.* **2014**, *28*, 1604–1616. [CrossRef]

36. Watson, F.G.R.; Becker, M.S.; Milanzi, J.; Nyirenda, M. Human Encroachment into Protected Area Networks in Zambia: Implications for Large Carnivore Conservation. *Reg. Environ. Change* **2014**, *15*, 415–429. [CrossRef]

37. Laurance, W.F. Does Research Help to Safeguard Protected Areas? *Trends Ecol. Evolut.* **2013**, *28*, 261–266. [CrossRef] [PubMed]

38. Geldmann, J.; Manica, A.; Burgess, N.D.; Coad, L.; Balmford, A. A Global-Level Assessment of the Effectiveness of Protected Areas at Resisting Anthropogenic Pressures. *Proc. Natl. Acad. Sci. USA* **2019**, *116*, 23209–23215. [CrossRef]

39. Kozlowski, J.; Vass-Bowen, N. Buffering External Threats to Heritage Conservation Areas: A Planner's Perspective. *Landsc. Urban Plan.* **1997**, *37*, 245–267. [CrossRef]

40. Palomo, I.; Martín-López, B.; Potschin, M.; Haines-Young, R.; Montes, C. National Parks, Buffer Zones and Surrounding Lands: Mapping Ecosystem Service Flows. *Ecosyst. Serv.* **2013**, *4*, 104–116. [CrossRef]

41. Sharma, U.R. An Overview of Park-People Interactions in Royal Chitwan National Park, Nepal. *Landsc. Urban Plan.* **1990**, *19*, 133–144. [CrossRef]

42. Weladji, R.B.; Moe, S.R.; Vedeld, P. Stakeholder Attitudes towards Wildlife Policy and the Bénoué Wildlife Conservation Area, North Cameroon. *Environ. Conserv.* **2003**, *30*, 334–343. [CrossRef]

43. Albers, H.J. Spatial Modeling of Extraction and Enforcement in Developing Country Protected Areas. *Resour. Energy Econ.* **2010**, *32*, 165–179. [CrossRef]

44. Bajracharya, B.; Uddin, K.; Chettri, N.; Shrestha, B.; Siddiqui, S.A. Understanding Land Cover Change Using a Harmonized Classification System in the Himalaya. *Mountain Res. Dev.* **2010**, *30*, 143–156. [CrossRef]

45. Abdullah, J.; Ahmad, C.B.; Jaafar, J.; Sa'ad, S.R.M. Stakeholders' Perspectives of Criteria for Delineation of Buffer Zone at Conservation Reserve: FRIM Heritage Site. *Proc. Soc. Behav. Sci.* **2013**, *105*, 610–618. [CrossRef]

46. Du, W.; Penabaz-Wiley, S.M.; Njeru, A.M.; Kinoshita, I. Models and Approaches for Integrating Protected Areas with Their Surroundings: A Review of the Literature. *Sustainability* **2015**, *7*, 8151–8177. [CrossRef]

47. Secretariat of the Convention on Biological Diversity. *CBD Protected Areas in Today's World: Their Values and Benefits for the Welfare of the Planet*; Secretariat of the Convention on Biological Diversity: Montreal, QC, Canada, 2008.

48. Ervin, J.; Sekhran, N.; Dinu, A.; Gidda, S.; Vergeichik, M.; Mee, J. *Protected Areas for the 21st Century: Lessons from UNDP/GEF's Portfolio*; UNDP: New York, NY, USA, 2010.

49. DeFries, R.; Karanth, K.K.; Pareeth, S. Interactions between Protected Areas and Their Surroundings in Human-Dominated Tropical Landscapes. *Biol. Conserv.* **2010**, *143*, 2870–2880. [CrossRef]

50. Nepal, S.K. Mountain Ecotourism and Sustainable Development: Ecology, Economics, and Ethics. *Mt. Res. Dev.* **2002**, *22*, 104–109. [CrossRef]

51. Hamilton, A.; Cunningham, A.; Byarugaba, D.; Kayanja, F. Conservation in a Region of Political Instability: Bwindi Impenetrable Forest, Uganda. *Conserv. Biol.* **2000**, *14*, 1722–1725. [CrossRef] [PubMed]

52. Bowles, I.; Rosenfeld, A.B.; Sugal, C.A.; Mittermeier, R.A. *Natural Resource Extraction in the Latin American Tropics*; National Geographic: Washington, DC, USA, 1998.

53. Khudzari, J.; Kurian, J.; Tartakovsky, B.; Raghavan, G.S.V. Bibliometric Analysis of Global Research Trends on Microbial Fuel Cells Using Scopus Database. *Biochem. Eng. J.* **2018**, *136*, 51–60. [CrossRef]

54. Zou, X.; Yue, W.L.; Vu, H. Le Visualization and Analysis of Mapping Knowledge Domain of Road Safety Studies. *Accid. Anal. Prev.* **2018**, *118*, 131–145. [CrossRef] [PubMed]

55. Chen, D.; Liu, Z.; Luo, Z.; Webber, M.; Chen, J. Bibliometric and Visualized Analysis of Emergy Research. *Ecol. Eng.* **2016**, *90*, 285–293. [CrossRef]

56. Geng, Y.; Chen, W.; Liu, Z.; Chiu, A.S.F.; Han, W.; Liu, Z.; Zhong, S.; Qian, Y.; You, W.; Cui, X. A Bibliometric Review: Energy Consumption and Greenhouse Gas Emissions in the Residential Sector. *J. Clean. Prod.* **2017**, *159*, 301–316. [CrossRef]

57. Kandel, P.; Chettri, N.; Chaudhary, S.; Sharma, P.; Uddin, K. Ecosystem Services Research Trends in the Water Tower of Asia: A Bibliometric Analysis from the Hindu Kush Himalaya. *Ecol. Indic.* **2021**, *121*, 107152. [CrossRef]

58. Fang, Y.; Yin, J.; Wu, B. Climate Change and Tourism: A Scientometric Analysis Using CiteSpace. *J. Sustain. Tour.* **2018**, *26*, 108–126. [CrossRef]

59. Chen, C.; Dubin, R.; Kim, M.C. Emerging Trends and New Developments in Regenerative Medicine: A Scientometric Update (2000–2014). *Exp. Opin. Biol. Ther.* **2014**, *14*, 1295–1317. [CrossRef] [PubMed]

60. United Nations. *UN List of Least Developed Countries*; UNCTAD: Geneva, Switzerland, 2017; Volume 5, pp. 8–11.

61. Vaidya, A.; Mayer, A.L. Critical Review of the Millennium Project in Nepal. *Sustainability* **2016**, *8*, 1043. [CrossRef]

62. Rotich, D. Concept of Zoning Management in Protected Areas. *Sage J.* **2012**, *2*, 173–183.
63. WWF. *The Greater One-Horned Rhinoceros Conservation Action Plan for Nepal (2006–2011)*; WWF: Washington, DC, USA, 2006.
64. Aryal, K.; Dhungana, R.; Silwal, T. Understanding Policy Arrangement for Wildlife Conservation in Protected Areas of Nepal. *Hum. Dimens. Wildlife* **2021**, *26*, 1–12. [CrossRef]
65. Bhattarai, B.R.; Wright, W.; Poudel, B.S.; Aryal, A.; Yadav, B.P.; Wagle, R. Shifting Paradigms for Nepal's Protected Areas: History, Challenges and Relationships. *J. Mt. Sci.* **2017**, *14*, 964–979. [CrossRef]
66. Heinen, J.T.; Kattel, B. Parks, People, and Conservation: A Review of Management Issues in Nepal's Protected Areas. *Popul. Environ.* **1992**, *14*, 49–84. [CrossRef]
67. Dhruba Bijaya, G.C.; Cheng, S.; Xu, Z.; Bhandari, J.; Wang, L.; Liu, X. Community Forestry and Livelihood in Nepal: A Review. *J. Anim. Plant Sci.* **2016**, *26*, 1–12.
68. Regmi, K.D.; Walter, P. Modernisation Theory, Ecotourism Policy, and Sustainable Development for Poor Countries of the Global South: Perspectives from Nepal. *Int. J. Sustain. Dev. World Ecol.* **2017**, *24*, 1–14. [CrossRef]
69. Heinen, J.T.; Yonzon, P.B. A Review of Conservation Issues and Programs in Nepal: From a Single Species Focus toward Biodiversity Protection. *Mt. Res. Dev.* **1994**, *14*, 61–76. [CrossRef]
70. Sharma, P.; Chettri, N.; Wangchuk, K. Human–Wildlife Conflict in the Roof of the World: Understanding Multidimensional Perspectives through a Systematic Review. *Ecol. Evolut.* **2021**, *11*, 11569–11586. [CrossRef]
71. Rana, S.K.; Rawal, R.S.; Dangwal, B.; Bhatt, I.D.; Price, T.D. 200 Years of Research on Himalayan Biodiversity: Trends, Gaps, and Policy Implications. *Front. Ecol. Evolut.* **2021**, *8*, 1–9. [CrossRef]
72. Lamsal, P.; Kumar, L.; Atreya, K.; Pant, K.P. Vulnerability and Impacts of Climate Change on Forest and Freshwater Wetland Ecosystems in Nepal: A Review. *Ambio* **2017**, *46*, 915–930. [CrossRef] [PubMed]
73. Zhong, L.; Yang, R.; Zhao, Z. Critical Review of English Literature for National Parks Based on Bibliometric Analysis. *Chin. Landsc. Archit.* **2018**, *07*, 23–28.
74. Explore the World's Protected Areas. Available online: https://www.protectedplanet.net/country/NPL (accessed on 15 December 2021).
75. The World Bank Group World Development Indicators. Available online: https://databank.worldbank.org/source/world-development-indicators (accessed on 29 May 2022).
76. DNPWC, Government of Nepal Ministry of Forests and Environment. Available online: https://dnpwc.gov.np/en/ (accessed on 15 November 2021).
77. Den Braber, B.; Evans, K.L.; Oldekop, J.A. Impact of Protected Areas on Poverty, Extreme Poverty, and Inequality in Nepal. *Conserv. Lett.* **2018**, *11*, 1–9. [CrossRef]
78. Nepal, S.K. Tourism in protected areas: The Nepalese Himalaya. *Ann. Tour. Res.* **2000**, *27*, 661–681. [CrossRef]
79. Shrestha, U.B.; Shrestha, S.; Chaudhary, P.; Chaudhary, R.P. How Representative Is the Protected Areas System of Nepal? *Mt. Res. Dev.* **2010**, *30*, 282–294. [CrossRef]
80. Rey-Martí, A.; Ribeiro-Soriano, D.; Palacios-Marqués, D. A Bibliometric Analysis of Social Entrepreneurship. *J. Bus. Res.* **2016**, *69*, 1651–1655. [CrossRef]
81. Albort-Morant, G.; Ribeiro-Soriano, D. A Bibliometric Analysis of International Impact of Business Incubators. *J. Bus. Res.* **2016**, *69*, 1775–1779. [CrossRef]
82. Gaviria-Marin, M.; Merigó, J.M.; Baier-Fuentes, H. Knowledge Management: A Global Examination Based on Bibliometric Analysis. *Technol. Forecast. Soc. Chang.* **2019**, *140*, 194–220. [CrossRef]
83. Jian, F.; Jingda, D. Comparison of Visualization Principles between Citespace and VOSviewer. *J. Libr. Inform. Sci. Agric.* **2019**, *31*, 31–37.
84. Aguillo, I.F. Is Google Scholar useful for bibliometrics? A webometric analysis. *Scientometrics* **2012**, *91*, 343–351. [CrossRef]
85. Ding, X.; Yang, Z. Knowledge Mapping of Platform Research: A Visual Analysis Using VOSviewer and CiteSpace. *Electron. Commerce Res.* **2020**, 1–23. [CrossRef]
86. Thelwall, M. Bibliometrics to Webometrics. *J. Inform. Sci.* **2008**, *34*, 605–621. [CrossRef]
87. Merigó, J.M.; Yang, J.B. A Bibliometric Analysis of Operations Research and Management Science. *Omega* **2017**, *73*, 37–48. [CrossRef]
88. Carvalho, M.M.; Fleury, A.; Lopes, A.P. An Overview of the Literature on Technology Roadmapping (TRM): Contributions and Trends. *Technol. Forecast. Soc. Chang.* **2013**, *80*, 1418–1437. [CrossRef]
89. Antonio-Rafael, R.-R.; José, R.-N. Changes in the Intellectual Structure of Strategic Management Research: A Bibliometric Study of the Strategic Management Journal. *Strategic Manag. J.* **2001**, *25*, 981–1004. [CrossRef]
90. Cornelius, B.; Landström, H.; Persson, O. Guidelines for Using Bibliometrics at the Swedish Research Council. 2015. Available online: https://www.vr.se/download/18.514d156f1639984ae0789dc2/1529480565499/Guidelines+for+using+bibliometrics+at+the+Swedish+Research+Council.pdf (accessed on 21 May 2022).
91. Xiang, C.; Wang, Y.; Liu, H. A Scientometrics Review on Nonpoint Source Pollution Research. *Ecol. Eng.* **2017**, *99*, 400–408. [CrossRef]
92. Medina-Mijangos, R.; Seguí-Amórtegui, L. Research Trends in the Economic Analysis of Municipal Solid Waste Management Systems: A Bibliometric Analysis from 1980 to 2019. *Sustainability* **2020**, *12*, 8509. [CrossRef]

93. Hoppen, N.H.F.; Vanz, S.A.d.S. Neurosciences in Brazil: A Bibliometric Study of Main Characteristics, Collaboration and Citations. *Scientometrics* **2016**, *109*, 121–141. [CrossRef]
94. Heinen, J.T.; Shrestha, S.K. Evolving policies for conservation: An Historical Profile of the Protected Area System of Nepal. *J. Environ. Plan. Manag.* **2006**, *49*, 41–58. [CrossRef]
95. Ojha, H.R.; Ghimire, S.; Pain, A.; Nightingale, A.; Khatri, D.B.; Dhungana, H. Policy without politics: Technocratic control of climate change adaptation policy making in Nepal. *Clim. Policy* **2016**, *16*, 415–433. [CrossRef]
96. Wang, Y.; Lu, Z.; Sheng, Y.; Zhou, Y. Remote Sensing Applications in Monitoring of Protected Areas. *Remote Sens.* **2020**, *12*, 1370. [CrossRef]
97. Martin, E.B. Ivory in Kathmandu. *Oryx* **1998**, *32*, 317–320. [CrossRef]
98. Fisscher, W.; Sulzer, W. Economical and Ecological Effects of Tourism in Langtang National-Park (Nepal). *Mitt. Osterr. Geogr. Ges.* **1994**, *136*, 225–242.
99. Thapa Karki, S. Do Protected Areas and Conservation Incentives Contribute to Sustainable Livelihoods? A Case Study of Bardia National Park, Nepal. *J. Environ. Manag.* **2013**, *128*, 988–999. [CrossRef]
100. Lamichhane, B.R.; Persoon, G.A.; Leirs, H.; Poudel, S.; Subedi, N.; Pokheral, C.P.; Bhattarai, S.; Thapaliya, B.P.; de Iongh, H.H. Spatio-temporal patterns of attacks on human and economic losses from wildlife in Chitwan National Park, Nepal. *PLoS ONE* **2018**, *13*, e0195373. [CrossRef]
101. Tessema, M.E.; Lilieholm, R.J.; Ashenafi, Z.T.; Leader-Williams, N. Community attitudes toward wildlife and protected areas in Ethiopia. *Soc. Nat. Resour.* **2010**, *23*, 489–506. [CrossRef]
102. Sunam, R.K.; Bishwokarma, D.; Darjee, K.B. Conservation Policy Making in Nepal: Problematising the Politics of Civic Resistance. *Conserv. Soc.* **2015**, *13*, 179–188. [CrossRef]
103. Karanth, K.K.; Nepal, S.K. Local residents perception of benefits and losses from protected areas in India and Nepal. *Environ. Manag.* **2012**, *49*, 372–386. [CrossRef]
104. Martino, D. Buffer zones around protected areas: A brief literature review. *Electron. Green J.* **2001**, *1*. [CrossRef]
105. Schwarz, H.; Revilla, M.; Weber, W. Memory Effects in Repeated Survey Questions Reviving the Empirical Investigation assumption of the Independent Measurements Assumption. *Survey Res. Methods* **2020**, *14*, 325–344.
106. WWF. Available online: https://www.wwfnepal.org/what_we_do/wildlife/snow_leopard/ (accessed on 25 April 2022).
107. DNPWC. *Department of National Parks and Wildlife Reserve, Ministry of Forests and Soil Conservation*; Annual Report; DNPWC: Kathmandu, Nepal, 2012.

 *land*

Article

Revealing Changes in the Management Capacity of the Three-River-Source National Park, China: An Application of the Best Practice-Based Evaluation Method

Xianyang Liu [1,2], Qingwen Min [1,2] and Wenjun Jiao [1,*]

[1] Institute of Geographic Sciences and Natural Resources Research, Chinese Academy of Sciences, Beijing 100101, China

[2] University of the Chinese Academy of Sciences, Beijing 100049, China

* Correspondence: jiaowj@igsnrr.ac.cn

Citation: Liu, X.; Min, Q.; Jiao, W. Revealing Changes in the Management Capacity of the Three-River-Source National Park, China: An Application of the Best Practice-Based Evaluation Method. *Land* **2022**, *11*, 1565. https://doi.org/10.3390/land11091565

Academic Editors: Le Yu, Rui Yang, Yue Cao and Steve Carver

Received: 6 August 2022
Accepted: 10 September 2022
Published: 14 September 2022

Publisher's Note: MDPI stays neutral with regard to jurisdictional claims in published maps and institutional affiliations.

Abstract: Management evaluation is increasingly required for national parks worldwide as it is an essential mechanism for improving management levels and achieving management objectives. The management capacity evaluation (MCE), an integral component of management evaluation, emphasizes the suitability of management measures. It helps identify the deficiencies in existing management measures and form feedback to improve them, thus increasing the overall management level of national parks. However, the existing MCE methods from international programs suffer from limited adaptability and are difficult to promote in other countries. In this research, we apply the best practice-based (BPB) method to the Three-River-Sources National Park (TNP), the first national park in China, to reveal the changes in its management capacity during the pilot period. The BPB method is new compared with other MCE methods, but is more adaptable to the current situation of China's national parks. Results show that TNP's comprehensive management capacity and the five aspects of management capacities improved effectively, which means the management measures adopted during the pilot phase were generally appropriate and practicable. Some management capacities, such as management organization, legal system construction, management planning, and natural resources confirmation and registration performed well or improved significantly during the pilot period, providing beneficial lessons for other national parks in China. Some management capacities, such as the ecological compensation scheme, monitoring and early warning system, and management team, are still deficient and should be prioritized for future improvement. The effectiveness and operability of the BPB method are validated in this research, as it provides a rapid and accurate diagnosis of TNP's management capacities and useful feedback for improving them. We submit that the BPB method not only contributes to the theoretical improvement of MCE methods, but also shows wider adaptability to different protected area types and countries.

Keywords: management capacity evaluation; management measures; national park; protected area; best practice; indicator system; Three-River-Source National Park

1. Introduction

Protected areas (PAs) are defined as an area of land and/or sea especially dedicated to the protection and maintenance of biological diversity, and of natural and associated cultural resources, managed through legal or other effective means [1]. They are of great significance in mitigating biodiversity loss, maintaining the crucial services provided, enhancing community health, and safeguarding national ecological security [2–4]. Since the first protected area was established in 1956, China built a system of PAs covering forest, grassland, wetland, marine, and desert ecosystem types, aiming to preserve rare and endangered species, natural relics, and natural landscapes [5]. As of 2018, there are more than 10 types of PAs in China, with more than 11,800 sites, covering a total area of about 18% of the land area and 4.6% of the sea area [6,7]. However, due to the unclear rights and responsibilities, the replication of administrative efforts, and the fragmentation of conservation expertise, these PAs suffer from varying degrees of management inefficiency [8]. To this end, China conducted the institutional reform in March 2018, and established a new protected area system comprised of three types: national parks, nature reserves, and natural parks [9].

National parks, comprising areas that showcase ecosystems characteristic of China, are defined as the mainstay of this new protected area system. National parks are important components of the global protected area system, not only having the function of providing high-quality ecological products, but also providing human society with public services, such as research, education, and recreation [10–12]. Since the Yellowstone National Park was established in the U.S. in 1872, more than 200 countries around the world participated in the construction of national parks, and a great amount of financial and non-financial resources are continuously invested in their construction [13]. However, it is increasingly questioned whether the input of various resources enhanced the management capacity and effectiveness of national parks and whether the management objectives of national parks were achieved as expected. Since the fourth World Parks Congress in 1992, these issues are featured prominently on the national park management agenda and became common concerns [14]. China also confronts these questions in the construction of national parks. Ten national park pilots were established since 2016, and the first batch of five national parks was officially recognized in October 2021 at the 15th meeting of the Conference of the Parties (COP15) to the United Nations Convention on Biological Diversity. During this period, the construction of national parks received extensive attention from governments at all levels, as well as obtained strong policy support and a large amount of capital and technology investments. However, whether and to what extent these investments improved the management capacity and efficiency of the national parks is not credibly answered.

In the management of PAs, management evaluation is recognized as an important mechanism to improve management practices, promote transparency for reporting, and create proper accountability [15]. International attention was drawn to the management evaluation of PAs since the 1970s [14]. After nearly 50 years of development, a composite management evaluation system that includes evaluations of effectiveness, capacity, threats and stresses, impacts, and biophysical characteristics was established [13,16–19]. The most widely used management evaluations are the management effectiveness evaluation and the management capacity evaluation. The management effectiveness evaluation (MEE) primarily refers to the responsiveness of management results to desired objectives [14,15]. The management cycle-based evaluation framework proposed by the International Union for Conservation of Nature (IUCN) became the cornerstone of a series of MEE methods. Several of these methods, such as rapid assessment and prioritization of protected area management (RAPPAM), the management effectiveness tracking tool (METT), and enhancing our heritage (EOH), are widely used in many PAs around the world [20–24]. MEE tends to reveal the effectiveness of management and conservation by measuring changes in the state of PAs, therefore being more applicable to mature PAs and mainly being used to make comparisons between PAs. Compared to MEE, the management capacity evaluation (MCE) focuses on the suitability of management measures, that is, whether management measures respond effectively to management needs [25]. The Parks in Peril Program, initiated by The Nature Conservancy (TNC), evaluates the management capacity of national parks from the perspectives of primary conservation action, long-term management, financial support, and local guarantee. This method was applied in 17 countries in Latin America and the Caribbean [26,27]. The Proyecto Ambiental Regional de Centro America program (PROARCA) proposed an MCE method with an indicator system comprised of social relations, administration, natural and cultural resources, political law, and finance. This method was applied in Panama, Honduras, El Salvador, Nicaragua, Guatemala, and Costa Rica in Central America [28]. MCE helps diagnose the soundness of management practices in an individual PA and is often used in the early stage of the PA's construction.

China started management evaluation practices in the 1990s, focusing on the MEE of nature reserves. Xue and Zheng explored the indicators and standards of MEE for nature reserves in China, and proposed an indicator system containing management conditions, management measures, a scientific research base, and management effectiveness [29]. Since then, several mature international methods were introduced to China, of which RAPPAM and METT are widely used. According to incomplete statistics, by the end of 2019 there were 2081 nature reserves with a total area of 1,238,500 km^2, which accepted MEE, accounting for 66.37% of the total PAs in China [30]. In addition, the management authorities developed specifications for the MEE of nature reserves, such as the *Technical regulations for the management effectiveness evaluation of nature reserves (LY/T 1726-2008)* and the *Standard for assessment of nature reserve management (HJ 913-2017)*. At this time, a mature MEE evaluation system for China's nature reserves is formed. However, due to the differences in conservation objectives and management needs, the existing MEE methods of nature reserves cannot be directly applied to national parks. Furthermore, the national parks of China are at the early stage of construction, so the MCE that emphasizes the suitability of management measures is more appropriate for them. It will help identify the deficiencies in existing management measures, improve the management processes, and increase the overall management level of national parks. Although the existing MCE methods from international programs can provide some references, their indicator systems and evaluation standards lack wider adaptability, making them difficult to promote in other countries, such as China.

In this research, we apply the best practice-based (BPB) method to the Three-River-Sources National Park (TNP), the first national park in China, aiming to reveal the changes in its management capacity during the pilot period. The BPB method summarized the best practices of national park management in different countries and proposed an indicator system and a set of evaluation standards for evaluating the management capacity of national parks, therefore having wider adaptability and a larger potential for promotion [31]. Compared with other MCE methods, this method is more adaptable to the current management situation of China's national parks, and is able to quickly identify the shortcomings in the management and the gaps at the international best level. However, the BPB method was rarely applied since it was proposed. Thus, its effectiveness and operability are lacking in validation. We submit that this research will not only help improve the management capacity of TNP and provide guidance for the management of other national parks in China, but will also test and validate the BPB method and contribute to the theoretical innovation and improvement of MCE methods.

2. Materials and Methodology

2.1. Best Practice-Based (BPB) Evaluation Method

As mentioned earlier, the MCE is an important component of protected area management evaluation and is particularly essential for current national park management in China. However, the existing MCE methods from international programs suffer from limited adaptability and are not well suited to the management needs of China's national parks. To this end, we apply and test the BPB method proposed by us in 2019 [31]. The most important feature of this method is that the evaluation indicators and standards are selected and determined based on the best practices of existing national parks in the world. To design the indicator system, we firstly made a systematic review of national park management practices in various countries, such as the U.S., Canada, the U.K., South Africa, Japan, South Korea, and Argentina, and obtained a summary of the best practices in worldwide national parks. Then we developed an indicator system with a total of 18 indicators in the five aspects and designed five criteria accordingly: institutional construction, guarantee mechanism, natural resources and ecosystem management, community management, and popularization and education (Table 1).

Table 1. Evaluation indicators of the best practice-based (BPB) method and their best standards.

Criterion	Indicator	Best Standards
Institutional construction	Management organization	An independent management organization is established in the national park with well-organized departments and a clear division of duties, which allows for efficient and orderly operation.
	Management team	The management team has excellent professional knowledge and comprehensive capacity and frequently participates in professional skills training.
	Management planning	The management planning can integrate multiple plans of the national park to meet its management needs to the largest extent and form a mechanism for dynamic adjustment and regular revision.
Guarantee mechanism	Financial support	The national park has sufficient financial investment, diversified and stable financing channels, and a sound capital management system.
	Legal system construction	The national park has a sound legal system, clear legal hierarchy, and professional enforcement team.
	Scientific research support	A dedicated team conducts long-term and steadily based scientific research, and the research results serve to construct the national park.
	Multi-stakeholder participation	The enterprises, social organizations, community residents, and other parties are involved in the management of the national park.
	Audit mechanism	An audit mechanism system that acts as a constraint has effective results.
Natural resources and ecosystem management	Natural resources inventory	The national park completed comprehensive natural and resource inventories and formed a complete resources database.
	Natural resources confirmation and registration	The national park completed the confirmation and registration of its natural resources.
	Ecosystem restoration	The national park implemented scientific and long-term ecological restoration initiatives, which resulted in significant effectiveness.
	Monitoring and early warning system	The national park has a complete monitoring and early warning mechanism, as well as the necessary facilities to monitor complete ecological elements and accurately warn of natural disasters.
Community management	Community organization construction	There is a community organization with a complete structure and standardized management, in which the community residents' interests can be covered and their recommendations can be presented regularly.
	Resident participation	A well-established community co-management mechanism allows community residents to participate in managing the national park in various ways.
	Ecological compensation scheme	The ecological compensation scheme is diversified and stable, providing flexible and diverse compensation methods that are satisfactory to the recipients.
Popularization and education	Recreation management	There are comprehensive recreation management regulations and standardized visitor management systems to meet the recreation needs of the public.
	Science popularization	By providing rich and colorful popular science activities, comprehensive science popularization facilities and exquisite science popularization materials, the national park realized an extensive publicity.
	Environmental education	The national park conducts a wide variety of environmental education activities to raise the environmental protection awareness of the community, visitors, and the general public.

Source: reference [31].

Based on the above indicator system, we applied the hierarchical analysis method and participatory evaluation process to determine the weight of each indicator. A total of 30 experts from the fields of ecology, environment, management, and planning were invited to make a judgment on the importance of each criterion, and the weights of the five criteria were then calculated (Table 2). The weights of institutional construction, guarantee mechanism, natural resources and ecosystem management, community management, and popularization and education were 0.380, 0.179. 0.212, 0.067, and 0.162, respectively. Next, the weights of each indicator under each criterion were calculated in the same way (Table 2).

Table 2. Weights of the evaluation indicators of the BPB method.

Criterion	Indicator	Weight
Institutional construction (0.380)	Management organization	0.354
	Management team	0.426
	Management planning	0.220
Guarantee mechanism (0.179)	Financial support	0.376
	Legal system construction	0.240
	Scientific research support	0.194
	Multi-stakeholder participation	0.122
	Audit mechanism	0.068
Natural resources and environment management (0.212)	Natural resources inventory	0.313
	Natural resources confirmation and registration	0.273
	Ecosystem restoration	0.193
	Monitoring and early warning system	0.221
Community management (0.067)	Community organization construction	0.422
	Resident participation	0.289
	Ecological compensation scheme	0.289
Popularization and education (0.162)	Recreation management	0.383
	Science popularization	0.217
	Environmental education	0.400

Source: reference [31].

To evaluate the management capacity of a national park, participants will be invited to score each indicator by judging the extent to which the corresponding management capacity meets the best standard. Each indicator is scored through a five-grade scale; 5, 4, 3, 2, and 1 represent fully compliant, relatively compliant, largely compliant, not very compliant, and not compliant, respectively, with the score of 100, 75, 50, 25, and 0 on a percentage scale. When calculating the comprehensive management capacity score for each national park, the scores in each aspect should be calculated first. The score of the management capacity in one aspect is calculated by summing the weighted score of each indicator (Equation (1)). Then the comprehensive management capacity score of the evaluated national parks is obtained by weighting the management capacity score in each aspect (Equation (2)).

$$S_c = \sum_{i=1}^{m} P_i S_i \tag{1}$$

where S_c represents the management capacity score in one aspect, P_i represents the weight of the i-th indicator in this criterion, and S_i represents the management capacity score of the i-th indicator in this criterion.

$$S = \sum_{c=1}^{j} P_j S_{jc} \ (j = 5) \tag{2}$$

where S represents the comprehensive management capacity score of the evaluated national park, P_j represents the weight of the j-th criterion, and S_{jc} represents the j-th S_c.

Both the single and comprehensive management capacity of the national park can be categorized into four grades of excellent, good, regular, and poor based on their scores (Table 3).

Table 3. Grades of the management capacities of national parks.

Grade	Excellent	Good	Regular	Poor
Scores	$90 \leq S \leq 100$	$75 \leq S < 90$	$60 \leq S < 75$	$S < 60$

Source: reference [31].

To make better use of the BPB method, we suggest following the below procedures: (1) defining the spatial and temporal scopes of the evaluation, (2) gathering background data and information, (3) applying the evaluation method, (4) analyzing the evaluation results, and (5) forming feedback to the park management.

- The first step is to identify the spatial and temporal scopes of the evaluation. The spatial boundary of the evaluated national park needs to be clarified, which is usually the entire area or a sub-zone. The time point of the evaluation needs to be determined, which is usually a specific year. If the changes in the management capacity within a certain period need to be evaluated, the baseline year and the evaluation year need to be selected.
- The second step is to gather and organize background information and data for each indicator, which is essential for preparing the evaluation and serves as a support for the subsequent phase of analyzing the evaluation results.
- The third step is to apply the evaluation method. The application of the BPB method needs such evaluation tools as questionnaires and scorecards, and the evaluation results are obtained through statistics.
- The fourth step is to analyze the evaluation results and look for the reasons behind the changes. This step helps to identify the reasons for high or low management capacity, as well as why it is improving or deteriorating.
- The final step is to provide feedback to the national park management. The findings are used to propose appropriate actions in order to enhance better management, including additional rectification of management measures, and adjustments to management priorities and resource allocation ratios.

2.2. Study Area

The Three-River-Source National Park (TNP) ($89°45'$–$102°23'$ E, $31°39'$–$36°12'$ N) is located in the southern part of Qinghai Province, China, and covers an area of 123,100 km^2. TNP serves as the source catchment area of the Yangtze River, Yellow River, and Lancang River, so it consists of three zones: the Yangtze River Source Zone, the Yellow River Source Zone, and the Lancang River Source Zone. The mountainous terrain in the park extends to an average altitude of over 4500 m, and alpine meadows and alpine grasslands are the primary ecosystem types [32]. TNP includes Zhiduo County, Qumalai County, Maduro County, Zaduo County, and the Hoh Xil nature reserve, comprised of 12 townships and 53 administrative villages (Figure 1). There are a total of 16,621 households and 64,000 residents in the park, most of whom are Tibetan. The economic development within the park is backward, and the industrial structure is single, primarily relying on traditional animal husbandry, with limited employment and income generation channels. Thus, there is a great incongruity between ecological conservation and economic development in the park.

In April 2016, TNP became China's first national park pilot, and in October 2021, it was formally certified as a national park. TNP became one of China's most nationally representative national parks by optimizing the conservation boundaries, coordinating the conflicts between conservation and development, etc. To evaluate its management capacity thoroughly, we define the entire area of TNP as the spatial boundary. We choose 2017 and 2021 as the baseline and evaluation years to quantify the changes in its management capacity.

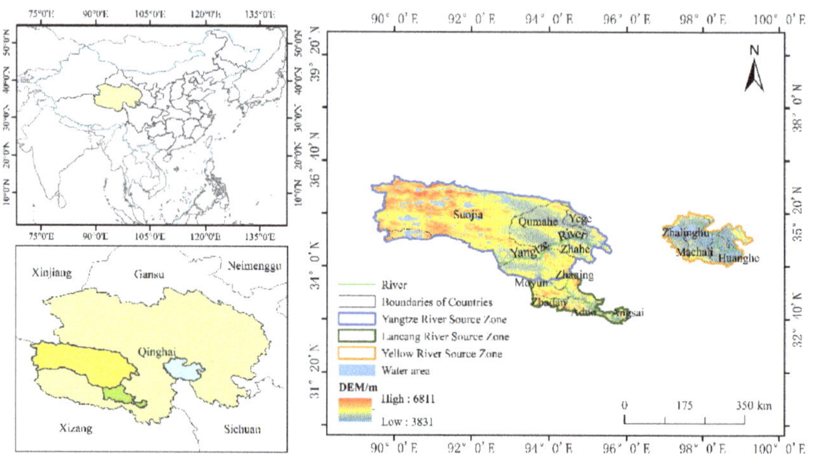

Figure 1. Location of the Three-River-Source National Park (TNP), China.

2.3. Data Collection

To collect background data and information, we conducted a field survey in TNP from 18 to 29 August 2018, and paid a return visit to the interviewees from 11 to 14 January 2022. The survey route in 2018 passed through the whole area of TNP, beginning in Xining City and continuing through Xinghai County, Mado County, Yushu City, and finally Zaduo County. Along the survey route, we conducted key person interviews and organized workshops with the staff of the TNP management authority, the three zone management committees, and some ecological protection stations.

Through interviews, we got first-hand information about management measures in management organization, the management team, scientific research support, community organization construction, science popularization, and environmental education in 2017. We also obtained statistical data, policy, and technical documents on management planning, financial support, monitoring and early warning system, natural resources inventory, and resident participation in 2017. In January 2022, we returned to the above interviewees to update these data for 2021. Furthermore, we collected information on legal system construction, ecosystem restoration, and environmental education between 2017 and 2021 by searching literature and websites. Based on the above process, we sorted out the major management measures taken by TNP according to the management criteria (Table 4).

We completed "the MCE Questionnaire of TNP" by inviting managers of TNP, including the staff from the National Park Authority, the management committees of three zones, and the ecological protection stations. We also distributed the questionnaire to experts who were engaged in soil and water restoration, ecological compensation and community co-management of TNP for a long time. A total of 79 questionnaires were returned for this research. Forty questionnaires were returned in 2018, of which, 30 were from managers and ten were from experts. Thirty-nine questionnaires were returned in 2022, of which 28 were from managers and 11 were from experts. During the data processing phase, the score for each indicator was calculated, with managers scoring 40% and experts scoring 60%. TNP's management capacity score of five aspects and comprehensive score were calculated by weighting and adding the scores for each indicator.

Table 4. Management measures adopted by TNP during the pilot period.

Criterion	Management Measures
Institutional construction	The TNP management authority was established in 2016 with a four-tier management structure: the management authority, the management committee of three zones, the management office, and the ecological protection station.
	The TNP management authority consists of ten departments with a staff number of 402 at the initial, which increased to 409 in 2022.
	A two-tier planning system was established, with an overall plan and several special plans.
	Once officially certified, the TNP adjusted its boundary in a timely manner by including the headwaters of the three rivers into the conservation scope.
Guarantee mechanism	The main funding source of TNP is the financial allocations at all levels. TNP also accepted several financial and in-kind donations from enterprises and social organizations.
	The regulations on the management of TNP promulgated by the Standing Committee of the Qinghai Provincial People's Congress became the primary basis for management.
	The TNP management authority established the "Legal Research Association of TNP", introduced a legal advisor system, and formulated 13 management measures.
	The National Park Police Headquarters was established, which is directly under the leadership of the TNP management authority, to carry out the investigation and prosecution of natural resources in national parks.
	An exclusive research support team, the Research Institute of TNP, was established to cooperate with renowned universities and research institutions at home and abroad to carry out scientific research from multiple perspectives.
Natural resources and environment management	The off-office auditing of natural resource assets of leading cadres of TNP was completed.
	The investigation and publication of the region's water, grassland, wetland, and forest resources in the region were completed.
	The integrated confirmation rights registration of water, forests, mountains, grasslands, wastelands, and mudflats resources of TNP were completed.
	Ecological protection works were implemented to achieve large-scale ecological restoration.
	TNP continuously increased the strength of enforcement and community popularization, raising the conservation awareness among community residents.
	A "sky-ground-air" monitoring platform was established, and the number of monitoring points was significantly increased.
Community management	TNP implemented a community eco-guard system, and improved the grant funding reward and performance appraisal mechanisms.
	A livestock insurance fund was established, and the accident compensation system was implemented.
	TNP carried out franchise management and drove herders to participate in ecological experience work.
Popularization and education	TNP adopted three strategies of science popularization: designing and using image logos, organizing various science popularization activities based on anniversaries and special festivals, and creating and publishing diversified science popularization works.
	TNP authorized Mado Yunxiang Nature Tours Company to carry out ecological experiences in the Yellow River Zone. By managing visitors strictly and conducting a booking and assessment mechanism, the ecological experiences achieved good results.
	The TNP management authority conducted environmental education activities for the community residents.

3. Results

3.1. Analysis of Single Management Capacity

We applied the BPB method to evaluate the management capacity of TNP in 2017 and 2021, respectively. Table 5 shows the scores of each management capacity of TNP in the two years.

Table 5. Scores of single management capacity of TNP in 2017 and 2021.

Criterion	Indicator	Scores	
		2017	2021
Institutional construction	Management organization	69.52	92.62
	Management team	76.38	84.30
	Management planning	66.76	89.04
Guarantee mechanism	Financial support	71.87	85.06
	Legal system construction	65.13	87.46
	Scientific research support	75.81	83.82
	Multi-stakeholder participation	62.52	77.64
	Audit mechanism	69.14	89.42
Natural resources and ecosystem management	Natural resources inventory	53.95	85.38
	Natural resources confirmation and registration	51.87	88.18
	Ecosystem restoration	69.53	90.02
	Monitoring and early warning system	66.33	73.36
Community management	Community organization construction	52.33	74.24
	Resident participation	69.07	83.96
	Ecological compensation scheme	69.67	76.62
Popularization and education	Recreation management	53.43	64.22
	Science popularization	68.52	89.58
	Environmental education	66.33	77.30

From Table 5, we can see that the scores of each management capacity of TNP were generally lower in 2017 compared with those in 2021. Among the 18 management capacities, only the management team and scientific research support, scoring 76.38 and 75.81, reached a good level; the four management capacities, namely natural resources confirmation and registration, community organization construction, recreation management, and natural resources inventory, were relatively poor, with scores of 51.87, 52.33, 53.43, and 53.95; and the remaining 12 management capacities were at a regular level (Figure 2). This indicates that the interviewees generally considered that the management capacities of TNP in 2017 were insufficient, and only the management teams and scientific research support were satisfactory.

The scores of each management capacity of TNP were generally higher in 2021 than those in 2017, with no management capacity performing poorly. Among the 18 management capacities, the management organization and ecosystem restoration reached an excellent grade with scores of 92.62 and 90.02, respectively; the three management capacities of recreation management, monitoring and early warning system, and community organization construction were relatively weak, with scores of 64.22, 73.36 and 74.24; and the remaining 13 management capacities were of a good grade (Figure 3). This suggests that the interviewees were generally satisfied with the management capacities of TNP in 2021, especially for the management organization and ecosystem restoration, while they considered the performance of recreation management, monitoring and early warning system, and community organization construction slightly inferior.

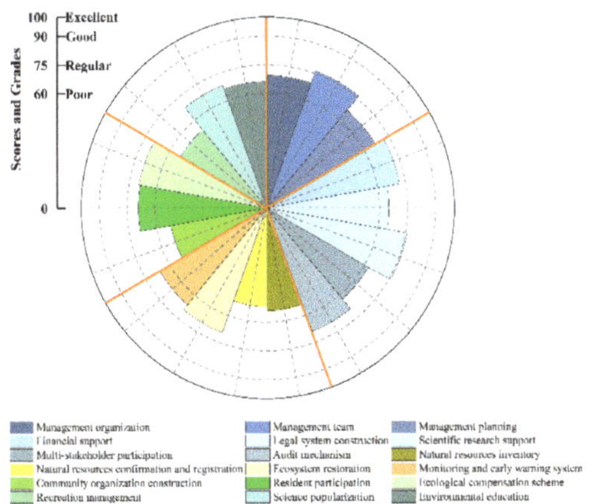

Figure 2. Grades of single management capacity of TNP in 2017.

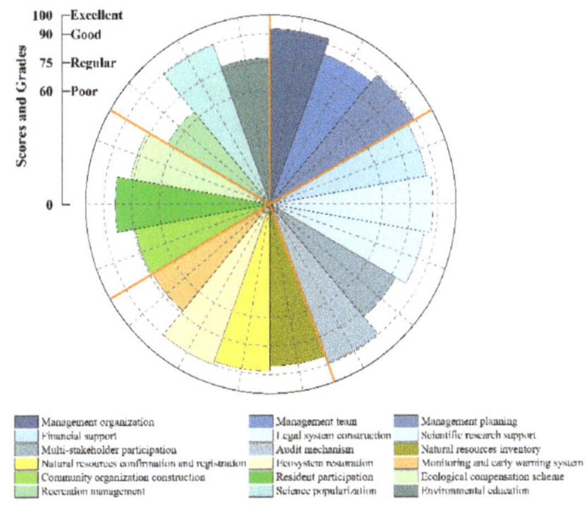

Figure 3. Grades of single management capacity of TNP in 2021.

By comparing the single management capacity between 2017 and 2021, we can see that all management capacities of TNP improved over the past four years. The most significant improvement is found in the management capacity score of natural resources confirmation and registration, which raised from 51.87 to 88.18, with an improvement of 70.00%. The management capacities of natural resources inventory and community organization construction also increased considerably. The management capacity score of natural resources inventory improved from 53.95 in 2017 to 85.38 in 2021, with an increase of 58.26%. The management capacity score of community organization construction increased from 52.33 in 2017 to 74.24 in 2021, showing an improvement of 41.87%. The scores of the other four management capacities increased by more than 30%, which are legal system construction (34.29%), management planning (33.37%), management organization (33.23%), and science popularization (30.74%). A minor improvement is found in the ecological compensation scheme, with the score increasing by 9.98%. Furthermore, three management capacities increased by less than 15%, namely, the management team (10.37%), scientific research support (10.57%), and monitoring and early warning system (10.60%). This indicates that the interviewees felt that the management capacities of TNP improved to different degrees in many aspects after four years of development. They considered that the TNP significantly improved most of the management capabilities. In contrast, the improvement in the management capabilities of the ecological compensation scheme, management team, scientific research support, and monitoring and early warning system is slightly less noticeable.

3.2. Analysis of Comprehensive Management Capacity

By adding up the weighted score of each indicator, we obtained the scores of the management capacities of TNP in the five aspects, and further, we got the comprehensive score. Table 6 shows the management capacity scores in different aspects and the comprehensive management capacity scores of TNP in 2017 and 2021.

Table 6. Comprehensive management capacity scores of TNP in 2017 and 2021.

	Institutional Construction	Guarantee Mechanism	Natural Resources and Ecosystem Management	Community Management	Popularization and Education	Comprehensive Management Capacity
2017	71.84	69.69	59.12	62.18	61.86	67.86
2021	88.29	84.79	84.40	77.74	74.95	84.96

The comprehensive management capacity score of TNP was 67.86 in 2017, which was a regular grade. The scores of the management capacities in the aspects of institutional construction, guarantee mechanism, natural resources and ecosystem management, community management, and popularization and education were 71.84, 69.69, 59.12, 62.18, and 61.86, respectively. The management capacities of TNP in the five aspects were relatively balanced in 2017, of which the natural resources and ecosystem management got the lowest score, but was close to the regular grade, and the other four were all at the regular grade.

The comprehensive management capacity score of TNP was 84.96 in 2021, ranking at a good grade. The scores of the management capacities in the five aspects were 88.29, 84.79, 84.40, 77.74, and 74.95, respectively. The strengths and weaknesses in the management capacities of TNP in 2021 were apparent. The institutional construction had the best performance, scoring close to the excellent grade, while the capacity in popularization and education was inferior, only ranking at the regular level.

By comparing the comprehensive management capacities between 2017 and 2021, we can see that the management capacities of TNP in all five aspects enhanced significantly over the past four years. The largest improvement is found in natural resources and ecosystem management, with its score rising from 59.12 to 84.40, showing an increase of 42.76%, and the grade rising from poor to good (Figure 4). The other four aspects all increased by more than 20%, of which the institutional construction, guarantee mechanism and community management increased from regular to good, and the popularization and education also immediately reached the good level. Specifically, the capacity of community management also improved over the past four years, with its score rising from 62.18 to 77.74, showing an increase of 25.02%. The score of institutional construction was the highest in both years, which raised from 71.84 to 88.29, with an increase of 22.90%, indicating that institutional construction plays a vital role in improving the management of TNP. The score of the guarantee mechanism of TNP raised from 69.69 in 2017 to 84.79 in 2021, showing an improvement of 21.67%. The score of popularization and education scores rose from 61.86 to 74.95, with an increase of 21.16%. As a result, the comprehensive management capacity score of TNP increased by 25.20%, with the performance pulled up from a regular grade to a good grade.

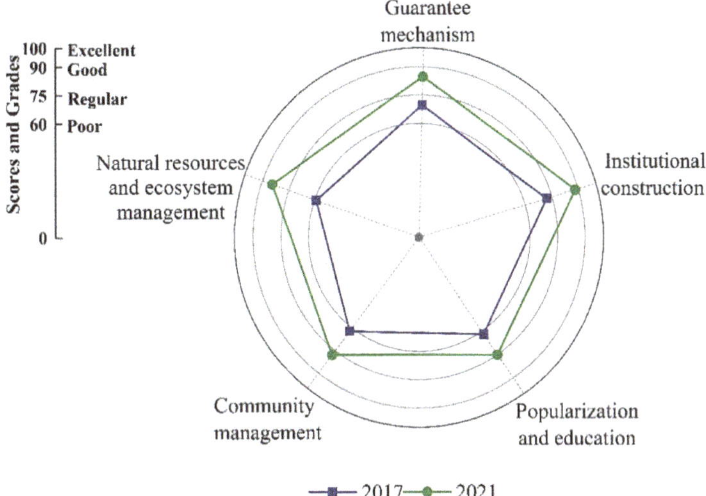

Figure 4. Grades of the comprehensive management capacities of TNP in2017 and 2021.

4. Discussion

4.1. The Beneficial Experiences from TNP's Management

All management capacities of TNP improved over the past four years. However, the reasons behind these improvements vary. Analyzing the reasons behind these improvements will help clarify the critical factors for increasing TNP's management capacities and thus provide management suggestions for other national parks.

Several management capacities achieved significant improvements, namely natural resources inventory and natural resources confirmation and registration, with scores increasing by 70.00% and 58.26%, respectively. This is highly due to the significant advancement in relevant work, such as resource surveys, unified verification, and registration. During the pilot period, the background situation of natural resources in TNP was thoroughly investigated, and their rights were uniformly confirmed and registered. In addition, TNP's capacity in community organization construction improved greatly, showing an increase of 41.87% in the score. With the appearance of community inquiries and supervision cards, residents gradually participated in the management of TNP, which may account for the significant enhancement of this management capacity. The capacity of audit mechanism also increased by 29.33%. This is mainly because of the completion of the off-office auditing of natural resource assets of leading cadres. It is demonstrated that the orderly implementation of related work is a vital way to improve the management capacities of national parks, especially in natural resources and ecosystem management.

Five management capacities increased to a certain degree because of the effective management measures taken by TNP. For example, the score of the legal system construction increased by 34.29%. In the absence of high-level legislation for national parks in China, such a marked improvement is partly due to the fact that a team of legal advisers provided strong support for the development of the regulations of TNP; and it is partly because an efficient and independent enforcement team ensured the implementation effectiveness of the regulations [33]. The score of the management plan increased by 33.37%. After analyzing relevant data and information, we argue that the reasons are multifaceted. Firstly, the overall plan of TNP was effectively implemented after the standardized preparation, validation, and approval, and it is now subject to third-party evaluation. Secondly, the social feedback and supervision mechanism for the planning of TNP is greatly improved. Thirdly, the sources of the three rivers were included after TNP was formally certified as a national park, so the boundaries of TNP were perfected in ecological integrity. This laid the foundation for more scientific and rational planning for TNP.

TNP's capacity in management organization greatly improved, with the score increasing by 33.23%. This is mainly because the management authority of TNP integrated related departments of the four counties within its boundary during the pilot period. The bureau for eco-environment and natural resources management and the bureau for natural resources and environmental enforcement were established in 2016 [34]. This realized the unified and efficient law enforcement of natural resources and environment in TNP and solved the problems of overlapped management, different standards, and cross functions to a certain extent. The score of science popularization is also raised, showing an increase of 30.74% in four years. There are three reasons for this: firstly, the image logo was put into use, letting more people know about TNP; secondly, a series of popular science works were created, spreading the values of TNP to the general public; and finally, a variety of publicity activities were carried out for local communities, raising their awareness of conservation. The score of ecosystem restoration also increased largely, with an improvement of 29.47%. Firstly, TNP already took measures to guide the behaviors of the residents, which raised their awareness of environmental protection effectively and reduced the ecological damage by residents fundamentally. Secondly, relying on large-scale ecosystem restoration projects with adequate investment, the ecosystem restoration of TNP achieved considerable progress.

4.2. The Deficiencies in TNP's Management Compared with Best Practices

Some management capacities of TNP, such as ecological compensation scheme, management team, scientific research support, and monitoring and early warning system, improved slightly in the past four years. The reasons are multifaceted. Exploring the factors for the slight improvement of these capacities can provide reasonable suggestions for improving TNP's management strategies and measures.

The score of the management team ranked first in 2017, but the improvement was only 10.37% over the past four years. The slight improvement was perhaps caused by the minor change in the quantity or quality of the management team in those four years. Firstly, the number of management staff did not increase significantly. Although TNP has the largest number of staff among the ten national park pilots, the area managed by each person in TNP is also the largest [35], which means that each person in TNP has more to do in a given time. With the adjustment of the boundary, the total area of TNP expanded by nearly half, so the workload of each person is even greater. Secondly, the number of management staff who have professional skills in grassland protection, as well as flora and fauna monitoring is still small. With the inclusion of river sources into TNP, more issues concerning zoning and management are brought about, requiring more professional teams to manage. To cope with the increased intensity and difficulty of management, we suggest that TNP further strengthen the management team by increasing the number of management staff and improving the quality of the professional team.

The score of scientific research support ranked second in 2017 while only increasing 10.57% during the past four years. As an area with the most concentrated plateau biodiversity and the most sensitive and fragile ecosystem, TNP attracted many scholars to conduct scientific research and produce fruitful scientific results. This is probably the reason why this capacity was outstanding in 2017. During the past four years, TNP formed a specialized research institution, signed cooperative agreements with other research institutions, actively participated in academic conferences and exchanged management practices with other PAs. However, due to a lack of systematic organization, transformation, and application of research results, these research practices are yet to fully exert the function of scientific research support, only bringing a slight increase to the related management capacity. In the future, TNP should strengthen the transformation and application of existing scientific research results and improve the relevance and applicability of subsequent research results.

The management capacities in ecological compensation scheme and monitoring and early warning system were under performing in both years. The score of ecological compensation scheme showed the slightest improvement of 9.98% during the past four years.

The eco-guard system is the main ecological compensation policy of TNP. With the number of eco-guards selected from communities expanded and the management mechanism improved, the capacity in ecological compensation scheme improved to a certain degree. However, two factors might limit the large improvement in this capacity Firstly, the range of compensation is enormous, while the funding is still deficient. As an area with more prominent ecological functions, TNP does not have much higher compensation standards than other areas, which results in the low motivation of protectors. Secondly, the existing compensation scheme is still imperfect [36]. The residents in TNP live in the upper reaches of the three rivers, and make sacrifices for the downstream areas, but they are not compensated by the downstream beneficiaries. To this end, we suggest that TNP seeks more funds from society and the market besides transfer payments from the central government, and establish a horizontal ecological compensation system for the watershed to solve the relationship of upstream conservation and downstream development.

The score of monitoring and early warning system only increased 10.60% in the past four years. Although the establishment of the Ecological Data Center improved the capacity of TNP in monitoring environmental elements, wildlife distribution, and ecosystem disasters, there are still several deficiencies in the management and utilization of monitoring data when compared with the best practices. For example, the data from different sources lack a unified storage and management mechanism, and the data are not fully shared with the public. International experiences show that effective data management and utilization are essential for national park monitoring. For example, all standards, background data, and monitoring data in Canada are recorded in the Information Center on Ecosystems Database [37]. In the U.S., the 32 networked eco-region units are required to regularly publish a series of resource summaries, data summary briefs, technical reports, trend analyses, and synthesis reports on the web [38]. Therefore, we suggest that TNP strengthen the management of the Ecological Data Center, including enhancing the collection of data from different sources, improving the organization and collation of data, and facilitating the transformation and integration of data.

Furthermore, some capacities, though increasing greatly in scores, still need further improvement. Take the capacity in community organization construction as an example. Although its score has increased significantly in the past four years, the construction of community organization still has gaps with the best practices, mainly due to the low participation of community residents in national park management decisions. One good practice comes from the Kakadu National Park and Parks Australia where the world's first community co-management agreement was signed in 1978 by residents. Since then, they established a National Park Management Committee with shared decision-making powers [39]. Another good practice is found in the Republic of Macedonia, where community residents became a more influential group in the management of national parks through the involvement of NGOs in environmental protection, educational seminars, field trips, and information dissemination [40]. Therefore, we suggest TNP increase the ways of residents' participation in park management, adopt residents' recommendations more fully, hold joint meetings regularly, and make consultation widely when making management decisions.

4.3. Strengths, Weaknesses and Applicability of the BPB Method

The BPB method is tested and validated in this research, and its application in TNP shows that the evaluation results can reflect the actual changes in management capacities. For example, TNP is significantly improved in institutional construction and natural resources and ecosystem management, and slightly improved in guarantee mechanism, community management, and popularization and education on the whole. It provides a rapid and accurate way for comprehensive management capacity evaluation of TNP and visualizes the management states of TNP in the form of evaluation scores. It also shows the management performance of TNP in the five aspects of institutional construction, guarantee mechanism, natural resources and ecosystem management, community management, and popularization and education. Applying the BPB method achieves a comprehensive diagnosis of TNP's management capacities and identifies the gap between its management status and the international best level. The comparison between the two years reflects the changes in the management capacities of TNP directly. The comparison and analysis of the evaluation results can help explain the reasons behind the changes and provide a reliable basis for improving TNP's management strategies. It is demonstrated that the BPB method can reveal the changes in the management capacities of national parks well and reflect the impact of various management measures on management capacities to a certain extent.

Compared to the existing MCE methods from international programs, the BPB method has broader adaptability. This adaptability is reflected in the indicator system, scoring method, and evaluation form. Firstly, the indicator system comes from a systematical summary of national park management experiences in different countries, which is well-rounded and applicable worldwide. In contrast, the indicator system proposed by The Parks in Peril Program only covers four aspects: conservation action, long-term management, financial support, and local guarantee, and lacks attention to conserving natural resources and ecosystems. Secondly, the BPB method uses a benchmark approach to invite interviewees to score each indicator, which improves the anchoring mechanism of the evaluation. Finally, compared with the objective quantitative evaluation with a long period and large resource input, the BPB method has a straightforward evaluation and a simple results processing process, which has advantages in timeliness and low investment. In addition, when applying this method to evaluate each indicator, we can compare the current management states with international best practices and provide feedback on current management measures to further improve management capacities.

The BPB method also has certain shortcomings. Firstly, this is a subjective evaluation method that relies on participants' judgment. Several studies suggest management evaluations that rely on expert knowledge and qualitative judgment may be more accurate than those relying on quantitative data or a mix of data types [41,42]. However, as the quantity and precision of national park monitoring data improves, it will become a future demand to explore accurate evaluation based on ecological monitoring results as evidence. Therefore, the BPB method and the objective quantitative evaluation can be used jointly and complement each other. Secondly, the evaluation indicator system is only two-leveled, which does not fully reflect every aspect of the national park management. How to refine the evaluation indicator system and measure the management capacity of national parks in an all-round way became the focus of future research. It should be noted, when designing the indicator system and setting the standards, we should consider the conservation needs and management objectives of a specific type of PA, and then make adjustments accordingly.

5. Conclusions

This research follows the BPB method and selects TNP, China's first national park system pilot and one of the first batch of national parks, as the case study. By evaluating the management capacity in 2017 and 2021 respectively, the changes in the management capacity of TNP during the pilot period have been explored and the appropriateness of related management measures has been revealed. Some management capacities, such as legal system construction, management planning, and natural resources confirmation and registration, performed well or improved significantly during the pilot period, providing beneficial lessons for other national parks in China. Some management capacities, such as ecological compensation scheme, monitoring and early warning system, and the management team, are still lacking and should be prioritized for future improvement. The BPB method is tested and validated in this research, showing a potential to be promoted to other PAs in China and even other countries. Not only is the effectiveness and operability of this method confirmed in this research, but its contribution to the theoretical improvement of MCE methods is also demonstrated.

Author Contributions: Conceptualization and methodology, X.L. and W.J.; investigation, data collection and analysis, X.L.; writing, reviewing and editing, X.L., W.J. and Q.M. All authors have read and agreed to the published version of the manuscript.

Funding: This research was funded by the Strategic Priority Research Program of the Chinese Academy of Sciences (XDA23100203) and the National Key R&D Program of China (2017YFC0506404).

Data Availability Statement: The data are not publicly available due to privacy or ethical restrictions.

Acknowledgments: The authors are grateful to the Three-River-Source National Park management authority for their assistance in developing this research. The authors would like to thank Shuaichen Yao for his help in data collection, thank Bojie Wang and Zhidong Li for their help in drafting the manuscript.

Conflicts of Interest: The authors declare no conflict of interest.

References

1. Petric, L.; Mandic, A. Visitor Management Tools for Protected Areas Focused on Sustainable Tourism Development: The Croatian Experience. *Environ. Eng. Manag. J.* **2014**, *13*, 1483–1495. [CrossRef]
2. Kolahi, M.; Sakai, T.; Moriya, K.; Makhdoum, M.F.; Koyama, L. Assessment of the Effectiveness of Protected Areas Management in Iran: Case Study in Khojir National Park. *Environ. Manag.* **2013**, *52*, 514–530. [CrossRef] [PubMed]
3. Chapman, C.A.; van Bavel, B.; Boodman, C.; Ghai, R.R.; Gogarten, J.F.; Hartter, J.; Mechak, L.E.; Omeja, P.A.; Poonawala, S.; Tuli, D.; et al. Providing health care to improve community perceptions of protected areas. *Oryx* **2015**, *49*, 636–642. [CrossRef]
4. Kelly, A.B.; Gupta, A.C. Protected Areas: Offering security to whom, when and where? *Environ. Conserv.* **2016**, *43*, 172–180. [CrossRef]
5. Peng, F.; Li, J.; Yuan, H.; Zhu, Y.; Li, B. *Research on Overall Scheme of Establishing National Park System*; China Environment Publishing Group: Beijing, China, 2019.
6. Tang, X.; Jiang, Y.; Liu, Z.; Chen, J.; Liang, B.; Lin, C. Top-level Design of the Natural Protected Area System in China. *For. Resour. Manag.* **2019**, *28*, 1–7. [CrossRef]
7. Ren, H.; Guo, Z. Progress and prospect of biodiversity conservation in China. *Ecol. Sci.* **2021**, *40*, 247–252. [CrossRef]
8. Li, J.; Wang, W.; Axmacher, J.C.; Zhang, Y.; Zhu, Y. Streamlining China's protected areas. *Science* **2016**, *351*, 1160. [CrossRef]
9. Wang, W.; Feng, C.T.; Liu, F.Z.; Li, J.S. Biodiversity conservation in China: A review of recent studies and practices. *Environ. Sci. Ecotechnol.* **2020**, *2*, 100025. [CrossRef]
10. Watkins, T.; Miller-Rushing, A.J.; Nelson, S.J. Science in Places of Grandeur: Communication and Engagement in National Parks. *Integr. Comp. Biol.* **2018**, *58*, 67–76. [CrossRef]
11. Vieira, F.A.S.; Bragagnolo, C.; Correia, R.A.; Malhado, A.C.M.; Ladle, R.J. A salience index for integrating multiple user perspectives in cultural ecosystem service assessments. *Ecosyst. Serv.* **2018**, *32*, 182–192. [CrossRef]
12. Kulczyk-Dynowska, A.; Bal-Domańska, B. The National Parks in the Context of Tourist Function Development in Territorially Linked Municipalities in Poland. *Sustainability* **2019**, *11*, 1996. [CrossRef]
13. Leverington, F.; Costa, K.L.; Pavese, H.; Lisle, A.; Hockings, M. A Global Analysis of Protected Area Management Effectiveness. *Environ. Manag.* **2010**, *46*, 685–698. [CrossRef] [PubMed]
14. Hockings, M.; Stolton, S.U.E.; Dudley, N. Management Effectiveness: Assessing Management of Protected Areas? *J. Environ. Policy Plan.* **2004**, *6*, 157–174. [CrossRef]
15. Hockings, M.; Cook, C.N.; Carter, R.W.; James, R. Accountability, reporting, or management improvement? Development of a state of the parks assessment system in New South Wales, Australia. *Environ. Manag.* **2009**, *43*, 1013–1025. [CrossRef]
16. de Oliveira Junior, J.G.C.; Campos-Silva, J.V.; Santos, D.T.V.; Ladle, R.J.; da Silva Batista, V. Quantifying anthropogenic threats affecting Marine Protected Areas in developing countries. *J Environ. Manag.* **2021**, *279*, 111614. [CrossRef]
17. Joppa, L.; Pfaff, A. Reassessing the forest impacts of protection: The challenge of nonrandom location and a corrective method. *Ann. N. Y. Acad. Sci.* **2010**, *1185*, 135–149. [CrossRef]
18. Munoz Brenes, C.L.; Jones, K.W.; Schlesinger, P.; Robalino, J.; Vierling, L. The impact of protected area governance and management capacity on ecosystem function in Central America. *PLoS ONE* **2018**, *13*, e0205964. [CrossRef]
19. Addison, P.F.E.; Flander, L.B.; Cook, C.N. Towards quantitative condition assessment of biodiversity outcomes: Insights from Australian marine protected areas. *J. Environ. Manag.* **2017**, *198*, 183–191. [CrossRef]
20. Nolte, C.; Agrawal, A. Linking management effectiveness indicators to observed effects of protected areas on fire occurrence in the Amazon rainforest. *Conserv. Biol.* **2012**, *27*, 155–165. [CrossRef]
21. Kurdoglu, O.; Cokcaliskan, B.A. Assessing the effectiveness of protected area management in the Turkish Caucasus. *Afr. J. Biotechnol.* **2011**, *10*, 17208–17222. [CrossRef]
22. Oyelowo, O.J.; Chima, D.U.; Oladoye, A.O. An assessment of the management of Osun Osogbo world heritage site. *J. Agric. For. Soc. Sci.* **2011**, *8*, 110–116. [CrossRef]
23. Stoll-Kleemann, S. Evaluation of management effectiveness in protected areas: Methodologies and results. *Basic Appl. Ecol.* **2010**, *11*, 377–382. [CrossRef]
24. Bencini, A.; Caneschi, A.; Carbonera, C.; Dei, A.; Gatteschi, D.; Righini, R.; Sangregorio, C.; van Slageren, J. Tuning the Physical Properties of a Metal Complex by Molecular Techniques: The Design and the Synthesis of the Simplest Cobalt-O-dioxolene Complex Undergoing Valence Tautomerism. *ChemInform* **2004**, *35*, 141–154. [CrossRef]

25. Nielsen, G. Capacity development in protected area management. *Int. J. Sustain. Dev. World Ecol.* **2012**, *19*, 297–310. [CrossRef]
26. Martin, A.S.; Rieger, J.F. *The Parks in Peril Site Consolidation Scorecard: Lessons from Protected Areas in Latin American and the Caribbean*; The Nature Conservancy: Arlington County, VA, USA, 2003; pp. 9–12.
27. Wright, R.G. Parks in Peril: People, Politics, and Protected Areas. *Q. Rev. Biol.* **2001**, *76*, 259. [CrossRef]
28. Hockings, M. Systems for assessing the effectiveness of management in protected areas. *Bioscience* **2003**, *53*, 823–832. [CrossRef]
29. Xue, D.; Zheng, Y. A study on evaluation criteria for effective management of the nature reserves in China. *Rural Eco-Env.* **1994**, *10*, 6–9.
30. Feng, B.; Li, D.; Zhang, Y.; Xue, Y. Progress and analysis on the management effectiveness evaluation of protected area based on Aichi Biodiversity Target 11th in China. *Biodivers. Sci.* **2021**, *29*, 150–159. [CrossRef]
31. Liu, X.; Min, Q.; Jiao, W.; He, S.; Liu, M.; Yao, S.; Zhang, B. Methodology of evaluating the management capacity of national parks based on best practices. *Acta Ecol. Sin.* **2019**, *39*, 8211–8220. [CrossRef]
32. Wu, J.; Wu, G.; Zheng, T.; Zhang, X.; Zhou, K. Value capturecapture mechanisms, transaction costs, and heritage conservation: A case study of Sanjiangyuan National Park, China. *Land Use Policy* **2020**, *90*, 104246. [CrossRef]
33. Tang, X. The establishment of national park system: A new milestone for the field of nature conservation in China. *Int. J. Geoherit. Parks* **2020**, *8*, 195–202. [CrossRef]
34. Su, H.; Wang, N.; Su, Y. The experience and its reference study of law enforcement system of Sanjiangyuan National Park pilot. *Biodivers. Sci.* **2021**, *29*, 304–306. [CrossRef]
35. Zhang, X.; Sun, G. Current Situation and Model Selection in the Construction of National Parks dministration. *J. Beijing For. Univ. (Soc. Sci.)* **2021**, *20*, 76–83. [CrossRef]
36. Wang, Y. Implementation and Suggestions of the Ecological Compensation Policy in the National Parks—Case Study of the Three-River-Source National Park Pilot. *Manag. World J.* **2020**, *7*, 22–26.
37. McGoldrick, D.J.; Clark, M.G.; Keir, M.J.; Backus, S.M.; Malecki, M.M. Canada's national aquatic biological specimen bank and database. *J. Great Lakes Res.* **2010**, *36*, 393–398. [CrossRef]
38. Fancy, S.G.; Gross, J.E.; Carter, S.L. Monitoring the condition of natural resources in US national parks. *Environ. Monit. Assess.* **2009**, *151*, 161–174. [CrossRef]
39. Oldekop, J.A.; Holmes, G.; Harris, W.E.; Evans, K.L. A global assessment of the social and conservation outcomes of protected areas. *Conserv. Biol.* **2016**, *30*, 133–141. [CrossRef]
40. Saška Petrova, S.B.-B.; Martin, Č. Landscapes of Flexibility: Negotiating the Everyday ǀ ǀ From inflexible national legislation to flexible local governance: Management practices in the Pelister National Park, Republic of Macedonia. *GeoJournal* **2009**, *74*, 589–598. [CrossRef]
41. Hockings, M.; Stolton, S.; Dudley, N.; James, R. Data credibility: What are the "right" data for evaluating management effectiveness of protected areas? *New Dir. Eval.* **2009**, *122*, 53–63. [CrossRef]
42. MacMillan, D.C.; Marshall, K. The Delphi process—An expert-based approach to ecological modelling in data-poor environments. *Anim. Conserv.* **2006**, *9*, 11–19. [CrossRef]

Article

Study on Functional Zoning Method of National Park Based on MCDA: The Case of the Proposed "Ailaoshan-Wuliangshan" National Park

Junze Liu [1], Xiaoyuan Huang [1,*], Huijun Guo [2], Zhuoya Zhang [1], Xiaona Li [1] and Mengxiao Ge [1]

[1] School of Geography and Ecotourism, Southwest Forestry University (SWFU), 650224 Kunming, China
[2] National Plateau Wetlands Research Center, Southwest Forestry University (SWFU), 650233 Kunming, China
* Correspondence: hxy21cn@swfu.edu.cn

Abstract: In a national park master plan, functional zoning plays a key role in developing differentiated zoning controls that achieve multiple park construction objectives. In this study, a geographical attribute code and basic zoning elements are developed for the proposed "Ailaoshan-Wuliangshan" National Park, followed by the development of spatial multi-criteria sets and weight sets to determine the suitability of the land. Next, we use a clustering algorithm and conflict unit prioritization to allocate space for multi-target units to get the preliminary zoning schemes, and then identify stable units and unstable units through sensitivity analysis. Ultimately, the functional zoning of the National Park was determined. According to the results, the proposed "Ailaoshan-Wuliangshan" National Park can be divided into nine types of 164 landscape units; the highest land suitability values of each zone showed the traits of differentiation and aggregation in spatial distribution; there are 97 stable units and 67 unstable units; approximately 62.83% and 37.17% of the total park area can be divided into core conservation area (primary sensitive area and secondary sensitive area) and general control area (ecological activity area and ecological control area). By implementing a comprehensive assessment and decision-making process, the defined functional zones are precise and simple to recognize on the ground, and they adhere to the area proportions needed by national standards. Furthermore, the functional zoning is clustered, which avoids the fragmentation of the zoning results causing difficulties in management, and serves as a point of reference for the functional zoning approaches used in other proposed national parks in China.

Keywords: national park; functional zoning; landscape unit; multi-criteria decision analysis

Citation: Liu, J.; Huang, X.; Guo, H.; Zhang, Z.; Li, X.; Ge, M. Study on Functional Zoning Method of National Park Based on MCDA: The Case of the Proposed "Ailaoshan-Wuliangshan" National Park. *Land* **2022**, *11*, 1882. https://doi.org/10.3390/land11111882

Academic Editors: Le Yu, Rui Yang, Yue Cao and Steve Carver

Received: 17 September 2022
Accepted: 20 October 2022
Published: 23 October 2022

Publisher's Note: MDPI stays neutral with regard to jurisdictional claims in published maps and institutional affiliations.

1. Introduction

In 2019, China put forward a policy to establish a system of nature reserves with national parks as the mainstay, and thus, the development of national parks in China has entered a new period [1]. According to the latest standards for establishing national parks in China, the main purpose of national parks is to protect natural ecosystems and to achieve scientific conservation and rational use of natural resources [2]. Worldwide, zoning designs are commonly used in order to balance conservation and development needs while making sure that integrated service functions of national parks can be fully realized [3], but the designs vary widely and have their own characteristics depending on the conflict between conservation and exploitation. For example, the United States has adopted a traditional zoning model with more refined sub-zones under each type of functional zoning to enhance management [4]. Germany's zoning plan embodies the idea of dynamic zoning [5]. New Zealand's national park zoning plan adopts a management zoning and special zoning approach [6]. Japan's national park zoning has a distinction between reflecting special areas and general areas [7]. According to the current planning and zoning scheme of China's national parks, they are generally divided into two control zones based on the protection level, core conservation areas and general control areas [8]. Some of the pilot national parks will, on the basis of the above-mentioned control zones, also carry more specific functional zones on the basis of the division of core areas, buffer areas, and experimental areas of the original types of protected areas [9].

Since 1983, foreign countries have put forward the concept of zoning and ideas and some basic principles and steps for the functional zoning of national parks. With the rapid development of national parks, the combination of qualitative and quantitative research on the functional zoning of national parks is becoming increasingly closer. Scholars from many countries have explored the construction of functional zoning evaluation index systems from both natural and social perspectives, such as ecological conservation perspective, Recreation Opportunity Spectrum (ROS), ecosystem health evaluation, stakeholder perspective, resource type or landscape type perspective, and animal behavior [10–15], and a variety of quantitative methods have been used for development planning and management implementation plans for national parks, including GAP analysis, landscape suitability assessment, spatial overlay, multivariate analysis, habitat distribution models, and condition value assessment methods [16–23]. The relevant functional zoning methods in China are currently diversified, but there are problems such as vague method descriptions, often unclear methods for graded zoning, and a lack of intuitive and operable zoning methods [24]. It is not conducive to the implementation of refined zoning controls. Thus, it is imperative to strengthen research on zoning methods for national parks.

Decisions about the functional zoning of national parks require the assessment of multiple land attributes against multiple objectives, which are inherently conflicting. In DSS (decision support systems), MCDA (Multi-Criteria Decision Analysis) is a method of comparing alternative courses of action based on multiple factors and identifying the most optimal path forward [25,26]. These methods employed include structuring decision problems, performing sensitivity analyses, increasing transparency, and enhancing the visual representation of results [27]. By using a multi-criteria decision analysis process, alternatives can be compared based on a set of clear criteria addressing the most relevant factors. Geneletti and Duren (2008) used MCDA for land suitability evaluation and completed an optimal adjustment of natural park zoning schemes through cluster analysis [22]. Randal et al. (2010) used MCDAS (a custom software application integrating GIS and MCDA) for management planning studies in forest landscapes [28]. Bereket et al. (2016) combined MCDA with GIS and developed a spatial zoning method for multipurpose marine protected areas through stakeholder consultation [29]. All of these studies mentioned above need to address the question of how to achieve optimal decision making in the context of spatial planning, with multiple objectives. As Linkov et al. argue [27], Multi-criteria decision analysis is well suited to participatory settings involving different objectives and different stakeholders. Therefore, this method is highly beneficial when applied to spatial planning.

In this study, MCDA is combined with GIS and applied for hierarchical functional zoning, using the proposed "Ailaoshan-Wuliangshan" National Park (AWNP) as an example. These areas cover a wide range of potentially conflicting conservation or identification targets with complex relationships. In the methods section, we explain the method of identifying landscape units and how to build up a set of evaluation criteria and combine them with stability tests to complete the assignment of landscape units. In the results section, we show the final scheme of the first-level zone and second-level zoning, reflecting the concept of refined and differentiated hierarchical zoning, alleviating related conflict issues and better balancing nature conservation and regional development. It is hoped that this will provide a reference for other national parks, especially those with a relatively fragmented spatial distribution, in terms of functional zoning methods.

2. Materials and Methods

2.1. Study Area

The proposed "Ailaoshan-Wuliangshan" National Park (AWNP) is located in the central part of Yunnan Province, which is the southern extension of the Hengduan Mountains and is in the area where four states (cities), namely Pu'er City, Yuxi City, Chuxiong Yi Autonomous Prefecture and Dali Bai Autonomous Prefecture, are connected. It is based on the results of the Yunnan Provincial Nature Reserve Consolidation and Optimization Plan carried out in 2020, which covers two national nature reserves, namely the Ailao Mountain National Nature Reserve and the Wuliang Mountain National Nature Reserve, as well as a number of provincial reserves and nature parks. From 2020 to 2022, the Yunnan Provincial Government commissioned the Kunming Branch of the Chinese Academy of Sciences and other institutions to conduct several scientific investigations and feasibility studies on the proposed AWNP, and completed the project declaration. The proposed AWNP's spatial distribution, unlike many national parks, tends to be linear and more fragmented, covering 1537.33 km^2 in total. Ailaoshan, Wuliangshan, and Konglonghe are the three areas of the park, while an ecological corridor links Ailaoshan with Wuliangshan (Figure 1).

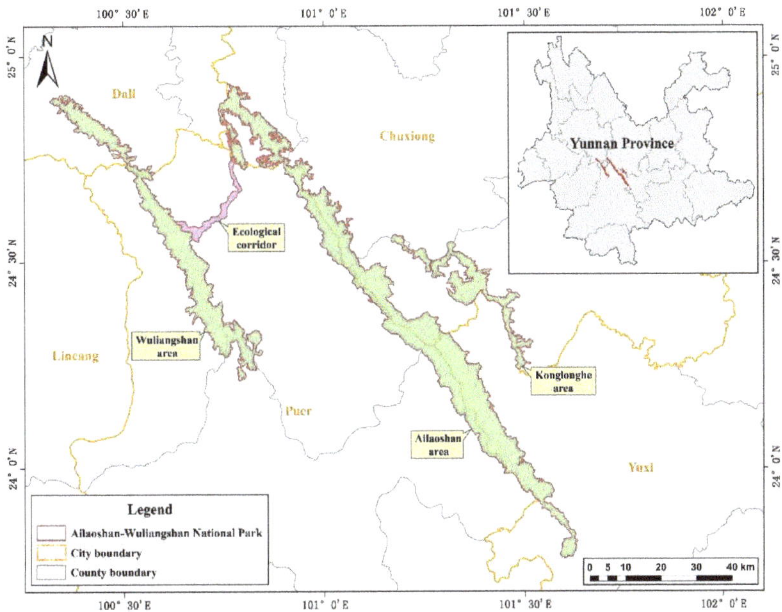

Figure 1. The study area location.

The proposed AWNP is rich in natural resources and is an invaluable corridor for tropical to temperate transitions, species migration and gene exchange in mainland Asia. It is one of eight migratory routes for migratory birds worldwide. This area is characterized by a high diversity of species, including 12 species of national class I protected animals, represented by the eastern black crested gibbon (*Nomascus nasutus*). It is also the main distribution area for eastern black crested gibbons in the world, holding over 90% of the extant population. There are also four national Class I protected plants in the proposed AWNP, and the forest preserves the largest area of montane evergreen broad-leaved forest in the subtropical region of China. The proposed AWNP region is rich in cultural resources, such as the Ancient Tea Horse Road, where the long-standing tea culture contributed to the economic development and cultural exchanges in ancient China. When the boundaries of the proposed national park were delineated, the villages in the area were divided on the periphery of the boundary, where ethnic groups such as the Yi, Hani and Yao live, creating a diverse ethnic culture. The proposed AWNP is, therefore, of great conservation and research value.

2.2. Data Sources and Processing

The data collected in this study mainly includes geographic information data, remote sensing image data and related textual information, as follows: land-use vector data (from the Second National Land Survey), soil-type vector data (from the World Soil Database, HWSD), vegetation cover-type vector data (from the Yunnan Provincial Forest Resources Class II Survey), and endangered animal habitat-range vector data (provided by the Southwest Forestry University team) within the proposed AWNP and its surrounding areas. Satellite imagery, DEM grid data, and vector data such as water system waters, settlements, tourist attractions, roads, traffic service points, and leisure and recreation points are downloaded through Bigemap GIS Office software. The vector boundary of the proposed national park is provided by the Kunming Branch of the Chinese Academy of Sciences. The textual material, including the AWNP construction proposal and other related drafts, was provided by the Yunnan Forestry Research and Planning Institute and the Southwest Forestry University team. The existing data was subsequently updated through the local Forestry and Grassland Bureau in collaboration with the Natural Resources Bureau and in conjunction with field research. In order to keep the relevant raster data consistent, all raster data was resampled to an image size of 30 m × 30 m for subsequent analysis.

2.3. Methods

We defined three zones for the analysis process in order to simplify the subsequent process and determine the final scheme: Zone A (core conservation areas), Zone B1 (ecological activity areas), and Zone B2 (ecological control areas), which are based on the Chinese "two-zone system" (core protection zone and general control zone). Specifically, Zone A is a first-level zone, which, in the final scheme, will be specifically divided into two second-level zones, primary sensitive areas and secondary sensitive areas. On the other hand, the B1 and B2 zones are secondary subzones under the general control zone. This is due to the more complex and multi-purpose nature of the general control area. By doing so, the secondary zoning plan for the general control zone can be made more accurate.

The methodology of this thesis consists of five stages (Figure 2). Firstly, the proposed AWNP area is divided into landscape units of the same nature. In phase two, a spatial multi-criteria analysis was carried out to complete the land suitability evaluation of the park. In the third phase, the spatial multi-target unit allocation is carried out, the conflicting units are identified and redistributed, and the preliminary partition results are generated. In the fourth stage, the stable and unstable units are identified through sensitivity analysis tests. Finally, the allocation of unstable units is completed and the first-level zoning and the second-level zoning are finalized.

2.3.1. Landscape Unit Delineation

Typically, a landscape unit (land unit) is defined as an area with similar geographical characteristics [30], such as topography, land use type, soil type, etc. As a parcel of land within a small and homogeneous geographical scale, rather than based on administrative or land-use boundaries. Consequently, landscape units of the same type were used as the basic zoning elements for this study.

Firstly, the four raster data of slope classification, elevation classification, land-use type and soil type were overlaid and processed using ArcGIS and ENVI to update the attribute codes (Table 1). A preliminary raster map of landscape units with both natural and socio-economic attributes was generated, comprising a total of 15,057 landscape units of 113 types. With too many landscape units the zoning results may be too fragmented, making the zoning scheme unreasonable and difficult to manage [31]. Therefore, we combine landscape units smaller than 100 hectares with the most similar neighboring units. This is supplemented by visual interpretation of remote sensing images to check and correct the boundaries of the landscape units. Ultimately, the proposed AWNP was divided into nine types of landscape units totaling 164 (Figure 3).

Table 1. Landscape unit attribute coding.

Data Type	Classification	Code	Data Types	Classification	Code
	537–1312 m	1		3–10%	1
	1312–1912 m	2	Slope	10–25%	2
Altitude	1912–2337 m	3		25–50%	3
	2337–2628 m	4		50–100%	4
	2628–3348 m	5		Arboreal forest	01
	LVg	01		Construction land	02
	LVk	02		Cultivated land	03
	LVj	03		Temporary land use	04
	PZg	04		Immature forest land	05
	PDd	05	Land use	Water area	06
Soil	CMd	06	type	Shrub land	07
	GLe	07		Barren hills and wasteland	08
	CMx	08		Unused land	09
	CMc	09		Harvested land	10
	LVh	10		Open woodland	11
	GLu	11			

Notes: 110101: Indicates tree woodland of type 537 m–1312 m above sea level, with a slope of 3–10% (gently sloping land) and a soil type LVg. Altitude classification: Divided into 5 levels according to the natural breakpoint method. Soil classification: The soil classification system used is FAO-90.

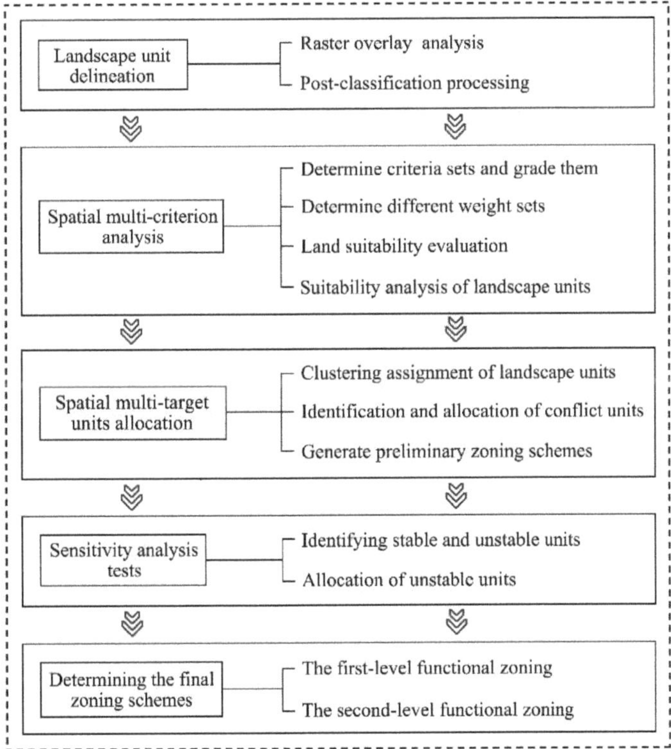

Figure 2. The technical flowchart of this study.

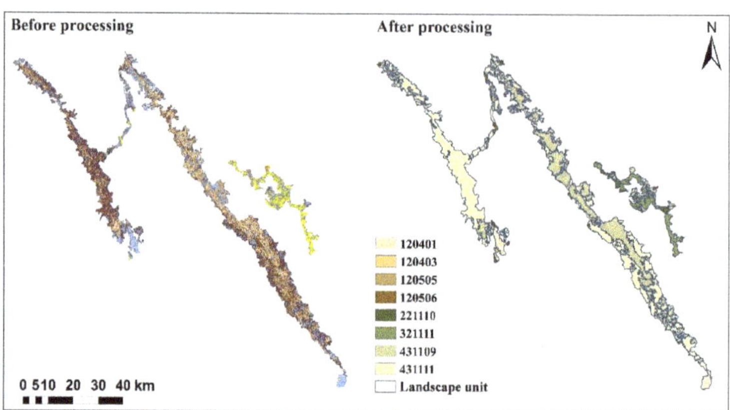

Figure 3. Landscape unit map for the proposed AWNP.

2.3.2. Spatial Multi-Criteria Analysis

To evaluate the relevance of different landscape units to different functional zones, two types of criteria, biotic and abiotic, need to be developed [32]. As Zone A is dominated by strict protection, the assessment of this zone is only relevant to ecological protection. As for Zone B1, it is dominated by landscape resources and distinguishes between agricultural landscapes with tourism value and areas within the proposed AWNP that are adjacent to tourist attractions. Unlike Zone B1, Zone B2 uses artificial facilities that have the greatest human impact for assessment. It focuses on the areas with the greatest human impact and requiring controlled restoration. To assess land suitability in each of the different zoning districts, three criteria were identified through consultation with experts (Table 2). Then, MCDA is used to group these criteria into suitability indices, which are assigned to each mapping unit for subsequent spatial multi-criteria analysis. We convert the suitability index from its original unit to a uniform and ordered value scale, scoring the criteria in descending order (1 to 5) [33], a step known as standardization [32]. In this process, the habitat index in Area A was graded according to the type of vegetation cover. As eastern black crested gibbons and other national Class I protected animals in the proposed AWNP area mainly inhabit evergreen broad-leaved forests, the evergreen broad-leaved forests were rated the highest suitability index (5 points), and other vegetation cover types were graded in descending order according to their biodiversity conservation value. The Agricultural Landscape Index for Zone B1, on the other hand, uses the current state of land use as the basis for grading. Terraces are rated highest in this index. Other land types are ranked in order of suitability, from largest to smallest, according to the degree of impact of human activity. Except for the two standard indices above, the rest are converted into Euclidean distance rasters based on the vector data they belong to, and their scores are inversely proportional to their distance from vector points or surfaces [34].

The Analytic Hierarchy Process (AHP) combined with the Experts Grading Method was used for weight evaluation [14]. Given that experts from different research directions may have different opinions on the importance of each criterion, three different sets of weights were determined for each group of criteria (Table 2). We conducted nine more spatial multi-criteria analyses by weighted sums since ecological conservation is the primary goal of national parks, and the core conservation area (Zone A) must be larger than 50% of the total park area [36]. By comparing the results generated by different weight sets and referring to relevant norms and expert opinions, weight set 1 was used for Area A, weight set 2 was used for Zone B1, and weight set 3 was used for Zone B2. In Zone B2, weight set 3 was selected (the proposed AWNP is intersected by a number of roads and has the greatest impact, while the other two criteria are primarily located on the periphery of the

AWNP). Based on these weight sets, maps of land suitability for AWNP in zones A, B1, and B2 were determined (Figure 4).

Table 2. Criterion sets and weight sets.

Criterion Sets	Zone A			Zone B1			Zone B2		
	VT	HEA	RWS	ATA	VAL	AL	TSF	RF	R
weight sets 1	0.4	0.4	0.2	0.4	0.4	0.2	0.4	0.4	0.2
weight sets 2	0.333	0.333	0.333	0.333	0.333	0.333	0.333	0.333	0.333
weight sets 3	0.275	0.275	0.450	0.275	0.275	0.450	0.275	0.275	0.450

Notes: VT = Vegetation type (Evergreen broad-leaved forest is 5 points, deciduous broad-leaved forest and bamboo forest are 4 points, warm coniferous forest is 3 points, shrub is 2 points, and non-woodland is 1 point). HEA = Habitat for endangered animals (Available habitat extent data are dominated by potential habitat and are subject to further research). RWS = River water sources. ATA = Available tourist attractions. VAL = Village architectural landscape (Mainly located on the periphery of the proposed AWNP boundary, only 2 villages are located within the park). AL = Agricultural landscape (5 points for arable land (terraces), 4 points for planted forest, 3 points for building land, and 1 point for other natural forest). TSF = traffic services facilities (Mainly located on the periphery of the proposed AWNP boundary). RF = Recreational facilities (All located on the periphery of the proposed AWNP boundary). R = Roads (According to the relevant specification [35], a straight line distance of 1 km on both sides of the road is converted by ArcGIS into an equidistant raster).

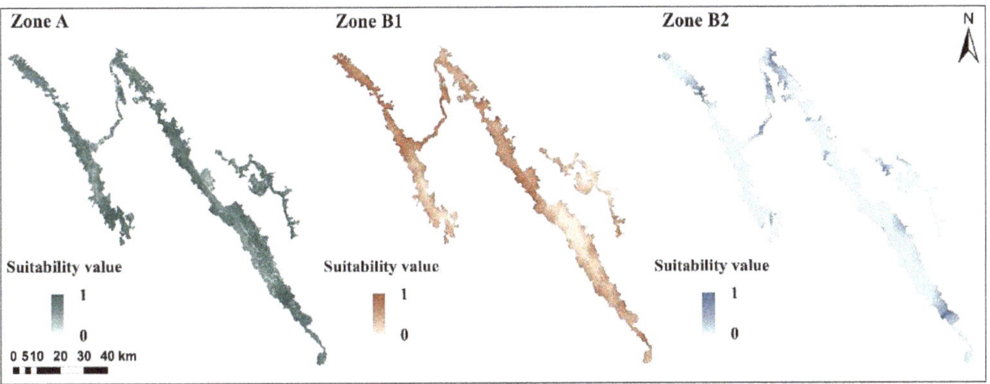

Figure 4. Land suitability maps for Zones A, B1, and B2.

In order to aggregate each of the three land suitability raster data into landscape units, we used ArcGIS to generate a fishing net and calculated the average value of land suitability in each landscape unit to obtain land suitability maps for all landscape units in Zone A, B1, and B2 (Figure 5). Subsequently, we calculated the average between the top 50% and the top 30% of all fishnet sites in each cell separately, in preparation for the subsequent execution of the sensitivity analysis [22].

2.3.3. Spatial Multi-Target Units Allocation

Considering that the proposed AWNP is a fragmented spatial distribution, we use k-means++ for the multi-target unit's allocation. It allows for a greater concentration of similar landscape units at spatial distances [37]. Firstly, the suitability atlas was obtained by applying the average of all fishing net points in each landscape unit. The average between the top 50% and the average between the top 30% were used as input elements. This was followed by a comparison of the results of the selection of the three functional divisions. When landscape units are selected for only one functional partition, they are assigned directly to that partition, while units selected for two or three functional partitions and units not selected by any partition are noted as "conflicting units" [22]. Through the above process, three sets of conflicting analyses of landscape unit allocations based on different mean values were completed (Figure 6).

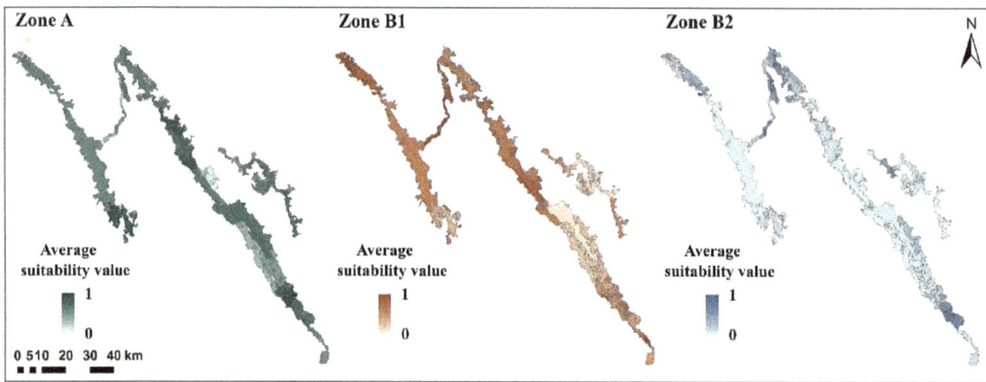

Figure 5. Landscape unit suitability maps for Zones A, B1, and B2.

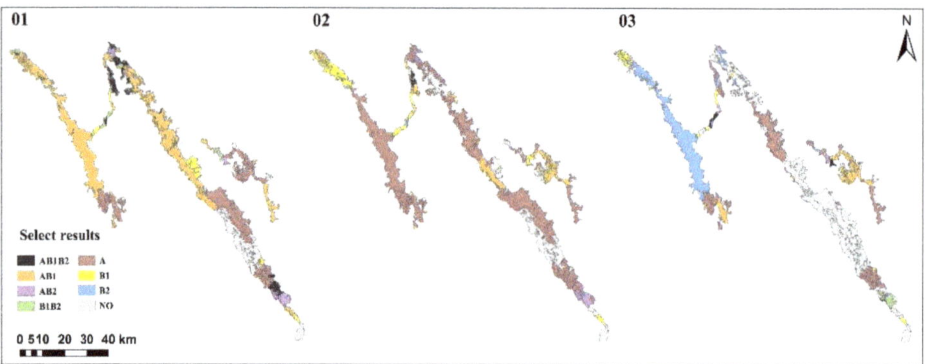

Figure 6. Landscape unit assignment conflict analysis (01, average; 02, average among best 50% cells; 03, average among best 30% cells).

Due to the fact that the land suitability maps, to which the three sub-areas belong, are based on different criteria, it is not possible to directly compare their base suitability values. Thus, by converting the base suitable values for each landscape unit to ordinal suitable values, they are made comparable. Since ecological conservation is the primary objective of the National Park, there is a need to balance the cultural and historical landscape with regional development coordination. Therefore, in the priority ranking, Area A has priority over Areas B1 and B2, and Area B1 has priority over Area B2. The allocation of all conflicting units is accomplished by successively meeting the needs of the higher objectives and then removing the needs of all remaining objectives [38]. After that, we obtained three preliminary zoning schemes (Figure 7).

By comparing the three zoning schemes, the sensitivity analysis of all landscape units was carried out, and it was found that the stable and unstable units, that is, landscape units that were not affected by the polymerization method at the time of partition distribution, participated in the landscape units of the zoning to which the polymerization method changed. As the protection of endangered wildlife is one of the most important objectives of the proposed AWNP, in order to determine the final zoning scheme, the unstable units containing the landscape units selected for Zone A were first overlaid with the potential habitat areas of endangered animals represented by the eastern black crested gibbon, and subsequently, the landscape units containing the intersecting parts of the two were assigned to Zone A, while the other unstable units were assigned to Zones B1 and B2, thus, completing the primary functional zoning. The stable units that each zoning district belongs to, since they are not sensitive to changes in aggregation methods, are considered the areas that best meet the criteria for that zoning district. Therefore, the stable units in Zone A are classified as primary sensitive areas. The areas that were formerly part of the conflict units and allocated to Zone A are classified as secondary sensitive areas. In contrast, the conversion of the original cardinal suitability values of the landscape units to ordinal suitability values has the greatest impact on zone B2, as the suitability map for this zone is generated using vector lines (roads) with vector points (surrounding villages and transport facilities) as standard elements. Within the proposed AWNP, the zone has a suitability value of mostly 0, with higher values concentrated in a few narrow-banded areas. Consequently, the values become smooth when aggregated to the landscape unit to which the overall suitability mean is applied. However, they become prominent when the other two aggregation methods are applied. In view of these characteristics, this study overlays the unstable units assigned to the general control area with the suitability map belonging to zone B2, and within each unstable unit, when the suitability value ≥ 0.5 and accounts for more than 50% of the area of the unit, the unit is classified as an ecological control area (zone B2), and when the standard is not met, it is classified as an ecological activity area (zone B1). By calculating, the remaining unstable unit allocation is completed and the secondary functional division is determined (Figure 8).

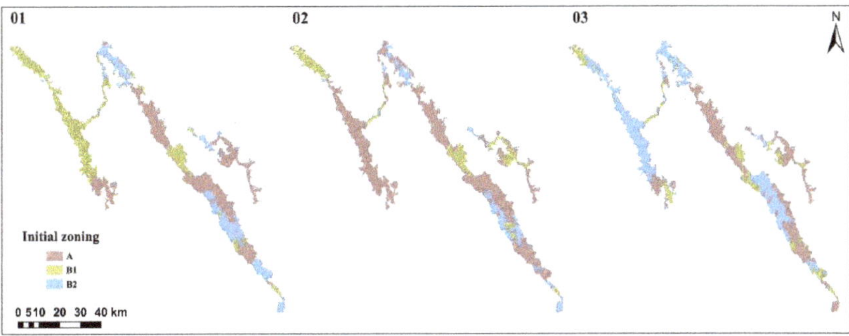

Figure 7. Comparison of preliminary zoning schemes (01, average; 02, average among best 50% cells; 03, average among best 30% cells).

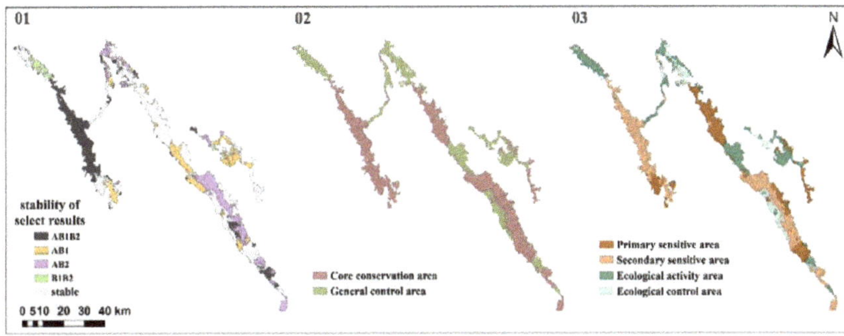

Figure 8. 01, Sensitivity analysis test; 02, First-level functional zoning; 03, Second-level functional zoning.

3. Results and Analysis

3.1. Characterization of Landscape Units

In the Wuliangshan area, located in the western part of the proposed AWNP, the number of landscape units subdivided is small, and the units are large and relatively intact, except in the northwestern part. The northern part of the Wuliang Mountains is a narrower, taller section of the entire Wuliang Mountains and has a more complex ecological environment. The canyon is part of a juxtaposition formed by the recent uplift of the mountains, disintegration of the plateau, and deep river cuts. In the other two areas, the number of landscape units is larger. The fragmented distribution of landscape units is most evident in the northern and southeastern parts of the Ailaoshan area and the central part of the Konglonghe area. The north of Ailao Mountain is narrow and tight, the ground is rugged, and the elevation is very different, and the southeast is a beaded landscape of gorges and basins. In the middle of the Konglonghe area, there are river valleys surrounded by high mountains, and multiple river terraces are visible (Figure 3).

3.2. Distribution Characteristics of Different Suitability Values

The highest land suitability values in Zone A are mainly found in the southeastern part of the Wuliangshan area, the north-central part of the Ailaoshan area and the southeastern part of the area. The first area is 1900–2700 m above sea level and the main vegetation type is broad-leaved evergreen forest. The second and third areas are at an altitude of 2200–3000 m. The vegetation type is mainly broad-leaved evergreen forests with warm coniferous forests, bamboo forests and shrubs. Most of these three areas overlap with endangered animal habitats. The highest land suitability values in Zone B1 are mainly found in the central and northwestern part of the Wuliangshan area, in the larger area in the central part of the Ailaoshan area and in the smaller area scattered in the northern and southeastern part of the Ailaoshan area and in the southeastern part of the Konglonghe area. These areas contain tourist attractions and are adjacent to villages, with some areas interspersed with complex site types. Despite the fact that land suitability values for Zone B2 were generated from linear and point vector data, three areas stand out as being of high suitability. They are located in a narrow area in the northern part of the Wuliangshan area and in the southeastern and northwestern parts of the Ailaoshan area adjacent to the ecological corridor. These areas have higher-level roads that pass through them and are near traffic service facilities (Figure 4).

The land suitability maps for all landscape units in Zones A, B1, and B2 provide a more visual and comprehensive comparison of the distribution of suitability values for different criteria within the proposed AWNP (Figure 5).

3.3. Zoning Assessment

By comparing the landscape unit allocation conflict analysis diagram (Figure 6), Zone B2 are most significantly affected after allocation using different suitability averages, as the suitability values for Zone B2 are generated from linear and point vector data. These values are smoothed out when the overall mean is applied to the landscape cells, but they become prominent when other methods are applied [22]. As a result, the distribution of conflict and non-conflict units has changed more significantly as a whole. When the overall average is applied, the conflict cells cover 71.68% of the area within the proposed AWNP. However, when the top 50% average and top 30% average are applied, the conflict cells cover 35.97% and 55.94% of the area within the proposed AWNP, respectively. The landscape units more consistently allocated to Area A are mainly located in the southeastern part of the Wuliangshan area and the southeastern part of the Ailaoshan area, while only small scattered landscape units are consistently allocated to Zones B1 and B2. The units that have not been selected by any of the sub-regions and are more stable are concentrated in two areas in the southeastern part of the Ailaoshan area. The area to the north is in the range of 1950–3348 m above sea level, making a very big difference in height, including the highest mountain peak in the Ailaoshan area. These two areas are far from water sources, have less overlap with potential habitats for endangered animals, and are far enough away from villages and tourist attractions that no roads cross them.

Comparing the three preliminary zoning scenarios for completing the allocation of conflict units (Figure 7), the portion of the stable allocation to Zone A, in addition to the two areas mentioned previously, is the southern part of the Konglonghe area. Most of the area is a low mountain valley between 537–1660 m above sea level, the lowest elevation in the proposed AWNP, and overall, part of the Dry-hot Valley adjacent to the Red River system. Due to the steep topography and dryness, the area is ecologically fragile with short channels, sparse water flow, and significant seasonal changes. The portion of the stable allocation to Zone B1 and Zone B2, with the exception of the south-central part of the Ailaoshan area, is the higher suitability value to which Zone B1 and Zone B2 each belong. Due to the small number of rural roads and science stations distributed in the south-central part of the Mourned Mountains area, the suitability values for Zone B2 are more prominently ranked when they are aggregated into larger landscape units, and are, therefore, consistently assigned to Zone B2.

Comparing the stable and unstable units, the stable units cover 41.24% of the proposed AWNP. In contrast, the unstable units fluctuating between zones A, B1, and B2 cover 23.29% of the park. These units are mainly located in the central part of the Wuliangshan area, the southeastern part of the Ailao Mountain area and the northwestern part of the Konglonghe area. These areas are where both potential habitats for endangered animals and tourist attractions are located. The vegetation type is mainly evergreen broad-leaved forest, adjacent to villages and roads. A total of 15% of the park consists of unstable units that fluctuate between zones A and B1. They are mainly located in the southeastern part of the Wuliangshan area, the central part of the Ailaoshan area, and the Konglonghe area. Most of these units are adjacent to rivers or water sources, as well as villages and contain a variety of land types. The unstable units fluctuating between zones A and B2 cover 17.24% of the AWNP, and they are mainly located in the south-central and northern parts of the Ailaoshan area, as well as in the northern scattered units of the Konglonghe area. There is a high ecological value to these units, but they are located closer to the road. Finally, only three units fluctuate between zones B1 and B2, which cover 3.23% of the park, the largest of which is located in the northwestern part of the Wulianghshan area, adjacent to the village and containing tourist attractions, but also with roads distributed about the periphery of the unit (Figure 8).

According to the final zoning results (Figure 8), the first-level zoning includes the Core Protection Zone and the General Control Zone. The core conservation area (Zone A) covers an area of 965.83 km^2 (62.83%) and the general control area (Zones B1 and B2) covers an area of 571.50 km^2 (37.17%). The core conservation area is divided into two subzones, namely the primary sensitive area and secondary sensitive area, based on ecological sensitivity and conservation priority. The general control area is divided into an ecological activity area (Zone B1) and an ecological control area (Zone B2). Within the core conservation area, the primary sensitive area covers 364.03 km^2, accounting for 37.71% of the core conservation area, and the secondary sensitive area covers 601.80 km^2, accounting for 62.29% of the core conservation area. Within the general control area, the ecological activity area covers 384.27 km^2, accounting for 67.24% of the general control area, and the ecological control area covers 187.23 km^2, accounting for 32.76% of the general control area.

4. Discussion

4.1. Scientific and Innovative

Taken as a whole, although the decision analysis process in this study is cumbersome, it is composed of a rigorous and orderly set of steps. It is possible to check and supplement the various data layers at any time, as well as to update the settings of the different criteria and targets. In addition, it is possible to carry out comparative analyses using the corresponding indicators and the different weights assigned to them. These findings confirm the assertions of Zhang et al. (2013) that this zoning method offers full flexibility and transparency [39]. From the initial analysis, the basic zoning elements of this study are landscape units with homogeneity, and the landscape units that complete the post-classification treatment are sufficiently large and representative of the overall national park space [30]. In contrast, if only grid cells are used for subsequent zoning studies, not only is the shape single and the area fixed, but also the boundaries of the cells are not easily and accurately identified on the ground, resulting in a final zoning scheme that is not suitable for practical application. In terms of intermediate processes, multi-criteria analysis and multi-objective allocation combined with sensitivity analysis tests show which areas are stable in the allocation process and which units need further study. It is also useful for national park managers to take a more comprehensive look at the impact of different criteria and prioritization on zoning outcomes. This will enable them to decide whether more information and data need to be collected on certain aspects. Geneletti and Duren's (2008) approach to zoning natural parks [22], while not suggesting how to determine the final zoning scheme, has helped us to understand the use of MCDA in conjunction with land suitability assessment and cluster analysis. We build on this approach and further propose how to determine the final zoning scheme and achieve a hierarchical zoning method. According to the final zoning scheme, both primary zoning schemes conform to the national norm of "two zones" [36] and secondary zoning schemes based on primary zoning, which fully take into account the multi-functional nature of national parks [1] and better balance the relationship between conservation and development.

4.2. Limitations and Future Research

From the analysis process and data: Firstly, the delineation of landscape units is based on natural and landscape features [30]. As the criteria and suitability indices are mainly determined by natural and landscape factors, it is reasonable to combine the suitability indices of zones A and B1 into these landscape units, while this may not be accurate and reasonable for zone B2, where the criteria and suitability indices are mainly determined by other factors. Thus, this point still needs further research. Secondly, for a more accurate land suitability assessment, existing data needs to be supplemented and updated, especially more comprehensive and precise data on the distribution of endangered species. Additionally, data such as pedestrian volume in the proposed AWNP area and the carrying capacity of the tourism infrastructure need to be collected and calculated and taken into account [40]. Furthermore, with regard to the allocation of conflict and unstable units, there is a need for further field research and the collection of various types of data from the relevant regions to complete the allocation of these units in a more scientific manner.

From the evaluation indicator system and zoning results: This study uses a mechanical method that emphasizes quantification, but national parks are not only natural spaces. If, as Hidle (2019) argues, only the state's interest in managing and controlling natural parks is considered at the expense of local stakeholders [41]. Then, there is a loss for both the park and the people, which can affect the subsequent balance of conservation and development objectives, especially for such ribbon and dispersed national parks, to the detriment of adjacent communities and visitors coming to experience the resources of the different geographical locations. Thus, as Eugenio et al. (2022) argue, the management of natural spaces cannot be considered a separate issue [42]. Although indicators related to social factors were used in this study, more social factors need to be included in other ways for the study of zoning methods. The system of evaluation indicators is supplemented by social surveys, for example, to increase the applicability of this method and make it more in line with the reality of the social-ecological system. In addition, we should combine the ROS theory with Manning's Managing Outdoor Recreation strategies and practices framework [43], and add secondary zoning, such as management service areas, to improve the ease of use, generalizability, and more comprehensive and rational achievement of zoning control in this study.

From the proposed ecological corridor: Within the ecological corridor that connects the Ailaoshan area with the Wuliangshan area, some areas of the ecological corridor are exposed to human disturbance due to the distribution of settlements and roads. Human disturbance may damage the restoration of potential ecological corridors [3]. Therefore, additional ecological corridors should be created at key locations that impede wildlife migration. This is an issue that cannot be ignored by national park authorities and still requires further research to address.

5. Conclusions

This study proposes a hierarchical zoning approach to explore how national parks can achieve finer and more differentiated zoning control and better balance conservation and development. This multi-use zoning approach takes into account the characteristics and conservation needs of different natural ecosystems, as well as the needs of society for the use of national park resources. We have improved and enhanced the zoning methodology of Geneletti and Duren (2008) [22] to further suggest how to determine the final zoning scheme and achieve hierarchical functional zoning. It provides a theoretical reference and methodological complement to the study of national park zoning methods. In addition, if extended to the area surrounding the park, the land suitability analysis and multi-target land allocation can be used to support the optimal adjustment of national park boundaries.

This method allows national park authorities and other stakeholders to understand the process of grading zones in a clear and transparent way. It is robust and flexible, and it is relatively easy to re-plan functional zoning even if new relevant policies are introduced in the future or the evaluation of the importance of a factor is changed. Sensitivity analysis helps managers, stakeholders, and the public to anticipate how well nature conservation, community development, and construction objectives will be implemented under different zoning scenarios, which avoids confusion in the communication process and helps park authorities to determine whether more data needs to be collected on certain aspects. In fact, according to our finalized first-level zone and second-level zoning scheme, it can provide a reference for the management agencies to develop zoning control measures. For example, in primary and secondary sensitive areas, different degrees of strict protection measures are implemented; in ecological activity areas, routes and designated areas are planned for ecological experience and science education activities; in ecological control areas, ecological restoration and ecological transformation of facilities are implemented [44,45].

Nevertheless, there is room for improvement in this study, as the data currently available is limited and there are some subjective assumptions in the planning of the zoning process. We also need to consider more social factors. In conclusion, the zoning method in this paper is able to combine theory with practice and hopefully contribute to the establishment of a nature reserve system with Chinese characteristics, with national parks as the mainstay.

Author Contributions: Conceptualization, J.L. and X.H.; methodology, J.L. and X.H.; software, X.H.; validation, J.L.; formal analysis, J.L. and X.H.; investigation, J.L., Z.Z. and X.H.; resources, H.G.; data curation, J.L. and X.H.; writing—original draft preparation, J.L. and M.G.; writing—review and editing, Z.Z., X.H. and M.G.; visualization, J.L.; supervision, Z.Z. and X.H.; project administration, H.G. and X.L.; funding acquisition, X.L. All authors have read and agreed to the published version of the manuscript.

Funding: This research was funded by the National Natural Science Foundation of China (Grant No. 31901322) and "A comprehensive study on supporting the construction of the Ailaoshan-Wuliangshan National Park" by the Kunming Branch of the Chinese Academy of Sciences (Grant No. 2019IB018).

Institutional Review Board Statement: Not applicable.

Informed Consent Statement: Not applicable.

Data Availability Statement: Not applicable.

Acknowledgments: The authors gratefully acknowledge the support of the funding. We also sincerely thank four reviewers and the journal editors.

Conflicts of Interest: The authors declare no conflict of interest.

References

1. Ma, B.; Zeng, W.; Xie, Y.; Wang, Z.; Hu, G.; Li, Q.; Cao, R.; Zhuo, Y.; Zhang, T. Boundary delineation and grading functional zoning of Sanjiangyuan National Park based on biodiversity importance evaluations. *Sci. Total Environ.* **2022**, *825*, 154068. [CrossRef]
2. Tang, X.; Jiang, Y.; Yang, R.; Zhao, Z.; Ouyang, Z.; Tian, Y.; Xu, W.; Chen, S.; Ma, S.; Liang, B.; et al. *GB/T 39737–2021; Specification for National Park Establishment*. Standardization Administration of the People's Republic of China: Beijing, China, 11 October 2021.
3. Wang, Y.; Yang, H.; Qi, D.; Songer, M.; Bai, W.; Zhou, C.; Zhang, J.; Huang, Q. Efficacy and management challenges of the zoning designations of China's national parks. *Biol. Conserv.* **2021**, *254*, 108962. [CrossRef]
4. National Park Service. General Management Plan, Yosemite [EB/OL]. Available online: https://www.nps.Gov/yose/learn/management/upload/YOSE_104_D1316B_-id338162.pdf (accessed on 8 October 2019).
5. Zhao, Z.; Peng, L. The Evolution and Development Trend of National Park Zoning Plan. *Landsc. Archit.* **2020**, *27*, 73–80. [CrossRef]
6. Tongariro National Park Management Plan [EB/OL]. Available online: https://www.doc.govt.nz/about-us/our-policies-and-plans/statutory-plans/statutory-planpublications/national-park-management/tongariro-nationalPark-management-plan/ (accessed on 8 October 2019).
7. *The History of the Nature Conservation Administration of the Department of Natural Protection: The 50th Anniversary of the Natural Park*; The Department of the Department of Natural Protection, First Law Publishing: Nieuw-Zuid-Wales, Australie, 1981; p. 73.
8. Liu, C. Analysis on Zoning Control in National Park. *J. Nanjing Tech. Univ.* **2020**, *14–30*, 111.
9. Zhang, Y.; Zhang, Z. Analysis of the construction and operation of national parks and zoning management models in China. *Manag. Adm.* **2022**, 143–148. [CrossRef]
10. Zhu, Z. A Preliminary Study on the Ecological Classification and Conservation Planning of Nanshan National Park. In Proceedings of the 2019 China Urban Planning Annual Conference, Beijing, China, 27–29 June 2019.
11. Shen, H.Q. Review of National Park Visitor Experience Indicators A Case Study of ROS, LAC and VERP. *Landsc. Archit.* **2013**, *5*, 23.
12. Coppedge, B.R.; Engle, D.M.; Fuhlendorf, S.D.; Masters, R.E.; Gregory, M.S. Landscape cover type and pattern dynamics in fragmented southern Great Plains grasslands, USA. *Landsc. Ecol.* **2001**, *16*, 677–690. [CrossRef]
13. Mulyana, A.; Moeliono, M.; Minnigh, P.; Indriatmoko, Y.; Limberg, G. Establishing Special Use Zones in National Parks. 2010. Available online: https://www.cifor.org/publications/pdf_files/infobrief/001-Brief.Pdf (accessed on 6 August 2022).
14. Fu, M.; Tian, J.; Ren, Y.; Li, J.; Liu, W.; Zhu, Y. Functional zoning and space management of Three-River-Source National Park. *J. Geogr. Sci.* **2019**, *29*, 2069–2084. [CrossRef]
15. Lin, Z.; Tu, W.; Hong, Y.; Huang, J.; Liu, J. Optimal design of functional zoning of nature reserves based multiple spatial characteristics. *J. For. Environ.* **2019**, *39*, 248–255. [CrossRef]
16. Wang, B.; Guan, W.B.; Jian-An, W.U. A Method for Assessing Regional Ecological Security Pattern to Conserve Biodiversity—GAP Analysis. *Res. Soil Water Conserv.* **2006**, *13*, 192–196.
17. Li, F.; Xu, M.; Liu, Q.; Wang, Z.; Xu, W. Ecological restoration zoning for a marine protected area: A case study of Haizhouwan National Marine Park, China. *Ocean Coast. Manag.* **2014**, *98*, 158–166. [CrossRef]
18. Yu, J.; Shen, Y.; Song, X.; Chen, X.; Li, S.; Shen, X. Evaluating the effectiveness of functional zones for black muntjac (Muntiacus crinifrons) protection in Qianjiangyuan National Park pilot site. *Biodivers. Sci.* **2019**, *27*, 5–12.
19. Ma, B.; Zeng, W.; Xie, Y. The functional zoning method for natural parks: A case study of Huangshan Scenic Area. *Acta Ecol. Sin.* **2019**, *39*, 8286–8298.
20. Tian, Q.; Huang, F.; Wang, K.; Wan, D.; Li, J.; Wang, H. Land ecological suitability evaluation of nature reserve: With Wanfo Mountain Nature Reserve in Hunan Province as an example. *J. Zhejiang Univ.* **2020**, *46*, 201–208.
21. Li, J.; Liu, X. Research of the Nature Reserve Zonation Based on the Least-cost Distance Model. *J. Nat. Res.* **2006**, *21*, 217–224.
22. Geneletti, D.; Duren, I.V. Protected area zoning for conservation and use: A combination of spatial multicriteria and multiobjective evaluation. *Landsc. Urban Plan.* **2008**, *85*, 97–110. [CrossRef]
23. Nandy', S.; Singh, C.; Das, K.K.; Kingma, N.C.; Kushwaha, S.P.S. Environmental vulnerability assessment of eco-development zone of Great Himalayan National Park, Himachal Pradesh, India. *Ecol. Indic.* **2015**, *57*, 182–195. [CrossRef]
24. Wang, Z.; Li, Y.; Hua, S.; Zhou, J.; Liu, W.; Liao, S. Functional Zoning of Potatso National Park by Ecological Protection Weighting. *J. Nanjing For. Univ.* **2021**, *45*, 225–231.
25. Zhang, J.; Wei, W.; Cheng, Y.; Zhao, B. Study on site selection of small and medium-sized city parks based on GIS suitability evaluation. *J. Nanjing For. Univ.* **2020**, *44*, 171–178.
26. Beinat, E.; Nijkamp, P. Land-use management and the path toward sustainability. In *Multicriteria Analysis for Land-Use Management*; Beinat, E., Nijkamp, P., Eds.; Kluwer Academic Publishers: Dordrecht, The Netherlands, 1998; pp. 1–13.
27. Esmail, B.A.; Geneletti, D. Multi-criteria decision analysis for nature conservation: A review of 20years of applications. *Methods Ecol. Evolut.* **2018**, *9*, 42–53. [CrossRef]
28. Randal, G.; Joan, E.; Rodolphe, D.; Brian, G.E. An approach to GIS-based multiple criteria decision analysis that integrates exploration and evaluation phases: Case study in a forest-dominated landscape. *For. Ecol. Manag.* **2010**, *260*, 2102–2114.
29. Bereket, T.H.; Fang, Q. Zoning for a multiple-use marine protected area using spatial multi-criteria analysis: The case of the Sheik Seid Marine National Park in Eritrea. *Mar. Policy* **2016**, *63*, 135–143.

30. Liu, G.H.; Liu, J.W.; Zhu, Y.H. Land-unit Based Flow Network Model in Small Catchment with GIS Support. *Prog. Geogr.* **2011**, *21*, 139–145.

31. Niekerk, A.V. A comparison of land unit delineation techniques for land evaluation in the Western Cape, South Africa. *Land Use Policy* **2010**, *27*, 937–945. [CrossRef]

32. Geneletti, D. A GIS-based decision support system to identify nature conservation priorities in an alpine valley. *Land Use Policy* **2004**, *21*, 149–160. [CrossRef]

33. Rincón, V.; Velázquez, J.; Gutiérrez, J.; Sánchez, B.; Hernando, A.; García-Abril, A.; Santamaría, T.; Sánchez-Mata, D. Evaluating European Conservation Areas and Proposal of New Zones of Conservation under the Habitats Directive. Application to Spanish Territories. *Sustainability* **2019**, *11*, 398. [CrossRef]

34. Boulad, N.; Hamidan, N. The use of a GIS-based multi-criteria evaluation technique for the development of a zoning plan for a seasonally variable Ramsar wetland site in Syria: Sabkhat Al-Jabboul. *Wetl. Ecol. Manag.* **2018**, *26*, 253–264. [CrossRef]

35. Cao, M.; Ma, Y.; Sun, Z.; Hu, H.; Zhu, H.; Sheng, C.; Li, Z.; Guo, H.; An, L.; Si, Z.; et al. *LY/T 2242-2014*; Technical Regulation for Biodiversity Impact Assessment of Construction Project in Nature Reserve. The State Forestry Administration of the People's Republic of China: Beijing, China, 21 August 2014.

36. Tang, X.; Jiang, Y.; Sun, H.; Chen, Y.; Peng, R.; Wu, C.; Cui, G.; Liang, B.; Liu, Z.; Chen, J.; et al. *GB/T 39736-2020*; Technical Regulations for the National Park Master Plan. Standardization Administration of the People's Republic of China: Beijing, China, 22 December 2020.

37. Arthur, D.; Vassilvitskii, S. K-Means++: The Advantages of Careful Seeding. In Proceedings of the Eighteenth Annual ACM-SIAM Symposium on Discrete Algorithms, SODA 2007, New Orleans, LA, USA, 7–9 January 2007.

38. Eastman, J.R.; Jiang, H.; Toledano, J. *Multi-Criteria and Multi-Objective Decision Making for Land Allocation Using GIS*; Springer: Dordrecht, The Netherlands, 1998.

39. Zhang, Z.; Sherman, R.; Yang, Z.; Wu, R.; Wang, W.; Yin, M.; Yang, G.; Ou, X. Integrating a participatory process with a GIS-based multi-criteria decision analysis for protected area zoning in China. *J. Nat. Conserv.* **2013**, *21*, 225–240. [CrossRef]

40. Zhou, K.; Liu, H.; Fan, J.; Yu., H. Environmental stress intensity of human activities and its spatial effects in the Qinghai-Tibet Plateau national park cluster: A case study in Sanjiangyuan region. *Acta Ecol. Sin.* **2021**, *41*, 268–279.

41. Hidle, K. How National Parks Change a Rural Municipality's Development Strategies—The Skjåk Case, Norway. *J. Rural Stud.* **2019**, *72*, 174–185. [CrossRef]

42. Eugenio, C.; Marilena, L.; Francisco, N. Protected Natural Spaces, Agrarian Specialization and the Survival of Rural Territories: The Cases of Sierra Nevada (Spain) and Alta Murgia (Italy). *Land* **2022**, *11*, 1166.

43. Moyle Brent, D. Managing outdoor recreation: Case studies in the national parks. *Ann. Tour. Res. A Soc. Sci. J.* **2013**, *41*, 253–254.

44. Zhong, L. Eco-friendly facilities: An important guarantee for the green development of national parks. *Tour. Trib.* **2018**, *33*, 8–9.

45. Jianghui, G.E.; Dexia, Z.A.N.G. Research on the Construction of Chinese National Park System Based on Sustainable Development Theory. *J. Landsc. Res.* **2015**, *7*, 13–16. [CrossRef]

land

Article

Interactive Effects on Habitat Quality Using InVEST and GeoDetector Models in Wenzhou, China

Xue Zhang [1], Lingyun Liao [1,*], Zhengduo Xu [1], Jiayu Zhang [1], Mengwei Chi [1], Siren Lan [1,2] and Qiaochun Gan [1]

[1] College of Arts College of Landscape Architecture, Fujian Agriculture and Forestry University, Fuzhou 350002, China; 1191775049@fafu.edu.cn (X.Z.); 3201726082@fafu.edu.cn (Z.X.); 1191775047@fafu.edu.cn (J.Z.); moviechi@fafu.edu.cn (M.C.); lkzx@fafu.edu.cn (S.L.); 1211775013@fafu.edu.cn (Q.G.)
[2] Engineering Research Center for Forest Park of National Forestry and Grassland Administration, Fuzhou 350002, China
* Correspondence: liaolingyun@fafu.edu.cn

Abstract: Global urbanisation has accelerated in recent years, especially in rapidly growing coastal cities, and the destruction of habitat and natural resources has intensified. Although much attention has been paid to the study of habitat quality, there are still gaps in our understanding of the factors that influence it and their interactions. In this study, the InVEST habitat quality evaluation model and the GeoDetector model were used to construct a framework for analysing the dynamic changes in habitat quality and their influencing factors from 1992 to 2015. Wenzhou City, Zhejiang Province, China, was selected as the study area. The new framework extends studies on habitat quality change to annual analysis and reduces the lag between the actual change and the mapping time. The interactions between natural and anthropogenic factors are explored, and the effects of different types of land use conversion on habitat quality are further discussed. The results show that: (1) During the study period, cultivated and construction land areas in Wenzhou City increased the most, and forest land area decreased the most. (2) Habitat quality in Wenzhou City was generally good during the study period, but it showed a declining trend from year to year, and the distribution of habitat quality decreased from west to east. (3) The interactions between land use change and annual precipitation change and those between land use change and population density change have the most significant impact on habitat quality. The conversion of forest land to cultivated land, conversion of water area to cultivated land, and conversion of forest land to building land have the greatest impact on habitat quality. The results of the study can provide recommendations for ecological restoration, optimal integration of protected areas, and provide a reference for the healthy and sustainable development of coastal regions.

Keywords: habitat degradation; LUCC; driving force; GeoDetector model; coastal city

Citation: Zhang, X.; Liao, L.; Xu, Z.; Zhang, J.; Chi, M.; Lan, S.; Gan, Q. Interactive Effects on Habitat Quality Using InVEST and GeoDetector Models in Wenzhou, China. *Land* **2022**, *11*, 630. https://doi.org/10.3390/land11050630

Academic Editor: Le Yu

Received: 8 March 2022
Accepted: 19 April 2022
Published: 24 April 2022

1. Introduction

Coastal cities represent intersections between land and sea, and are characterised by their special geographical location, differential natural resources, and good economic foundations [1]. Land reclamation and urbanisation can have a significant impact on coastal ecological health in terms of issues such as habitats for plant and animal communities, and changes in soil properties [2]. Habitat quality refers to the provision of a suitable living environment for individuals or populations in an ecosystem [3] and reflects the biodiversity to a certain extent [4]. High quality of habitat is the basis of ecosystem services, providing humans with significant economic benefits and cultural values [5]. However, according to the Global Assessment of Biodiversity and Ecosystem Services, one million of the eight million species found globally are now threatened by extinction due to human activities [6]. Comparing with inland cities, the habitat quality of coastal cities faced more pressure from urban development and climate change [7,8] Assessment of the spatiotemporal evolution of habitat quality and exploration of the factors affecting its change are critical to the construction of a regional ecological security framework [9], and to spatial and layout planning in coastal cities.

How is habitat quality measured? Different methods are developed for the assessment of habitat quality at different scales [10]. At a small scale, the distribution and abundance of major species in the region are determined based on field survey data [11]. For example, Loffler and Fartmann [12] measured regional habitat quality based on vegetation structure, analysed the effects of grassland landscape and habitat quality on Orthoptera insects, and proposed that improving regional habitat quality is critical to strengthening the conservation and management of Orthoptera insects. However, it is difficult to study all types of provinces and cities with large regional areas using this method. To extract environmental factors at a large and medium scale, the InVEST model [13] and the remote sensing ecological index (RSEI) [14] are usually used to rapidly evaluate habitat quality. Many previous studies have used the InVEST model to explore the correlation between habitat quality and land use changes in different regions [15]. However, most current studies discuss habitat quality changes over different time periods at intervals of five or ten years, but annual habitat quality changes have not been studied.

Although much attention has been paid to the study of habitat quality, there are still gaps in our understanding of the factors that influence it and their interactions. The determination of factors affecting changes in habitat quality provides the basis for the protection of the habitat biodiversity [16], and it is important to solve certain key problems for the promotion of habitat protection [17]. In recent studies, the influencing factors have been classified into two groups [18]: natural factors and anthropogenic factors. Natural factors, such as elevation, do not change significantly in the short term; however, they indirectly limit human activities. Differences in topography, precipitation, and vegetation distribution affect the spatial component of various habitats [19]. Anthropogenic factors, such as population density, gross domestic product (GDP), and land use changes, are the most direct manifestations of human activities [20]. The results of recent studies show that land use changes are the main factor controlling changes in habitat quality [21]. However, it remains unclear which types of land use changes have the most significant effects on habitat quality. In real life, changes in habitat quality are due to interactions among various influencing factors, which form a relatively complex network. The relative importance of each influencing factor changes depending on the time, policy, and region.

Ordinary least squares regression and geographically weighted regression analysis are the most important methods used to study the correlations between habitat quality and influencing factors [22]. These two methods can be used to analyse the spatial heterogeneity of similar geographical attributes and local clustering [23]. These effects can be explained by the position, relation, and weight of spatial distance of influencing factors based on a linear model [24]. However, these methods only involve linear interpolation and have certain limitations, although they all are based on linear regression models [25]. In most studies, county-level administrative units or different grid units are used, but the effects of different influencing factors at different scales are ignored [26]. The nonlinear GeoDetector model, which is a group of statistical methods that can be used to determine geospatial heterogeneity and its driving forces [27], was used in this study to analyse the correlations between changes in habitat quality and influencing factors at different scales and in different regional zones. The advantage of the GeoDetector model is that it can analyse both numerical and qualitative data. It can be used to determine the interactions among different factors [28] and to quantitatively analyse the interactions between two factors [29].

Wenzhou City located in the east of Zhejiang Province, China, is a prefecture-level city that is designated as one of the three major representative cities after the reform and opening up of China. It has developed rapidly and is often regarded as a typical example of the privatisation and corporatisation of Chinese cities in the context of market-oriented reforms and open experiments [30]. However, rapid economic development often comes at the cost of damaging the environment, putting enormous pressure on the ecological environment due to urban expansion, changes in land use, and uneven spatial allocation of resources and environment. This study chose Wenzhou City as the research area and analysed annual changes in its habitat quality and the factors influencing it, to provide a reference for ecological environmental protection and land use in rapidly developing Chinese coastal cities, and to promote the pace of urban transformation to encourage the construction of the ecological civilisation of China. This study focuses on the following questions:

1. What is the overall habitat quality situation in the urban area of Wenzhou City?
2. How has habitat quality in Wenzhou changed from 1992 to 2015, year by year?
3. What are the main factors contributing to changes in habitat quality?
4. How do natural and anthropogenic factors interact with each other?

The aim of this study is to construct a framework for analysing the dynamic changes in habitat quality and their influencing factors and explore interactive effects between natural factors and anthropogenic factors on habitat quality. Based on the InVEST habitat quality evaluation model, the annual assessment of habitat quality is conducted. In addition, the GeoDetector model is applied in the study to explore the interaction between the factors that influence the change of habitat quality. The continuous evaluation of habitat quality and the effects of various factors may provide a scientific basis for the ecology evaluation in territorial spatial planning, optimal integration of protected areas, and suggest the direction for future planning measures and development.

2. Methodology

2.1. Study Area

Wenzhou City is on the south-eastern coast of China. It is surrounded by mountains on three sides and faces the sea on one side. The territory is rich in natural resources and known as 'seven mountains, two waters, and one field'. The topography of Wenzhou City is trapezoidal from the southwest to the northeast and the landforms can be divided into low mountainous areas in the west, low mountainous and hilly basins in the centre, plain and tidal flat areas in the east, and coastal island areas. The region is rich in forest resources, with forest coverage of 60.03% [31]. The forest ecosystem is the most extensive ecosystem type found in Wenzhou City. Wenzhou City has a dense river network and a developed surface water system, including the Oujiang, Feiyun, and Ao rivers [32]. According to the Wenzhou Bureau of Statistics [33], the land and sea areas of Wenzhou City are 12,083 and 8649 km^2, respectively. Wenzhou City has 12 counties and districts under its jurisdiction, with a permanent population of 9.3 million people. In recent years, Wenzhou has seen rapid urbanisation, with the city's gross domestic product (GDP) standing at 1.322 billion yuan in 1978; surpassing 100 billion yuan in 2002; and reaching 687.09 billion yuan in 2020, with rapid economic growth.

2.2. Methods

2.2.1. Description of the Habitat Quality Model

The habitat quality model was selected as the submodule of the InVEST model. Habitat quality maps were generated by combining land use data and information about biodiversity threat factors to determine the regional environmental quality [34]. Habitat quality scores were calculated as follows:

$$Q_{ab} = M_b\left(1 - \frac{G_{ab}^z}{G_{ab}^z K^z}\right) \tag{1}$$

where Q_a represents the habitat quality score of habitat pixel a in habitat type b; M represents the habitat suitability of habitat type b; and G and K are the default model parameters, which are 2.5 and 0.5, respectively.

The degree of the degradation of the habitat at this location is:

$$G_{ac} = \sum_{e=1}^{E} \sum_{f=1}^{F} \left(\frac{W_e}{\sum_{e=1}^{E} w_e} \right) e_c i_{eac} \beta_a S_{be} \tag{2}$$

where G_{ac} denotes the habitat degradation degree of habitat pixel a in habitat type c; e is the threat source for the habitat; and f is the grid of the threat source e.

The stress effect of $e_{(ef)}$ in grid f on the habitat in grid a is i_{eac}:

$$i_{eac} = 1 - \frac{G_{ac}}{G_{max}} \text{(linear decay)}$$
$$i_{eac} = \exp\left(-\frac{2.99}{G_{max}} G_{ac}\right) \text{(exponential decay)} \tag{3}$$

where G_{ac} is the distance between pixel a of the habitat and pixel c of the threat source; W and G_{max} are the weight and maximum influence range of the threat source E, respectively; β represents the effects of local conservation policies; and S represents the relative sensitivity of each habitat to different threat sources.

The land-use was derived from MODIS Land Cover/Dynamics (MCD12) data [35]. The original land-use and cover classification includes six categories: agricultural land, forest land, grassland, wetland, construction land, and other land uses, as well as 22 subcategories from the original spatial dataset [36]. Based on the actual situation in Wenzhou City, six categories were used in this study: cultivated land, forest land, grassland, aquatic environments, construction land, and bare land. Cultivated land, construction land, and bare land were considered to be the primary habitat threat factors, because all three are disturbed by human activities, the natural conditions of bare land, bare rock, and stony land in Wenzhou City are poor, and bare land causes dust pollution in sunny days and erosion in rainy days.

Information about threat sources and habitat sensitivity parameters were obtained from Haiyan [37] and related studies and were used as input for the habitat quality module of the InVest model (Tables S1 and S2).

2.2.2. GeoDetector Model

The GeoDetector model is composed of four detectors: factor detector, risk detector, interaction detector, and ecological detector. In this study, the interaction and factor detectors were used to explore the factors affecting the spatial differentiation of habitat quality.

Factor detector

The calculation is as follows [28]:

$$q = 1 - \frac{\sum\limits_{a=1}^{A} N_a \sigma_a^2}{n\sigma^2} \tag{4}$$

where q is the degree to which an impact factor explains the distribution of the habitat quality; A is the stratification of the impact factors; N_a and n are the habitat quality of layer a and the whole region, respectively; and σ^2_h and σ^2 are the variance of layer a and the study area. If the explanatory power of the influencing factor q is within the range of (0–1), the closer q is to 1, and the stronger the influence of this factor on the spatial differentiation of the habitat quality. Otherwise, the influence is weaker.

Interaction detector

Based on the comparison of the q values of single factors, the sum of two single factor q values, and the q values of the double-factor interaction, the interaction detector can be divided into five classes based on the highs and lows of these three values. If $q(Y_1 \cap Y_2) < \text{Min}[q(Y_1), q(Y_2)]$, the interaction is decreasing nonlinearly. If $\text{Min}[q(Y_1), q(Y_2)] < q(Y_1 \cap Y_2) < \text{Max}[q(Y_1), q(Y_2)]$, the interaction decreases linearly by a single factor. If $q(Y_1 \cap Y_2) > Max[q(Y_1), q(Y_2)]$, the interaction is a double-factor enhancement. If $q(Y_1 \cap Y_2) = q(Y_1) + q(Y_2)$, the interaction is independent. If $q(Y_1 \cap Y_2) > q(Y_1) + q(Y_2)$, the interaction is a nonlinear enhancement.

Selection of the best GeoDetector parameters

As a spatial statistical method, the GeoDetector model analyses the correlations among different factors based on grid data. The sizes of different aggregation areas or spatial distribution, i.e., the scale and partition effects, must be considered to produce different results [23]. Firstly, the effect of scale on the GeoDetector model results was analysed to determine the best parameters. Based on previous studies, the data resolution, and research area, five grid scales were selected: 1000×1000 pixel, 2000×2000 pixel, 3000×3000 pixel, 4000×4000 pixel, and 5000×5000 pixel. Secondly, different types of independent variables were used to determine the effect of partitioning on the GeoDetector model. Different classification methods are generally used for the discretisation of numerical independent variables. In this study, the natural breakpoint and manual classification methods were adapted to test and select the best model.

Selection of the spatial distribution of influencing factors

The factors affecting changes in habitat quality are diverse and complex. Based on previous studies, the natural environment is in innate existence, and its transformation is caused by human interference. Therefore, the effects of changes in habitat quality can be divided into natural factors and anthropogenic factors. Based on the literature and on existing data for the study area, six influencing factors were selected for analysis in this study (Table 1).

In this study, the indexes that represent natural factors include the elevation, annual mean precipitation, and normalised vegetation index. Elevation influences the composition of the vegetation structure and indirectly affects the habitat selection of animals [38], and thus the habitat quality. Annual average precipitation is an important factor that affects the growth and reproduction of organisms as well as the food resources of animals and plants [39]. The normalised difference vegetation index (NDVI) can reflect the growth status and coverage degree of vegetation. It is the most direct method that can be used to model the advantages and disadvantages of regional ecological environments [40]. Anthropogenic factors include urbanisation, population density, and land use/cover change (LUCC). Night-time light data can be used to measure the overall urbanisation intensity [41]. Rapid urbanisation leads to a gradual imbalance in ecosystem functions [42]. The population density reflects the distribution of the population. Where habitat quality is good, there tends to be less human disturbance. LUCC can reflect the human economic activities and is an embodiment of human activities [43]. The digital elevation model (DEM), precipitation, and population density data were obtained from the Resources and Environmental Science Data Center, Chinese Academy of Sciences (Beijing, China; http://www.resdc.cn/, (accessed on 5 March 2021)). The NDVI data originated from the Geospatial Data Cloud (http://www.gscloud.cn/#page6, (accessed on 5 March 2021)). Night-time light data were derived from the Chinese Long-term Series Annual Artificial Night-time Light dataset (1984–2020) [44].

Table 1. Summary of factors affecting habitat quality based on the literature and factors used for the GeoDetector model.

Category	Factors Affecting Habitat Quality, Based on the Literature	Factors Used for the Geodetector Model
Natural factors	Topographic factors [5]	Elevation
	Climate [45]	Annual average precipitation changes
	Vegetation coverage [46]	NDVI changes from 1992 to 2015
Anthropogenic factors	Urbanisation [42]	Changes of night-time light data from 1992 to 2015
	Population [41]	Population changes from 1992 to 2015
	LUCC [47,48]	Land use transformation from 1992 to 2015

2.2.3. Improvement of the Logistic Multiple Regression Model for the Analysis of Influencing Factors

SPSS Statistics 26.0 (IBM Corp, Armonk, NY, USA) statistical software was used to re-analyse the same data, using logistic regression model analysis. Based on sample data, the model generates regression coefficients for each variable. The correlations between dependent and independent variables in the model can be determined and discussed based on these coefficients. The subordinate variables are 0 and 1, where 0 indicates no change in the habitat quality [19]. If q is the occurrence probability of the event and the value range is 0–1, 1−q is the probability that the event does not occur, which can be calculated using a logistic function:

$$Q = \frac{\exp(\beta_0 + \beta_1 N_1 + \beta_2 N_2 + \ldots + \beta_n N_n)}{1 + \exp(\beta_0 + \beta_1 N_1 + \beta_2 N_2 + \ldots + \beta_n N_n)} \tag{5}$$

where $N_1, N_2, \ldots, N_n$ are factors that affect habitat quality change, such as the elevation, precipitation, and land use change. The constants in the β_0 equation, that is, β_1, β_2, and so forth, are partial regression coefficients of the β_n logistic regression that represent the degree of the influence of the independent variables.

2.2.4. Principal Component Analysis

Principal component analysis (PCA) is a statistical analysis method that converts multiple indicators into a small number of comprehensive indicators [49]. In this study, various land use changes were used as indicators for the PCA, to obtain the characteristic roots of the matrix and corresponding variance contribution rate. PCA was undertaken using SPSS Statistics 26.0 statistical software. Subsequently, the score of each component factor was calculated using the linear expression and a comprehensive score was obtained using the variance contribution rate as the weight [50]. Finally, the comprehensive score of each principal component was converted to a percentage system and used as the evaluation score, which more accurately reflects the effects of various land use changes on habitat quality.

3. Results

3.1. Spatiotemporal Patterns of Land Use in Wenzhou City from 1995 to 2015

From 1992 to 2015, the proportions of forest land and cultivated land in Wenzhou City remained high. Cultivated land and construction land areas increased slowly, whereas forest land, grassland, bare land, and aquatic environments decreased gradually. From 1992 to 2015, the cultivated land area increased the most (77,652.98 ha). It increased rapidly from 1992 to 2000 and remained at 57,000 ha in 2015. The construction land area increased by 45,851.43 ha; it expanded continuously and, after 2000, the construction area increased more rapidly (Figure 1).

The forest land area decreased to 112,313.87 ha. The forest land area decreased from 1992 to 2000 and has remained at ~560,000 ha since then. The grassland area increased to a maximum of 35,137.44 ha in 2001, then decreased to 30,000 ha and remained at a relatively stable level. The bare land area slightly decreased, whereas the aquatic environments remained unchanged.

3.2. Spatiotemporal Changes in the Habitat Quality

The value range of habitat quality is (0–1). The larger the value is, the higher the habitat suitability [41]. To explain the changes in habitat quality in Wenzhou City, the results of the calculations of the 24 phases of the habitat quality index were divided into five ranges: 0–0.2, 0.2–0.4, 0.4–0.6, 0.6–0.8, and 0.8–1. The quality of the habitat was then graded according to the following five levels: extremely poor, poor, medium, good, and excellent. Based on these results, it was found that there is no good habitat in Wenzhou City.

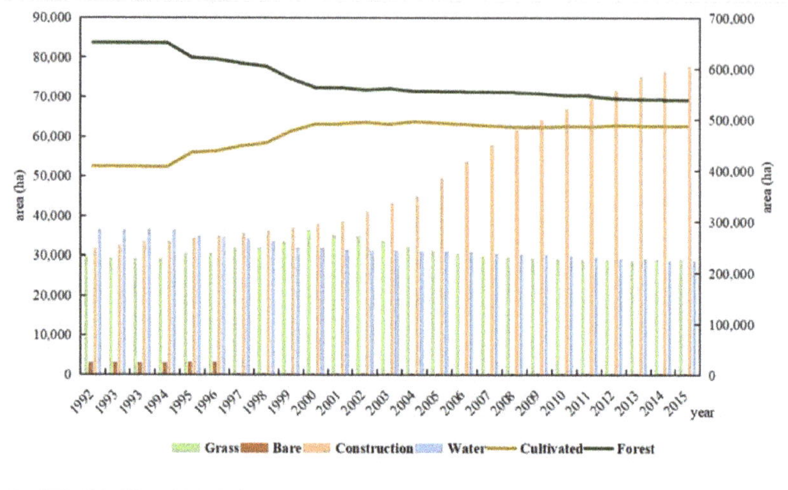

Figure 1. Changes of the forest land and cultivated land areas in Wenzhou City from 1992 to 2015 (left ordinate: grassland, bare land, construction land, and aquatic environments; right ordinate: forest land and cultivated land).

The habitat quality in Wenzhou City was relatively good from 1992 to 2015. The area of excellent habitat quality accounted for the largest proportion of the whole city. However, the area of excellent habitat decreased from 1992 to 2003 and then stabilised at ~610,000 ha. The area of medium habitat quality increased from 406,615.78 ha in 1992 to 491,415.81 ha in 2005 and then remained at the same level. The area of extremely poor or poor habitat quality was smaller. The area of extremely poor habitat quality increased from 37,334.59 ha in 1992 to 83,121.74 ha in 2015. During 1996, the area of poor habitat quality sharply decreased by 2894.69 ha and reached 197.83 ha in 1997. Subsequently, the area of poor habitat quality slightly fluctuated, but did not exceed 307.99 ha (Figure 2).

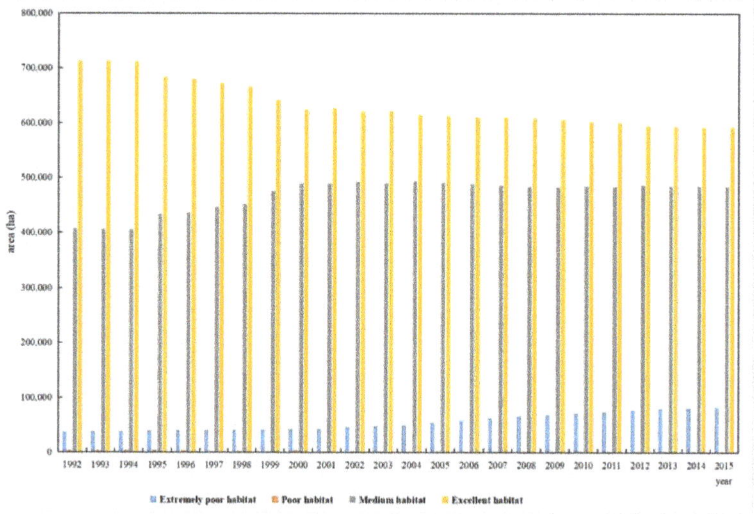

Figure 2. Statistical chart of habitats of different qualities in Wenzhou City from 1992 to 2015.

The spatial distribution of habitat quality in Wenzhou City considerably differs. Overall, habitat quality decreases from the west to the east; that is, the average habitat quality in the inland region is better than in coastal counties and cities (Figure 3). Excellent habitat is mainly located in Yongjia, Taishun, and Wencheng counties, followed by Longwan District and Dongtou District. The prime quality habitats decreased in all counties and cities. The habitats in Cangnan County, Ruian City, Yueqing City, and Pingyang County decreased significantly. Yongjia County had a medium habitat quality. Ruian City, Yueqing City, Cangnan County, Pingyang County, Taishun County, and Wencheng County have experienced rapid growth for 24 years. Ruian City rapidly grew from 1994 to 2002 and then stabilised. In 2005, it slowly declined to the 2015 level of Cangnan County. Lucheng District, Longwan District, Longgang City, and Dongtou District have fewer areas of medium habitat quality. The area of poor habitat quality decreased in all counties. It declined rapidly from 1996 to 1997, especially in the Longwan District, which had the poorest habitat quality. The area of extremely poor habitat quality increased in all counties and cities. The rate of increase started to accelerate after 2000, and this trend was noticed particularly in Yueqing City and Ruian City. The area of extremely poor habitat quality in the Ouhai District rapidly increased from 2000 to 2010, ranking third after the Lucheng District. Taishun County and Wencheng County had the smallest areas and slowest growth rates.

Figure 3. Changes in habitat quality in Wenzhou City from 1992 to 2015.

3.3. Relative Level of Habitat Degradation

The degree of habitat degradation ranges from zero to one. The larger the value is, the higher the habitat suitability [41]. To explain the changes in the habitat degradation in Wenzhou City, habitat degradation has been reclassified in GIS into five stages: little degradation (0–0.2), mild degradation (0.2–0.4), moderate degradation (0.4–0.6), high degradation (0.6–0.8), and extremely high degradation (0.8–1).

The land area that experienced habitat degradation in Wenzhou City was relatively low from 1992 to 2015. The areas of little degradation accounted for the largest proportion of the whole city. These decreased over the 24 years, whereas areas with the other four types of degradation increased. Areas of moderate degradation grew the most rapidly. In 2015, these areas accounted for approximately ten times their area in 1992. There were no areas with a relatively high degree of degradation from 1992 to 1995, but they began to appear after this time. After 2004, the growth rate of areas with relatively high degree of degradation accelerated and reached ~11,241.60 hectares in 2015. There were no areas of extremely high degradation from 1992 to 2006. These began to appear after 2007 and reached 902.05 hectares in 2015 (Figure 4).

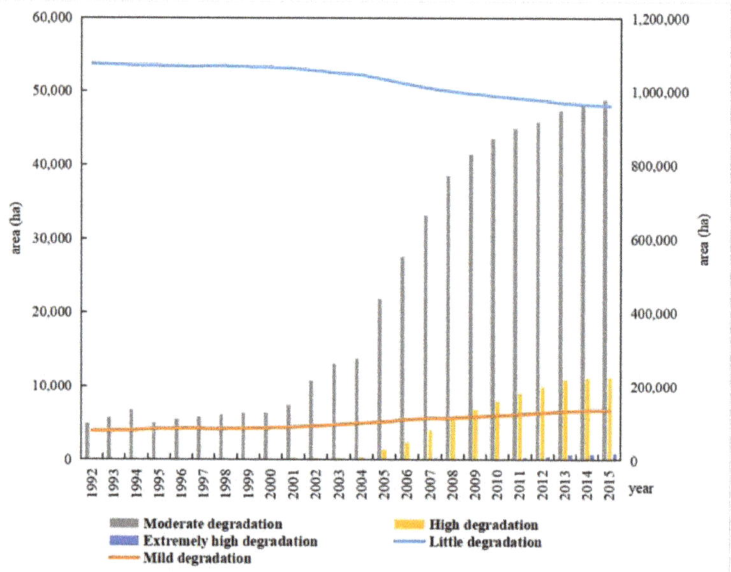

Figure 4. Statistical chart of the changes in the areas of habitat degradation in each county of Wenzhou City from 1992 to 2015.

Coastal areas were the main areas that experienced habitat degradation in Wenzhou City, whereas the degree of degradation of inland areas was relatively low. Areas of little degradation were concentrated in Yongjia County, Taishun County, and Wencheng County; there was a lower proportion of such areas in in Longgang City and the Longwan, Dongtou, Lucheng, and Ouhai districts. The areas of mild degradation increased, except in the Ouhai and Longwan districts. The area of mild degradation in Leqing City remained stable from 1992 to 2000 and then sharply increased. The area of mild degradation was the largest in 2015. In 1992, Pingyang County experienced the least degradation. Over the past 24 years, its rapid growth surpassed that of many other counties and cities; it ranks second. The areas of moderate degradation in each county and city increased by different amounts over the past 24 years. The Ouhai District showed slow growth from 1992 to 2000 and rapid increase from 2000 to 2007. The area of moderate degradation then slightly decreased to 10106.41 ha. After 2004, the growth rate of Ruian City accelerated. In 2012, it surpassed that of Ouhai District. The county accounted for the largest area of moderate degradation in Wenzhou City. After 2000, the area of relatively high degradation was zero. Subsequently, each county and city started to grow. The Ouhai District had the fastest growth rate. It accounted for an area of ~5254.77 hectares in 2015. Before 2006, there were no areas of extremely high degradation in Wenzhou City, but these increased by different degrees in the Lucheng, Longwan, and Ouhai districts and in Yongjia County. The Ouhai District had a rapid rate of growth; the area of extremely high degradation area was ~354.01 hectares in 2015. Before 2011, the area of extremely high degradation in the Lucheng District was ~157.34 hectares. Subsequently, the area exceeded that of the Ouhai District and accounted for the largest area of high degradation area in Wenzhou City (~445.16 hectares; Figure 5).

Figure 5. Changes in areas with moderate habitat degradation in each county in Wenzhou City from 1992 to 2015.

3.4. Analysis of Factors That Affected Changes in Habitat Quality

The natural breakpoint method was used to match the natural and anthropogenic factors with habitat quality in Wenzhou City. The results were imported into the GeoDetector model to calculate the Q value of each factor, to illustrate its influence on the spatial distribution of habitat quality. The result shows that the Q value of each influencing factor can be ranked from large to small as follows: land use change (0.696); elevation (0.211); night-time light change (0.144); NDVI change (0.120); precipitation change (0.097); and population density change (0.068). Therefore, land use change is the key factor that affected the changes in habitat quality in Wenzhou City. As the areas of cultivated land and construction land in Wenzhou City has risen, the intensity of land use has significantly increased. This increased the possibility of changes in land use type from habitat to non-habitat. The effect of elevation is more notable than that of other natural factors. Wenzhou City is surrounded by mountains on three sides and the sea on one side. Landform can be divided into low mountainous areas in the west, low mountainous and hilly basins in the centre, plain and tidal flat areas in the east, and coastal island areas. Plains and hilly areas are conducive to the development of human activities, whereas mountains have a higher elevation and rugged terrain that is not conducive to human activities. The change in night-time light ranks third. In recent years, the rate of urbanisation has accelerated, which significantly contributed to the changes in habitat quality. The increase in the regional economic GDP of Wenzhou City attracts migrants and promotes urbanisation. The transportation networks and infrastructure were significantly improved, but this increased the possibility of a decline in habitat quality.

Analysis of the factor interactions (Figure 6) shows that when compared with single factor interaction, the results of the interactions between each impact factor and other factors are enhanced by varying degrees. Land use ∩ precipitation change (0.733) and population density change (0.730) have the largest values, followed by the land use ∩ night-time light change (0.723), land use elevation ∩ DEM (0.715), and land use ∩ NDVI (0.715). The population density change ∩ precipitation change (0.211) showed the least interaction. There is a notable interaction between land use change and other factors. The land use type significantly affected the distribution patterns of different ecosystems. In addition, the increasing frequency of human activities led to an accelerated transformation of land use types, which increased the ecological pressure of surrounding towns and sped up changes in habitat quality. The interaction among anthropogenic factors was stronger than among natural factors, although natural factors are the foundations of habitat quality; they change slightly with weak intensity. Obvious changes in natural factors such as elevation would not occur without the disturbance of human activities. Therefore, the pairwise interaction among the anthropogenic factors is stronger than that among natural factors or between natural factors and social-economic factors.

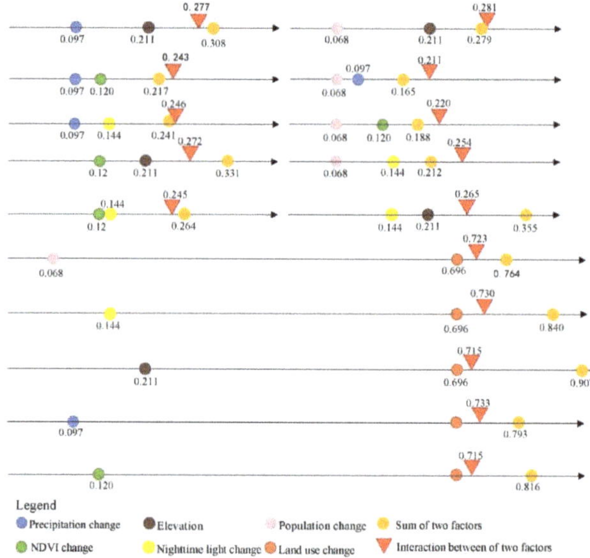

Figure 6. Results of interaction detection of factors influencing habitat quality in Wenzhou City.

4. Discussion

4.1. Analysis of the Habitat Quality in Wenzhou City

From 1992 to 2015, the overall habitat quality in Wenzhou City was good. Habitat quality slightly decreased from west to east. This is mainly due to the rich ecological resources in the western part of Wenzhou City. There are contiguous forest lands in the western part of Wenzhou City with high forest coverage. Economic development is slower in the western region than in the eastern region. According to Pan et al. [51], higher habitats in Zhejiang Province are mostly concentrated in the mountainous and hilly areas of Wenzhou City, which is consistent with the findings of this paper. In 1994, the State Council issued and implemented the Regulations of the People's Republic of China on Nature Reserves [52]. In the same year, Wenzhou City council established the Zhejiang Wuyanling National Nature Reserve in Taishun County to protect its subtropical forest ecosystem and rare plants and animals, such as pheasants and macaques [53]. In 1997, Wenzhou City established its first national forest park, Zhejiang Yandang Mountain National Forest Park. By the end of 2015, the number of natural protected areas in Wenzhou City significantly increased. It had 32 protected areas at and above the provincial level. During this period, the forest area of Wenzhou City increased by 33,000 hectares and the forest coverage rate increased by 20.9% [33]. Therefore, the quality of excellent habitats in Wenzhou City is well protected, and the quality of habitats in the western region is higher than in the eastern region. In addition, during 1995–2015, urban expansion in Wenzhou City was mainly concentrated in urban centres and coastal areas [54], with the largest area of land occupied by construction land, followed by forest land. The findings of our study are consistent with previous studies mentioned above. The habitat quality in Wenzhou City has been declining year by year in recent years, especially in coastal areas, where it is emergent to establish protected areas or take other effective conservation measures to change the degradation status.

4.2. Analysis of the Factors Affecting the Change in the Habitat Quality

Based on the results discussed in Section 3.3, the social and economic factors are the key factors that have affected changes in habitat quality in Wenzhou City. Based on the Wenzhou City 1992–2015 Yearbook [55], the land area, population, and GDP of Wenzhou City increased by 281 km^2, 1.35 million, and 417.619 billion yuan, respectively, mainly in Ouhai District, and in Yueqing City and Ruian City. Based on the adjustment of the land use structure in Wenzhou City (1997–2010) [56], the largest increase in construction land was mainly concentrated in Yueqing City and in the Ouhai and Lucheng districts, which further suggests that areas of extremely poor and poor habitat quality in the Ouhai District, Ruian City, and Yueqing City accounted for the largest proportion of the whole city. Regions with relatively slow population and GDP growth were mainly distributed in Yongjia County, Taishun County, Wencheng County, and the Dongtou District, which corresponded to the regions with the most extensive distribution of superior habitat. LUCC, population density, and economic growth have a notable spatial aggregation effect and the driving forces of social and economic growth in Wenzhou City weaken gradually from the east to the west.

In addition, natural factors play an important role in changes in the habitat quality. The geographical environment restricts the development of human activities in the western region and forest ecosystems are well protected from construction by the creation of protected areas, and the relatively high annual rainfall ensures a suitable environment for the growth of flora and fauna, creating areas rich in species diversity. Therefore, ecological restoration at a national scale should consider designating areas with a long history of excellent habitat quality within the biodiversity protection red line, and strictly controlling the human destruction of ecosystems within the ecological red line. For areas where habitat destruction has been greater and serious fragmentation has occurred in recent years, a combination of biological and engineering remediation measures should be used to actively promote the restoration of natural habitats. It is important to build the protected areas network, with protected areas as ecological sources, and important rivers where land and sea meet as ecological corridors to maintain the stability of habitat quality. The results of Section 3.2 of this study can be used to help optimal integration of the boundaries of protected areas, and ensure ecological measures are taken together to optimise the ecological network pattern, enhance biodiversity, promote the restoration of natural ecosystems, and create a high-quality ecological hinterland.

4.3. Effect of the Different Type of Land Use Change on the Habitat Quality

PCA was carried out using SPSS to extract eight components with a cumulative variance contribution rate of 84.414%, based on the principle that the characteristic root should be greater than one and the cumulative variance contribution rate should reach $\geq$ 80% [57]. Based on an analysis of the land use change types and eight principal components, it was determined that the main factors that affected habitat quality changes are cultivated land and construction land. The factors that contributed the most to the changes in the habitat quality can be ordered as follows, from large to small: conversion of forest land to cultivated land (0.889), aquatic environments to cultivated land (0.866), forest land to construction land (0.742), grassland to forest land (0.734), aquatic environments to forest land (0.715), and cultivated land to forest land (0.704). Additional data are shown in Tables S3 and S4.

Therefore, the conversion of forest land to cultivated land had the greatest effect on the deterioration of the habitat quality. Based on Section 3.1, the area of forest land decreased sharply around 2000, whereas the area of cultivated land rapidly increased. This was principally due to the promulgation of the Land Management Law in 1998, which stipulated the implementation of a compensation system for occupied farmland and the implementation of measures to protect the balance between the occupation of, and compensation for, occupied farmland for non-agricultural construction. Under this policy, people receive economic benefits from land reclamation, and as a result forest land was converted to cultivated land. Subsequently, the state implemented a policy of 'returning farmland to forest' in 2002 to protect and improve the ecological environment. From 2010 to 2014, Wenzhou City enacted the Balance of Arable Land, an act that allocated an equivalent amount and quality of arable land to the local government to supplement the amount of arable land occupied by construction, at the municipal level [58], and cultivated land and forest areas stabilised. Combined with the trend in habitat quality changes shown in Figure 3, it can be concluded that the quality of prime habitat rapidly declined from 1998 to 2000. This decline slowed after 2002. After the Balance of Arable Land was enacted from 2012 to 2014, the area with a prime habitat quality remained stable. Therefore, the effect of relevant policies on habitat quality changes has been validated by this study.

The conversion of aquatic environments to cultivated land also led to a deterioration in habitat quality. The explosive growth of the marine economy of Wenzhou City has led to an increase in land reclamation at many locations and aquatic environments have gradually decreased. Therefore, changes in areas with an excellent and intermediate habitat quality slowed down in Wenzhou City, whereas the areas of extremely poor habitat quality continued to increase, mainly in the eastern coastal areas. The eastern coastal area, which is the core of the economic construction and development of Wenzhou City, is mainly characterised by plains and hills. Frequent reclamation projects and other activities have led to the deterioration of the habitat quality in coastal areas, water pollution, and the continuous disappearance of natural ocean shorelines and harbours. The coastal waters, canals, plain river networks, and urban inland rivers within its boundaries are severely polluted. The eutrophication of water bodies is significant and wetland functions are severely threatened [59].

4.4. Comparison of the Logistic Binary Regression and GeoDetector Models

Logistic regression model analysis was carried out on the same data using SPPS. The contribution of land use changes was the largest, but the p value was greater than 0.05. The Exp(B) values of elevation, rainfall change, population change, and night-time light change were the same (see Figure 7). In contrast, the contribution of NDVI change was the smallest. An increase in the NDVI will likely mean a decrease in habitat quality deterioration, i.e., these two factors are negatively correlated. The Exp(B) values of multiple influencing factors were the same as those obtained using the GeoDetector model. However, the results obtained using the GeoDetector model explained the influence of multiple factors in more detail. In addition, the effect of NDVI change ranked fourth in Section 3.3, which slightly differs from the results obtained from the regression analysis. NDVI reflects the degree of vegetation coverage, which is the basis of the ecological environment and directly affects the quality of the habitat. Its influence is normally greater than of rainfall, as has been confirmed by numerous studies [29,60] The changes in habitat quality as a result of the combined action of these two factors cannot be compared. To conclude, the GeoDetector model is superior to the logistic binary regression model. Its operation is simpler, and collinearity and correlations among various factors are not considered.

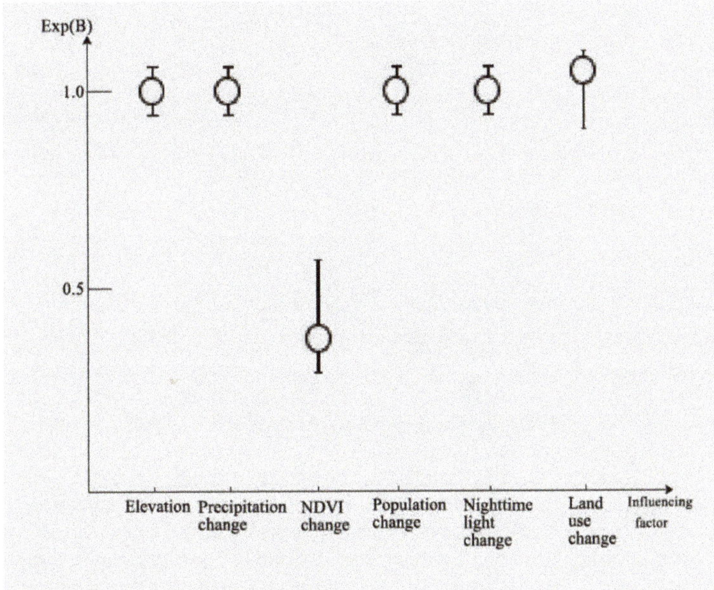

Figure 7. Model estimation of the influencing forces of the changes in habitat quality in Wenzhou City.

4.5. Overall Analysis

Coastal cities are at the forefront of economic development, but they are also habitats for many important plants and animals, including birds; land use changes have led to large-scale losses of natural coastal wetlands [61]. Consequently, many fast-growing coastal cities are under great ecological pressure. In this study, the annual changes in the habitat quality from 1992 to 2015 were analysed, which reduced the lag between the actual change and mapping time compared with analyses of the total change over a five-year period, because the changes analysed could have occurred at any point in this period (Figure 8). For example, the area with a medium habitat quality in Taishun County increased from 1992 to 2000. However, it first decreased and then increased during this period. The increase occurred between 1998 and 2000; therefore, the lag time of the change is two years. Thus, the use of a simple and efficient method to track the annual changes in habitat quality provides time for the mitigation of the habitat degradation in rapidly developing coastal cities of China.

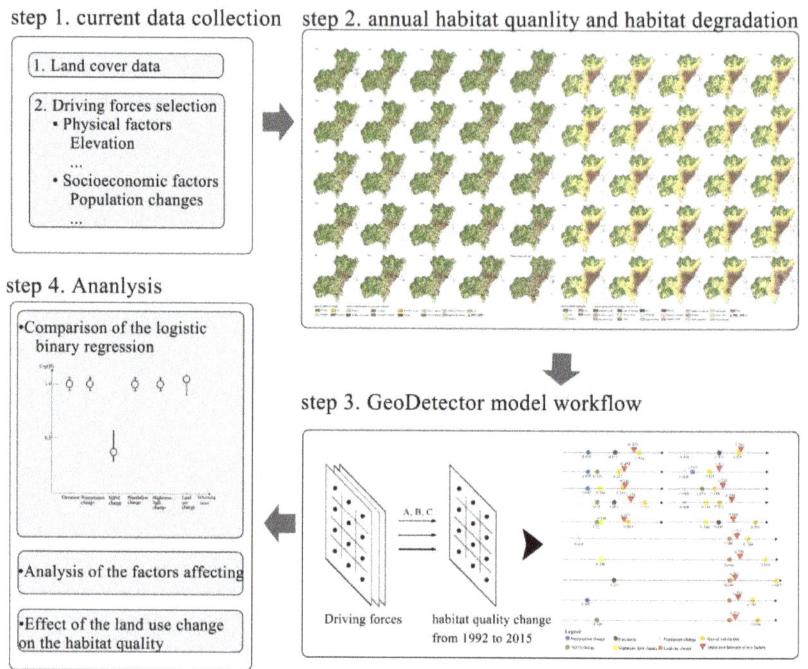

Figure 8. The specific working steps of this article.

The GeoDetector model revealed the factors that influence changes in habitat quality and then helped determine measures for changing the natural condition of coastal areas [37]. For example, the most influential factor in the changes in habitat quality in Wenzhou City was land use change, particularly the conversion of forest land and aquatic environments into arable land. Consequently, improvements to ecological environment quality in Wenzhou City should focus on increasing urban green space coverage, improving the layout of green space in urban areas, improving water resource conservation and wetland protection, preventing reclamation of aquatic environments, protecting water resources, reducing water pollution, reducing reclamation projects including shore zone restoration and beach restoration, and enhancing the landscape effect and ecological service value of water bodies.

China's territorial spatial planning includes the evaluation of the carrying capacity of resources and environment and the suitability of territorial spatial development, among which the evaluation of the carrying capacity of resources and environment summarises the characteristics of regional resource and environmental endowments and analyses their strengths and weaknesses according to three different functional directions: ecological protection, agricultural production, and urban construction. This research framework is simple to operate and can be applied in areas where data collection is difficult, to analyse regional ecological conservation by modelling habitat quality, to use habitat degradation to illustrate the direction of town construction, and to explore the factors influencing ecological change, providing a scientific basis for recent implementation measures in territorial spatial planning.

In this study, due to limitations in the collection of endemic species data information, it was not possible to validate the distribution of plants and animals against the quality of habitats simulated by the InVEST model. In the next step of the study, species distribution data could be used as a basis for habitat definition, and a plant ecology approach could be used as a basis for the model, potentially reducing the subjectivity of the results. Habitat quality changes are generally caused by multiple factors. In this study, GeoDetector models were used to explore the interaction between these two factors. However, the interaction between these factors is complex and uncertainties affect the results. In this study, when selecting climate change factors, rainfall change was chosen as an indicator due to the availability of data, but the indicators of climate change could be further refined, e.g., temperature change could be added to the influencing factors, as extreme weather may also influence the choice of sites for human activities, etc. The human factors also should be refined, the addition of policies and government interventions should be considered, and the mechanisms driving the multi-factor coupling should be analysed.

The intensity and interaction between the drivers of habitat quality may change in the future according to some studies, while the main driver may remain the same in the future [21]. This study aims to investigate the factors influencing past changes in habitat quality and to provide scientific guidance for the implementation of next steps in the short and medium term. Subsequent studies can forecast land use changes in the study area and provide a scientific basis for longer-term coastal urban planning in the future.

5. Conclusions

The land use structure in coastal cities, as the pacesetters of national major development strategies, has changed significantly and different land use structures have affected the regional ecological environment. The InVEST model is parameterised by the use of expert knowledge, so that the framework can be extended for use in different coastal cities. In addition, analysis of influencing factors using the GeoDetector model showed that habitat quality in coastal cities is impacted by natural, human, and social factors. Cross-probing between two random factors shows different degrees of enhancement. Significant interactions between land use and other factors were noted in the study area. Land-use types have an important influence on the distribution patterns of different ecosystems and, in combination with increasingly frequent anthropogenic activities, force an accelerated rate of land type change. This increases the ecological pressure on the surrounding towns and accelerates changes in habitat quality. Therefore, attention should be paid to the social and economic development of future cities, the construction of environmentally friendly land use patterns, and targeted action measures that protect and mitigate ecological threats. This study demonstrates that this research framework provides a simple, efficient, and low-cost decision support tool that can provide time to mitigate habitat degradation in the rapidly growing coastal cities of China and provide a scientific basis for ecological civilisation and spatial ecological restoration of the country, leading to a robust and integrated approach to land use planning and management.

Supplementary Materials: The following supporting information can be downloaded at: https://www.mdpi.com/article/10.3390/land11050630/s1, Table S1: Weights of the threat data; Table S2: Sensitivities of the habitat types to each threat; Table S3: Characteristic roots and variance contribution rate derived from the principal component analysis; Table S4: Factors of each component.

Author Contributions: Conceptualization, X.Z. and L.L.; Data curation, Z.X. and J.Z.; Formal analysis, X.Z.; Funding acquisition, L.L. and S.L.; Investigation, L.L., J.Z. and M.C.; Methodology, X.Z. and Z.X.; Project administration, M.C. and S.L.; Software, Z.X.; Supervision, L.L.; Validation, J.Z.; Writing—original draft, X.Z., L.L. and Z.X.; Writing—review & editing, L.L., M.C., S.L. and Q.G.All authors have read and agreed to the published version of the manuscript.

Funding: The study was funded by Distinguished Youth Program of Fujian Agriculture and Forestry University [No. XJQ201932] and the Interdisciplinary Integration Guidance Project of the College of Landscape Architecture, Fujian Agriculture and Forestry University [No. YSYL-xkjc-3].

Institutional Review Board Statement: Not applicable.

Informed Consent Statement: Not applicable.

Acknowledgments: The study was also supported by the Wenzhou Natural Resources and Planning Bureau. The authors very much appreciate their contributions.

Conflicts of Interest: The authors declare no conflict of interest.

References

1. Xu, Y. Development strategy of China's coastal cities for addressing climate change. *Clim. Chang. Res.* **2020**, *16*, 88–98.
2. Zhang, Y.; Chen, R.; Wang, Y. Tendency of land reclamation in coastal areas of Shanghai from 1998 to 2015. *Land Use Policy* **2020**, *91*, 104370. [CrossRef]
3. Van Horne, B. Density as a misleading indicator of habitat quality. *J. Wildl. Manag.* **1983**, *47*, 893–901. [CrossRef]
4. Johnson, M.D. Measuring habitat quality: A review. *Condor* **2007**, *109*, 489–504. [CrossRef]
5. Aneseyee, A.B.; Noszczyk, T.; Soromessa, T.; Elias, E. The InVEST Habitat Quality Model Associated with Land Use/Cover Changes: A Qualitative Case Study of the Winike Watershed in the Omo-Gibe Basin, Southwest Ethiopia. *Remote Sens.* **2020**, *12*, 1103. [CrossRef]
6. Global Assessment Report on Biodiversity and Ecosystem Services 2019. Available online: https://ipbes.net/global-assessment (accessed on 12 February 2021).
7. Zhang, X.; Huang, X.; Zhao, X. Calculation of Ecology Service Value of Land Use Change in Jiangsu Coastal Area. *Res. Soil Water Conserv.* **2015**, *22*, 252–256.
8. Liu, C.; Yang, M.; Hou, Y.; Xue, X. Ecosystem service multifunctionality assessment and coupling coordination analysis with land use and land cover change in China's coastal zones. *Sci. Total Environ.* **2021**, *797*, 149033. [CrossRef]
9. Peng, J.; Xu, F.X.; Wu, J.; Deng, K.; Hu, T. Spatial Differentiation of Habitat Quality in Typical Tourist City and their Influencing Factors Mechanisms: A Case Study of Huangshan City. *Resour. Environ. Yangtze Basin* **2019**, *28*, 13.
10. Liang, X.; Yuan, L.; Ning, L.; Song, C.; Cheng, C.; Wang, X. Spatial pattern of habitat quality and drivingfactors in Heilongjiang Province. *J. Beijing Norm. Univ.* **2020**, *12*, 56.
11. Fellman, J.B.; Hood, E.; Dryer, W.; Pyare, S. Stream physical characteristics impact habitat quality for Pacific salmon in two temperate coastal watersheds. *PLoS ONE* **2015**, *10*, e0132652. [CrossRef]
12. Löffler, F.; Fartmann, T. Effects of landscape and habitat quality on Orthoptera assemblages of pre-alpine calcareous grasslands. *Agric. Ecosyst. Environ.* **2017**, *248*, 71–78. [CrossRef]
13. Zheng, Y.; Zhang, P.T.; Tang, F. The Effects of Land Use Change on Habitat Qualaty in Changli County Based on InVST Model. *Chin. J. Agric. Resour. Reg. Plan* **2018**, *39*, 121–128.
14. Wei, L.; Lan, S.; Xiong, H.; Shen, Q.; Lu, D.; Chen, X. Habitat Quality Evaluation of Wuyi Mountain National Nature Reserve in 1988–2018. *J. Southwest For. Univ.* **2021**, *41*, 93–102.
15. Zhang, M.; Zhang, F.; Li, X. Evaluation of Habitat Quality Based on InVEST Model: A Case Study of Tongzhou District of Beijing, China. *Landsc. Archit.* **2020**, *27*, 95–99.
16. Fahrig, L. Effects of habitat fragmentation on biodiversity in China. *J. Ecol.* **2017**, *36*, 2605–2614.
17. Sun, X.; Jiang, Z.; Liu, F.; Zhang, D. Monitoring spatio-temporal dynamics of habitat quality in Nansihu Lake basin, eastern China, from 1980 to 2015. *Ecol. Indic.* **2019**, *102*, 716–723. [CrossRef]
18. Chen, T.; Feng, Z.; Zhao, H.; Wu, K. Identification of Ecosystem Service Bundles and Driving Factors in Beijing and its Surrounding Areas. *Sci. Total Environ.* **2019**, *711*, 134687. [CrossRef]
19. Yan, S.; Wang, X.; Cai, Y.; Li, C.; Yan, R.; Cui, G.; Yang, Z. An Integrated Investigation of Spatiotemporal Habitat Quality Dynamics and Driving Forces in the Upper Basin of Miyun Reservoir, North China. *Sustainability* **2018**, *10*, 4625. [CrossRef]
20. Ju, H.; Zhang, Z.; Zuo, L.; Wang, J.; Zhang, S.; Wang, X.; Zhao, X. Driving forces and their interactions of built-up land expansion based on the GeoDetector–a case study of Beijing, China. *Int. J. Geogr. Inf. Sci.* **2016**, *30*, 2188–2207. [CrossRef]
21. Li, S.P.; Liu, J.L.; Lin, J.; Fan, S.L. Spatial and temporal evolution of habitat quality in Fujian Province, China based on the land use change from 1980 to 2018. *J. Appl. Ecol.* **2020**, *31*, 4080–4090.
22. Liu, Z.F.; Tang, L.N.; Qiu, Q.Y.; Xiao, L.; Xu, T.; Yang, L. Temporal and spatial changes in habitat quality based on land-Use change in Fujian province. *Acta Ecol. Sin.* **2017**, *37*, 4538–4548.
23. Song, Y.; Wang, J.; Ge, Y.; Xu, C. An optimal parameters-based GeoDetector model enhances geographic characteristics of explanatory variables for spatial heterogeneity analysis: Cases with different types of spatial data. *GIScience Remote Sens.* **2020**, *57*, 593–610. [CrossRef]
24. Brunsdon, C.; Fotheringham, A.S.; Charlton, M.E. Geographically weighted regression: A method for exploring spatial nonstationarity. *Geogr. Anal.* **1996**, *28*, 281–298. [CrossRef]
25. Liu, C.; Wu, X.; Wang, L. Analysis on land ecological security change and affect factors using RS and GWR in the Danjiangkou Reservoir area, China. *Appl. Geogr.* **2019**, *105*, 1–14. [CrossRef]
26. Xue, X.; Wang, X.; Duan, H.; Yang, L.; Xie, Y. Analysis on Spatio-temporal Evolution of Habitat Quality in Qilian Mountains Based on Land Use Change. Bull. *Soil Water Conserv.* **2020**, *40*, 278–284.
27. Wang, J.F.; Li, X.H.; Christakos, G.; Liao, Y.L.; Zhang, T.; Gu, X.; Zheng, X.Y. GeoDetectors-based health risk assessment and its application in the neural tube defects study of the Heshun Region, China. *Int. J. Geogr. Inf. Sci.* **2010**, *24*, 107–127. [CrossRef]
28. Wang, J.F.; Xu, C.D. Geodetector: Principle and prospective. *Acta Geogr. Sin.* **2017**, *72*, 116–134.
29. Zhu, Z.; Kasimu, A. Spatial-temporal evolution of habitat quality in Yili Valley based on GeoDetector and its influencing factors. *Chin. J. Ecol.* **2020**, *39*, 3408–3420.

30. Lin, S.; Li, Z. City profile: Wenzhou-A model city of transitional China. *Cities* **2019**, *95*, 102393. [CrossRef]
31. Wenzhou Municipal People's Government. Natural Resources. Available online: http://www.wenzhou.gov.cn/art/2019/3/22/art_1633785_31546755.html (accessed on 11 September 2021).
32. Jian, H. A brief history of water resources in Wenzhou. *Zhejiang Water Conserv. Sci. Technol.* **1996**, *23*, 60–64.
33. Wenzhou Bureau of Statistics. 2020 Wenzhou Statistical Yearbook. Available online: http://wztjj.wenzhou.gov.cn/art/2020/11/9/art_1467318_58725689.html (accessed on 11 September 2021).
34. Sharp, R.; Douglass, J.; Wolny, S.; Arkema, K.; Bernhardt, J.; Bierbower, W.; Chaumont, N.; Denu, D.; Fisher, D.; Glowinski, K.; et al. InVEST 3.10.2.post21+ug.gb784d7e User's Guide. 2020. Available online: https://storage.googleapis.com/releases.naturalcapitalproject.org/invest-userguide/latest/index.html (accessed on 7 March 2022).
35. Sulla-Menashe, D.; Friedl, M. MCD12C1 MODIS/Terra+Aqua Land Cover Type Yearly L3 Global 0.05Deg CMG V006. Available online: https://lpdaac.usgs.gov/products/mcd12c1v006/ (accessed on 7 March 2022).
36. UCL-Geomatics. Land Cover CCI Product User Guide (Version 2.0). 2017. Available online: https://www.esa-landcover-cci.org/?q=webfm_send/84 (accessed on 20 August 2020).
37. Zhu, C.; Zhang, X.; Zhou, M.; He, S.; Gan, M.; Yang, L.; Wang, K. Impacts of urbanization and landscape pattern on habitat quality using OLS and GWR models in Hangzhou, China. *Ecol. Indic.* **2020**, *117*, 106654. [CrossRef]
38. Liao, Y.; Wang, X.Y.; Zhou, J.M. Suitability assessment and validation of giant panda habitat based on GeoDetector. *J. Geoifor. Sci.* **2016**, *18*, 767–778.
39. Jia, L.; Yankuo, L.; Lujun, M.; Guangyong, X.; Fangkai, Y.; Yan, H.; Peng, X. Effects of wintering site climatic conditions on the population size of white spoonbills in Poyang Lake Nature Reserve. *J. Ecol.* **2014**, *34*, 5522–5529.
40. Qiu, C.; Hu, J.; Yang, F. Analysis of conservation effectiveness of nature reserves based on NDVI in Yunnan Province. *Acta Ecol. Sin.* **2020**, *40*, 7312–7322.
41. Wu, J.; Li, X.; Luo, Y.; Zhang, D. Spatiotemporal Effects of Urban Sprawl on Habitat Quality in the Pearl River Delta from 1990 to 2018. *Sci. Rep.* **2021**, *11*, 13981. [CrossRef]
42. Bai, L.; Xiu, C.; Feng, X.; Liu, D. Influence of urbanization on regional habitat quality: A case study of Changchun City. *Habitat Int.* **2019**, *93*, 102042. [CrossRef]
43. Liu, J.; Meng, P.; Gong, X. Accelerating the New Round Amendment of Land Administration Law: Reviews from the Workshop of "Promoting the Amendment of Land Administration Law in Accordance with the Constitution". *China Land Sci.* **2015**, *29*, 16–21.
44. Zhang, L.; Ren, Z.; Chen, B.; Gong, P.; Fu, H.; Xu, B. A Prolonged Artificial Nighttime-light Dataset of China (1984–2020). Available online: https://data.tpdc.ac.cn/en/data/e755f1ba-9cd1-4e43-98ca-cd081b5a0b3e/?q= (accessed on 7 March 2022).
45. Huang, J.; Tang, Z.; Liu, D.; He, J. Ecological response to urban development in a changing socio-economic and climate context: Policy implications for balancing regional development and habitat conservation. *Land Use Policy* **2020**, *97*, 104772. [CrossRef]
46. Zhong, Q.; Ma, J.; Zhao, B.; Wang, X.; Zong, J.; Xiao, X. Assessing spatial-temporal dynamics of urban expansion, vegetation greenness and photosynthesis in megacity Shanghai, China during 2000–2016. *Remote Sens. Environ.* **2019**, *233*, 111374. [CrossRef]
47. Sharma, R.; Nehren, U.M.; Rahman, S.A.; Meyer, M.; Rimal, B.; Seta, G.A.; Baral, H. Modeling Land Use and Land Cover Changes and Their Effects on Biodiversity in Central Kalimantan, Indonesia. *Land* **2018**, *7*, 57. [CrossRef]
48. Li, X.; Zhou, Y.; Asrar, G.R.; Mao, J.; Li, X.; Li, W. Response of vegetation phenology to urbanization in the conterminous United States. *Glob. Chang. Biol.* **2017**, *23*, 2818–2830. [CrossRef] [PubMed]
49. Pearson, K. Principal components analysis. *Lond. Edinb. Dublin Philos. Mag. J. Sci.* **1901**, *6*, 559. [CrossRef]
50. Wang, Y.; Wang, J.; Yao, Y.B.; Wang, J.S. Evaluation of Drought Vulnerability in Southern China Based on Principal Component Analysis. *Ecol. Environ. Sci.* **2014**, *23*, 1897–1904.
51. Pan, Y.; Bao, H.; Huang, L.; Zhu, J. Characterizing spatiotemporal dynamics and impacts of coastal urbanization on habitat quality in response to coastal booms. *Shanghai Land Resour.* **2020**, *41*, 18–24.
52. Gao, J.X.; Xu, M.J.; Zou, C.X. Development Achievement of Natural Conservation in 70 Years of New China. *China Environ. Manag.* **2019**, *11*, 25–29.
53. Wenzhou Local History. Nature Reserve. Available online: http://www.wenzhou.gov.cn/art/2019/3/29/art_1633803_31791118.html (accessed on 11 September 2021).
54. Gao, C.; Feng, Y.; Tong, X.; Jin, Y.; Liu, S.; Wu, P.; Ye, Z.; Gu, C. Modeling urban encroachment on ecological land using cellular automata and cross-entropy optimization rules. *Sci. Total Environ.* **2020**, *744*, 140996. [CrossRef]
55. Wenzhou Bureau of Statistics. 1992–2015 Statistical Yearbook. Available online: http://wztjj.wenzhou.gov.cn/col/col1467318/index.html (accessed on 11 September 2021).
56. Wenzhou City Land and Resources Bureau Net. Wenzhou City Land Use Master Plan (1997–2010). Available online: http://www.mnr.gov.cn/gk/ghjh/201811/t20181101_2324679.html (accessed on 11 September 2021).
57. Chen, Y.; Huang, L.; Zhou, M. Assessment of Water Quality of the Main Stream of Tuojiang River based on Principal Component Analysis Method. *Sichuan Environ.* **2021**, *40*, 133–137.
58. Lin, L.; Jia, H.; Pan, Y.; Qiu, L.; Gan, M.; Lu, S.; Deng, J.; Yu, Z.; Wang, K. Exploring the Patterns and Mechanisms of Reclaimed cultivated land Utilization under the Requisition-Compensation Balance Policy in Wenzhou, China. *Sustainability* **2017**, *10*, 75. [CrossRef]

59. Peng, X.; Xu, R.; He, Y. Analysis on coastline and coastal wetland changes in Yueqing bay in recent 30 years. *Mar. Environ. Sci.* **2019**, *38*, 68–74+83.
60. Li, Z.; Xu, Y.; Sun, Y.; Wu, M.; Zhao, B. Urbanization-Driven Changes in Land-Climate Dynamics: A Case Study of Haihe River Basin, China. *Remote Sens.* **2020**, *12*, 2701. [CrossRef]
61. Jiang, T.T.; Pan, J.F.; Pu, X.M.; Wang, B.; Pan, J.J. Current status of coastal wet-lands in China: Degradation, restoration, and future management. *Estuar. Coast. Shelf Sci.* **2015**, *164*, 265–275. [CrossRef]

Article

Spatio-Temporal Variation of the Ecosystem Service Value in Qilian Mountain National Park (Gansu Area) Based on Land Use

Lili Pu [1], Chengpeng Lu [2,*], Xuedi Yang [1] and Xingpeng Chen [1,2]

[1] College of Earth and Environmental Sciences, Lanzhou University, Lanzhou 730000, China
[2] Institute of County Economic Development & Rural Revitalization Strategy, Lanzhou University, Lanzhou 730000, China
* Correspondence: lcp@lzu.edu.cn

Citation: Pu, L.; Lu, C.; Yang, X.; Chen, X. Spatio-Temporal Variation of the Ecosystem Service Value in Qilian Mountain National Park (Gansu Area) Based on Land Use. *Land* 2023, 12, 201. https://doi.org/10.3390/land12010201

Academic Editors: Rui Yang, Yue Cao, Steve Carver and Le Yu

Received: 2 December 2022
Revised: 5 January 2023
Accepted: 6 January 2023
Published: 7 January 2023

Abstract: The value of ecosystem services and service capabilities continue to improve, and the way to form a path of resource industrialization development has become one of the important directions of sustainable development. This paper mainly takes the construction of national parks as a major opportunity and explores the temporal and spatial changes in the value of ecosystem services in Qilian Mountain National Park (Gansu area) and the construction path of the industrial system of national park construction. The total value of ecosystem services was calculated using a comprehensive index of the degree of land use, land contribution rate, ecological service value, equivalent factor of economic value, and the improved value coefficient of farmland ecological services, and then the Sensitivity index was used to reveal the dependence of the value of ecosystem services on the value index over time. The results showed the following: (1) Human disturbance factors in Qilian Mountain National Park (Gansu area) are weak, and the land use of Qilian Mountain National Park (Gansu Area) was mainly grassland, followed by unused land, forest land, and glacial snow, with the change in glacial snow cover being the largest. (2) The ecosystem of Qilian Mountain National Park (Gansu area) is strong, and the contribution rate of forest land, construction land, unused land, and glacial snow cover in Qilian Mountain National Park (Gansu Area) was positive, while cultivated land, grassland, and water area were negative. Among them, glacial snow cover contributed the most at 10.4723 the ecological barrier function plays a stable role. (3) The ecosystem service value (ESV) in Qilian Mountain National Park (Gansu Area) showed a fluctuating growth trend on the whole, showing the characteristics of high northwest and low southeast, among which the total value of grassland was the largest, the value of unused land was the smallest with the largest increase range, and the increase in water area was the smallest. (4) Qilian Mountain National Park (Gansu Area) is mainly based on regulated services, followed by support services, supply services, and cultural services, all showing a clear growth trend, increasing by 181.77%, 183.90%, 196.19%, and 170.38%, respectively. With the development of low-carbon economy and circular economy as the main idea, we aim to build a national park industrialization development path of direct product supply, indirect product supply, and basic guarantee.

Keywords: ecosystem services value; land use intensity; land use change; sensitivity analysis; Qilian Mountain National Park (Gansu Area)

1. Introduction

The continuous satisfaction of economic and social service functions by ecosystem services is an important basic prerequisite for achieving their continued function [1]. In 2015, China promulgated and implemented the "Opinions of the Development and Reform Commission on the Key Work of Deepening Economic System Reform in 2015" to carry out a "pilot national park system" in nine provinces, including Sichuan, Hainan, and Guangdong, and in 2021, China officially established the first batch of national parks, which included

Sanjiangyuan, giant pandas, Northeast tigers and leopards, Hainan tropical rainforest, and Wuyi Mountain, covering an area of 230,000 square kilometers, covering nearly 30% of the terrestrial areas of the national key protected wild animal and plant species. Due to the coupling characteristics of the natural and cultural landscapes of China's natural resources themselves [2], more attention is being paid to the attributes of cultural characteristics and the needs for integrated development, such as ecosystem service functions, social functions, and premium functions, in the process of their development [3]. The proposal national park construction explores the shift from the ecological protection system dominated by nature reserves to the nature reserve system with national parks as the main body, providing a typical development model for the overall protection of the global natural system and paying attention to the important role of ecological assets [4]. As an important ecological barrier typical of western China [5], the Qilian Mountains play an important role in helping to maintain the balance of the oasis ecosystem in the Hexi Corridor [6] and cultural symbols [7,8], and the way to better highlight the service characteristics in the protection system dominated by national parks has become an urgent problem.

Ecosystem services refer to the environmental conditions and utilities formed and maintained by ecosystems for human survival and development, and all the benefits directly or indirectly obtained by human beings from the ecosystem, including four aspects of supply services, regulation services, support services, and cultural services [9]. The research on the value of China's ecosystem services has been carried out by Xie Gaodi [10] to develop the "China terrestrial ecosystem service value equivalent factor table". It provides a basis for calculating regional ecosystem values and is widely used, and the coordination between ecosystem services is constantly weighed [11]. The main types of ecosystems are farmland, forests, grasslands, wetlands, oceans, and cities [12], which can provide people with systematic service functions—that is, the various utilities that humans obtain from the ecosystem [13]. Similarly, they provide a variety of services to humans, directly or indirectly, and have been widely discussed in the academic community [14]. For example, Costanza first assessed global natural capital in 1997, mainly using ecosystem goods and services [15]. De Groot et al. defined ecosystem functioning as the ability of natural processes and their components to provide goods and services that meet direct or indirect human needs [16]. Since the United Nations Millennium Assessment (2005), which pointed out that ecosystem services refer to the benefits that people receive from ecosystems, ecosystem services science has made many advances in developing the core concepts and methods [17]. The research and development of ecosystems continue to deepen, and the importance of the development of economy [18], society [19], and urban ecosystem service value prediction continues to increase [20], which not only plays an important role in the construction of national parks [21] but also in human development, such as cultural development [22] and landscape value [23]. In 2021, the United Nations officially adopted the new framework of environmental–economic accounting–ecosystem accounting (SEEA-EA) to further promote sustainable economic and social development. In the study of ecosystem service value in China, it has been proposed that ecological equivalent factors [10] rely on continuous optimization and in-depth calculation of ecosystem value. In 2020, the Ministry of Ecology and Environment and the Research Center for Eco-Environmental Sciences of the Chinese Academy of Sciences jointly compiled a technical guide for accounting for the terrestrial ecosystem product (GEP) and then extending the function and value of recreation services to the ecosystem [24], which continues to enrich the research on the value system of ecosystem services with a focus on counties [25]. Similarly, with the transformation and development of China's economy and society, more attention should be paid to connotative development and cross-regional ecological economic linkage development [26], and the role of the vegetation index in ecosystems should be fully utilized [27].

Ecosystem service function and ecological sensitivity are important contents of ecological protection evaluation [28], and the process of national park construction not only pays attention to the supply capacity of the ecosystem itself but also divides national parks into strictly protected areas, ecological conservation areas, traditional use areas, and scientific and educational recreation areas [29], and also pays more attention to the reuse of other extended functions such as cultural aesthetics. Some scholars have made calculations based on GEP (gross ecosystem product), demonstrating that the ecological value is the most prominent [30]. The Qilian Mountains are ecologically fragile and sensitive areas, and ecological restoration is more difficult [31], but the way to further realize the service value of the ecosystem as a national park, better serve the local economy, and society to play a better role and form a benign interaction with the ecosystem has become an urgent problem to be solved. As such, the systematic protection of national parks as the main body has become a typical case demonstration.

This paper mainly relies on the importance and resource characteristics of ecological economic development, taking Qilian Mountain National Park (Gansu area) as an example. First, the ecosystem service value equivalent factor was used to analyze the changes in ecosystem service value from 2000 to 2019 and enrich the application research of ecosystem service value equivalent factor. Second, combined with the economic development of the Qilian Mountains and its surrounding areas, highlight the characteristics shared by the people of national park construction, build a national park industrialization development path of direct product supply, indirect product supply and basic guarantee, and put forward countermeasures and suggestions for national park construction. We also hoped to provide a typical case for the development of terrestrial ecosystems around the world.

2. Overview of the Study Area

Qilian Mountain National Park (Gansu Area) covers an area of 34,400 km^2, accounting for 68.5% of the total area, involving the seven counties (districts) of Subei Mongol Autonomous County, Aksai Kazakh Autonomous County, Sunan Yugur Autonomous County, Minle County, Yongchang County, Tianzhu Tibetan Autonomous County, and Liangzhou District, including Qilian Mountain National Nature Reserve, Yanchiwan National Nature Reserve, Tianzhu Three Gorges National Forest Park, Horseshoe Temple Provincial Forest Park, Binggou River Provincial Forest Park, and other protected areas. The terrain is basically high in the south and low in the north, located in a cold area with a plateau continental climate and rich natural environment. It consists mainly of Qinghai spruce forest, shrub forest, and a small number of Qilian cypress, birch, and aspen forests, grassland meadow steppe, desert steppe, and alpine grassland. The vegetation growth in the area is good, and the forest coverage rate reaches 28.8% [32] (Figure 1).

As of 2019, The 7 counties (districts) of Qilian Mountain National Park (Gansu Area) have a land area of 1232.2 square kilometers and a population of 1460.3 thousand, the GDP totaled 7.974 billion USD, the investment in fixed assets was 5.338 billion USD, and the added value of the primary, secondary, and tertiary output was 1.964, 1.641, and 4.369 billion USD, respectively (According to the information released by the National Bureau of Statistics of China, the conversion of US dollars and RMB is based on the average exchange rate of US dollars and RMB in 2020—that is, 1 US dollar to 6.8974 yuan) (Table 1).

Table 1. Statistics of major indicators of Qilian Mountain National Park (Gansu Area) in 2019.

County (District)	Major Indicators						
	Area [1]	Population [2]	GDP	Value of the Primary	Value of the Secondary	Value of the Tertiary Output	Fixed Investment
	sq. km.	tp	Billion USD	Billion USD	Billion USD	Billion USD	Billion USD
Subei mongolian prefecture	667	15.1	0.236	0.016	0.099	0.121	0.6
Akesai kazak autonomous county	314	11.0	0.149	0.012	0.046	0.091	0.416
Minle county	37	192.5	0.851	0.275	0.163	0.413	0.793
Yongchang county	74	177.6	1.13	0.263	0.303	0.564	0.451

Table 1. Cont.

County (District)	Major Indicators						
	Area [1]	Population [2]	GDP	Value of the Primary	Value of the Secondary	Value of the Tertiary Output	Fixed Investment
	sq. km.	tp	Billion USD	Billion USD	Billion USD	Billion USD	Billion USD
Tianzhu tibetan autonomous county	71	151	0.663	0.17	0.127	0.366	0.552
Liangzhou district	49	885.3	4.559	1.127	0.788	2.644	2.368
Sunan Yugur Autonomous County	202	27.8	0.386	0.101	0.115	0.17	0.158
Total	1232.2	1460.3	7.974	1.964	1.641	4.369	5.338

1. sq. km.: Square kilometer. The data are mainly from the official websites of seven county (district) governments.
2. Population data are the seventh national census. "tp" represents "thousand people".

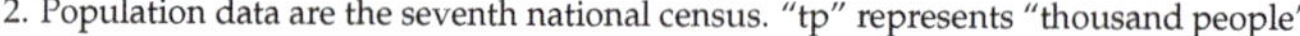

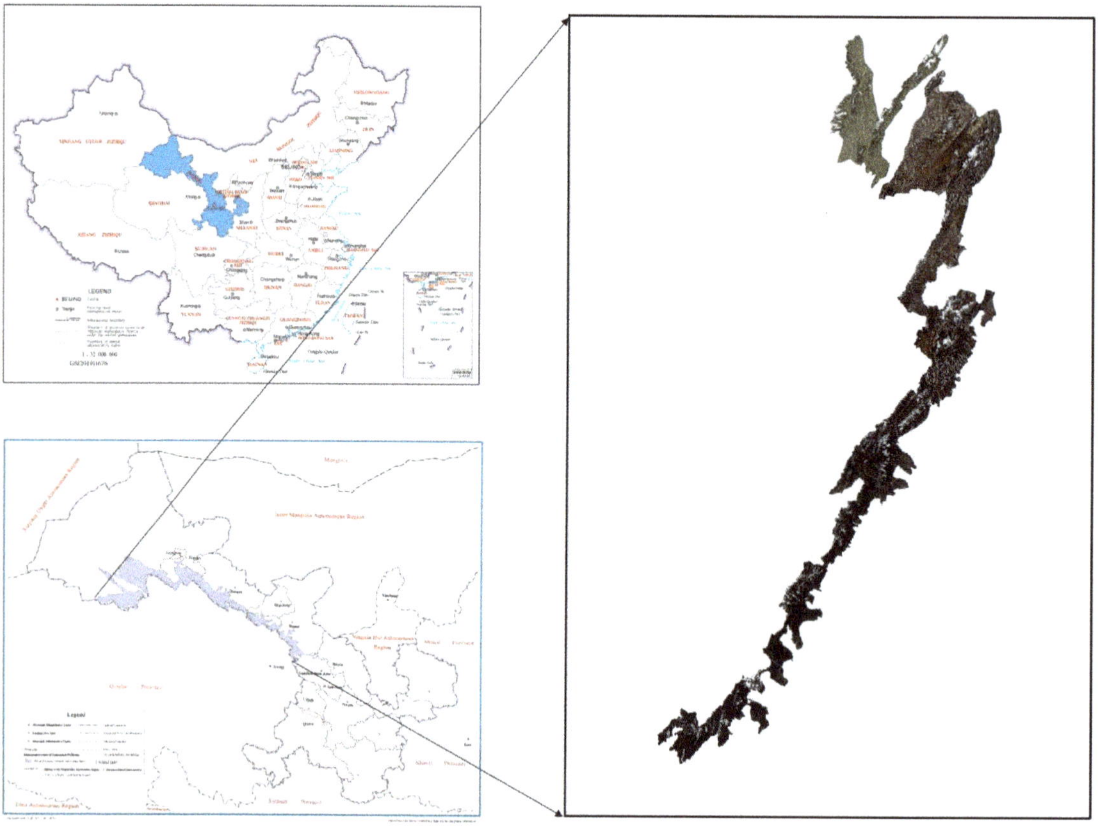

Figure 1. Location of the study area.

3. Materials and Methods

3.1. Data Sources and Processing

The remote sensing monitoring dataset of land cover change in China (CNLUCC) provided by the Data Center for Resources and Environmental Sciences of the Chinese Academy of Sciences from 2000 to 2019 was provided by the Data Center for Resources and Environment Science of the Chinese Academy of Sciences, and this paper analyzed the land use changes according to the first-level classification method of land use type of the system, namely, arable land, forest land, grassland, water, construction land, and unused land. The data on grain crop output and sown area came from the Gansu Development Yearbook, while the grain price data were from the Summary of National Agricultural Product Cost and Benefit Data. These data are widely used in the study of the value of ecosystem services in China [33,34] (Figure 2).

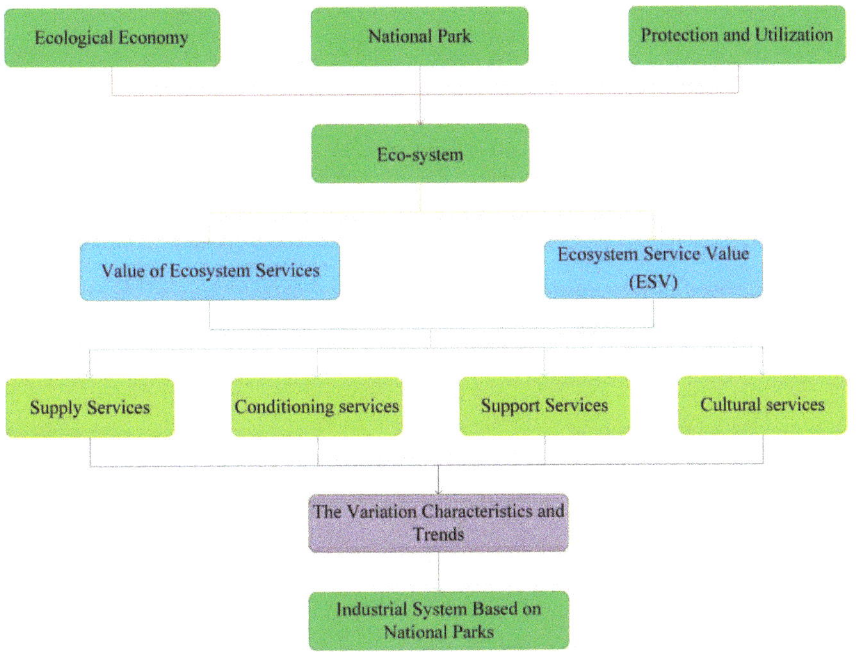

Figure 2. Flow chart. Based on the background of the construction of a community with a shared future for man and nature with Chinese characteristics, relying on the construction of the main body of national parks, highlighting the relationship between ecological economic development and ecosystem protection and utilization, calculating the value of ecosystem services through the equivalent factor of ecosystem services, maximizing the benefits of the four major ecosystem service functions of supply, regulation, support and culture, and analyzing the changing characteristics and trends of the four, and then putting forward countermeasures and suggestions for the construction of an industrial system dominated by national parks.

3.2. Research Methods

3.2.1. Analysis of Degree of Land Use and Change Characteristics

1. Composite Index of Land Use

The comprehensive index of the degree of land use (L) reflects the degree of human development and utilization of regional land and is an important indicator to measure the depth and breadth of regional land use. Its formula is expressed as [35]:

$$I = \sum_{n-1}^{n} (L_i \cdot P_i) \cdot 100\%, \tag{1}$$

where I represents the comprehensive index of land use intensity, L_i represents the land use intensity grade of the class I land use type, and P_i represents the proportion of class I land use type to the total land area.

In order to quantify the influence of each land use type on the change of the comprehensive index of land use intensity, the contribution rate of land type use intensity was introduced, and the calculation method is as follows [35]:

$$R_i = \frac{I_{ib} - I_{ia}}{I_{ia}} = \frac{L_i \cdot (P_{ib} - P_{ia})}{L_i \cdot P_{ia}}, \tag{2}$$

where I_{ib} and I_{ia} are the land use intensity index for the class I land use types b year and a year, respectively. P_{ib} and P_{ia} refer to ratio of the type I land use type to the total land area, respectively. L-i. denotes the land use intensity rating of the class I land use type. R_i is the contribution rate of the land use intensity composite index of class I land use type from a to b years, where a negative value means that its contribution makes the land use intensity composite index smaller, while a positive value indicates that its contribution makes the land use intensity composite index larger. The larger the absolute value of R_i, the greater the contribution of class I land use types to the change of the overall land use intensity composite index—that is, the greater the impact.

2. Analysis of land use change characteristics

The land use transfer matrix is the basis for analyzing the direction of regional land use change, which can reveal the structural characteristics and transfer direction of land use changes [36]. The rate of land use change can be expressed in terms of land use dynamics. A single land use dynamic degree can visually reflect the intensity of change in various land types [37].

$$K = \frac{U_b - U_a}{U_a} \times \frac{1}{T} \times 100\%, \tag{3}$$

where K is the dynamic degree of a certain land use type. Ua and U_b represent the area of a land use type at the beginning and end of the study period, respectively. T is the study period for a land type.

3.2.2. Approaches to Valuing Ecosystem Services

Referring to the research results of Xie Gaodi [38], Sutton and Costanza [39], and others, the economic value of the national ecosystem ecological service value equivalent factor was calculated, and the proposed equivalent factor table defines the economic value of the annual natural food yield of farmland, with a national average yield of 1 hm^2 being 1 [40] and the value equivalent factor of other ecosystem services being a relative quantity, which refers to the contribution of the ecological service relative to the farmland food production service.

The economic value of grain production can be calculated as [35]:

$$E_c = \frac{1}{7} T_a \cdot T_b, \tag{4}$$

where E_c is the economic value of grain production. T_a is the average grain benchmark yield (kg/hm^2) in the study area study area. T_b is the unit price of grain in the study area. 1/7 refers to the natural ecosystem without human input in the unit area, and the economic value provided by the natural ecosystem without human input is 1/7 provided by existing farmland [35]. According to the biomass factor table of farmland ecosystem in different provinces in China [10], the biomass factor of farmland ecosystem in Gansu was 0.42, and the value coefficient of farmland ecological service in Qilian Mountain area was 0.85 after adjustment according to the actual situation.

The service value coefficient of each ecological service function can be calculated as follows [41]:

$$VC_{ij} = E_c \cdot f_{ij}, \tag{5}$$

where VC_{ij} is the coefficient of the jth ecological service value of the ith land use type (dollar/hm$^2 \cdot$a), and f_{ij} represents the equivalent factor of the jth ecological service value of the ith land use type. From 2000 to 2019, the average grain output of Qilian Mountain National Park (Gansu Area) was 66,009.02 kg/hm^2, and in 2019, the average grain price of the seven counties (districts) of Qilian Mountain National Park (Gansu Area) was 4.26 USD/hm^2, while the value of ecosystem services in Qilianshan National Park (Gansu Area) was calculated as 38,587.18 USD/hm^2. Furthermore, the value of ecosystem services in the study area was calculated [41]:

$$ESV = \sum_{i=1}^{n}(A_k \times VC_k) \qquad ESV_f = \sum_{i=1}^{n}\left(A_k \times VC_{jk}\right), \qquad (6)$$

where ESV and ESV_f are the total value of ecosystem services and the functional value of the f-service, respectively. A_k represents the area of land use type k (hm^2). VC_k and VC_{jk} are the ecosystem service value coefficient and the f-service function value coefficient for land use type k, respectively.

3.2.3. Sensitivity Analysis

This paper used the Coefficient of Sensitive (CS) index commonly used in economics to reveal the dependence of the value index on the change of ecosystem service value over time, so as to reduce the uncertainty of the results. According to CS, to better verify the stability of the change trend and characteristics of the total value of ecosystem services in Qilian Mountain National Park (Gansu area) from 2000 to 2019. In this paper, CS was calculated by increasing or decreasing the ecological service value coefficient VC by 50% for each land use type [42].

$$CS = \left| \frac{(ESV_j - ESV_i)/ESV_i}{\left(VC_{jk} - VC_{ik}\right)/VC_{ik}} \right|, \qquad (7)$$

where VC_{ik} and VC_{jk} represent the value coefficient of ecological services per unit area of Category k ecosystems before and after adjustment. ESV_i and ESV_j represent the total value of ecological services before and after the adjustment, respectively. CS is the sensitivity of the value coefficient of each ecosystem service in the study area. If $CS > 1$, ESV is elastic to VC, the accuracy of the value coefficient is poor, and the confidence is low. If $CS < 1$, ESV is not elastic to VC and the results are credible.

4. Results

4.1. Change Characteristics of Land Use Degree

4.1.1. Land Use Change Characteristics

From 2000 to 2019, Qilian Mountain National Park (Gansu Area) was mainly divided into four phases of arable land, forest land, grassland, water, unused land, construction, and glacier five types of land use types. Specifically, there were mainly the following aspects:

In the study periods of 2000, 2005, 2010, 2015, and 2019, different land types in Qilian Mountain National Park (Gansu Area) changed to varying degrees according to the remote sensing monitoring dataset of land cover change in China, mainly as follows: The area of unused land continued to increase, and the area of forest land, glacial snow cover, and construction land fluctuated and increased. The fluctuation of cultivated land and grassland area decreased. Specifically, the proportion of unused land increased from 35.21% (1,075,888.26 hm^2) in 2000 to 36.43% (1,122,641.01 hm^2) in 2019. The proportion of forest land increased from 5.18% (158,374.26 hm^2) in 2000 to 11.37% (350,380.26 hm^2) in 2019, the proportion of glacial snow area increased from 0.29% (8892.72 hm^2) in 2000 to 3.34% (102,892.68 hm^2) in 2019, the proportion of construction land increased from 0.0015% in 2000 (46.08 hm^2) to 0.0048% (147.60 hm^2) in 2019, the proportion of cultivated land decreased from 0.31% (9324.27 hm^2) in 2000 to 0.28% (8717.04 hm^2) in 2019, and the proportion of grassland area decreased from 58.4668% (1,786,700.07 hm^2) in 2000 to 48.34% in 2019 (1,489,829.58 hm^2) (Table 2).

Table 2. Changes in land use area and proportion of Qilian Mountain National Park (Gansu Area) from 2000 to 2019 (units: hm^2, %).

Land Use Types	Area and Proportion					2000–2019 Rate of Change/%	Dynamics of Single Land Use/%
	2000	2005	2010	2015	2019		
Farmland	9324.27 0.31%	12,627.63 0.40%	9274.59 0.30%	9231.30 0.30%	8717.04 0.28%	−6.51	−0.02
Forestland	158,374.26 5.18%	457,176.42 14.66%	164,074.86 5.26%	319,009.05 10.2883%	350,380.26 11.37%	121.24	0.30
Grassland	1,786,700.07 58.47%	1,166,820.21 37.42%	1,613,574.54 51.72%	1,468,309.32 47.35%	1,489,829.58 48.34%	−16.62	−0.04
Water	16,695.90 0.55%	17,494.11 0.56%	6078.24 0.19%	113,096.97 3.65%	7452.81 0.24%	−55.36	−0.14
Built-up area	46.08 0.0015%	58.41 0.0019%	22.59 0.0007%	88.47 0.0029%	147.60 0.0048%	220.31	0.55
Unused land	1,075,888.26 35.21%	1,389,957.39 44.57%	1,219,935.60 39.11%	1,103,931.09 35.60%	1,122,641.01 36.43%	4.35	0.01
Glacial snow	8892.72 0.29%	74,350.26 2.38%	106,632.45 3.42%	87,024.51 2.81%	102,892.68 3.34%	1057.04	2.64

From the perspective of land use structure, grassland was the main one, followed by unused land, forest land, and glacial snow cover, with annual average area ratios of 48.66%, 38.18%, 9.35%, and 2.45%, respectively, while the annual average area ratios of cultivated land and construction land were 0.32% and 0.0023%, respectively. In terms of change rate and up, the change range was 1057.04%, 220.31%, and 121.24%, and the dynamic degree of single land use was 2.64%, 0.55%, and 0.30%, respectively (Figure 3).

4.1.2. Land Use Change Characteristics

In this paper, with reference to the land use intensity grading method [35,43], the use intensity of the land use type in the study area was divided into five levels and assigned the corresponding index in Formula (1), with the specific land use degree detailed in Table 3.

Table 3. Assignment table for land use intensity ratings.

	Unused Land (Glacial Snow)	Water	Forestland (Grassland)	Farmland	Built-Up Area
Degree of land use	1	2	3	4	5

According to the actual situation of the study area and the division of land use intensity grades in existing studies, this paper divided them into five levels, assigned them to the grades, and obtained the land use intensity index and its changes in the four phases of Qilian Mountain National Park (Gansu Area) in 2000, 2005, 2010, 2015, and 2019 (Table 4).

Table 4. Land use intensity and rate of change in Qilian Mountain National Park (Gansu Area).

	Land Use Intensity Index	Amount of Change in the Land Use Intensity Index	Rate of Change in Land Use Intensity
2000	2.2877	—	—
2005	2.0594	−0.2283	−9.98%
2010	2.1506	0.0912	4.43%
2015	2.1984	0.0478	2.22%
2019	2.2052	0.0069	0.31%

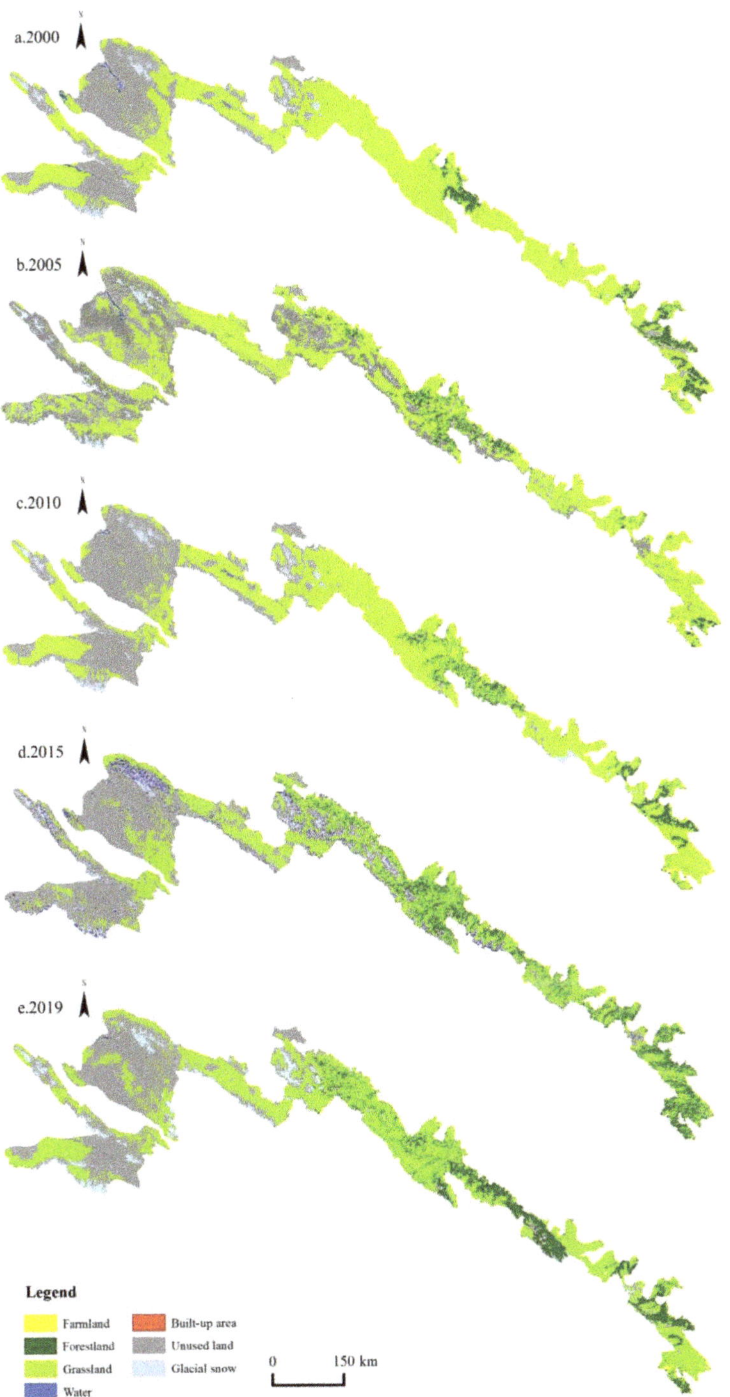

Figure 3. Land use status of Qilian Mountain National Park (Gansu Area) from 2000 to 2019. a-e represent the current status of land use in 2000, 2005, 2010, 2015, and 2019, respectively.

The variation range of Qilian Mountain National Park (Gansu Area) was large during the study period, and the change range of each study period was very different, but the land use intensity index was very low. Analysis of the results calculated according to Formulas (2) and (3), The land use intensity indices and their changes for the five periods 2000, 2005, 2010, 2015, and 2019 are shown in Table 5.

Table 5. Contribution rate of land use intensity by land type in Qilian Mountain National Park (Gansu Area).

	Farmland	Forestland	Grassland	Water	Built-Up Area	Unused Land	Glacial Snow
2000–2005	0.3271	1.8288	−0.3600	0.0268	0.2421	0.2660	7.1931
2005–2010	−0.2658	−0.6412	0.3824	−0.6527	−0.6134	−0.1226	0.4337
2010–2015	0.0014	0.9561	−0.0845	17.7203	2.9402	−0.0896	−0.1789
2015–2019	−0.0500	0.1050	0.0208	−0.9337	0.6784	0.0231	0.1895
2000–2019	−0.0731	1.1936	−0.1732	−0.5574	2.1760	0.0346	10.4723

From the study period from 2000 to 2019, the contribution rate of forest land, construction land, unused land, and glacial snow cover was positive, while for cultivated land, grassland, and water area, it was negative. The contribution rate of glacial snow cover was 10.4723, and the contribution rate of construction land and forest land was also relatively large and positive, indicating that during the study period, a very small portion of arable land, grassland, and water areas in Qilian Mountain National Park (Gansu Area) was developed or developed into glacial snow, construction land, or woodland. Specifically:

First, from 2000 to 2005, the contribution rate of arable land, forest land, water area, construction land, unused land, and glacial snow cover was positive, only grassland contributed negatively. The contribution rate of glacial snow cover was the largest at 7.1931, and the contribution rate of forest land and cultivated land was 1.8288 and 0.3271, respectively. This shows that from 2000 to 2005, grassland was developed or developed into arable land, forest land, water, construction land, unused land, or glacial snow.

Second, from 2005 to 2010, the contribution rate of grassland and glacial snow cover was positive, while the contribution rate of cultivated land, forest land, water area, construction land, and unused land was negative, and the contribution rate of water area was the largest and negative. This indicates that from 2000 to 2005, arable land, forest land, water areas, construction land, and unused land were developed or developed into grassland or glacier snow.

Third, from 2010 to 2015, the contribution rate of arable land, forest land, water area, and construction land was positive, while the contribution rate of grassland, unused land, and glacial snow cover was negative, and the contribution rate of water area was the largest and positive. This illustrates that grassland, unused land, and glacial snow cover were developed or developed into arable land, forest land, water area, or construction land.

Fourth, from 2015 to 2019, the contribution rate of forest land, grassland, construction land, unused land, and glacial snow cover was positive, while the contribution rate of cultivated land and water area was negative, and the contribution rate of construction land was the largest and positive. This shows that, from 2015 to 2019, cultivated land and water areas were developed or developed into forest land, grassland, construction land, unused land, or glacier snow.

4.2. The Value of Ecosystem Services

4.2.1. The Temporal Variation Characteristics of the Total Value of the Service

From the perspective of the total value of ecosystem services of land types, the total value of ecosystem services in Qilian Mountain National Park (Gansu Area) from 2000 to 2019 showed a trend of fluctuation with an increase, with an overall increase of 990.2085 billion USD according to Formulas (4)–(6).

First, the total value of services increased from 542.1147 billion USD in 2000 to 3521.2048 billion USD in 2015, an increase of 182.66%, and then dropped to 1532.3232 billion USD in 2019, a decrease of 54.68%, showing a clear inverted "U" growth trend (Table 6).

Table 6. The total value and proportion of ecosystem services of each land type in Qilian Mountain National Park (Gansu Area) and the corresponding changes.

	Total Value of the Service and Percentage									
	2000		2005		2010		2015		2019	
Farmland	0.4844	0.09%	1.2263	0.14%	1.9925	0.10%	2.2898	0.06%	1.3488	0.09%
Forestland	46.7240	8.62%	252.0863	27.94%	200.1601	9.91%	449.3289	12.76%	307.8546	20.08%
Grassland	455.8161	84.08%	556.3551	61.65%	1702.1868	84.26%	1788.3872	50.79%	1131.9449	73.87%
Water	37.9062	6.99%	74.2339	8.22%	57.0635	2.82%	1225.9076	34.81%	50.3932	3.29%
Unused land	0.0007	0.01%	0.0016	0.01%	0.0013	0.01%	0.0061	0.01%	0.0062	0.01%
Glacial snow	1.1833	0.21%	18.4907	2.04%	58.6723	2.90%	55.2853	1.57%	40.7754	2.66%
Total	542.1147	100%	902.3941	100%	2020.0764	100%	3521.2048	100%	1532.3232	100%

	Amount of Change				
	2000–2005	2005–2010	2010–2015	2015–2019	2000–2019
Farmland	0.7417	0.7664	0.2972	−0.9411	0.8644
Forestland	205.3623	−51.9262	249.1688	−141.4742	261.1306
Grassland	100.5390	1145.8315	86.2005	−656.4423	676.1288
Water	36.3277	−17.1704	1168.8439	−1175.5144	12.4869
Unused land	0.0009	−0.0003	0.0046	0.0003	0.0057
Glacial snow	17.3075	40.1814	−3.3868	−14.5100	39.5920
Total	360.2794	1117.6823	1501.1284	−1988.8816	990.2085

Second, from the perspective of the total value, the total value of all types of land types showed an increasing trend, with the largest total value for grassland, the smallest and largest increase for unused land, and the smallest increase for water areas. First, the total value of grassland was the largest and showed an increasing trend, increasing from 455.8161 to 1131.9449 billion USD, an increase of 148.33%, with the average proportion being 70.9316%. Second, the total amount of unused land was the smallest, but its increase was the largest—that is, from 0.0007 billion USD in 2000 to 0.0062 billion USD in 2019, an increase of 853.95%. Third, the total value of water areas increased the least, from 37.9062 billion USD in 2000 to 50.3932 billion USD in 2019, an increase of 32.94% (Table 6).

4.2.2. Spatial Variation Characteristics of the Total Service Value

From the perspective of the spatial total ecosystem service value of land type, the total ecosystem service value of Qilian Mountain National Park (Gansu Area) from 2000 to 2019 showed the characteristics of high northwest and low southeast values. In 2000, it was mainly high in the Western Arctic, while other regions were mainly moderately distributed. In 2005, it was mainly high in the west and arctic, while other regions were low and very low. In 2010, it was dominated by extremely high in the northwest, and in 2015, it was basically the same as in 2010. In 2019, the northwest was dominated by extremely high, the middle region was dominated by very low, and the southeast region was dominated by medium and low values (Figure 4).

4.2.3. The Function of the Service and the Changing Characteristics of the Value of the Individual Service

The service functions of Qilian Mountain National Park (Gansu Area) were analyzed from the perspective of supply, regulation, support, and cultural services, mainly based on regulation services, followed by support, supply, and cultural services, all showing obvious growth trends, increasing by 181.77%, 183.90%, 196.19%, and 170.38%, respectively (Table 7).

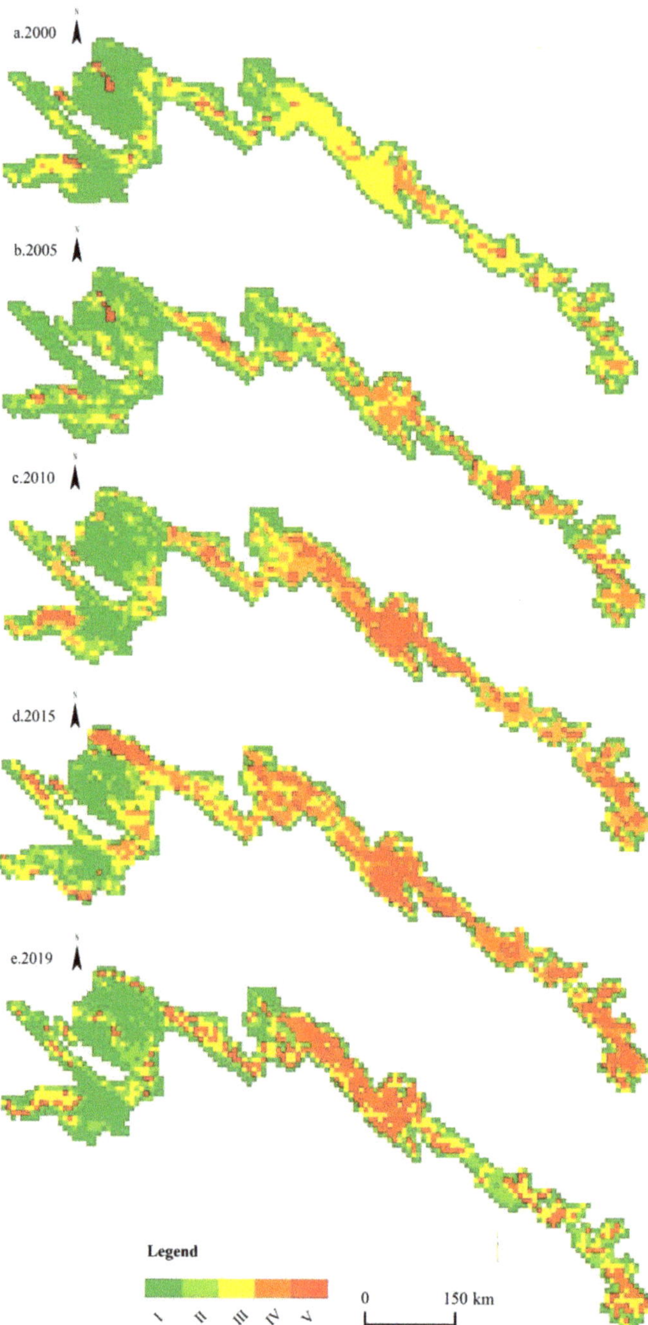

Figure 4. Value characteristics of land ecosystem services in Qilian Mountain National Park (Gansu area), 2000–2019. (**a–e**) represent the value characteristics of land ecosystem services in 2000, 2005, 2010, 2015, and 2019, respectively. I–V mainly represent the intensity ranking of the total value of ecosystem services from low to high, classified according to the five-level natural fracture method in ArcGIS software.

Table 7. Total value and proportion of ecosystem services by land type in Qilian Mountain National Park (Gansu Area) and the corresponding changes (billion USD %).

Service Features	Supply Services	Conditioning Services	Support Services	Cultural Services
2000	34.8308	356.9154	125.0125	25.3560
	6.42	65.84	23.06	4.68
2005	60.1541	599.3683	203.6830	39.1886
	6.67	66.42	22.57	4.34
2010	137.2245	1323.3005	466.6521	92.8994
	6.79	65.51	23.1	4.6
2015	238.6677	2489.8058	642.5233	150.2079
	6.78	70.71	18.25	4.27
2019	103.1654	1005.6898	354.9110	68.5569
	6.73	65.63	23.16	4.47
2000–2019	68.3346	68.3346	68.3346	68.3346
Amount/rate of change	196.19%	181.77%	183.90%	170.38%

From 2000 to 2019, the single service functions of Qilian Mountain National Park (Gansu Area) were mainly based on climate, water, and soil regulation and remained basically stable, accounting for an average of 24.04% and 25.63%. Meanwhile, soil conservation, diversity, gas regulation, environmental purification, aesthetic landscape, raw material production, water supply, food production, and nutrient cycling accounted for 10.87%, 10.34%, 8.95%, 8.20%, 4.47%, 2.57%, 2.47%, 1.64%, and 0.82%, respectively. In terms of the proportion of total ecological service value, the proportion of total ecological service value basically maintained a growth trend, except for air purification, water and soil regulation, soil conservation, diversity, and aesthetic landscape, while the proportion of other individual service functions showed a growth trend (Table 8).

Table 8. Total value of ecosystem services by land type in Qilian Mountain National Park (Gansu Area) and the corresponding proportion (billion USD %).

Service Features	Individual Service Features	2000	2005	2010	2015	2019	Amount of Change (2000–2019)	Rate of Change (2000–2019)
Supply services	Food production	9.6633	14.6791	36.1903	45.2447	26.4925	16.8292	174.16%
		1.78%	1.63%	1.79%	1.28%	1.73%		
	Raw material production	14.6270	24.1152	55.0894	70.2096	42.1374	27.5105	188.08%
		2.70%	2.67%	2.73%	1.99%	2.75%		
	Water supply	10.5405	21.3599	45.9448	123.2134	34.5355	23.9950	227.65%
		1.94%	2.37%	2.27%	3.50%	2.25%		
	Subtotal	34.8308	60.1541	137.2245	238.6677	103.1654	68.3346	196.19%
Conditioning services	Gas conditioning	51.1063	83.3408	193.1937	245.3344	146.7329	95.6266	187.11%
		9.43%	9.24%	9.56%	6.97%	9.58%		
	Climate comfort	136.0364	227.8777	516.8104	644.0117	398.0173	261.9808	192.58%
		25.09%	25.25%	25.58%	18.29%	25.97%		
	Clean-up operation	45.9109	74.8265	170.1298	260.4256	129.0871	83.1762	181.17%
		8.47%	8.29%	8.42%	7.40%	8.42%		
	Soil–water regulation	123.8618	14,713.76	30,566.97	92,427.52	22,889.19	14,345.9491	167.92%
		22.85%	23.64%	21.94%	38.06%	21.66%		
	Subtotal	356.9154	599.3683	1323.30	2489.806	1005.69	648.7744	181.77%

Table 8. *Cont.*

Service Features	Individual Service Features	2000	2005	2010	2015	2019	Amount of Change (2000–2019)	Rate of Change (2000–2019)
Support services	Soil conservation	62.2532	101.1645	234.1863	297.6786	177.9187	115.6654	185.80%
		22.85%	23.64%	21.94%	38.06%	21.66%		
	Nutrient cycling	4.6870	7.6643	17.6358	22.5078	13.4347	8.7477	186.64%
		0.86%	0.85%	0.87%	0.64%	0.88%		
	Diversity	58.0724	94.8543	214.8300	322.3370	163.5577	105.4854	181.64%
		10.71%	10.51%	10.63%	9.15%	10.67%		
	Subtotal	125.0125	203.6830	466.6521	642.5233	354.9110	229.8985	1.8390
Cultural services	Aesthetic landscape	25.3560	39.1886	92.8994	150.2079	68.5569	43.2010	170.38%
		4.68%	4.34%	4.60%	4.27%	4.47%		
	Subtotal	25.3560	39.1886	92.8994	150.2079	68.5569	43.2010	170.38%

4.3. Sensitivity Analysis

According to the sensitivity analysis in Formula (7) of the 50% increase in the value coefficient of ecological services, the sensitivity index of different land use types was very different, but there was little difference between different years of the same type, and the sensitivity index was less than 1. Among them, grassland had the largest sensitivity index, while arable land had the lowest sensitivity index. The total value of ecosystem services in the study area was not elastic to the value coefficient, so the value coefficient used in this calculation was suitable for Qilian Mountain National Park (Gansu Area), and the results are credible (Table 9).

Table 9. Ecosystem service value sensitivity index by land type in Qilian Mountain National Park (Gansu Area).

	Amount of Change				
	2000	2005	2010	2015	2019
Farmland (VC ± 50%)	0.00089	0.00136	0.00099	0.00065	0.00088
Forestland (VC ± 50%)	0.08619	0.27935	0.09909	0.12761	0.20091
Grassland (VC ± 50%)	0.84081	0.61653	0.84263	0.50789	0.73871
Water (VC ± 50%)	0.06992	0.08226	0.02825	0.34815	0.03289
Unused land (VC ± 50%)	0.00000	0.00000	0.00000	0.00000	0.00000
Glacial snow (VC ± 50%)	0.00000	0.00000	0.00000	0.00000	0.00000

5. Discussion

(1) First, the land use of Qilian Mountain National Park (Gansu Area) was mainly grassland from 2000 to 2019, followed by unused land, forest land, and glacial snow, with an annual average area ratio of 48.66%, 38.18%, 9.35%, and 2.45%, respectively, during which the largest variation of glacial snow cover occurred. From 2000 to 2020, the area of water bodies increased significantly, and the desert area decreased significantly in Sanjiangyuan National Park [44], while the changes in Qilian Mountain National Park (Gansu Area) were mainly glacial snow, construction land, and forest land, reflecting that Qilian Mountain National Park (Gansu Area) has low human interference factors and obvious originality and integrity characteristics. It shows that the construction and self-repair ability of the ecosystem of Qilian Mountain National Park (Gansu Area) continue to improve, provide high-quality system service resources for the construction of the national park, provide original natural landscape, provide a more intuitive landscape system for further exerting its ecosystem service value, which is conducive to the development of a green industrial

system based on sightseeing and tourism, and lay the foundation for the optimization of the ecosystem for the construction of the national park.

(2) Second, the contribution rate of forest land, construction land, unused land, and glacial snow cover in Qilian Mountain National Park (Gansu Area) from 2000 to 2019 was positive, while that of cultivated land, grassland, and water area was negative. Among them, the contribution rate of glacial snow cover was 10.4723, and the contribution rate of construction land and forest land was relatively large and positive. Conversely, the grassland and water bodies of Sanjiangyuan National Park contributed greatly to the ecological environment of the park [45], and the evolution of land use types was related to the value of ecosystem services. This shows that a very small portion of arable land, grassland, and water areas in Qilian Mountain National Park (Gansu Area) has been developed or developed into glacial snow, construction land, or woodland, which reflects the integrity of the system. This poses a new challenge to how to realize the protection of ecosystem integrity in the construction of national parks, not only focusing on strengthening the authenticity and integrity protection of natural ecosystems in the process of national park construction but also putting forward a more severe test for the path of utilization.

(3) Third, the ESV in Qilian Mountain National Park (Gansu Area) from 2000 to 2019 showed a fluctuating growth trend on the whole, demonstrating the characteristics of high northwest and low southeast values, and showed opposite spatial characteristics with the characteristics of a high value in the northeast and a low value in the northwest of the ecosystem service value of the Yangtze River Source Park and Lancang River Source Park, The Yellow River Source Park presented the characteristics of a high value in the west and a low value in the east. [46]. Qilian Mountain National Park (Gansu Area) had the largest total grassland value, the smallest unused land value, and the smallest increase in water area. Regulating services, followed by support services, supply services, and cultural services, all showed a clear growth trend, increasing by 181.77%, 183.90%, 196.19%, and 170.38%, respectively. It shows that as an important ecological barrier in the western region of China, the Qilian Mountains play an important role in the regulation of the overall environment, and at the same time, with the opportunity of developing eco-tourism in the western region, the cultural service function of the Qilian Mountains ecosystem is well played, and the construction results of the national park are shared by the whole people.

(4) Fourth, this article took the comprehensive services of Qilian Mountain National Park (Gansu Area) as the mainstay, giving full play to the four major service functions of supply, regulation, support, and cultural services, maintaining ecological security, ensuring ecological regulation functions, providing products for a good living environment, establishing a sound long-term ecological compensation mechanism to help provide financial guarantee for the park [47], taking the development of low-carbon economy and circular economy as the main idea, and building direct product supply. The industrialization development path of national parks with indirect product supply and basic guarantee appropriately develops the construction of direct market and life-oriented product systems for agricultural production, forestry services, animal husbandry production, and fishery production according to the characteristics of the region. Agricultural production mainly relies on the natural conditions of the region to develop the production of wheat, corn, vegetables, fruits, and other green agriculture, meet the basic needs of the region, and the most suitable development of large-scale agricultural seed production and production base. Develop a forestry service system focusing on forestry breeding and renewal and better realize the breeding and renewal of forest land. In turn, high-quality natural ecosystems are used to develop circular pastoral production and suitable fishery production. In addition, it extends and cultivates business systems such as accommodation and catering, leisure vacation, culture and art, and fitness and leisure activities, such as the development of campsite products, ecological catering services, and other green and ecological tertiary industry service systems. Similarly, effective conversion mechanisms for ecosystem goods and markets should be combined and considers the use of carbon sink compensation mechanisms and ecological banks in the process of consumption or marketization of these

products. Through the systematic and intelligent sustainable use of the ecosystem of Qilian Mountain National Park (Gansu Area), we can better help the construction of national parks and become an area jointly built and shared by the people. In particular, the construction of national parks is more prominent in the construction of the people's sharing mechanism for construction results, paying attention to the integrity of the ecosystem and paying more attention to the realization of its added value and maximizing its benefits. Build a mechanism for mutual coordination and unification of direct product supply, indirect product supply and basic security system, and explore the construction of a sustainable industrial system with the goal of human and natural communities. Combined with the actual situation and industrial characteristics of China's national park construction, fully tap the cultural supply capacity of ecosystem services, provide direct product supply, such as agriculture, forestry, animal husbandry, and fishery, explore diversified indirect product supply, such as accommodation and catering, leisure vacation, culture and art, fitness and leisure activities, and more direct, systematic, and intelligent protection and supervision systems, so as to realize the effective docking of product supply and sustainable utilization (Figure 5).

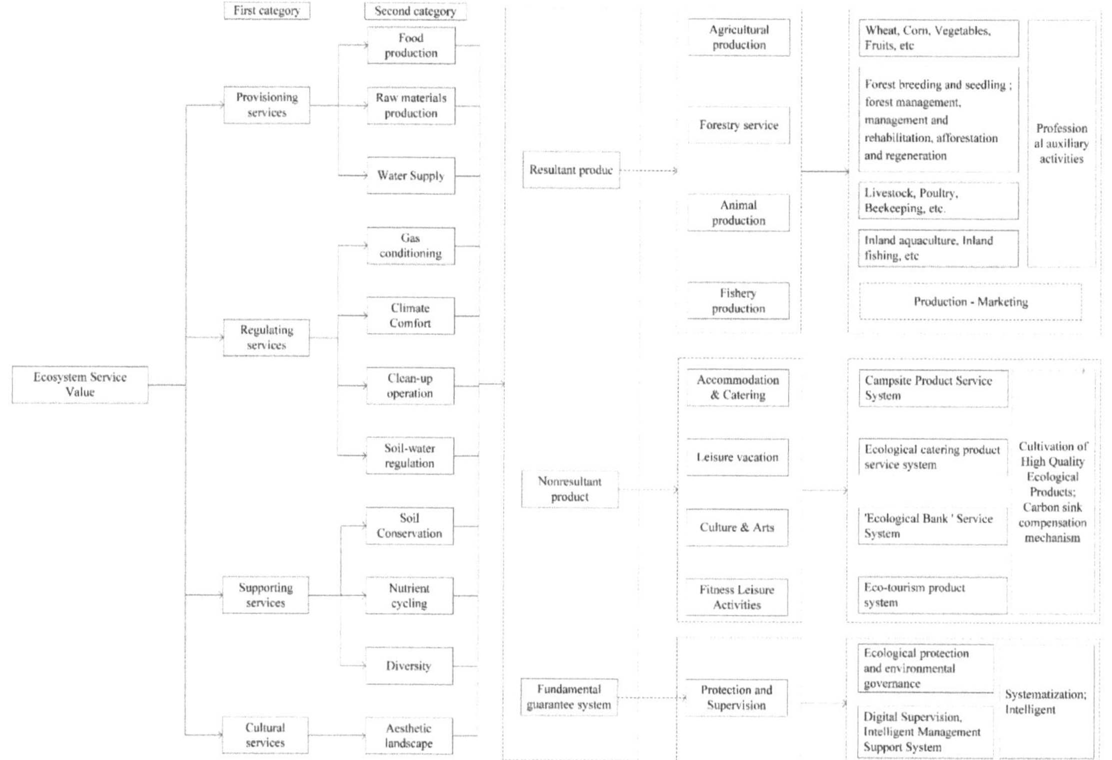

Figure 5. Industrial system construction diagram.

6. Conclusions

Based on the equivalent factor of ecosystem services, this paper calculated the ecosystem value of Qilian Mountain National Park (Gansu Area) by assetization, which provides theoretical support for its market-oriented development. From 2000 to 2019, the land use of Qilian Mountain National Park (Gansu Area) was mainly grassland, during which the largest change in glacial snow cover occurred. The value of ecosystem services in 2019 was 1532.32 billion USD, showing a clear inverted "U" growth trend, taking the development of the low-carbon and circular economies as the main idea, and putting forward the path of marketization or industrialization development of national parks. However, it is mainly based on the analysis of economic equivalent factors, and more GEP and other methods should be used in the process of method selection for in-depth research and exploration, and the research area is mainly selected for the study of Qilian Mountain National Park (Gansu Area), and the comparative study with Qinghai Area and Qilian Mountain National Park should be considered, and the actual development of the industry in the region should be explored in depth.

This paper studied the calculation of the total value of ecosystem services in Qilian Mountain National Park (Gansu Area) from 2000 to 2019, which needs to be combined with the new framework of the environmental–economic accounting–ecosystem accounting (SEEA EA) officially adopted by the United Nations in 2021 and China 2020. The annual compilation of the gross ecosystem product (GEP) calculation guide further deepens the value research of ecosystems, and the value comparison of different internal regions should also be studied in depth.

Author Contributions: Conceptualization, L.P., X.C. and C.L.; methodology, L.P. and C.L.; software, L.P. and X.Y.; validation, X.C. and L.P.; resources, L.P.; data curation, L.P. and X.Y.; writing—original draft preparation, L.P. and X.Y.; writing—review and editing, L.P., X.Y. and C.L. All authors have read and agreed to the published version of the manuscript.

Funding: Open Project of Institute of County Economy Developments & Rural Revitalization Strategy, Lanzhou University (xyjj2019007).

Data Availability Statement: Not applicable.

Conflicts of Interest: The authors declare no conflict of interest.

References

1. Yang, Q.; Wang, X.H. Evaluating relationship between water projects and ecological protection from the respect of Dujiang Weir. *China Water Resour.* **2020**, *3*, 22–24.
2. Xu, L.L.; Yu, H. Research progress of international landscape evaluation, protection, and utilization of national parks and implications to China. *Resour. Sci.* **2022**, *44*, 1520–1532. [CrossRef]
3. Wang, S.K.; Jiao, Y. On realization of the concept of 'benefit to all' in legislation on national parks. *J. Southeast Univ. (Philos. Soc. Sci.)* **2021**, *23*, 50–59.
4. Qiu, X.; Xiao, Y.; Shi, L.; Wang, H.M.; Liu, Y.H.; Sun, H.L. Assessment of ecological conservation benefit in the Inner Mongolia based on ecological assets. *Acta Ecol. Sin.* **2022**, *42*, 5255–5263.
5. Lv, Z.X.; Zhao, T.W. Pluralistic Co-governance System Construction of Qilian Mountain National Park. *J. Northwest Minzu Univ. (Philos. Soc. Sci.)* **2021**, *4*, 82–88.
6. Chang, Z.Q.; Xu, B.L.; Wang, J.Y.; Chang, X.X.; Wang, Y.L. Forest Grazing Ecosystem and the Sustainable Management in Qilian Mountains. *J. Northwest For. Univ.* **2001**, *S1*, 66–69.
7. Wang, Y.Z.; Zeng, G. Research on the construction of the Silk Road Cultural Heritage Corridor from the spatial perspective—A case of Gansu section. *World Reg. Stud.* **2022**, *31*, 862–871.
8. Dong, S.C.; Li, B.; Xue, M.; Li, Z.H. Comparative analysis and discussion on tourism development between Gansu and six neighboring provinces and regions. *Gansu Soc. Sci.* **2008**, *1*, 240–243.
9. Daily, G.C. *Nature's Service: Societal Dependence on Natural Ecosystems*, 4th ed.; Island Press: Washington, DC, USA, 1997; pp. 49–68.
10. Xie, G.D.; Xiao, Y.; Zhen, L.; Lu, C.X. Study on ecosystem services value of food production in China. *Chin. J. Eco-Agric.* **2005**, *3*, 10–13.
11. Wen, Y.L.; Zhang, X.L.; Wei, J.H.; Wang, X.L.; Cai, Y.J. Temporal and spatial variation of ecosystem service value and its trade-offs and synergies in the peripheral region of the Poyang Lake. *Sci. Geogr. Sin.* **2022**, *42*, 1229–1238.
12. Yang, Y.T.; Yin, Y.; Jiang, K.D. Ecological security pattern planning of coastal zone in Maoming city from land-sea integration perspectives. *Mar. Environ. Sci.* **2022**, *41*, 705–713.
13. Ouyang, Z.Y.; Wang, X.K.; Miao, H. A primary study on Chinese terrestrial ecosystem services and their ecological-economic values. *Acta Ecol. Sin.* **1999**, *19*, 19–25.
14. Frelichova, J.; Vackar, D.; Lorencova, E. Integrated assessment of ecosystem services in the Czech Republic. *Ecosyst. Serv.* **2014**, *8*, 110–117. [CrossRef]
15. Costanza, R.; d'Arge, R.; DeGroot, R.; Farber, S.; Grasso, M.; Hannon, B.; Limburg, K.; Naeem, S.; ONeill, R.V.; Paruelo, J.; et al. The value of the world's ecosystem services and natural capital. *Nature* **1997**, *387*, 253–260. [CrossRef]

16. DeGroot, R.S.; Wilson, M.A.; Boumans, R.M. A typology for the classification, description and valuation of ecosystem functions, goods and services. *Ecol. Econ.* **2002**, *41*, 393–408. [CrossRef]

17. Small, N.; Munday, M.; Durance, I. The challenge of valuing ecosystem services that have no material benefits. *Glob. Environ. Chang.-Hum. Policy Dimens.* **2017**, *44*, 57–67. [CrossRef]

18. Grabowski, J.H.; Brumbaugh, R.D.; Conrad, R.F.; Keeler, A.G.; Opaluch, J.J.; Peterson, C.H.; Piehler, M.F.; Powers, S.P.; Smyth, A.R. Economic Valuation of Ecosystem Services Provided by Oyster Reefs. *Bioscience* **2012**, *62*, 900–909. [CrossRef]

19. Ding, M.M.; Liu, W.; Xiao, L.; Zhong, F.X.; Lu, N.; Zhang, J.; Zhang, Z.H.; Xu, X.L.; Wang, K.L. Construction and optimization strategy of ecological security pattern in a rapidly urbanizing region: A case study in central-south China. *Ecol. Indic.* **2022**, *136*, 108604. [CrossRef]

20. Morshed, S.R.; Fattah, M.A.; Haque, M.N.; Morshed, S.Y. Future ecosystem service value modeling with land cover dynamics by using machine learning based Artificial Neural Network model for Jashore city, Bangladesh. *Phys. Chem. Earth* **2022**, *126*, 103021. [CrossRef]

21. Sutton, P.C.; Duncan, S.L.; Anderson, S.J. Valuing Our National Parks: An Ecological Economics Perspective. *Land* **2019**, *8*, 54. [CrossRef]

22. Chan, K.M.A.; Satterfield, T.; Goldstein, J. Rethinking ecosystem services to better address and navigate cultural values. *Ecol. Econ.* **2012**, *74*, 8–18. [CrossRef]

23. Castro, A.J.; Verburg, P.H.; Martin-Lopez, B.; Garcia-Llorente, M.; Cabello, J.; Vaughn, C.; Lopez, E. Ecosystem service trade-offs from supply to social demand: A landscape-scale spatial analysis. *Landsc. Urban Plan.* **2014**, *132*, 102–110. [CrossRef]

24. Zeng, Y.X.; Zhong, L.S.; Yu, H.; Zhou, B. The structure of recreational ecosystem services: A case study of the Sanjiangyuan National Park. *Acta Ecol. Sin.* **2022**, *42*, 5653–5664.

25. Li, J.H.; Huang, L.; Cao, W.; Wu, D. Accounting of gains and losses of ecological assets in counties of key ecological function regions in Yangtze River Delta. *J. Nat. Resour.* **2022**, *37*, 1946–1960. [CrossRef]

26. Guo, F.Y.; Tong, L.J.; Qiu, F.D.; Li, Y.M. Spatio-temporal differentiation characteristics and influencing factors of green development in the eco-economic corridor of the Yellow River Basin. *Acta Geogr. Sin.* **2021**, *76*, 726–739.

27. Lu, C.P.; Hou, M.C.; Liu, Z.L.; Li, H.J.; Lu, C.Y. Variation Characteristic of NDVI and its Response to Climate Change in the Middle and Upper Reaches of Yellow River Basin, China. *IEEE J. Sel. Top. Appl. Earth Obs. Remote Sens.* **2021**, *14*, 8484–8496. [CrossRef]

28. Wang, S.L.; Qu, Y.B.; Zong, H.N.; Zhang, Y.J.; Guan, M.; Zhang, Y. Research on multi-dimensional decomposition and conduction path of territory spatial pattern at the municipal level. *J. Nat. Resour.* **2022**, *37*, 2803–2818. [CrossRef]

29. Wu, D.T.; Wang, X.; Liu, H.H.; Hu, J.; Wang, G.Q. Building a National Park System Based on World Natural Heritage Sites. *Study Nat. Cult. Herit.* **2022**, *7*, 61–70.

30. Fan, Z.L.; Li, N.; Li, W.M. Assessment of ecosystem service value in Qilian Mountain National Park, Gansu Province. *Contemp. Econ.* **2021**, *12*, 70–77.

31. Wang, Z.Y.; Shi, P.J.; Zhang, X.B.; Yao, L.T.; Tong, H.L. Grid-scale-based ecological security assessment and ecological restoration: A case study of Suzhou district, Jiuquan. *J. Nat. Resour.* **2022**, *37*, 2736–2749.

32. Su, J.D.; Pu, J.L.; Li, G.X.; Zhao, X.J. Study on ecosystem benefit evaluation of Qilian Mountains National Nature Reserve of Gansu Province. *Ecol. Sci.* **2021**, *40*, 89–94.

33. Zhang, C.Y.; Bai, Y.P.; Yang, X.D.; Li, L.W.; Liang, J.S.; Wang, Q.; Chen, Z.J. Identification of ecosystem service bundles in Ningxia Plain under multi-scenario simulation. *Geogr. Res.* **2022**, *41*, 3364–3382.

34. Zhang, N.L. Dynamic Changes of Land Use and Ecosystem Services Value in Altai Mountains. *Chin. J. Soil Sci.* **2022**, *53*, 1286–1294.

35. Wang, J.; Zhou, S.K.; Meng, F.L.; Zhang, L. Impacts of Land Use Evolution on the Evaluation of Ecosystem Service Value and Temporal Variation in the Lugu Lake Basin. *J. West China For. Sci.* **2022**, *51*, 34–42.

36. Cheng, G.B.; Ju, X.Q. Response of Ecosystem Service Value to Land Use Change Based on RS and GIS Technology: Taking Urumqi City Circle as an Example. *J. Nat. Resour.* **2021**, *37*, 169–175.

37. Zhu, H.; Li, X. Discussion on the Index Method of Regional Land Use Change. *Acta Geogr. Sin.* **2003**, *58*, 643–650.

38. Tan, Z.; Guan, Q.Y.; Lin, J.K.; Yang, L.Q.; Luo, H.P.; Ma, Y.R.; Tian, J.; Wang, Q.Z.; Wang, N. The response and simulation of ecosystem services value to land use/land cover in an oasis, Northwest China. *Ecol. Indic.* **2020**, *118*, 106711. [CrossRef]

39. Xie, G.D.; Zhen, L.; Lu, C.X.; Xiao, Y.; Chen, C. Expert Knowledge Based Valuation Method of Ecosystem Services in China. *J. Nat. Resour.* **2008**, *5*, 911–919.

40. Sutton, P.C.; Costanza, R. Global estimates of market and non-market values derived from nighttime satellite imagery, land cover, and ecosystem service valuation. *Ecol. Econ.* **2002**, *41*, 509–527. [CrossRef]

41. Xie, G.D.; Lu, C.X.; Cheng, S.K. Progress in Evaluating the Global Ecosystem services. *Resour. Sci.* **2001**, *6*, 5–9.

42. Zhang, F.; Yusufujiang, R.S.L.; Aierken, T.E.S. Spatio-temporal change of ecosystem service value in Bosten Lake Watershed based on land use. *Acta Ecol. Sin.* **2021**, *41*, 5254–5265.

43. Huang, M.Y.; Fang, B.; Yue, W.Z.; Feng, S.R. Spatial differentiation of ecosystem service values and its geographical detection in Chaohu Basin during 1995–2017. *Geogr. Res.* **2019**, *38*, 2790–2803.

44. Zhuang, D.F.; Liu, J.Y. Study on the Model of Regional of differentiation of Land Use Degree in China. *J. Nat. Resour.* **1997**, *2*, 10–16.

45. Wang, Z.F.; Xu, J. Impacts of land use evolution on ecosystem service value of national parks: Take Sanjiangyuan National Park as an example. *Acta Ecol. Sin.* **2022**, *42*, 6948–6958.
46. Zheng, D.F.; Hao, S.; Lv, L.T.; Xu, W.J.; Wang, Y.Y.; Wang, H. Spatial-temporal change and trade-off/synergy relationships among multiple ecosystem services in Three-River-Source National Park. *Geogr. Res.* **2020**, *39*, 64–78.
47. Wu, P.; Zhang, H. Assessment of Appropriate Standards for Ecological Compensation in Sanjiangyuan National Park: Based on the Perspective of Value Supply of Ecosystem Services. *Qinghai Soc. Sci.* **2022**, *1*, 50–58.

 *land*

Article

Constructing a Model of Government Purchasing of Ecological Services: Evidence from China's Northeast Tiger and Leopard National Park

Hongge Zhu [1], Yutong Zhang [1], Yaru Chen [2,*], Menghan Zhao [1] and Cao Bo [1]

[1] School of Economics and Management, Northeast Forest University, Heilongjiang 150040, China
[2] Development Research Center, National Forestry and Grassland Administration, Beijing 100714, China
* Correspondence: chenyaru09@126.com

Abstract: The harmonious coexistence of man and nature is the primary goal of the establishment of national parks. Creating an ecological service supply model that takes into account the efficiency of ecological services, the fairness of residents' livelihoods, and the reasonable distribution of rights and responsibilities is an important way of achieving that goal. China's Northeast Tiger and Leopard National Park (NTLNP) is a typical national park with state-owned forest land as the main body. Before the establishment of the national park, state-owned forest enterprises (SOFEs) and local government forest departments (LGFDs) were always the undertakers of ecological services. Issues such as the distribution of rights and responsibilities between the NTLNP Administration, SOFEs, and LGFDs and the livelihood of forest workers need to be resolved urgently. This study takes the NTLNP as the study area and constructs a model of government purchasing of ecological services. The main results show the following: (1) The driving factors of the government purchasing of ecological services are increasing the workload of ecological services, the need for workforce transfer, and the optimization of subsidy standards. (2) In the construction of the responsibility system, the NTLNP Administration is the purchaser, SOFEs and Protection Stations are the undertakers, and groups such as third-party institutions and the public are the Supervisors and Evaluators. (3) Setting the purchase price in 2022 at CNY 47,654.44 per person while maintaining an average annual growth rate of 6.10% will match the per capita wage income level of urban workers nationwide in 2035. Based on the research results, it is proposed that payment for ecosystem services (PES) and ecological compensation (EC) have mature research paradigms in solving the problems of efficiency and fairness, but government purchasing of ecological services is a more appropriate policy tool in terms of arranging rights and responsibilities. This study attempts to construct a model of government purchasing of ecological services in order to provide a useful reference for national parks with state-owned land as the main body.

Keywords: Northeast Tiger and Leopard National Park (NTLNP); government purchasing of ecological services; payment for ecosystem services (PES); ecological compensation (EC); state-owned forest enterprises (SOFEs)

Citation: Zhu, H.; Zhang, Y.; Chen, Y.; Zhao, M.; Bo, C. Constructing a Model of Government Purchasing of Ecological Services: Evidence from China's Northeast Tiger and Leopard National Park. *Land* **2022**, *11*, 1737. https://doi.org/10.3390/land11101737

Academic Editor: Rui Yang

Received: 31 August 2022
Accepted: 5 October 2022
Published: 8 October 2022

Publisher's Note: MDPI stays neutral with regard to jurisdictional claims in published maps and institutional affiliations.

1. Introduction

For national parks with state-owned land as the main body, the government is usually the provider of ecological services. Across the world, land in most national parks is owned by the central or federal government [1]. The provision of ecological services by the government is one of the best ways to solve the problem of positive externalities of ecological services. However, the government still faces many problems in providing ecological services. First, the government is not only the provider but also the producer of ecological services and faces the problem of low efficiency [2]. For example, national park management departments are scattered, and the management objectives of various departments are mixed, resulting in weakened protection power; national parks directly managed by local governments are prone to the problem of focusing on development and ignoring protection [3]. Second, the government has strict ecological protection responsibilities. If the traditional production and lifestyle of residents in the national park is restricted and alternative approaches have not been formed, strict protection will have a negative impact on residents' income [4–6]. Third, there is the problem of distribution of rights and responsibilities between the government and multi-stakeholders after the establishment of the national park [7,8]. Therefore, it is of great functionality and research significance to explore an ecological service supply model that takes into account the efficiency of supplying ecological services and improves residents' livelihood and the sensible distribution of rights and responsibilities.

China's Northeast Tiger and Leopard National Park (NTLNP) is a typical national park dominated by state-owned land, the main protection targets are forest ecosystems with Siberian tigers and Siberian leopards as flagship species. From the early days of the founding of the People's Republic of China in 1949 to the establishment of the NTLNP, the ecological services within the NTLNP have mainly been undertaken by the state-owned forest enterprises (SOFEs) (also known as Forest Bureau) and the local government forest departments (LGFDs). The workers of SOFEs and LGFDs who are engaged in afforestation, tending, and management are the most direct producers of ecological services. After the establishment of the NTLNP, problems related to the supply of ecological services gradually surfaced. First, the wage income level of the workers who engaged in ecological services in the NTLNP is lower than that of workers in the same province and across the country, and significantly lower than that of on-the-job workers in the forest and grass industry[1]. By comparing the income sources of sample worker families engaged in ecological services inside and outside the NTLNP, it was found that the wage income level of workers and families engaged in ecological services within the NTLNP is lower than that of worker families engaged in ecological services outside the park [9]. Second, because the protection and management of various natural resource assets in the park and the control of land and space use are all performed by the NTLNP Administration [10], the distribution of rights and responsibilities between the NTLNP Administration, SOFEs, and LGFDs must be clearly defined. Therefore, the NTLNP urgently needs to construct an ecological services supply model that embeds ecological, social, and management goals.

Government purchasing of public services is a means to improve government administrative efficiency and the quality of public services. It is widely used in promoting social justice and improving the environment [11]. This study is based on the first-hand data obtained from an investigation of the NTLNP in 2020, based on the theory of government purchasing of public services. The government purchasing of ecological services model is constructed with the five following components: institutional environment analysis, driving factor analysis, responsibility system construction, purchase price strategy, and the whole process evaluation chain. It attempts to address the following three core questions: (1) Why purchase?—Analyze the drivers of government purchases of ecological services. (2) How to purchase?—Clarify the distribution of rights and responsibilities of multiple stakeholders in the government's purchasing of ecological services. (3) How much?—Develop pricing strategies for government purchases of ecological services. Solving the above problems is of great significance for constructing a government purchasing ecological services model that can be used as a reference.

The government purchasing of ecological services is still in the exploratory stage in terms of theoretical system construction and practical operation. Compared with the literature, the marginal contribution of this study is mainly reflected in the following three aspects: (1) Based on the research question, we explore the model of government purchasing of ecological services, and further clarify the elements of the model. The current international concept that is most similar to the government purchasing of ecosystem services is government-funded payment for ecosystem services (PES) [12]; China's Sloping Land Conversion Program and Natural Forest Protection Program are representative of such PES projects [13]. Current research themes focus on the effect evaluation after the implementation of the project [14–16]. There are no studies specifically addressing normative processes for government-funded PES projects. (2) From a theoretical perspective, most existing studies are based on the idea of ecological economics and take the Coase Theorem and Pigou Theory as their theoretical foundation. The research perspective is how to incentivize and compensate producers of ecological services. The research on PES represented by Wunder has formed a mature analytical framework [17]. It is applicable to situations where the beneficiaries of ecosystem services are easily defined. There is a lack of research on government-funded PES, which is difficult for users to identify and define. This study focuses on the theoretical basis of public economics. The research is based on the creation of a policy tool that embeds ecological, social, and management objectives. It is more applicable to the issue of the supply mode of ecological services in national parks under state-owned property rights. (3) In terms of research method, based on the minimum wage standard method and the opportunity cost method, this study formulates a pricing plan for the government purchasing of ecological services. This plan includes aspects ranging from meeting the basic living needs of ecosystem service producers to covering their opportunity costs. The connotation of the price of government purchasing of ecological services and the standard of EC are similar. The calculation methods of EC standards can be roughly divided into two types: based on the results of ecological services [18–20] and the opportunity cost of, or willingness to pay, personnel engaged in ecological services [21–23]. However, there is no precedent for formulating compensation standards from the perspective of meeting the decent living needs of ecological services practitioners. There is also no decision-making range and development space that can be used as a reference for the price setting of the government's purchasing of ecological services.

The rest of this study is structured as follows. Section 2 provides the research background, research area, theoretical analysis framework, data, and methods. Section 3 elaborates the research results for five components: institutional environment analysis, driving factor analysis, responsibility system construction, purchase price strategy, and the whole process evaluation chain. Section 4 provides a discussion of the results, and Section 5 sets out conclusions and policy implications.

2. Materials and Methods

2.1. Research Background

Based on the experience of developed countries in the management of nature reserves over the past 100 years, China proposes to establish a national park system. Since 2016, China has successively launched 10 national park system pilots, including Sanjiangyuan, Wuyishan, and Siberian Tiger and Leopard, involving 12 provinces. The ownership by the whole population of natural resource assets in the national parks is exercised by the central government and provincial governments at different levels. Among them, the ownership of natural resource assets owned by the whole population in the NTLNP is directly exercised by the central government.

After a five-year pilot period, in October 2021, China announced the official establishment of the first batch of five national parks: Sanjiangyuan, Giant Panda, Siberian Tiger and Leopard, Hainan Tropical Rainforest, and Wuyi Mountain (Table 1). The protected area is about 230,000 km^2, covering 30% of China's terrestrial national key protected wildlife species. Ecological services such as wildlife monitoring and protection, forest tending and patrolling, grazing prohibition and restoration of grasslands, and publicity and education of conservation concepts within the park require a lot of human capital and capital injection.

Table 1. Overview of China's officially established national parks.

National Park	Geographical Location and Area	Land Tenure	Land Use Type	Ecological Service Supply Method
Siberian Tiger and Leopard National Park	42°31′06″N~44°14′49″N 129°5′0″E—131°18′48″E Total area 14,926 km^2	State-owned 13,644 km^2 (91.41%) Collective 1282 km^2 (8.59%)	Woodland 1431 km^2 (95.92%) Arable land 545 km^2 (3.65%)	The national park administration established ecological public welfare posts; SOFE workers, farmers, poor households, etc. undertake ecological services.
Giant Panda National Park	28°51′03″N~34°10′07″N 102°11′10″E~108°30′52″E Total area 27,134 km^2	State-owned 19,378 km^2 (71.41%) Collective 7,756 km^2 (28.59%)	Woodland 23,231 km^2 (85.61%) Arable land 1809 km^2 (6.67%)	The national park administration jointly handled by NGOs, local communities, public welfare foundations, SOFEs, and other stakeholders to provide ecological services [24].
Wuyishan National Park	27°31′20″N~27°55′49″N 117°24′13″E~117°59′19″E Total area 1001 km^2	State-owned 335 km^2 (33.4%) Collective 667 km^2 (66.6%)	Woodland 956 km^2 (95.50%) Garden 18 km^2 (1.80%)	The national park administration undertakes ecological services and implements unified management of collective forest land. One is that the national park service purchases the ownership of the prohibited trees. The second is to implement "separation of two rights" national park agency exercising access rights, use natural forest logging subsidy and scenic spot ticket income as EC funds [25].
Sanjiangyuan National Park	32°22′36″N~36°47′53″N 89°50′57″E~99°14′57″E Total area 123,100 km^2	–	Grassland 86,832 km^2 (73.58%) Rivers, lakes, and wetlands 29,843 km^2 (25.29%) Woodland 495.2 km^2 (0.42%)	The national park administration has established ecological management and protection posts, and implemented "one post for one household", and poor households living on pastures undertake ecological services [26].
Hainan National Park	18°33′16″~19°14′16″N 108°44′32″E~110°04′43″E Total area 4402 km^2	State-owned 3553 km^2 (80.7%) Collective 849 km^2 (19.3%)	Woodland 4020 km^2 (91.30%) Garden 178 km^2 (4.04%)	The national park administration undertakes ecological services, switching between collective land and state-owned land in the park. Taking the natural village as a unit, the replacement of land ownership between the place emigrated to and the place emigrated from shall be carried out, all the land collectively owned by the peasants who moved from the land will be transferred to the state, and the original state-owned land at the move-in place is determined to be collectively owned by farmers [27].

The NTLNP is a typical national park dominated by state-owned land. The land area of key state-owned forest areas[2] The Jilin and Heilongjiang provincial governments hand over the responsibility of the owner of natural resource assets within the NTLNP to the NTLNP Administration. The NTLNP is managed according to the vertical management system of "Administration Bureau–Management Sub-bureau–Protection Station". The NTLNP Administration is an agency directly under the State Forest and Grassland Administration (SFGA), co-located with the Commissioner's Office of the SFGA in Changchun. The management branch is co-located with the SOFEs and the LGFDs. All township governments and state-owned forest farms within the scope of the national park will establish Protection Stations, exercising the responsibilities of ecological services within their respective jurisdictions (Figure 1).

Table 2. The area and number of workers of the management branch included in the NTLNP.

Management Branch	Affiliated Unit	Area			Number of People		
		Total Area (km^2)	Included in the Park Area (km^2)	Proportion (%)	Total Number of People	Number of People in NTLNP	Proportion (%)
Hunchun Bureau	Changbai Mountain Forest Industry Group	4051	2719	67.10%	1754	1226	69.90%
Tianqiaoling Bureau	Changbai Mountain Forest Industry Group	2035	1992	97.90%	3689	2791	75.66%
Wangqing Bureau	Changbai Mountain Forest Industry Group	3042	2952	97.00%	4142	3327	80.32%
Daxinggou Bureau	Changbai Mountain Forest Industry Group	1272	594	46.70%	2454	1884	76.77%
Suiyang Bureau	Longjiang Forest Industry Group	5165	2563	49.60%	4366	1172	26.84%
Muling Bureau	Longjiang Forest Industry Group	2675	679	25.40%	3,928	403	10.26%
Dongjingcheng Bureau	Longjiang Forest Industry Group	4180	712	17.00%	6228	586	9.41%
Hunchun Municipal Bureau	Hunchun Municipal Government	1403	659	47.00%	147	34	23.13%
Wangqing County Bureau	Wangqing County Government	3289	1229	37.40%	1706	1049	61.49%
Dongning Municipal Bureau	Dongning Municipal Government	3065	513	16.70%	507	149	29.39%

2.2. Study Area

The study focused on an analysis of the jurisdictions of seven SOFEs included in the NTLNP. The total area of these is 12,211 km^2, accounting for 81.81% of the total area of NTLNP. The NTLNP is located in the southern part of Laoyeling, where the two provinces of Jilin and Heilongjiang meet in China. It is the connecting area of the border between China, Russia, and North Korea. Covering 6 counties (cities), 17 townships, and 105 administrative villages in Jilin and Heilongjiang provinces, with a total area of 14,926 km^2. The physical coordinates are 42°31′06″N-44°14′49″N, 129°5′0″E-131°18′48″E (Figure 2).

Historically, when large virgin forests were not developed, Siberian tigers and leopards were distributed across the mountain plains of northeast China. With the construction of the Chinese Eastern Railway in northeast China and war aggression, forest resources in Northeast China have been severely damaged, the number of Siberian tigers has also dropped sharply from thousands to about 500 [28]. In the early days of the founding of the People's Republic of China in 1949, the forest resources of the state-owned forest areas in northeast China were important strategic materials. The high-intensity exploitation and utilization of wood seriously threatens the integrity of the ecosystem, resulting in the near extinction of the Siberian tiger and leopard in China. In the 1980s, the northeast state-owned forest area fell into a crisis of recoverable resources and an economic crisis for SOFEs. In 1998, major floods broke out in the Yangtze River, Songhua River, Nen River, and other basins. To date, key state-owned forest areas in northeast China have begun to implement natural forest protection projects, which have greatly reduced timber production. In 2015, the commercial logging of natural forests will be completely stopped, the function of SOFEs has changed from producing timber to protecting natural forests, the role of forest workers has also changed from "lumber jacks" to "forest rangers".

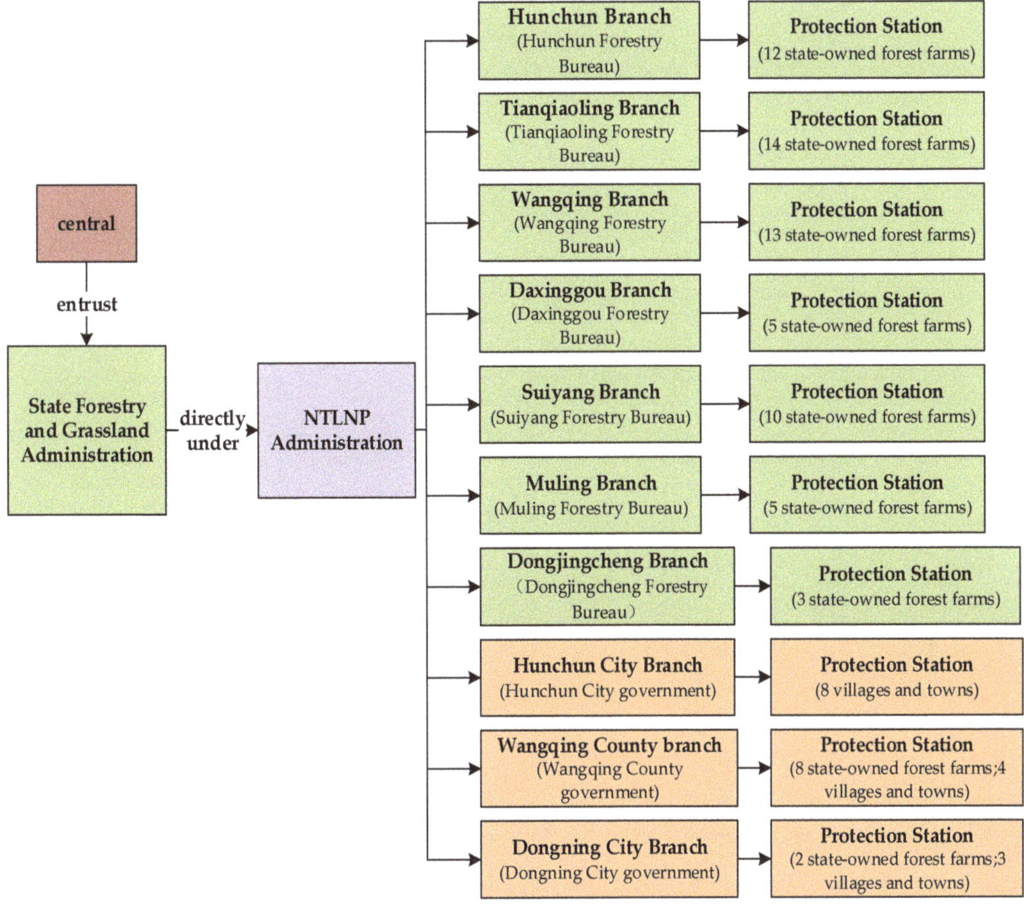

Figure 1. Organizational structure of the vertical management system of the NTLNP.

2.3. Theoretical Analysis Framework

The government purchasing of public services originated in the western administrative reform movement in the late 1970s, aiming to improve the efficiency of government administration and the quality of public services. International scholars generally believe that the connotation of government procurement of public services is similar to outsourcing public service contracts. The representative point of view, as suggested by Savas, is that "Government purchasing of public services means that the government provides public services by signing contracts with the private sector or the non-profit sector" [29]. The practice of purchasing public services by the Chinese government originated in 1998 when the Shanghai Pudong New Area Social Development Bureau purchased elderly care services from the Shanghai Young Men's Christian Association to improve the efficiency of management of the civic leisure center. Subsequently, the research of Chinese scholars in the field of government purchasing of public services has gradually increased. In discussion on the definition of government purchasing of public services based on China's national conditions, the most common view, as suggested by Wang, is that "The government will hand over the public service matters undertaken by itself to professional enterprises or social organizations through direct funding or public bidding, and finally pay the service fee according to the quantity and quality of public services provided by the undertakers" [30]. The concept of government purchasing of public services, a theoretical system based on public goods theory, new public management theory, new public service theory, governance theory, and transaction cost theory, has gradually formed. Based on the abovementioned classical public economics theory and following the logical sequence of "institutional environment analysis–driving factor analysis–responsibility system construction–purchase price strategy–whole process evaluation chain", we developed a theoretical analysis framework (Figure 3).

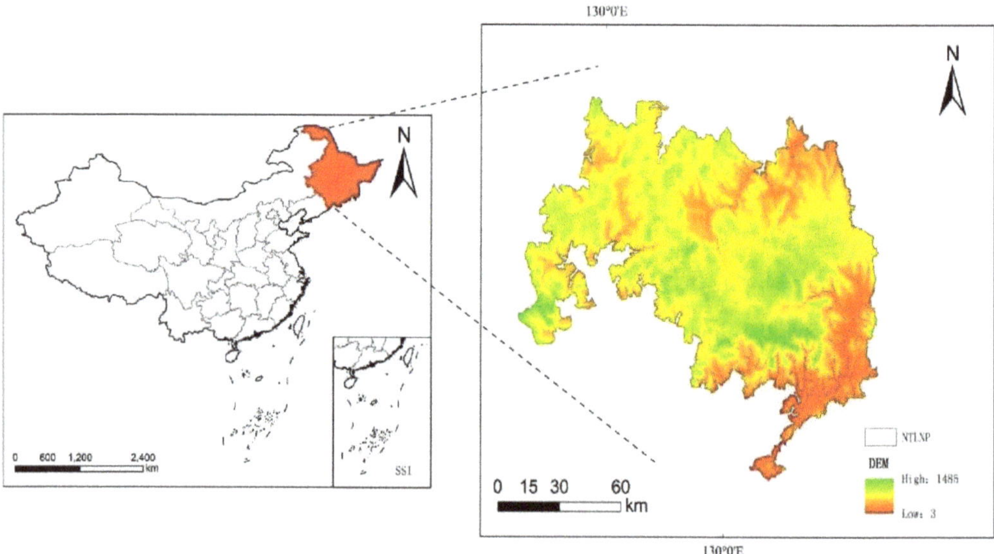

Figure 2. Study area.

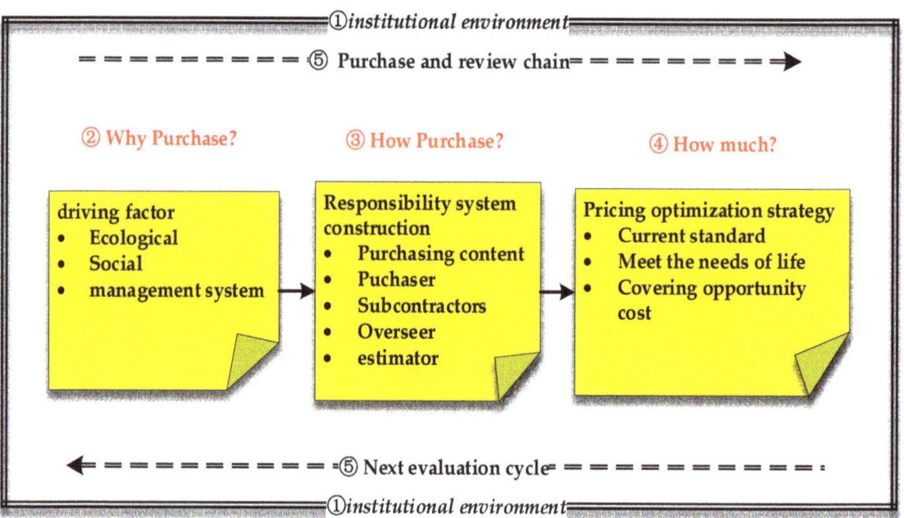

Figure 3. Theoretical analysis framework.

- Institutional environment.

First, government purchasing of services is embedded in a certain institutional environment [31], which is a prerequisite for the smooth implementation of government purchasing of ecological services. Research experience shows that different political parties in some countries have different political preferences for public services providers. For example, Gradus analyzed the production models of waste collection in all cities in the Netherlands and concluded that conservative liberals preferred government purchase or privatization, while Christian Democratic parties held a negative attitude toward privatization [32]. Ferris also suggested that the American Republican Party may be more in favor of the government purchasing of public services [33]. For China, government purchasing of public services is one of the most important governance tools, and policy documents issued by administrative agencies at all levels provide strict implementation and constraints for the government to purchase public services. For example, government procurement-related documents issued by the Ministry of Finance of the People's Republic of China will regulate the content and boundaries of government purchasing of public services, and national park development plans formulated by relevant departments such as the SFGA will clearly encourage the types of ecological services in which the government purchase public services. Therefore, giving priority to the institutional environment in which the NTLNP is located is the guarantee of practical advancement and the basis for theoretical research.

- Why purchase?

Second, the theory of public goods is the theoretical starting point for the government purchasing of ecological services, and satisfying the public interest is the primary motivation for the government purchasing of ecological services. All natural resource assets within the NTLNP are owned by everyone; ecosystem services (such as clean air, clean water) formed by strict ecological protection in national parks have the attributes of public goods or quasi-public goods. This ecological benefit has a positive externality to the surrounding residents and even the whole country [34]. As the beneficiaries of national park ecosystem services are difficult to define, the government, as the main supplier of ecological services, can make up for market failures and improve the efficiency of resource allocation. However, government organizations that integrate supply and production often lead to inefficiencies in the supply of public services or even bureaucratic systems due to scale imbalances, while government purchasing of services from social institutions is a cheaper and more flexible solution [35]. Ostrom proposed two ways for the government to provide public services: The first is produced by the government's own officials and workers; at this time, the government is both the provider and producer of public services. The second is to pay the funds to social institutions, which provide professional services to citizens. In this case, the government is the provider of public services, while the social institutions are the producers [36]. Based on this, Ostrom regarded the interests and needs of citizens as the primary motivation for the government purchasing of public services. Wang has a similar view and regards the lack of administrative resources and the limited financial capacity of local governments as the basic motivation for the government to purchase public services [30]. Lin pointed out that the more complex the social population, the greater the residents' demand for public services [37]. The NTLNP is located in the border area of northeast China, where the economic development level is underdeveloped and a large number of agricultural and forestry people are distributed in the area. Therefore, clarifying the ecological services needs of the public, especially the ecological services needs of multiple stakeholders within the NTLNP, is the logical starting point for constructing a government purchasing of ecological services model.

- How purchase?

Third, new public management theory, new public service theory, and governance theory are the theoretical basis for the construction of the responsibility system in the government purchasing of ecological services. A perfect responsibility system needs to take into account social needs and administrative efficiency by building a collaborative platform for multiple stakeholders. Therefore, the central aim of this study is to reasonably allocate the rights and responsibilities of the purchaser, undertakers, supervisors, and evaluators. The practical exploration of developed countries is the root of the evolution of the theory of government purchasing of public services, from new public management theory to new public service theory and governance theory, which reveals research trends shifting from market value priority to public value priority [38]. In the process of the new public management movement, new public management theory, which advocates improving the quality and efficiency of public service supply through market forces, emerged as the times required [39]. However, overemphasizing market efficiency would be out of touch with public needs, leading to the destruction of fairness, justice, and citizenship [40,41]. Therefore, Denhardt proposed new public service theory, the active participation of the public and the negotiation between the government, with the public being regarded as the prerequisites for the effective operation of government purchasing of public services. However, the government has reached cooperation agreements with social organizations in the form of service outsourcing, and it is becoming increasingly difficult to realize the coordination and orderly management of the public services system [42]. Governance theory has gradually become a theoretical guide to coordinate the power–responsibility relationship between participating subjects [43]. Ostrom proposed that a good governance structure can help reduce the cost of social governance. Governance structures include relationships within and between organizations, the relationship between organizations includes the relationship between the government and other social organizations as well as between its own branches. Public services reform under the guidance of governance theory pays more attention to the construction of coordination systems between departments behind public services outsourcing. In this study, the NTLNP involved multiple stakeholders such as the Administration Bureau, SOFEs and LGFDs, community residents, and social organizations. Clarifying the rights and responsibilities of multiple stakeholders is the core link of constructing a government purchasing of ecological services model.

- How much?

Fourth, transaction cost theory aims to optimize the price of government purchasing of ecological services. Relevant research on the minimum wage theory and the semi-market theory provides innovative ideas for designing the purchase price range. Williamson divides transaction costs into search costs, information costs, bargaining costs, decision-making costs, monitoring costs, and default costs. He also proposes that comparing transaction costs and organizational costs can help the government choose whether to produce services by itself or purchase services from social institutions [44]. Reducing the transaction cost is the motivation for the government to optimize the purchase price of ecological services. This study focuses on the purchase price in the bargaining cost and divides the purchase price of the government purchasing of ecological services into three stages: The first stage is the current wage level of workers engaged in ecological services within the NTLNP. The second stage is to formulate a benchmark price to meet the living needs of ecological services producers (ecological management workers) based on the minimum wage theory. In the third stage, the purchase price to compensate the opportunity cost of ecological management workers is formulated based on semi-market theory. Marx's minimum wage theory is the representative theoretical basis for the establishment of the minimum wage standard. The minimum wage should consist of the value of the means of subsistence necessary for the owner of the labor force to sustain themselves and their descendants [45]. The government purchasing of ecological services is similar to EC. The relevant theories of ecological compensation standard can be used as the theoretical basis for the price of government purchasing of ecological services. Among them, the theory of semi-market theoretical value is the core theory of ecological compensation standard [46]. The semi-market theory can establish the standard of EC when the establishment of the ecosystem service function market is difficult; the main purpose is to determine the compensation standard from the two aspects of market supply and demand, such as opportunity costing. This study integrates the classical theories of public economics, labor economics, and ecological economics to set the price of government purchasing of ecological services, aiming to meet the needs of the public and reduce government transaction costs.

- Purchase and review chain

Fifth, new public service theory, transaction cost theory, and governance theory are the theoretical foundations for designing the whole process evaluation chain. The whole process evaluation chain runs through the government purchasing of ecological services. Satisfaction of public demand and efficiency of fund use are the main components of the evaluation chain, and third-party evaluation is the primary evaluation method. Huang proposed that government departments should form a comprehensive and transparent institutional framework for purchasing public services that transcends departmentalism, and make the public participate in the demand assessment, design, and acceptance of purchasing services [47]. This study attempts to construct the whole process evaluation chain of the government purchasing of ecological services according to the characteristics of the NTLNP. The analysis framework is based on the whole process evaluation chain of government purchasing of public services constructed by Jiang, which includes project evaluation, process evaluation, and result evaluation. The chain also includes evaluation procedures and evaluation methods.

2.4. Data and Methods

2.4.1. Research and Data

The data sources consist of primary data obtained from field research and statistical yearbook data. The survey data include two parts: the symposium survey data and the questionnaire survey collection data. The survey was conducted from September to December 2020.

(1) Symposium survey data. The research object of the symposium is the NTLNP Administration and its 7 administrative branches. Participants are management cadres responsible for natural resource management, comprehensive management (financial planning, personnel management, etc.), and ecological protection and restoration in each branch. The symposium survey consists of three parts: (i) types and tasks of ecological services undertaken by ecological management and protection positions within the NTLNP; (ii) the composition and number of personnel included in the NTLNP and engaged in ecological services; (iii) salary composition and current standards for personnel assigned to the NTLNP and engaged in ecological services.

(2) Questionnaire data. The data of the questionnaire come from the "NTLNP Resident Livelihood Survey Project" launched in 2020. The survey uses a combination of computer-assisted interview techniques (CAPI) and computer-assisted telephone interview techniques (CATI) to conduct structured interviews. A multi-stage random sampling technique was used for sample selection. First, the level of SOFEs includes seven SOFEs included in the NTLNP. Secondly, in each SOFE, according to the list of forest farms and communities, 2 forest farms on the mountain and 1 community under the mountain were equally selected. Finally, in each sample forest farm and community, 10 worker families were randomly selected as sample households according to the household registration list. The structured interview consists of two parts: (i) demographic and sociodemographic characteristics of the respondents; (ii) living standard of the families of forest workers. The survey obtained the survey data of 209 sample worker families from 7 SOFEs, among which are 78 worker families engaged in ecological services. This study focuses on these 78 worker families and their family members. Unfortunately, due to COVID-19, we have not been able to visit more households engaged in ecological services.

(3) Statistical Yearbook Data. The regional average income level and minimum wage data come from the *Heilongjiang Statistical Yearbook 2020* [48], the *Jilin Statistical Yearbook* [49], and the *China Forest and Grassland Statistical Yearbook 2020* [50].

2.4.2. Research Methods

Qualitative research and quantitative research were used to analyze the different sub-problems of the central problem of "constructing the model of government purchasing of ecological services". First, in order to answer the question of "the driving factors of the government purchasing of ecological services", qualitative research was used to conduct textual analysis of the symposium data obtained from field research. From this, we can interpret the changes in the types and tasks of ecological services in the NTLNP, the composition and number of personnel assigned to the NTLNP and engaged in ecological services, and the salary composition of personnel assigned to the NTLNP and engaged in ecological services with current standards. At the same time, quantitative research was used to analyze the sample data obtained by the questionnaire survey, and the current wage income level of the workers who are assigned to the NTLNP and engage in ecological services was calculated. Second, in order to address the issue of "the distribution of rights and responsibilities of multiple stakeholders in the government purchasing of ecological services", qualitative research was used to analyze the text of policy documents. From this, the specific requirements of the central government for the participants of the government purchasing of ecological services can be obtained, and the responsibility system for the government to purchase ecological services was constructed in combination with the multiple stakeholders involved in the NTLNP. Finally, in order to answer the question of "the price at which the government purchases ecological services", the sample data obtained from the questionnaire survey were used to formulate a price strategy through quantitative research.

The pricing strategy is divided into three parts: The first part is the current income level of the personnel engaged in ecological services obtained according to the survey data. The second part refers to and improves the minimum wage standard calculation method specified in *Order No. 21 of the Ministry of Labor and Social Security of the People's Republic of China*; thus, a benchmark price that can fully meet the living needs of the ecological service workers in NTLNP was calculated. The proportion method was calculated by multiplying the per capita consumption expenditure of residents of poor households based on the survey data of urban households by the dependency coefficient of the employed person and the ratio of wage income to disposable income, plus an adjustment coefficient [51]. Appropriate adjustments were made to the selection of indicators according to the purpose of the research. The reason for the adjustment is that this study measures the price of government purchasing of ecological services, and the price should meet or improve the basic living standards of workers engaged in ecological services, not the minimum wage standard for poor households. Therefore, the indicators of poor households were replaced by the indicators of the sample workers engaged in ecological services in the NTLNP, so as to calculate the benchmark of the purchase price of ecological services through the adjusted proportion method. The calculation formula is as follows:

$$M = C \times S \times B \tag{1}$$

where M is the annual salary standard of ecological service workers in NTLNP; C is the per capita expenditure of sample worker families engaged in ecological services in the NTLNP; S is the per capita support coefficient of sample workers engaged in ecological services; and B is the ratio of the per capita wage income to the per capita total income of sample worker families engaged in ecological services in the NTLNP.

The third part, with reference to the opportunity cost method [52], takes the calculation result of the proportion method as the benchmark price to meet the basic living needs of workers, and calculates that it will catch up with the average annual growth rate of the per capita wage income of urban workers in 2035. The difference in wage income level can reflect the economic losses borne by the workers engaged in ecological services in the NTLNP due to their restricted development rights. In November 2020, the *Proposal of the Central Committee of the Communist Party of China on Formulating the Fourteenth Five-Year Plan for National Economic and Social Development and the Vision for 2035* proposed the goal of doubling the total economic volume or per capita income by 2035. The calculation formula is as follows:

$$P_1 = P_0 \times (1 + i)^n \tag{2}$$

where P_1 represents the target salary income value that should be achieved, P_0 represents the benchmark price of government purchasing of ecological services, i is the average annual growth rate of wages, and n is the number of years required for the increase.

2.4.3. Sample Description

Among the surveyed worker families, there are 78 worker families engaged in ecological services, with an average family size of 3 people and a family support coefficient per capita of 1.82. The wage income of these 78 families from SOFEs accounts for 86.88% of the total wage income, and the wages of SOFEs are the main source of family livelihood. The average wage income for SOFEs was CNY 41,056.46, and the average wage income for all workers was CNY 47,254.54 yuan. The total household income is slightly larger than the total household expenditure, at CNY 64,028.94 and CNY 62,352.06, respectively (Table 3).

Table 3. Family characteristics of ecological service workers.

Survey Item	Mean	Std	Scope	Frequency (Sample = 78)	Percentage (%)
Family members	2.62	0.81	1	5	6.41
			2	29	37.18
			3	37	47.44
			4–5	7	8.97
Support coefficient	1.82	0.66	1	12	15.38
			1–2	55	70.52
			3	9	11.54
			4	2	2.56
Salary from Forest Bureau (CNY)	41,056.46	17,650.14	≤30,000	19	24.36
			30,001–60,000	45	57.69
			60,001–90,000	12	15.39
			≥90,001	2	2.56
Income from salary (CNY)	47,254.54	20,450.83	≤30,000	16	20.51
			30,001–60,000	41	52.56
			60,001–90,000	16	20.52
			≥90,001	5	6.41
Total income (CNY)	64,028.94	23,430.11	≤30,000	4	5.13
			30,001–60,000	35	44.87
			60,001–90,000	25	32.05
			≥90,001	14	17.95
Total outcome (CNY)	62,352.06	42,803.34	≤30,000	10	12.82
			30,001–60,000	41	52.56
			60,001–90,000	14	17.95
			≥90,001	13	16.67

There are 81 workers engaged in ecological services in 78 worker families, and most of these workers are male, with 66 male workers (81.48%) and 15 female workers (18.52%). The workers engaged in ecological services are mainly middle-aged groups, with an average age of 46.07, of which 53 (65.43%) are 45-59 years old. The average number of years of education for workers is 11.47, and most of them have a high-school and junior high-school education. The average wage income of workers engaged in ecological services from SFEs is CNY 33,496.35, the number of workers with wages between CNY 30,001 and 40,000 is 49 (60.49%), and the number of workers ≤ CNY 30,000 is 24 (29.63%) (Table 4).

Table 4. Characteristics of sample workers engaged in ecological services.

Survey Item	Mean	Std	Scope	Frequency (Sample = 81)	Percentage (%)
Gender	0.81	0.39	Male	66	81.48
			Female	15	18.52
Age	46.07	7.64	25–35	12	14.81
			36–44	16	19.76
			45–59	53	65.43
Education	11.47	2.42	≤ 9	30	37.04
			10–13	31	38.27
			≥ 14	20	24.69
Salary from Forest Bureau (CNY)	33,496.35	5932.37	$\leq 30,000$	24	29.63
			30,001–40,000	49	60.49
			$\geq 40,001$	8	9.88

3. Results

3.1. Institutional Environment Analysis

The plans and measures related to government purchasing of services issued by the Ministry of Finance of the People's Republic of China and the SFGA have created a favorable institutional environment for the NTLNP to explore government purchasing of ecological services. In December 2017, the SFGA and the provincial governments of Jilin and Heilongjiang jointly formulated the *General Plan for the Northeast Tiger and Leopard National Park (2017–2025) (Draft for Comment)* (referred to here as the "Plan") [28] to encourage the government to purchase ecological services to solve the problems of ecological protection and improve people's livelihoods. The 2020 Ministry of Finance of the People's Republic of China Order No. 102 *Administrative Measures for Government purchasing of Services* (referred to as the "Measures") formulated detailed implementation rules for government purchasing of services, providing action guidelines. The details are as follows.

First, the *Plan* proposes to "explore ways of purchasing services to manage and protect state-owned natural resource assets". The specific method is to set up ecological public welfare posts, and give priority to the state-owned forest areas, forest farm reform, and diversion of workers, farmers who have abandoned farmland and prohibit grazing, and poor people who have been filed and registered, so that they can benefit from participating in the ecological protection and operation of the NTLNP. The *Plan* clarifies the three functions of ecological public welfare posts, namely, field patrol, forest tending, and resource monitoring. For the NTLNP, the government purchasing of ecological services has the functions of ecological protection and improvement of people's livelihoods.

Second, the *Measures* clearly define the government purchasing of services, the identity requirements of participating subjects, the restricted scope of purchase content, the implementation conditions of purchase activities, and the supervision and management responsibilities. It provides a reference for the construction of a government purchasing of ecological services model in NTLNP. First, the core of the definition of government purchasing of ecological services is that the government acts as a purchaser and supervisor of ecological services and social institutions as undertakers of ecological services. The government pays the corresponding fees according to the quantity and quality of ecological services to provide high-quality ecological services to the public. Second, the purchaser must be a state agency at all levels, such as the NTLNP Administration. The undertaker must be a legally established enterprise or social organization, excluding public institutions. SOFEs and Protection Stations are the best choices for this role. Third, the specific scope and content of the government purchasing of ecological services shall be managed by an instructive catalogue, and ecological services such as field patrol, forest tending, and resource monitoring within the NTLNP have been included in the government-purchased services guidance catalogue. Fourth, the government purchasing of ecological services should highlight public welfare and prioritize projects that are related to the livelihood of national park residents and are conducive to transforming government functions and improving financial performance. The purchaser implements the performance management of the purchase project and conducts performance evaluation on the implementation of the purchased services on a regular basis. Fifth, the purchaser and they should consciously accept the supervision of finance, audit, society, and service objects.

3.2. Driving Factor Analysis

The increase in the content and tasks of ecological services is the first driving factor for the government to purchase ecological services. Based on the data obtained from the symposium, the contents and tasks of ecological services after the establishment of NTLNP were summarized and the following conclusions were drawn: After the establishment of the NTLNP, the content of ecological services has been significantly expanded and the intensity and difficulty of the task have increased. In particular, management and protection services have expanded from forest resource management and protection to wildlife protection and resource monitoring.

The transfer of participants in ecological services is the second driving factor for the government purchasing of ecological services. Based on the data obtained from the symposium, the identity types of ecological services personnel were summarized, and the following conclusions were drawn: Forest workers engaged in forest tending and management in the forest departments of 7 SOFEs and 3 LGFDs are the main producers of ecological services in NTLNP. Filed and registered poor households in rural areas are secondary subjects engaged in ecological services, and community residents and one-time resettlement personnel[3] are supplementary. There are a large number of people engaged in ecological services in the NTLNP, and their identities are mixed. Ecological services participants originally belonging to SOFEs and LGFDs can only be transferred to the NTLNP by government purchasing of ecological services. It is not only helpful to solve the problem of connecting the vertical management system and the SFE management system, but it is also helpful to lift the poor population out of poverty.

Optimizing the subsidy standard for ecological services and improving the income level of forest workers are the third driving factor for the government purchasing of ecological services. According to the family data of workers and families of 7 SFEs engaged in ecological services in the NTLNP collected through questionnaires, the following conclusions are drawn: First, the wage income level of frontline workers engaged in ecological services is significantly lower than that of the urban employed population in Heilongjiang, Jilin, and the whole country. The prevailing wage system ignores the need to improve the livelihoods of workers. Specifically, the survey data involved 81 front-line workers of SFEs engaged in ecological services; in 2019, the per capita wage income was 33,496.35 yuan, and the average wage income of the urban employed population[4] in Jilin, Heilongjiang, and the whole country was CNY 36,307.87, 41,597.86, and 49,020.14, respectively. Therefore, the salary level of forest workers in the NTLNP is significantly different from the regional and national average levels. Second, the wages of forest workers come from financial subsidies issued by the central government according to the amount of afforestation, tending, and management tasks that SOFEs are responsible for each year. The subsidy standard is CNY 7500/ha for artificial arbor forest; CNY 4500/ha for replanting, reconstruction, and cultivation; CNY 1800/ha for forest tending; and CNY 75/ha for forest management. The management and protection tasks of forest workers are arranged according to the area, and the management and protection wages are hourly wages, which comprise file wages + seniority allowance + other various allowances. Piece rate wages are implemented for afforestation and tending, and the unit price for afforestation or tending is determined on the basis of the number of afforestation or tending plants per Mu. Management and care wages are the main source of income for front-line workers, but according to the current management and care wages, the wage level of young workers with the same workload is significantly lower than that of middle-aged workers with long working years. At the same time, it is unfair to SFEs with small areas and many personnel that the central government issues the total amount of subsidies based on the area under management and protection. Therefore, NTLNP undertakes the dual mission of ecological protection and improving the livelihoods of residents. Relying on the way the government purchasing of ecological services to reassess the wages of ecological management and protection workers is an important way to promote a decent life for producers of ecological services.

3.3. Responsibility System Construction

Participants in the government purchasing of ecological services include purchaser, undertakers, evaluators, and supervisors (Figure 4). The purchaser establishes a relationship with the undertaker by purchasing the ecological services provided; evaluators establish relationships with purchaser and undertakers by evaluating the performance of the whole process of government purchasing of ecological services; the supervisors establish contact with the other three subjects through the supervision of the purchasing of ecological services and the evaluation of the behavior.

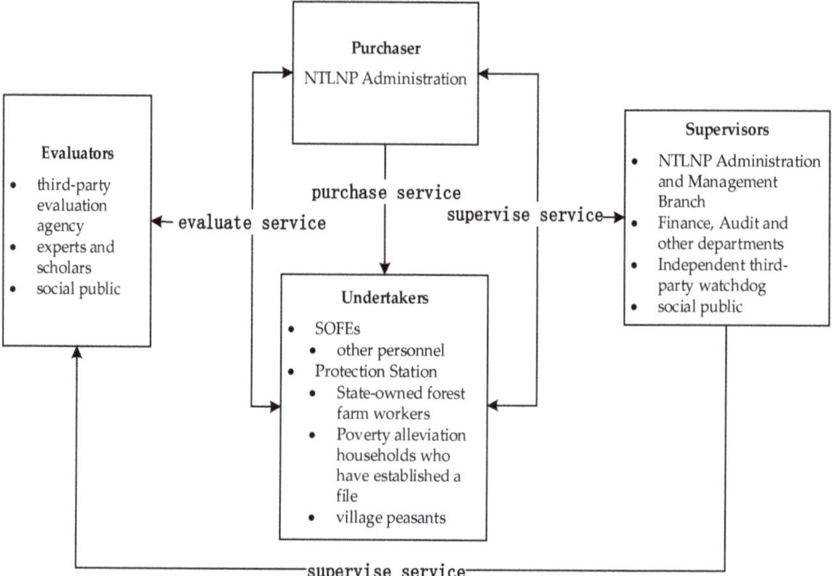

Figure 4. Construction of government purchasing ecological services responsibility system.

(1) Purchaser—NTLNP Administration.

(2) Undertakers - SOFEs and Protection Station. (i) The NTLNP involves 7 SOFEs from the Changbai Mountain Forest Group (Wangqing, Hunchun, Tianqiaoling, and Daxinggou) and the Longjiang Forest Group (Suiyang, Muling, and Dongjingcheng). It is a public welfare enterprise with forest management as its main business. At the same time, most of the forest farms established by SFEs are divided into the NTLNP as a whole, which is an important component of the Protection Stations. In addition, after the commercial logging of natural forests was completely stopped, SOFEs generated a large number of people who were transferred and diverted, and they took on ecological services through SOFEs. (ii) The NTLNP also involves 11 state-owned forest farms under the jurisdiction of three counties and cities of Hunchun City, Wangqing County, and Dongning City, which should establish Protection Stations as the main body of purchasing services. Registered and on-the-job workers from local state-owned forest farms are regarded as the main body of the ecological management and protection public welfare positions. In addition, according to the requirement that government should pay attention to public welfare in purchasing

services, the Protection Stations also provide ecological services to households that have been released from poverty and the peasants in the villages under their jurisdiction.

(3) Evaluators—Third-party evaluation agencies, experts and scholars, and the general public. The talent, technology and resource advantages of third-party institutions can ensure the fairness, impartiality and professionalism of the evaluation results. During the evaluation process, experts and scholars present professional opinions to ensure the scientific nature of the evaluation. As the direct beneficiaries of ecological services, the public can evaluate the effect of public services through information feedback mechanisms such as satisfaction surveys.

(4) Supervisors—NTLNP Administration and Management Branch, as well as financial and auditing departments, independent third-party monitoring agencies, and the public. First, the government carries out supervision and inspection of the purchase behavior, which is called internal supervision. The supervision carried out by finance, audit, and other relevant departments in accordance with their functions is called external supervision; this is called government supervision and constitutes the main supervision subject of the government's purchasing of ecological services. Second, the public, as the supervisor, can supervise and provide feedback on whether the purchasing subject is fair and impartial, whether there is delay in fund allocation, and other behaviors in the purchasing process.

3.4. Purchase Price Strategy

According to data from the 2020 survey on the livelihood of residents around the NTLNP, 78 family members who were engaged in the ecological services of the NTLNP were screened. For these 78 households, the per capita expenditure, average dependency coefficient, per capita wage income and per capita total income were calculated. The results are as follows: the per capita expenditure of the family is CNY 24,519.66; the average maintenance coefficient is 1.82; the per capita wage income is CNY 19,088.29; the total per capita income is CNY 21,349.89. The benchmark price for the government purchasing of ecological services calculated by the proportion method is CNY 39,898.56.

Taking the benchmark price measured by the proportion method as the baseline, it is estimated that the average annual growth rate of the wage income of the urban employed population in 2035 will catch up with the average wage income of the urban employed population in China, and it is estimated that the value of the government purchasing of ecological services should reach it in 2022 under this growth rate. In 2020, the average wage income of the urban employed population nationwide was CNY 51,438.13; the benchmark price for purchasing ecological services in the NTLNP tracks the average wage income of the national employed population, that is, by 2035, it will reach the goal of twice the wage income (CNY 102,876.26) of the current national urban employed workers, providing that it maintains a growth rate of 6.10%. Under this growth rate, the price of ecological services purchased by the government will reach CNY 47,654.44 in 2022.

Based on the above results, the government's price strategy for purchasing ecological services is divided into three stages. (1) Initial stage: according to the survey data, the average wage income of 81 forest workers in the NTLNP is CNY 33,496.35 (Figure 5). The benchmark price for the government purchasing of ecological services is CNY 39,898.56 per person. (2) Development stage: the proposed price for the government purchasing of ecological services in 2022 is CNY 47,654.44 per person. (3) Flat stage: to catch up with the per capita wage income level of urban workers nationwide in 2035, it is necessary to maintain an average annual growth rate of 6.10%.

Compared with the growth rate of the wage income of workers in the province where the NTLNP is located and China's forest and grass industry, it is relatively easy to achieve a growth rate of 6.1% for the price of ecological services purchased by the government. From 2015 to 2019, the average annual growth rate of the average wage income of urban on-the-job workers in Heilongjiang Province was 6.8%, and from 2015 to 2019, the average annual growth rate of the average wage income of on-the-job workers in the national forest and grass system was 7.7%. Compared to other industries in China, it is also easy to achieve a growth rate of 6.1% in the price of government purchasing of ecological services. From 2015 to 2019, the average annual growth rates of wage income of on-the-job workers in the five industries of manufacturing, construction, transportation, education, and health and social work were 7.2%, 6.1%, 7.1%, 8.0%, and 8.7%, respectively. Therefore, it is feasible for the government purchasing of ecological services with an average annual growth rate of 6.1% to catch up with the wage income level of urban workers across the country.

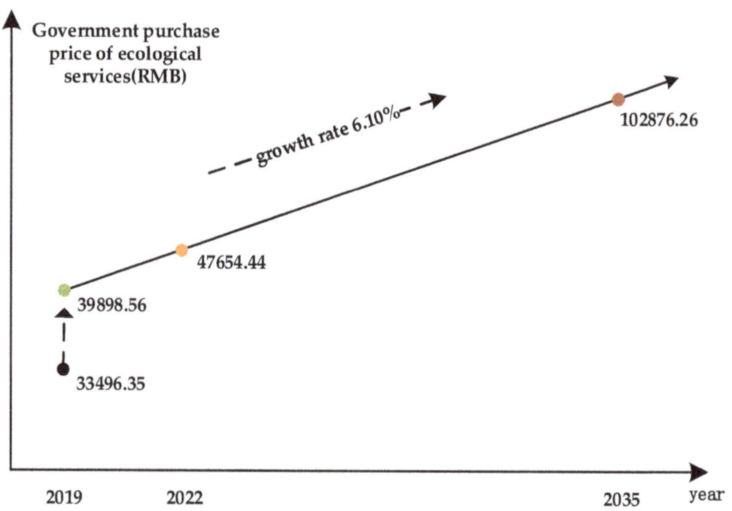

Figure 5. The price strategy of government purchasing of ecological services.

During the pilot period of the NTLNP, the funds of each branch mainly came from the forest reform and development funds in the special funds of the central fiscal year, and a small part of the funds came from the investment of non-profit organizations. National park financial subjects and special accounts have not yet been established [10]. In order to ensure the implementation of the government purchasing of ecological services, it is a prerequisite to establish a special account for national parks and financial items, and the budget items include the cost of purchasing ecological services. Fundraising can be divided into three parts. First, the forestry reform and development funds issued by the central government to the seven SOFEs can be partially transferred to the special account for national parks. Second, encourage local residents or enterprises to engage in production and operation activities through franchising in the ecological experience area or surrounding areas and use the franchise income to feed back to ecological services. Finally, raise funds for the whole society and allow large donations of enterprises or individuals to set up special funds under the name.

3.5. The Whole Process Evaluation Chain

The government purchasing of ecological services should establish a whole-process evaluation chain based on the purchase chain. The performance evaluation system includes project evaluation, process evaluation, and result evaluation, and the evaluation chain also includes evaluation procedures and evaluation methods (Figure 6). (1) Project evaluation. Set performance goals, demonstrate conduct requirements, and ensure the efficiency and quality orientation of ecological services projects. The evaluation indicator system should be set up according to the budget preparation, and the index system should include indicators such as the efficiency and quality of ecological services, and the satisfaction of the public. (2) Process evaluation. In the implementation process of government purchasing of ecological services, it is necessary to supervise the efficiency of budget execution, supervise the performance of contracts, and focus on evaluating the quality of services. (3) Results evaluation. First, disclose project information and evaluation results in a timely manner and accept social supervision. Second, establish a mechanism compatible with incentives and punishments, so that ecological management workers have the motivation to improve their abilities. Third, investigate the responsible subject through the accountability mechanism, and make timely rectification in the follow-up work. (4) Evaluation procedure. First, the pre-event, in-process and post-event evaluation should be connected to effectively prevent the information asymmetry in the purchase process through the evaluation procedure. Second, regular inspection is combined with random inspection, and the completion of the responsibilities of ecological public welfare positions should check irregularly. Third, the mid-term acceptance check is combined with the final acceptance check, and the results of the mid-term acceptance check are the main reference for the final acceptance check. (5) Evaluation method. First, expand the scope of third-party evaluation and adopt the evaluation mechanism of experts and third-party institutions. Second, optimize the evaluation index system and establish a quantitative index system for the three major benefits of ecology, economy, and society. Third, establish a dynamic tracking and monitoring mechanism to monitor the entire process of purchasing services in real time.

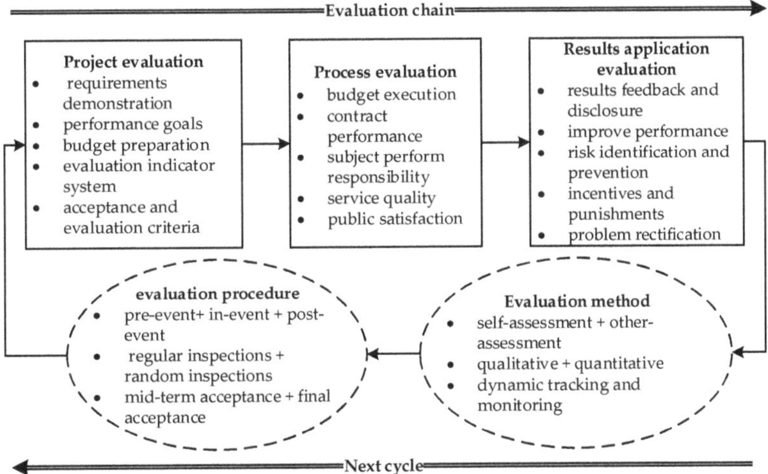

Figure 6. The whole process evaluation chain of government purchasing of ecological services.

3.6. Purchase Costs Analysis

Cost savings is an important factor for government to consider when providing ecological services. Based on the transaction cost theory, analyzing the costs of the two modes of direct government provision and purchasing provision, and comparing the financial burden under the two modes are the key stages of government decision-making.

The most important parts of the costs of the government's direct provision of ecological services are: The first is the labor costs, including the costs of additional posts created by the NTLNP Administration to provide ecological services and the costs of personnel management. The second is the costs of sites and materials, including the costs of establishing protection sites within the national park, and purchasing materials required for protection. The third is the costs of information, including the costs of the NTLNP Administration to search for grass-roots information and coordinate powers and responsibilities with SOFEs and state-owned forest farms. Fourth, the management costs of the NTLNP Administration itself, including office expenses, postal and telecommunication expenses, and transportation expenses.

The most important parts of the costs of the government purchasing of ecological services are: The first is the purchase costs of the NTLNP Administration, which includes the salaries of the employees discussed in this article, and the management costs of the ecological caretaker organized by the protection stations. The second is the third-party fees, which involves the costs of third-party supervision and evaluation. The third is the management costs of the NTLNP Administration itself, including office expenses, postal and telecommunication expenses, and transportation expenses.

Comparing the costs of direct provision of ecological services by the NTLNP Administration and of the purchasing of ecological services under the condition of providing the same quantity and quality of ecological services, several conclusions can be drawn: (1) In terms of labor costs, the costs of government provision is similar to that of government purchases. When provided directly, a large number of ecological management workers need to be recruited, and the wages and management costs of these workers must be paid. When purchasing services, it is necessary to pay the salaries of ecological protection workers and the management fees for the participation of residents in the protection stations. (2) In terms of sites and material costs, government purchases costs tend to be zero. When directly provided, a management sites needs to be established in the park and management equipment needs to be purchased. When purchasing services, SOFEs and state-owned forest farms have ready-made protection sites and professional ecological protection equipment, and no additional purchases are required. (3) In terms of information cost, the government purchase cost tends to zero. When providing direct information, it is necessary to search for basic social and ecological information in the park and to coordinate power and responsibility relations with SOFEs and state-owned forest farms. when purchasing services, SOFEs and state-owned forest farms have a similar grasp of the ecological and social conditions in the park. They have professional ecological management teams and technologies without additional information costs. (4) In terms of third-party fees, government purchases need to be paid to experts, scholars, auditing departments and other third-party institutions for supervision and evaluation, while direct provision does not require third-party fees. (5) In terms of the management cost of the national park itself, the cost of government provision is similar to that of government purchases.

Based on the above analysis, it is concluded that government purchases are more cost-effective than government provision, especially in terms of sites and material costs and information costs. SOFEs and state-owned forest farms have "natural" advantages, which can reduce the financial expenditure of the NTLNP Administration.

4. Discussion

4.1. Links between Ecosystem Services and Ecological Services

Ecological services in this study mean the ecological protection services provided by humans in the process of generating ecosystem services. In the literature, there are rich research results on ecosystem services based on the theoretical background of ecology or economics [53–55]. The definition of ecosystem services is generally understood as follows: "Ecosystem services are the benefits that humans derive from ecosystems", which was proposed by MAE [56]. The definition describes the flow of individual services from ecosystems to humans. However, recent research on ecosystem services seems to have reached a consensus that ecosystem services are not produced by ecosystems independent of humans but by human interactions with ecosystems [57–60]. In order to explain the role of humans in ecosystems, Comberti proposed the concept of "Services to Ecosystems" to reflect human actions to maintain and improve ecosystems. On the basis of the original "single service flow from ecosystem to human", the service flow from human to ecosystem is added, and a closed loop of a reciprocal relationship between human and ecosystem is constructed [61]. The concept of ecological services in this study is similar to the abovementioned "Services to Ecosystems", which reflect the important role of humans in ecosystems. The government purchasing of ecological services is a way of providing such services, and it is suitable for national parks with state-owned land as the main body, and such a national park is used as the study area, and the ecological services at this time belong to the category of public services.

4.2. A Policy Tool That Takes into Account Efficiency and Fairness and Can Clarify the Rights and Responsibilities of Stakeholders

Driven by realistic problems such as the increase in the content and tasks of ecological services in NTLNP, the connection between the old and new management systems, and the low subsidy standards for ecological services, this study proposes the government purchasing of ecological services as a new policy tool. The connotation of government purchasing of ecological services is similar to "government-funded" PES [62] and government-funded EC [63]; all three are applicable to situations where it is difficult for users to identify and define the ecosystem when it is a public good, and government intervention is an inevitable approach. In order to improve the efficiency of natural resource management and alleviate poverty, PES and EC have been widely adopted by the international community, especially developing countries, and have achieved remarkable results. In the study of PES and EC, watersheds, forests, grasslands, wetlands, biodiversity, and habitats are the main areas of research [13,64]. The research content revolves around key links such as conceptual connotation, theoretical basis, purchasers, and sellers (subject and object of compensation), payment standard (compensation standard), and effect evaluation [19,65–67]. Although the research frameworks of PES and EC are becoming more mature, and a mutually beneficial situation between efficiency and fairness has been achieved in practice [68], the relationship between rights and responsibilities among relevant stakeholders is a blind spot for research.

The issues involved in the government purchasing of public services include efficiency, fairness, and the relationship between the rights and responsibilities of the stakeholders. Wang divides government purchasing of public services into three models according to the relationship between the purchaser and the undertakers: the independent relationship–competitive purchasing model, the independent relationship–non-competitive purchasing model, and the dependent relationship–non-competitive purchasing model. Whereas the SOFEs and LGFD specializing in ecological services existed before the establishment of the NTLNP Administration, in this study, the NTLNP Administration is the purchaser, the SOFEs and Protection Stations are the directional undertakers, and the forest workers are the most direct producers of ecological services. The model of government purchasing of ecological services is independent and non-competitive. The government purchasing of ecological services clarifies the relationship of rights, responsibilities, and interests between stakeholders on the basis of taking into account the efficiency of supplying ecological services and the livelihoods of forest workers. This is the main reason for constructing the model of government purchasing of ecological services instead of following the existing research framework of PES or EC.

Due to the severe aging of ecological management and care workers, they retire at an average rate of 5% every year, and the ecological management and protection positions are recruited in the form of only decreasing and not increasing. After the retirement of existing ecological management and conservation workers, the government purchasing of ecological services model with community organizations as the undertakers is a future research direction. The research aim is to investigate the willingness of community residents and social organizations to participate and to carry out evolutionary game analysis on the multi-stakeholders of government purchasing of ecological services.

4.3. Enterprises, NGOs, Community Residents and Indigenous People Are the Main Undertakers of Ecological Services in National Parks

Based on the natural resources and management system background of state-owned forest areas in northeast China, a model of government purchasing of ecological services for the NTLNP, which is also applicable to other national parks in China. Among the five national parks officially established in China, the Giant Panda National Park was established in the state-owned forest areas of northwest and southwest China. Before the establishment of the national park, SOFEs and state-owned forest farms were the main body of ecological services supply in this area. The general situation of its natural resources and management system is similar to that of the northeast state-owned forest areas. Sanjiangyuan National Park and The Hainan Tropical Rainforest National Park belong to mountainous and totropical national parks. There are a large number of farmers and herdsmen, forestry workers, and even urban residents living in and outside these national parks [69]. The government purchasing of ecological services is an important means to solve the problems of the connection of management systems and the livelihood of community residents. For the Giant Panda National Park, the Giant Panda National Park Administration can be the purchaser of ecological services, 15 SOFEs and protection stations can be the undertakers of ecological services, state-owned forest farm workers, people who have been released from poverty and resettlement of immigrants Residents can participate in ecological services through protection stations. For the Sanjiangyuan National Park, it is impossible to complete such a large-scale ecological environmental protection work with only the management team of more than 400 people from the Sanjiangyuan National Park Administration [26]. It must rely on the strength of local herdsmen. Therefore, the three management committees under the Sanjiangyuan National Park Administration (Sanjiangyuan, Yellow River and Lancang River Source Management Committee) [70] can be the purchasers of ecological services. The surrounding herdsmen voluntarily participate in ecological services through the protecion stations. For Hainan Tropical Rainforest National Park, there are almost 40 ethnic groups, mainly Li and Miao, in the park [71]. Tropical rainforests provide the material base and living conditions for minority residents to thrive, and it is very important to guide residents to participate in the ecological services of national parks. Hainan Tropical Rainforest National Park Administration can be the main purchaser of ecological services, and community residents can participate in ecological services through the protecion stations.

From an international perspective, some national parks with state-owned land as the main body have begun to use the government purchasing of ecological services, and these cases provide lessons for other national parks. For example, about half of Melbourne's water sources are located in Kinglake National Park, Yarra Ranges National Park and Baw Baw National Park. The Victorian Government of Australia protects these forest water sources by purchasing the ecological services of Melbourne Water Company [72]. The Serbian government has entrusted the management of the Secovlje Salina National Park to a private mobile phone company through a combination of concessions and government purchases. The company's revenue comes from both concessions and the government's annual budget. On the one hand, the government franchises it for sea salt production; on the other hand, the government purchasing of its ecological services for national parks [72]. These experiences show that for national parks with fewer residents, it is more efficient for the government purchasing of ecological services from state-owned or private enterprises and NGOs. For national parks with more community residents or aboriginal people, such as China's national parks, residents' participation in national park ecological services through protection stations or community organizations is the best choice to solve the contradiction between protection and development.

5. Conclusions and Policy Implication

5.1. Conclusions

This study takes China's NTLNP as the research area and constructs a government purchasing of ecological services model based on the classical theory of government purchasing of public services. The government purchasing of ecological services model is divided into five components: institutional environment analysis, driving factor analysis, responsibility system construction, purchase price strategy, and the whole process evaluation chain. Among them, the driving factors, responsibility system, and purchase price of the government purchasing of ecological services are the focus of the study. In the literature, the text analysis method was used to analyze the relevant policy texts and the symposium data obtained from field research. From this, we obtained the incentives and constraints for the government purchasing of public services in the current policies, as well as the status of ecological services tasks, composition, number, and salary standards of the NTLNP. We adopted the revised minimum wage standard method and opportunity cost method to formulate the purchase price strategy. Comparing the two environmental policy tools, PES and EC, the government purchasing of ecological services is more suitable for national parks with state-owned land as the main body, and the ability to clarify the rights and responsibilities of multiple stakeholders is the unique feature of the government purchasing of ecological services. At the same time, formulating purchase price strategies from the perspective of meeting the decent living needs of ecological service participants and narrowing the gap with the average wage income level of urban workers is more conducive to achieving efficiency and fairness. The core conclusions of the study are as follows.

First, the driving factors for the government purchasing of ecological services are: (1) the content and tasks of ecological services in the NTLNP have increased; (2) ecological services workers in the forest sector need to be transferred; (3) subsidies for ecological services need to be improved. First, since the management and protection services in the NTLNP have expanded from forest resource management to wildlife protection and resource monitoring, if the NTLNP Administration establishes a special agency to engage in ecological services, it will inevitably increase government financial expenditure. The government purchasing of services is an ecological services supply method that not only saves costs but also provides professional services. Secondly, there are a large number of participants engaged in ecological services in the NTLNP, and their identities are diverse. Through the process of government purchase, all the participants engaged in ecological services from the forest department can be transferred to the NTLNP. Finally, the per capita wage income of the sample workers engaged in ecological services is CNY 33,496.35, which is significantly lower than the average wage income level of urban employed population in Jilin, Heilongjiang, and the whole country. Public services related to improving people's livelihoods are the priority items for the government purchasing of services. Increasing the wage income level of ecological services participants is an important way to improve the livelihood of residents in national parks.

Second, this research constructed a responsibility system for the government purchasing of ecological services under the institutional environment related to government purchasing of services. Among them, the purchaser is the NTLNP Administration, which uniformly exercises the management and protection responsibilities of state-owned natural resource assets in the region. Undertakers are seven SOFEs and three Protection Stations; the registered and on-the-job workers directly transferred by the SOFEs are the main body that performs ecological services, and a small number of one-time resettlement personnel, registered and off-duty personnel, urban workers, and other transferable personnel will undertake ecological services through the SOFEs. The registered and on-the-job workers of local state-owned forest farms are the main body of ecological services, and the poor people in the villages and towns under their jurisdiction undertake ecological services through the Protection Stations. The evaluators are third-party evaluation institutions, experts, and scholars, and the general public. The supervisors are the NTLNP Administration, the management branch, the financial and auditing departments, independent third-party supervision agencies, and the public.

Third, we divided the government's price strategy for purchasing ecological services into three stages: (1) Initial stage. According to the livelihood data of the sample forest workers in the NTLNP obtained through the survey, the benchmark price for the government purchasing of ecological services is calculated by the proportion method to be CNY 39,898.56 per person. (2) Development stage. It is estimated that the price of ecological services purchased by the government in 2022 was CNY 47,654.44 per person. (3) Flat stage. To catch up with the level of per capita wage income of urban workers nationwide by 2035, it is necessary to maintain an average annual growth rate of 6.10%.

Fourth, the whole process evaluation chain of the government purchasing of ecological services was designed, comprising three main parts: project evaluation, process evaluation, and result application evaluation. The evaluation procedures and methods are also described in the chain. In the project evaluation stage, it is necessary to set up an evaluation system that includes indicators such as ecological service efficiency and quality, and social public satisfaction. In the evaluation of the process, it is necessary to supervise the efficiency of budget execution, supervise the performance of contracts, and focus on evaluating the quality of ecological services. In the application for the evaluation of results, the project information and evaluation results are made transparent, and an incentive and punishment mechanism is established.

Fifth, compare the financial burdens of government provision and government purchases based on transaction cost theory. The costs of government purchases are lower than those of government provision, especially in terms of sites and material costs, and information costs.

5.2. Policy Implications

According to the research results, this study puts forward the policy implications of implementing the government purchasing of ecological services in stages. The purchase content expands from traditional ecological resource management to intelligent management. The purchase method extends from economic payment to franchising. The participant subjects change from administrative to market-oriented and diversified. The fund utilization expands from EC to ecological product value. Finally, it will realize the coordinated development and evolution of ecological, economic, and social benefits from guaranteeing ecological benefits.

The first stage combines the ecological protection goal of the government purchasing of ecological services with the consolidation of poverty alleviation achievements, so as to achieve the coordinated development of ecological and social benefits. We completely strip the social management functions undertaken by the SOFEs, and at the same time, increase the efforts to withdraw and merge forest farms, and gradually establish a management system of "Administration Bureau–Management Sub-bureau–Management and Protection Station". The Protection Stations will be established with SFEs and LGFDs as the main body to undertake the purchasing of ecological services. The border villages in the park realize "one household, one post", responsible for natural resource management and protection, ecological experience, environmental education services, ecological protection projects, and ecological monitoring.

In the second stage, the government purchasing of ecological services includes a combination of economic payment and franchise, so as to realize the transformation from ecological protection compensation to ecological product value. Government-funded franchising is also an option for the government purchasing of ecological services, and the undertakers of ecological services are encouraged to exploit their own advantages to carry out franchising. As an enterprise unit, SOFEs have established the cultivation and breeding bases of cattle, forest frogs, forest medicine, and black fungus through franchising, and carried out moderate-scale intensive management. Priority will be given to diverting workers from the reform of SFEs, farmers who have returned farmland to forests and banned grazing, and poor people who have been filed and registered. The implementation of single or joint household contracting is explored for forest workers' families to undertake ecological management and protection or under-forest planting and breeding and broaden the channels for local residents to increase their income.

In the third stage, the government purchases ecological services to transform them into intelligent ecological management and protection, using integrated monitoring technology to realize a new type of intelligent ecological management and conservation system. An early warning and response mechanism is established to greatly improve the efficiency of management and protection. It is important to build the brand system of "NTLNP", and incorporate various ecological products such as forest food, forest health care, nature education, and ecological research into the brand scope. Brand cultivation and protection are strengthened, and the premium of ecological products is increased. The franchise system is improved, and an interest linkage mechanism and a profit feedback mechanism are established. The moderate-scale operation and brand value-added income will feed back to the ecological management and social services public welfare positions. Finally, the government changes from government purchase to user payment, realizes the diversification and marketization of purchasers, and further realizes multiple ecological, economic, and social benefits.

Author Contributions: Conceptualization, H.Z. and Y.Z.; methodology, Y.Z. and C.B.; software, M.Z.; validation Y.C. and H.Z.; formal analysis, Y.Z.; investigation, H.Z. and C.B.; resources, Y.C.; data curation, Y.Z.; writing—original draft preparation, H.Z. and Y.Z.; writing—review and editing, Y.Z.; visualization, M.Z.; supervision, C.B.; project administration, H.Z.; funding acquisition, Y.C. All authors have read and agreed to the published version of the manuscript.

Funding: This research was supported by the State Forest and Grassland Administration Project of "Research on Ecological Services Purchased by the Government of the Northeast Tiger and Leopard National Park", grant number JYCL-2020-00046.

Data Availability Statement: Not applicable.

Acknowledgments: The authors are particularly grateful to all researchers and institutes for providing data for this study. The authors are also very grateful to the editors and reviewers for their comments and suggestions for improving this study.

Conflicts of Interest: The authors declare no conflict of interest.

Notes

[1] A survey found that in 2019, the per capita wage income of sample workers engaged in ecological services in the NTLNP was CNY 33,496.35. In the same year, the average wage income of the urban employed population in Jilin, Heilongjiang, and the entire country was CNY 36,307.87, 41,597.86, and 49,020.14, respectively. According to the "China Forest and Grassland Statistical Yearbook 2020", the average annual salary of workers in the China's forest and grassland system in 2019 was CNY 67,782.

[2] In 1949, the early days of the founding of New China, in order to meet the demand for wood in national economic construction, the government of China invested in the establishment of state-owned forest areas and a number of state-owned forest farms. In the nine provinces and regions with rich forest resources in the northeast, southwest, and northwest, 138 SFEs have been established, which specialize in timber harvesting. Among them, 87 SOFEs in Heilongjiang, Jilin, and Inner Mongolia constitute the key state-owned forest areas. State-owned forest farms are public institutions dedicated to afforestation and forest management and protection set up in concentrated state-owned barren mountains and wasteland suitable for forest. There are 4855 state-owned forest farms in 31 provinces. accounts for 83.28% of the total area of the NTLNP. It involves 7 SOFEs and 3 LGFDs; the former is a public welfare state-owned enterprise, while the latter is a public institution (Table 2). The NTLNP is the first national park in China that is directly handled by the central government.

[3] From 2000 to 2020, China implemented two phases of Natural Forest Protection Projects, from cutting down logging tasks to complete logging bans. The increase of surplus personnel in forest enterprises and the heavier burden of enterprises have led to the layoff of some forest workers, the state issues a one-time resettlement fee for these groups. A small number of one-time resettlement personnel will also participate in the forest operations of SFEs during the afforestation and tending season, and their wages are calculated according to the piece-rate system.

[4] The statistical yearbook does not include the average wage level of the urban employed population in Jilin, Heilongjiang, and the whole country. Therefore, this information is obtained indirectly through the calculation of (urban per capita disposable wage income $\times$ total urban population)/urban employment.

References

1. Yang, R. China's National Park Governance System: Principles, Goals and Paths. *Bio. Sci.* **2021**, *29*, 269–271.
2. Wedel, K.R. Government Contracting for Purchase of Service. *Soc. Work.* **1976**, *5*, 101–105.
3. Peng, L. Analysis on the Problems of Protected Area System in China and the Countermeasure. *China Landsc. Archit.* **2017**, *33*, 108–113.
4. Kideghesho, J.R.; Røskaft, E.; Kaltenborn, B.P. Factors Influencing Conservation Attitudes of Local People in Western Serengeti, Tanzania. *Biodivers. Conserv.* **2007**, *16*, 2213–2230. [CrossRef]
5. Vedeld, P.; Jumane, A.; Wapalila, G.; Songorwa, A. Protected Areas, Poverty and Conflicts. *For. Policy Econ.* **2012**, *21*, 20–31. [CrossRef]
6. Liu, Y.; Zou, X.; Chen, J.; Pan, T. Impacts of Protected Areas Establishment on Pastoralists' Livelihoods in the Three-River-Source Region on the Qinghai–Tibetan Plateau. *Land Use Policy* **2022**, *115*, 106018. [CrossRef]

7.   Zhou, D.; Wang, Z.; Lassoie, J.; Wang, X.; Sun, L. Changing Stakeholder Relationships in Nature Reserve Management: A Case Study on Snake Island-Laotie Mountain National Nature Reserve, Liaoning, China. *J. Environ. Manag.* **2014**, *146*, 292–302. [CrossRef]
8.   Musakwa, W.; Gumbo, T.; Paradza, G.; Mpofu, E.; Nyathi, N.A.; Selamolela, N.B. Partnerships and Stakeholder Participation in the Management of National Parks: Experiences of the Gonarezhou National Park in Zimbabwe. *Land* **2020**, *9*, 399. [CrossRef]
9.   Zhu, H.G.; Zhao, M.H. Research on the impact of national parks on the livelihood and income structure of different types of forest workers' families—Taking the Siberian Tiger and Leopard National Park as an example. *J. Agric For. Eco. Manag.* **2022**, *21*, 78–86.
10.   Chen, Y.R.; Han, J.K.; Qin, L.N.; Yang, H.C. Research on the problems and development paths of the pilot system of the Northeast Tiger and Leopard National Park. *Environ. Prot.* **2019**, *47*, 61–65.
11.   Hafsa, F.; Darnall, N.; Bretschneider, S. Social Public Purchasing: Addressing a Critical Void in Public Purchasing Research. *Public. Admin. Rev.* **2021**, *82*, 818–834. [CrossRef]
12.   Sattler, C.; Matzdorf, B. PES in a Nutshell: From Definitions and Origins to PES in Practice—Approaches, Design Process and Innovative Aspects. *Ecosyst. Serv.* **2013**, *6*, 2–11. [CrossRef]
13.   Salzman, J.; Bennett, G.; Carroll, N.; Goldstein, A.; Jenkins, M. The global status and trends of Payments for Ecosystem Services. *Nat. Sustain.* **2018**, *1*, 136–144. [CrossRef]
14.   Bennett, M.T. China's Sloping Land Conversion Program: Institutional Innovation or Business as Usual? *Ecol. Econ.* **2008**, *65*, 699–711. [CrossRef]
15.   Long, K.; Omrani, H.; Pijanowski, B.C. Impact of Local Payments for Ecosystem Services on Land Use in a Developed Area of China: A Qualitative Analysis Based on an Integrated Conceptual Framework. *Land Use Policy* **2020**, *96*, 104716. [CrossRef]
16.   Yost, A. Mechanisms behind Concurrent Payments for Ecosystem Services in a Chinese Nature Reserve. *Ecol. Econ.* **2020**, *11*, 106509. [CrossRef]
17.   Wunder, S. *Payments for Environmental Services: Some Nuts and Bolts*; CIFOR: Bogor, Indonesia, 2005.
18.   Zhou, C.; Ding, X.H.; Li, G.P.; Wang, H.Z. Research on the ecological compensation standard of the water source area of the middle route of the South-to-North Water Diversion Project—From the perspective of ecosystem service value. *Resour. Sci.* **2015**, *37*, 792–804.
19.   Sheng, W.; Zhen, L.; Xie, G.; Xiao, Y. Determining Eco-Compensation Standards Based on the Ecosystem Services Value of the Mountain Ecological Forests in Beijing, China. *Ecosyst. Serv.* **2017**, *26*, 422–430. [CrossRef]
20.   McDonald, J.A.; Helmstedt, K.J.; Bode, M.; Coutts, S.; McDonald-Madden, E.; Possingham, H.P. Improving Private Land Conservation with Outcome-based Biodiversity Payments. *J. Appl. Ecol.* **2018**, *55*, 1476–1485. [CrossRef]
21.   Li, X.G.; Miao, H.; Zheng, H.; Ouyang, Z.Y.; Xiao, Y. The Application of Opportunity Cost Method in Determining Ecological Compensation Standard—Taking the Central Mountainous Area of Hainan as an Example. *J. Ecol.* **2009**, *29*, 4875–4883.
22.   Hu, Z.T.; Liu, D.; Kong, D.S.; Jin, L.S. Estimation of grazing prohibition subsidy standard in grassland ecological compensation based on opportunity cost method. *J. Arid. Environ.* **2017**, *31*, 63–68.
23.   Liu, D.; Hu, Z.T.; Jin, L.S. Research on fallow compensation standard in groundwater overexploitation area based on farmers' willingness to pay. *Chinese J. Popul. Resour. Environ.* **2019**, *29*, 130–139.
24.   Li, S.; Feng, J.; Li, B.B.; Lv, Z. Experience and Challenges of the Pilot System of Giant Panda National Park. *Bio. Sci.* **2021**, *29*, 307–311.
25.   He, S.Y.; Su, Y. Wuyishan pilot experience and suggestions for improvement: Difficulties in the protection of national parks in the southern collective forest area and a way out for reform. *Bio. Sci.* **2021**, *29*, 321–324.
26.   Zhao, X.; Zhu, Z.Y.; lv, Z.; Xiao, L.Y.; Mei, S.N.C.; Wang, H. Community-based protection: Reflections on the public welfare post of ecological management and protection in Sanjiangyuan National Park. *Bio. Sci.* **2018**, *26*, 210–216.
27.   Long, X.; Du, Y.; Hong, X.; Zang, R.; Yang, Q.; Xue, H. Hainan Tropical Rainforest National Park Pilot Experience. *Bio. Sci.* **2021**, *29*, 328–330. [CrossRef]
28.   State Forestry Administration. People's Government of Jilin Province.; People's Government of Heilongjiang Province. General Plan for the Northeast Tiger and Leopard National Park (2017–2025) (Draft for Comment). Available online: https://www.forestry.gov.cn/uploadfile/main/2018-3/file/2018-3-9-599430e5ec1249bab08927453227ff14.pdf (accessed on 20 September 2022).
29.   Savas, E.S. *Privatization and Public-Private Partnerships*; Academia: San Fracisco, CA, USA, 2000.
30.   Wang, P.Q.; Salamon, L.M. *Research on the Government's Purchase of Public Services from Social Organizations: An Analysis of Chinese and Global Experiences*; Peking University Press: Beijing, China, 2010.
31.   Lv, F. Heterogeneous governance and the predicament of grassroots government purchasing services—Taking the government purchasing service project in S Street as an example. *J. Manag. World.* **2021**, *37*, 147–158.
32.   Gradus, R.; Dijkgraaf, E.; Wassenaar, M. Understanding Mixed Forms of Refuse Collection, Privatization, and Its Reverse in the Netherlands. *Int. Public. Manag. J.* **2014**, *17*, 328–343. [CrossRef]
33.   Ferris, J.; Graddy, E. Contracting out: For what? With whom? *Public Admin. Rev.* **1986**, 332–344. [CrossRef]
34.   Wang, Y.F. On the legal path of ecological compensation in national parks in my country. *Environ. Protect.* **2018**, *46*, 56–59.
35.   Dorwart, R.A.; Schlesinger, M.; Pulice, R.T. The promise and pitfalls of purchase-of-service contracts. *Psychiatr. Serv.* **1986**, *37*, 875–878. [CrossRef] [PubMed]
36.   Ostrom, V.; Robert, L.B.; Ostrom, E. *Local Government in the United States*; Peking University Press: Beijing, China, 2004.

37. Lin, M.W. Government Procurement of Public Services: An Integrative Analytical Framework. *J. Beijing Inst. Technol.* **2017**, *19*, 91–98.
38. Peng, J. From market value priority to public value priority—Progress, deficiencies and prospects of research on government purchasing responsibility. *J. Financ. Serv. Res.* **2018**, *1*, 43–52.
39. Osborne, D.; Gaebler, T. *Reforming Government: How Entrepreneurship Is Reforming the Public Sector*; Shanghai Translation Publishing House: Shanghai, China, 1996.
40. Terry, L.D. Administrative Leadership, Neo-Managerialism, and the Public Management Movement. *Public Admin. Rev.* **1998**, *58*, 194–200. [CrossRef]
41. Denhardt, R.B.; Denhardt, J.V. The new public service: Serving rather than steering. *Public Admin. Rev.* **2000**, *60*, 549–559. [CrossRef]
42. Huang, X.C. Rethinking the growth conditions of Chinese social organizations: A general theoretical perspective. *Socio Stud.* **2017**, *32*, 101–124.
43. Christensen, T.; Lægreid, P. Post New Public Management Reforms-Exploring the "Whole-of-Governmenty" Approach to Public Reform. In *Rethinking the Reform Questio*; Cambridge Scholars Publishing: England, UK, 2007; Volume 24, pp. 24–45.
44. Williamson, O.E. Markets and Hierarchies: Analysis and Antitrust Implications: A Study in the Economics of Internal Organization. *Account. Rev.* **1975**.
45. Marx, K.; Engels, F. *Marx & Engels Collected Works*; Progress Publishers: Moscow, Russia, 1975; Volume 2, pp. 1838–1842.
46. Li, X.G.; Miao, H.; Zheng, H.; Ouyang, Z.Y. The main methods of determining ecological compensation standards and their applications. *Acta Ecol. Sin.* **2009**, *29*, 4431–4440.
47. Jiang, A.H.; Yang, Q. Research on the "whole process" performance evaluation of government procurement of public services. *J. Cent. Univ. Financ. Eco.* **2020**, *3*, 3–9.
48. Statistics Bureau of Heilongjiang Province. *Heilongjiang Statistical Yearbook 2020*; China Statistics Press: Beijing, China, 2020; pp. 37–82.
49. Statistics Bureau of Jilin Province. *Jilin Statistical Yearbook 2020*; China Statistics Press: Beijing, China, 2020; pp. 39–54.
50. State Forestry and Grassland Administration. *China Forest and Grassland Statistical Yearbook 2020*; China Science Press: Beijing, China, 2020; pp. 166–173.
51. Wu, Z.; Guan, J.; He, J. An empirical study on the calculation of the minimum wage standard—A dynamic combination calculation based on the objective weighting of the CRITIC-entropy weight method. *Mod. Eco. Sci.* **2019**, *41*, 103–117.
52. Gong, F.; Chang, Q.; Wang, F.; Liu, X. An empirical study on the ecological compensation standard of grassland in Inner Mongolia. *J. Arid. Land Resour Environ.* **2011**, *25*, 151–155.
53. Costanza, R.; d'Arge, R.; de Groot, R.; Farber, S.; Grasso, M.; Hannon, B.; Limburg, K.; Naeem, S.; O'Neill, R.V.; Paruelo, J.; et al. The Value of the World's Ecosystem Services and Natural Capital. *Nature* **1997**, *387*, 253–260. [CrossRef]
54. Daily, G.C. *Nature's Services: Societal Dependence on Natural Ecosystems (1997)*; Yale University Press: New Haven, CT, USA, 2017; pp. 454–464.
55. Boyd, J.; Banzhaf, S. What Are Ecosystem Services? The Need for Standardized Environmental Accounting Units. *Ecol. Econ.* **2007**, *63*, 616–626. [CrossRef]
56. Toth, F.L. *Ecosystems and Human Well-Being: A Framework for Assessment*; Island Press: Washington, DC, USA, 2003.
57. Fischer, A.; Eastwood, A. Coproduction of Ecosystem Services as Human–Nature Interactions—An Analytical Framework. *Land Use Policy* **2016**, *52*, 41–50. [CrossRef]
58. Jones, L.; Norton, L.; Austin, Z.; Browne, A.L.; Donovan, D.; Emmett, B.A.; Grabowski, Z.J.; Howard, D.C.; Jones, J.P.G.; Kenter, J.O.; et al. Stocks and Flows of Natural and Human-Derived Capital in Ecosystem Services. *Land Use Policy* **2016**, *52*, 151–162. [CrossRef]
59. Palomo, I.; Felipe-Lucia, M.R.; Bennett, E.M.; Martín-López, B.; Pascual, U. Disentangling the Pathways and Effects of Ecosystem Service Co-Production. In *Advances in Ecological Research*; Woodward, G., Bohan, D.A., Eds.; Academic Press: Cambridge, MA, USA, 2016; Volume 54, pp. 245–283.
60. He, S.Y.; Su, Y.; Wang, L.; Cheng, H.G. Constructing a social situation analysis tool to promote the coordination of community resource use and protection goals in protected areas—Practice in the pilot area of Wuyishan National Park. *Acta Ecol. Sin.* **2019**, *39*, 3861–3870.
61. Comberti, C.; Thornton, T.F.; Wyllie de Echeverria, V.; Patterson, T. Ecosystem Services or Services to Ecosystems? Valuing Cultivation and Reciprocal Relationships between Humans and Ecosystems. *Global. Environ. Chang.* **2015**, *34*, 247–262. [CrossRef]
62. Engel, S.; Pagiola, S.; Wunder, S. Designing Payments for Environmental Services in Theory and Practice: An Overview of the Issues. *Ecol. Econ.* **2008**, *65*, 663–674. [CrossRef]
63. Jin, L.S.; Chu, Z.L.; Zou, C.G. The role of different types of ecological compensation in ecological protection and restoration of mountains, rivers, forests, fields, lakes and grasses. *Acta Ecol. Sin.* **2019**, *39*, 8709–8716.
64. Liu, G.H.; Wang, X.H.; Wen, Y.H.; Xie, J.; Zhang, Y.F.; Hua, Y.Y.; Zhu, Y.Y.; Hao, C.X. Research progress and practice mode of ecological compensation in my country in the past 20 years. *China J. Environ. Manag.* **2021**, *13*, 109–118.
65. Wunder, S. The Efficiency of Payments for Environmental Services in Tropical Conservation. *Conserv. Biol.* **2007**, *21*, 48–58. [CrossRef]
66. Tacconi, L. Redefining Payments for Environmental Services. *Ecol. Econ.* **2012**, *73*, 29–36. [CrossRef]

67. Liu, D.; Hu, Z.T.; Jin, L.S. A review of research on the analytical framework of ecological protection compensation. *Acta Ecol. Sin.* **2018**, *38*, 380–392.
68. Liu, J.; Li, S.; Ouyang, Z.; Tam, C.; Chen, X. Ecological and Socioeconomic Effects of China's Policies for Ecosystem Services. *Proc. Natl. Acad. Sci. USA* **2008**, *105*, 9477–9482. [CrossRef] [PubMed]
69. Rui, Y.; Xiaoli, S.; Keping, M. Recommendations on building up China's National-Park-centric Protected Area System. *Bio. Sci.* **2019**, *27*, 137–139. [CrossRef]
70. People's Government of Qinghai Province. Overall Planning of Sanjiangyuan National Park. Available online: http://www.gov.cn/xinwen/2018-01/17/5257568/files/c26af29955e141bda0d736a673dac4c5.pdf (accessed on 23 September 2022).
71. State Forestry and Grassland Administration. Hainan Tropical Rainforest National Park Planning (2019–2025). Available online: https://www.forestry.gov.cn/html/main/main_4461/20200423094840466465936/file/20200423094937861802994.pdf (accessed on 23 September 2022).
72. Borrini, G.; Dudley, N.; Jaeger, T. Governance of Protected Areas: From understanding to action. In *Best Practice Protected Area Guidelines Series No.20*; IUCN: Gland, Switzerland, 2013; pp. 46–47.

 *land*

Article

Evaluation of Cultural Ecosystem Service Functions in National Parks from the Perspective of Benefits of Community Residents

Peng Wang [1], Nan Li [2], Yating He [1] and Youjun He [1],*

[1] Research Institute of Forestry Policy and Information, Chinese Academy of Forestry, Beijing 100091, China
[2] International Center for Bamboo and Rattan, National Forestry and Grassland Administration, Beijing 100102, China
* Correspondence: hyjun163@163.com

Abstract: The ecosystem of national parks bears some cultural features. How the cultural ecosystem service functions are perceived by the public and how the cultural ecosystem service functions shape the public's cognition have become urgent scientific questions. This paper performs a case analysis on the Qianjiangyuan National Park System Pilot Area, a representative national park in China, which clarifies the main types of cultural ecosystem service functions from the perspective of the landscape aesthetics benefits of community residents, and analyze the varied impacts of demographics on functional cognition. On this basis, the entropy weight method was adopted to evaluate the importance of each function. Fuzzy comprehensive evaluation was employed to assess the composite level of the cultural service functions. The results show that: (1) the community residents value the benefits brought by the national park the most in terms of the ecological improvement function, and the situation is consistent across the four towns/townships; by contrast, the community residents attach the least importance to the benefits in terms of system governance function. (2) Except for the years of local residence, the community residents' cognition of different cultural ecosystem service functions may vary significantly, owing to factors like gender, age, education level, occupation, and annual mean income. (3) Concerning the importance of functional indices, the importance scores of the natural experience functions, humanistic concern functions, and social service functions are 0.3286, 0.3503, and 0.3211, respectively. The community residents had a moderate to high level of cognition for the cultural ecosystem service functions (3.99). The different types of functions can be sorted by effectiveness as: the social service functions (4.11) > natural experience functions (4.03) > humanistic concern functions (3.86). The research results provide a reference for improving the management level of national parks, and ease the increasingly prominent contradiction between people and land.

Keywords: national park; cultural ecosystem service; community resident; function evaluation; landscape

Citation: Wang, P.; Li, N.; He, Y.; He, Y. Evaluation of Cultural Ecosystem Service Functions in National Parks from the Perspective of Benefits of Community Residents. *Land* **2022**, *11*, 1566. https://doi.org/10.3390/land11091566

Academic Editors: Zhonghua Gou, Rui Yang, Yue Cao, Steve Carver and Le Yu

Received: 19 July 2022
Accepted: 12 September 2022
Published: 14 September 2022

Publisher's Note: MDPI stays neutral with regard to jurisdictional claims in published maps and institutional affiliations.

1. Introduction

More and more people perceive the traditional way of natural protection negatively, mostly owing to the overlook of cultural factors in the protection process [1]. The philosophy of binary opposition contributes greatly to the frequent occurrence of social, economic, and ecological crises in the international community [2]. Under the influence of this philosophy, the national parks of the United States (US) initially excluded indigenous residents from the natural protection program, treating them as an obstacle to ecological protection [3,4]. That is why the indigenous Indians were driven away from the Yellowstone National Park [5,6]. This kind of protection model disregards humanistic factors, separates humans from nature, and leads to the failure of the traditional way of natural protection. Rotherham called this protection model "cultural severance" and proved that the model accelerates ecological destruction [1]. For a long time, human factors were considered detrimental to the stability of the ecosystem. National parks (or protected areas) are often regarded as a system completely independent of humans. Thus, all human interference needs to be eliminated during natural protection [4]. This thinking mode, which separates nature from culture and detaches material from spirit, adds twists and turns to the construction of national parks, especially those in densely populated areas [7]. In fact, except for a few that are in absolute wildernesses, the ecosystem in most national parks (including other types of protected areas) constantly interacts and merges with the human society. As a result, the ecosystem of national parks bears some cultural features and exhibits as a social ecosystem that integrates humans and nature. Such a social ecosystem is a concentrated embodiment of the authenticity of the ecosystem of different national parks [8,9].

The cultural features of a national park are deeply rooted in the local eco-environment, forming a sense of the place unique to locals, and, in return, affects the ecosystem. Against this backdrop, governments around the world have started incorporating cultural factors into natural protection. For example, Japan formed the concept of community governance in the field of natural protection as early as the 1980s. National parks in Japan, a narrow and densely populated country, face a complicated land ownership. Due to the dense population in national parks, there are often complex relationships between property rights, financial rights, industries, and management. To sort out the relationships, the Japanese government issued the *Japans System of Natural Park (Zoning-System)*, which manages and zones park land from the perspective of resource preservation and sustainable utilization. This system is mainly implemented by signing landscape protection agreement with residents. For landowners, the original residents can sign agreements to obtain tax benefits and reduce the cost of land management, thereby reducing the burden of land management [10,11]. In the US, the National Park Service included public participation in multiple links of national parks, ranging from establishment, planning, decision making, management, to operation, and passed the Civic Engagement and Public Involvement and the National Environmental Policy Act (NEPA). These legislations allow the public to participate in at least three phases of national park development: scope delineation, environmental impact assessment (EIA) drafting, and EIA finalization [12,13]. Taking the Yellowstone National Park as an example, one fourths of the scientific research programs ratified each year are completed by foundations and other social organizations [14,15]. India, the world's most populous country, faces similar socioeconomic pressures as China does in the management of national parks. To cope with the pressures, the Indian government has established community reserve management committees and introduced joint community co-ownership programs in an attempt to solve the contradiction between ecological protection and community economic development [16]. Some indigenous Indians naturally worship forests. The concept of "sacred forest" stems from the ancient tradition of nature conservation. In ancient times, indigenous people ringfenced a specific area as a sacred forest, and protected it as a holy land [17]. Their efforts subtly contribute to the protection of forest resources and the maintenance of local biodiversity. In 2006, India enacted the *Forest Act*, which assured that local communities could manage nearby forests. In recent years, the ancient term of "scared forest" has gained prominence among ecological researchers [18,19].

There are diverse research methods and technologies for cultural ecosystem services which involve multiple disciplines [20]. The mainstream approaches include a questionnaire survey, participatory mapping, geographic information, and social media photographs. The evaluation methods for cultural ecosystem services include index system evaluation and value assessment. Plieninger et al. carried out a questionnaire survey on the perception patterns of cultural services among German respondents of different social and demographic backgrounds, and concluded that the respondents tend to associate cultural services and local places with personal happiness [21]. Brown et al. performed participatory mapping to identify areas of significant conservation value in New Zealand [22]. Raymond (2009) mapped the distribution of community values in Australia, and identified the areas threatening the ecosystem services [23]. With the orderly advancement of the national park system reform in China, many theorists and practitioners have shifted their attention to national parks. The problems in the system construction of national parks are explored from multiple angles, namely, biological diversity monitoring, ecological compensation, and endangered species protection, as well as planning and evaluation [24–26]. On the whole, the evaluation methods for cultural ecosystem services are still being explored, without forming mature research methods or paradigms. The research of cultural ecosystem services is closely related to social sciences. At present, ecological experts attach great importance to the regulation and support services of ecosystem services, while paying little attention to the cultural services related to human perception. As a result, the ecosystem assessment in key areas is not objective enough. The systematic study of cultural ecosystem services needs inspiration from disciplines focusing on human well-being, public health, and psychological change, such as social science, psychology, and behavioral science.

Ecosystem services refer to the conditions and efficacies of the natural environment, which are formulated and maintained by the ecosystem, are essential to human survival, and represent all the benefits obtained by humans from the ecosystem [27]. The research focus of ecosystem services is shifting from the accounting of service values to the coupling between humans and their well-being [28–30]. Ecosystem services mainly include the supply service, regulation service, supporting service, and cultural service. Among them, the cultural service has the closest bond to humans. In this background, the cultural ecosystem service functions both intermingle and conflict with the supply, regulation, and supporting services. The cognition of the cultural ecosystem service functions directly bears on the success of the pilot program of the national park. China began to construct national parks in 2015. National parks are established in batches; they are managed by levels and controlled by zones. Following the participatory community management, the original residents are encouraged and supported to engage in environmentally friendly business activities and participate in the management of national park affairs. The difficulty of national park management lies in the trade-off and synergy between the various ecosystem services such as service functions like the ecological supply, regulation and support, and cultural functions like the aesthetic service and recreation. These services are provided to different stakeholders, which affects their behavior. The overemphasis on a particular ecosystem service will definitely affect and damage other ecosystem services [31]. The scientific management of national parks hinges on clear ecosystem services, which are reflected in many national park management policies. Ecosystem services are important indices in many current standards and codes of China, including the *National park functional zoning specification* (LY/T 2933—2018), *Technical specification for the national park master plan* (GB/T 39736-2020), *Specification for monitoring of the national park* (B/T 39738-2020), and *Specification for assessment of the national park* (GB/T 39739-2020). Their importance is reflected primarily in the links between background surveys, analysis, and evaluation, as well as scheme comparison. Based on ecological protection, it is highly necessary to study the cognition and evaluation of the cultural ecosystem services of national parks from the perspective of the community residents.

For the national park system pilot area's cultural ecosystem, the background is not clear, the service function evaluation methods are not perfect, and a lack of cultural values means a realization mechanism to solve practical problems is required. The purpose of this study is to ensure a cultural ecosystem service of the cognitive evaluation is applied to the national park service management, to promote the consideration of cultural factors in planning decisions, and to satisfy the current social growing demand for a better life. Hence, this research is urgently needed to answer the following scientific questions: how are cultural ecosystem service functions perceived by the public? How do cultural ecosystem service functions shape public cognition? What should we do to promote the refined management decision making of national parks through the cognition of cultural services? The research results promote the integration between multiple disciplines and expand the breadth and depth of applying landscape science to the research on cultural services of national park ecosystems. It is suffice to say that our research lays a scientific basis for the rational planning, construction management, policy formulation, protection, and utilization of national parks after the completion of the pilot system reform.

In the light of the above analysis, this paper performs a case analysis on the Qianjiangyuan National Park System Pilot Area (QNPSPA), a representative national park in China, clarifies the main types of cultural ecosystem service functions from the perspective of the landscape aesthetics benefits of the community residents, and scientifically evaluates the importance and overall level of each function, laying the basis for improving the management level of national parks and the ecological welfare of residents. Note that the community residents in the study area, who are generally poorly educated and aged, may understand the concept of cultural ecosystem services differently, as their cognition is strongly affected by direct perception of landscapes. To facilitate field surveys and interviews, our surveys and analyses on the cultural ecosystem service functions of the QNPSPA were carried out mainly from the angle of landscape aesthetics benefits, which are easily felt by the residents. This research perspective was determined in reference to Hatan et al. (2020) and Booth et al. (2017), who also evaluated cultural ecosystem service functions from the angle of landscape aesthetics [32,33].

2. Materials and Methods

2.1. Classification of Cultural Ecosystem Service Functions

Cultural ecosystem services appeared along with ecosystem services in the mid to late 1960s. In the 1990s, cultural services gradually attracted attention from scholars [20]. In 1997, Costanza defined cultural services as the aesthetic, artistic, educational, and scientific values of an ecosystem [34]. In the 21st century, with the publication of the Millennium Ecosystem Assessment report, the definition of cultural ecosystem services was expanded to include human well-being. The classification of cultural services was extended from recreation to more areas, such as aesthetic value, recreation and ecotourism, spirituality and religion, inspiration, sense of place, cultural heritage, social relations, and education, to name but a few. Chan et al. believed that the cultural ecosystem services of natural resources bring non-material benefits to humans, e.g., experience and ability [35]. Russell et al. regarded cultural ecosystem services as the spiritual and cultural well-being that ecosystems contribute to humans through immaterial processes [36]. Considering the actual situation of the study area, and the opinions of experts in the relevant fields, the QNPSPA cultural ecosystem services were divided into natural experience functions, humanistic concern functions, and social service functions, from the perspective of the landscape aesthetics benefits of the community residents.

Specifically, the natural experience functions include the ecological improvement, wilderness protection, and system governance functions, highlighting that the national park ecosystem protects landscape ecology for the community residents. The humanistic concern functions cover the spiritual worship, folk culture popularization, and art inspiration functions, stressing that the national park ecosystem provides the community residents with the landscape art functions, social service functions, involving living environment improvement, science education, and health care functions, reflecting on the fact that the national park ecosystem offers social or ecological public services to the community residents.

Our evaluation system for the cultural ecosystem service functions of national parks consists of a goal layer, a criteria layer, and an index layer. To evaluate the cognition and functions, the indices were transformed into quantitative indices by Likert quantification standards. The cognition was measured by a 3-point scale, with strongly agree and strongly disagree being assigned 3 points and 1 point, respectively. The functions were measured by a 5-point scale, with very high score and very low score being assigned 5 points and 1 point, respectively. The evaluation index system is presented in Table 1.

Table 1. Evaluation index system.

Goal Layer (A)	Criteria Layer (B)	Index Layer (C)
Cultural ecosystem service functions (A)	Natural experience functions (B1)	Ecological improvement function (C1) Wilderness protection function (C2) System governance function (C3)
	Humanistic concern functions (B2)	Spiritual worship function (C4) Folk culture popularization function (C5) Art inspiration function (C6)
	Social service functions (B3)	Living environment improvement function (C7) Science education function (C8) Health care function (C9)

Unlike the functions of ecosystem regulation services and support services, the function of natural experience is a non-material benefit from natural resources to humans [35]. The function of ecological improvement refers to the landscape services that community residents obtain from natural resources, such as air, water, land, forests, and organisms. The function stresses that people achieve ecological well-being through a natural experience of the ecosystem [36]. The function of wilderness protection refers to the concrete protection of the landscape in the natural ecosystem and highlights a human cognition and appreciation of wild natural values, namely virgin forests, and cultural values in areas with little human interference. The function of system governance reflects the integrity of the elements and processes of the national park ecosystem, and the governance features of living communities. It involves mountains, rivers, forests, fields, lakes, and grasslands and emphasizes the cultural values of ecosystem diversity and integrity.

On the humanistic concern functions, spiritual worship refers to the landscape cultural functions formed through the pious appreciation and worship of natural landscapes, mainly including holy mountains, divine trees, and feng-shui forests; the folk culture popularization function stands for the cultural identity with historically significant material and non-material products, such as historical sites, famous ancient trees, legends, and the long tradition of ecological protection; and the art inspiration function refers to the art aesthetic function of the national park landscapes as it is often said that the national park is as beautiful as a landscape painting.

On the social service functions, the living environment improvement function refers to the landscape attributes of the national park in terms of the beautification and greening of the environment; the science education function stands for the science promotion and education functions of the national park, as well as the social values generated by these functions, relative to the residents; the health care function, focusing on the health care of the natural environment in the national park, mainly refers to the delightful feeling generated through the regulation of mental and physical health, as well as the resulting cultural services.

2.2. Study Area and Sample Selection

The QNPSPA is one of the first ten system pilot areas (SPAs) for Chinese national parks. Located in the west of the Zhejiang Province, it is a 252 km^2 area at the junction of three provinces, namely, Zhejiang, Jiangxi, and Anhui (Figure 1). The QNPSA is home to 9744 people living in four towns/townships: Suzhuang Town (1030 people in 383 households), Qixi Town (2621 people in 659 households), Hetian Township (2068 people in 587 households), and Changhong Township (3825 people in 1044 households).

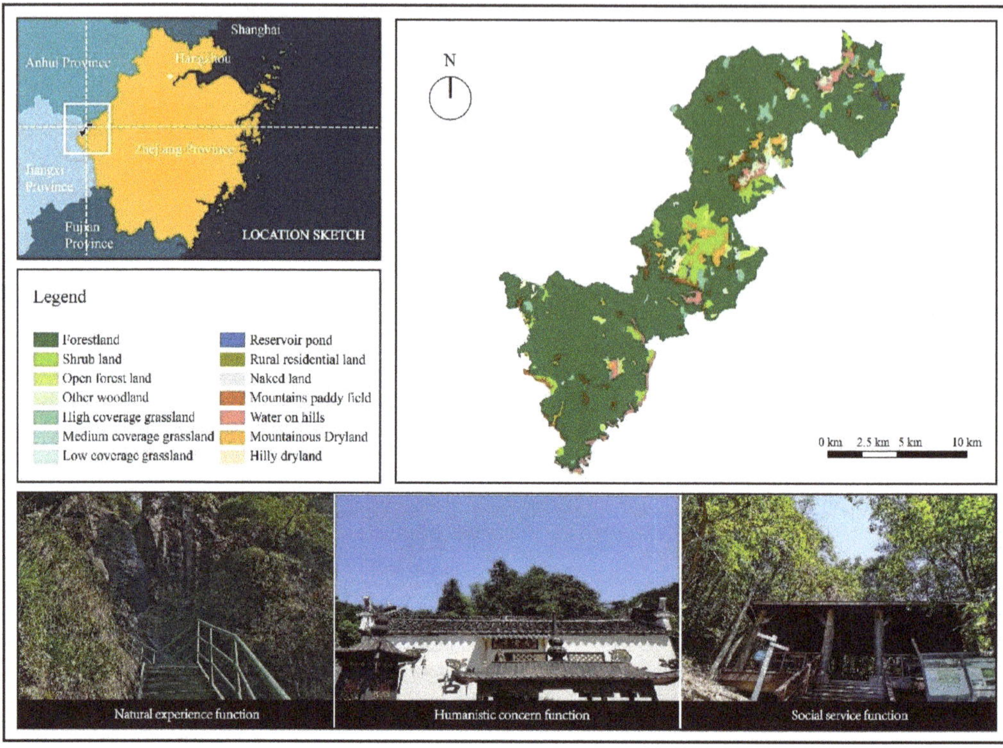

Figure 1. Land use map and cultural ecosystem service functions of the study area. Note: land use data are from the Resources and Environmental Science and Data Center, Chinese Academy of Sciences (https://www.resdc.cn/, accessed on 3 July 2022).

The QNPSPA carries the typical features of **the** collective forest areas in southern China: a high proportion of collective forest lands and a complex ownership. The industrial structure in the area is relatively simple; the residents mainly make a living by selling agricultural and forestry products, or working as migrant workers. The primary agricultural crops are rice and corn, while the dominant product of economic forests is tea seed oil. The economic development of the community relies on the production of bamboo and wood, tea, and other agricultural and forestry by-products. Agritainment and other forms of leisure tourism are still in their infancy. Most of the leisure tourism projects in the area are operated spontaneously on a small scale. According to the *Overall Plan for Qianjiangyuan National Park System Pilot Area (2016–2025)*, the Qianjiangyuan National Park Administration Bureau will develop community industries in an orderly manner and reasonably guide the industrial upgrading process. The communities within the boundaries of the national park will be allowed to carry out the following activities: eco-agriculture, eco-forestry, and rural tourism. These activities involve such industries as organic tea production, freshwater fish farming, creative agriculture, camellia oleifera economic forest, moso bamboo, and rural tourism. Among them, the primary industries like tea production, freshwater fish farming, and camellia oleifera economic forest are relatively large.

Drawing on the previous research [32,33], the empirical questionnaire mainly involves personal conditions, the cognition of the cultural service functions, and value evaluation (Appendix A). Three pre-surveys were carried out from August to September, 2020. The residents are mostly middle-aged and elderly. The pre-surveys found that the residents did not fully understand the concepts of cultural ecosystem services, landscape aesthetics, and functional values (Appendix B, Figure A1). Therefore, the residents were surveyed one after another through face-to-face interviews.

Based on the community and population data provided by Qianjiangyuan National Park Administration Bureau, the research team conducted a field investigation of the 19 administrative villages in the 4 towns/townships (i.e., Suzhuang, Changhong, Hetian, and Qixi) within the QNPSPA, namely, Hengzhong, Yucun, Tangtou, Xixi, Maotan, Suzhuang, Gutian, Xiachuan, Zhenzikeng, Kukeng, Gaosheng, Lulian, Tianfan, Longkeng, Liyangtian, Renzongkeng, Shangcun, Zuoxi, and Qixi. Owing to ecological migration and the relocation policies, there was no permanent resident in Gaosheng.

The research team was led by the staff of Qianjiangyuan National Park Administration Bureau, the staff of the law enforcement office in the relevant towns/townships, the cadres of the said administrative villages, and the forest rangers, and explained the details of the questionnaire survey to community residents in public venues like cultural halls, ancestral temples, village committee offices, and party and public service centers. The residents were invited to fill out the questionnaire. In addition, household visits were paid by the research team under the guidance of village cadres and forest rangers.

A total of 531 questionnaires were released, and 457 (86.06%) effective responses were obtained, including 145 from Suzhuang Town, 79 from Changhong Township, 124 from Hetian Township, and 109 from Qixi Town (Table 2). On the whole, the survey results of this study are basically consistent with the sample survey results conducted before the establishment of the national park.

Table 2. Statistics on valid samples.

Towns/ Townships	Administrative Villages	Village Code	Number of Households	Population	Sample Size
Suzhuang Town	Hengzhong	HZ	174	578	25
	Yucun	YC	44	118	18
	Tangtou	TT	— —	— —	22
	Xixi	XX	— —	— —	9
	Maotan	MT	— —	— —	40
	Suzhuang	SZ	38	106	20
Changhong Township	Gutian	GT	137	428	11
	Xiachuan	XC	438	1487	25
	Zhenzikeng	ZXK	229	815	34
	Kukeng	KK	377	1523	20
Hetian Township	Gaosheng	GS	— —	— —	— —
	Lulian	LL	154	493	46
	Tianfan	TF	140	496	31
Qixi Town	Longkeng	LK	293	1079	47
	Liyangtian	LYT	110	337	26
	Renzongkeng	RZK	192	634	19
	Shangcun	SC	219	670	9
	Zuoxi	ZX	119	612	28
	Qixi	QX	119	368	27
Total	19		2783	9744	457

Note: The dash line in the column of population indicates that the administrative village does not fall in the QNPSPA, yet the land owned by the village collective (e.g., farmlands and forests) belong to that area. The residents of these villages were also surveyed.

2.3. Entropy Weight Method (EWM)

The predecessors held that indices for the cognition of cultural services contribute differently to the composite score of the cultural ecosystem service functions [37]. Before comprehensive evaluation, it is necessary to assign a proper weight to each index. The EWM, an objective weighting approach for composite index evaluation, eliminates the effects of subjective human factors and outshines the traditional subjective weighting methods in terms of reliability. The EWM is primarily based on the information volume of each index. Entropy can be regarded as a measure of uncertainty. The greater the information volume, the lower the uncertainty of the index, the smaller the entropy, and the larger the index weight. The inverse is also true. The EWM can be implemented in the following steps:

Step 1. Data normalization

The original data x_{ij} are nondimensionalized through the normalization of the deviance, producing the initial matrix for comprehensive evaluation $Y = (y_{ij})_{n \times m}$ $(0 \leq i \leq m, 0 \leq j \leq n)$. Under the j-th index, the index weight $z_{ij}(0 \leq z_{ij} \leq 1)$ of the i-th resident can be calculated by:

$$z_{ij} = y_{ij} / \sum_{i=1}^{m} y_{ij} \tag{1}$$

On this basis, the proportion matrix $Z = (z_{ij})_{m \times n}$ is established for the survey data. The information entropy e and information utility d are computed for each index of cognition of cultural services. The information entropy e_j of the j-th cognition index can be calculated by:

$$e_j = -K \sum_{i=1}^{m} z_{ij} In(z_{ij}) \tag{2}$$

where $K = 1/(In(m))$ is a constant. The information utility d_j of the j-th cognition index depends on the difference between the entropy e_j and 1. The greater the d_j, the larger the weight of that cognition index. The information utility d_j can be calculated by:

$$d_j = 1 - e_j \tag{3}$$

Step 2. Index weighting

The greater the information utility d_j, the larger the weight of the index of cognition of cultural services, and the more prominent the contribution of the index to the composite cognition. The weight of the j-th cognition index can be calculated by:

$$w_j = d_j / \sum_{j=1}^{n} d_j \tag{4}$$

Step 3. Composite score calculation

The composite score U is obtained by weighted summation. The greater the U, the better the effect of the samples. Let w_j be the weight of the j-th index. Then, the composite score U can be calculated by:

$$U = \sum_{i=1}^{n} y_{ij} w_j \times 100 \tag{5}$$

2.4. Fuzzy Comprehensive Evaluation (FCE)

Utilizing the membership theory of fuzzy mathematics, the FCE is a comprehensive evaluation method for quantitative analysis. For the QNPSPA, the cognition evaluation of the cultural ecosystem service functions is a fuzzy task. Based on fuzzy mathematics, the qualitative evaluation for the cultural ecosystem service functions of the QNPSPA was transformed into quantitative evaluation, which is more pertinent and systematic than traditional approaches like analytic hierarchy process (AHP). Inspired by existing studies, the cultural ecosystem service functions of the QNPSPA were evaluated through primary FCE and the overall cognitive value of these functions was assessed by secondary FCE [37].

The primary FCE includes the following steps:

Step 1. Setting up the FCE index set

The sets of primary indices are established as $B_1 = \{C_1, C_2, C_3\}$, $B_2 = \{C_4, C_5, C_6\}$, and $B_3 = \{C_7, C_8, C_9\}$; the set of secondary indices is established as $A = \{B_1, B_2, B_3\}$.

Step 2. Setting up the comment set

The comment set can be established as $V = \{V_1, V_2, V_3, V_4, V_5\}$: {strongly high, slightly high, neutral, slightly low, strongly low}.

Step 3. Setting up the FCE matrix

The weight set vector K is calculated for each index by the EMV. Then, m residents are invited to evaluate the index set A, forming a fuzzy mapping. The relevant results are summarized into the FCE matrix R:

$$R = \begin{bmatrix} r_1 \\ r_2 \\ \vdots \\ r_m \end{bmatrix} = \begin{bmatrix} r_{11} & r_{12} & \cdots & r_{1n} \\ r_{21} & r_{22} & \cdots & r_{2n} \\ \vdots & \vdots & \vdots & \vdots \\ r_{m1} & r_{m2} & \cdots & r_{mn} \end{bmatrix} \tag{6}$$

where r_{ij} is the degree of comments $V_1, V_2, \cdots, V_5$ made by each resident on each index ($0 \leq i \leq m$, $0 \leq j \leq n$). According to the principle of maximum membership, the maximum r_{ij} is set to 1.

Step 4. Setting up the primary FCE set

The primary FCE set is derived from the weight set vector K and the FCE matrix R. For the three types of cultural service functions, the primary FCE sets S_{B1}, S_{B2}, and S_{B3} can be calculated by:

$$S_{Bi} = K_{Bi} * R_{Bi} = (b_{1i}, b_{2i}, \ldots b_{ni}) \tag{7}$$

where $*$ is the generalized fuzzy synthetic operation.

By the principle of maximum membership in FCE, the comment set V_j corresponding to maximum b_j is the optimal result of our primary FCE. Let $*^{\wedge}$ be the generalized fuzzy AND operation; $*^{\vee}$ be the fuzzy OR operation. Then, b_j can be calculated by:

$$b_j = (a_1 *^{\wedge} r_{1j}) *^{\vee} (a_2 *^{\wedge} r_{2j}) *^{\vee \cdots} *^{\vee} (a_m *^{\wedge} r_{mj}) \tag{8}$$

The evaluation is one-sided, when only S_{Bi} is taken as the evaluation index. For comprehensiveness, S_{Bi} was sorted out to obtain the secondary index S_B for secondary FCE. The secondary FCE includes the following steps:

Step 1. Setting up the secondary judgement matrix

Based on the secondary fuzzy index set A, S_{B1}, S_{B2}, and S_{B3} can be organized into the secondary judgement matrix S_B:

$$R_B = \begin{bmatrix} S_{B1} \\ S_{B2} \\ S_{B3} \end{bmatrix} = \begin{bmatrix} b_{11} & b_{12} & b_{13} \\ b_{21} & b_{22} & b_{23} \\ b_{31} & b_{32} & b_{33} \end{bmatrix} \tag{9}$$

Step 2. Setting up the secondary FCE set

The latter part of the secondary FCE is consistent with that of the primary FCE. The secondary FCE set S_B can be established as:

$$S_B = K_B * R_B = (b_1, b_2, \cdots, b_n) \tag{10}$$

Since the QNPSPA cultural ecosystem service functions are fuzzy, the idea of fuzzy mathematics was drawn to comprehensively consider the fuzzy comment subsets S_B and S_{Bi}. In this way, the cognitive aesthetic values of the QNPSPA were depicted quantitatively, making the evaluation more realistic. Specifically, the level was determined for each comment in the comment set V. The column vectors of the lines were compiled into the score set $N = (N_1, N_2, N_3, N_4, N_5)^T = (5, 4, 3, 2, 1)^T$ corresponding to the comment set. Based on the score set N, the levels can be solved through the inner product operation of the vectors:

$$S_B \cdot N = \sum_{j=1}^{n} b_j \cdot N_j \tag{11}$$

where S_B is a fuzzy comment subset; N is the score set. Note that the specific levels P are real numbers. In this paper, the result of secondary FCE S_B is normalized such that $0 \leq b \leq 1$ and $\sum b_j = 1$. Thus, the value of real number P is the weighted mean of the secondary FCE set S_B as the weight vector is relative to N_1, N_2, N_3, N_4, and N_5. In other words, the value of real number P reflects the comprehensive information from the secondary FCE set S_B and the score set N, laying the basis for solving the actual composite score for the QNPSPA cultural ecosystem service functions.

3. Results

3.1. Demographics

To ensure the data quality of the questionnaire survey, the sample data were subjected to reliability and validity tests on IBM SPSS Statistics. It was calculated that the Cronbach's alpha (0.805) was greater than 0.8, and the Kaiser–Meyer–Olkin (KMO) statistic (0.873) fell between 0.8 and 0.9 and achieved significance at the level of 95%, and even 99%. Referring to the standards of reliability and validity, the survey data on the community residents in the QNPSPA are of good internal consistency and structural validity.

As shown in Table 3, more male residents (50.33%) were surveyed than females (49.67%). The middle-aged (41–55) group was the largest age group among the respondents (39.82%), followed by the middle-aged and elderly (56–70) (32.39%). The smallest group was young people of 25 and below (2.84%). Among the respondents, 45.30% were either illiterate or graduates of primary schools; 34.79% and 19.91% had graduated from junior high schools and senior high schools and above, respectively. In terms of occupation, most of the respondents were farmers (62.58%). The second largest group (20.35%) worked in individual service industries, such as agritainment, homestays, and sales. Quite a few respondents worked in factories (7.44%) or worked in other cities (5.91%). The majority of the respondents earned CNY 50,000 and below. Notably, 188 (41.14%) of the respondents had an annual mean income of CNY 20,000 and below; 35.89% had an annual mean income of CNY 30,000–CNY 50,000; only 3.50% earned an average of CNY 160,000–CNY 300,000; and 1.53% earned CNY 310,000 and above, respectively, each year. The respondents (87.53%) had largely been living in the study area for 20 years or more. Only 2.19% had been living there for five years or less.

Table 3. Demographics of the respondents.

Demographics		Number (People)	Proportion (%)
Gender	Male	230	50.33
	Female	227	49.67
Age	≤25	13	2.84
	26–40	57	12.47
	41–55	182	39.82
	56–70	148	32.38
	≥71	57	12.47
Education level	Primary school and below	207	45.30
	Junior high school	159	34.79
	Senior high school and secondary technical school	67	14.66
	Higher vocational school and junior college	21	4.60
	Ordinary college and above	3	0.66
Occupation	Farmers	286	62.58
	Individual service workers	93	20.35
	Enterprise employees	34	7.44
	Migrant workers	27	5.91
	Students	8	1.75
	Others	9	1.97

Table 3. *Cont.*

Demographics		Number (People)	Proportion (%)
Annual mean income	≤CNY 20,000	188	41.14
	CNY 30,000–CNY 50,000	164	35.89
	CNY 60,000–CNY 150,000	82	17.94
	CNY 160,000–CNY 300,000	16	3.50
	≥CNY 310,000	7	1.53
Years of local residence	5 years and below	10	2.19
	6–10 years	23	5.03
	11–20 years	24	5.25
	21 years and above	400	87.53

3.2. Cognition of Cultural Ecosystem Services

As shown in Figure 2, more residents (91.68%) perceived the ecological improvement function in the QNPSPA than the other aesthetic values. More than 80% of the responds perceived the living environment improvement function (89.72%) and art inspiration function (86.43%). The following aesthetic values were cognized by more than 70% of the respondents: art inspiration function (78.77%), spiritual worship (75.05%), health care (75.05%), and wilderness protection (74.84%). System governance (69.37%) and folk culture popularization (68.71%) were perceived by over 60% of the respondents.

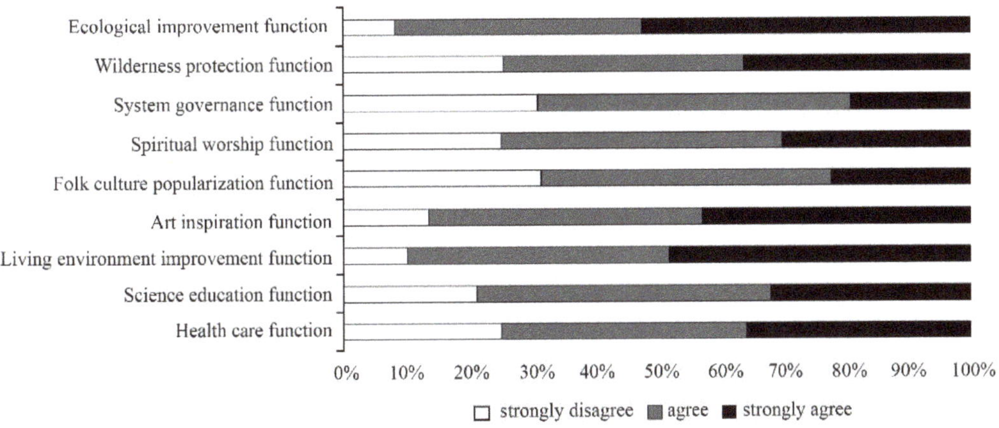

Figure 2. Proportion of community residents with different cognition degrees of cultural service functions.

Over 50% of the respondents strongly agreed that the QNPSPA had the ecological improvement function. Over 40% strongly agreed that the QNPSPA had the living environment improvement function and art inspiration function. Over 30% perceived wilderness protection, health care, the art inspiration function, and spiritual worship significantly. Over 20% found the folk culture popularization very prominent. In addition, over 10% strongly perceived the system governance function.

The QNPSPA covers four towns/townships in different functional zones. The different management methods lead to variations in how the residents of different towns/townships perceive cultural ecosystem services. The cognition of cultural ecosystem services in Suzhuang Town, Hetian Township, Changhong Township, and Qixi Township is displayed in Figure 3.

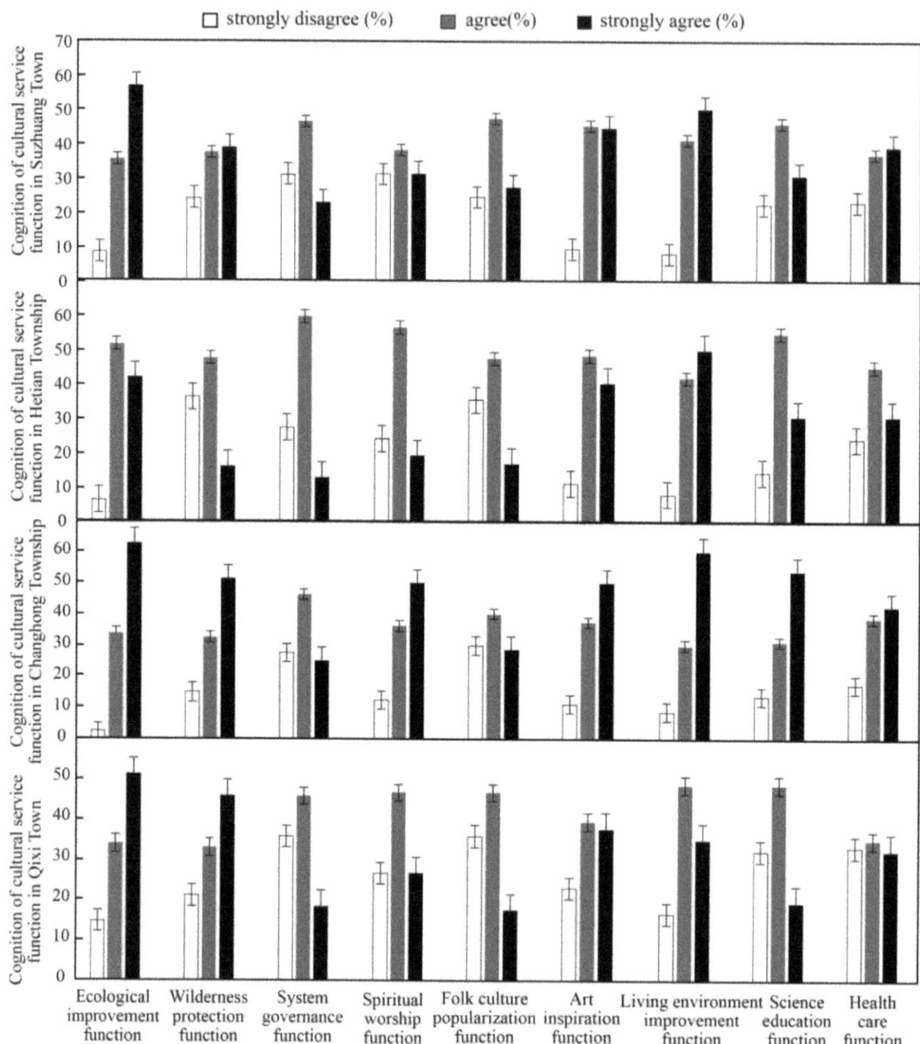

Figure 3. Cognition of cultural service function in different towns.

Among the residents of Suzhuang Town, over 90% held the view that the QNPSPA had ecological improvement, art inspiration, and living environment improvement functions. Around 75% believed that the area had wilderness protection, folk culture popularization, and science education functions. Sixty-eight percent claimed that the area had system governance and spiritual worship functions. Every cultural ecosystem service of the QNPSPA was recognized by more than 60% among the residents of Hetian Township. The ecological improvement function won the most beholders (93.55%), while that of wilderness protection was perceived by the fewest respondents (63.71%). Every cultural ecosystem service of the QNPSPA was recognized by more than 40% among the residents of Changhong Township. The ecological improvement function was cognized by the largest group of respondents (97.47%), while that of folk culture popularization was perceived by the smallest group (69.62%). Every cultural ecosystem service of the QNPSPA was recognized by more than 50% among the residents of Qixi Town. More than 80% of the respondents agreed that the QNPSPA boasts the functions of ecological improvement and living environment improvement. By contrast, system governance and folk culture popularization were the least perceived functions, but the recognizers still took up more than 60% of the respondents.

As shown in Table 4, the cultural ecosystem services of the QNPSPA can be ranked by the community residents' cognitive score as: the ecological improvement function (2.44) > living environment improvement function (2.38) > art inspiration function (2.29) > wilderness protection/science education/health care function (2.11) > spiritual worship function (2.05) > folk culture popularization function (1.91) > system governance function (1.89). In general, the residents of different administrative villages had a high cognition of ecological improvement, living environment improvement, and art inspiration functions, and a low cognition of folk culture popularization and system governance functions. The ecological improvement function was the most cognized cultural service function in all the villages of Suzhuang Town, Changhong Township, and Qixi Town, while living environment improvement was the most perceived function among the residents of Hetian Township. The system governance function was the least perceived function among those living in Suzhuang Town and Changhong Township, while folk culture popularization was that among the residents in Hetian Township and Qixi Town.

Table 4. Community residents' cognitive scores of cultural service functions.

Cultural Service Function	Suzhuang	Hetian	Changhong	Qixi	Total
Ecological improvement	2.48	2.35	2.61	2.37	2.44
Wilderness protection	2.14	1.8	2.37	2.25	2.11
System governance	1.92	1.85	1.97	1.83	1.89
Spiritual worship	2.00	1.95	2.38	2.00	2.05
Folk culture popularization	2.03	1.81	1.99	1.82	1.91
Art inspiration	2.35	2.29	2.39	2.15	2.29
Living environment improvement	2.42	2.42	2.52	2.18	2.38
Science education	2.08	2.16	2.41	1.87	2.11
Health care	2.16	2.06	2.25	1.99	2.11

The cognitive scores of the cultural service functions in 18 administrative villages (residents of Gaosheng had been entirely relocated) were subjected to cluster analysis. The resulting spatial distribution of the cultural service functions cognized in different villages is displayed in Figure 4.

The living environment improvement function was perceived as neutral and strongly high in 17 administrative villages; the ecological improvement and wilderness protection functions were cognized as neutral and strongly high in 15 administrative villages; the folk culture popularization and science education functions were cognized as neutral and strongly high in 14 administrative villages; the health care function was cognized as neutral and strongly high in 13 administrative villages; the system governance and art inspiration functions were cognized as neutral and strongly high in 11 administrative villages; and the spiritual worship function was cognized as neutral and strongly high in 10 administrative villages. The spiritual worship function received a strongly low cognition in more administrative villages (eight) than any other function, followed by the system governance and art inspiration functions, each of which received a strongly low cognition in only seven villages. Overall, at the spatial pattern level, the residents in different communities (administrative villages) differed significantly in the cognition of different cultural service functions.

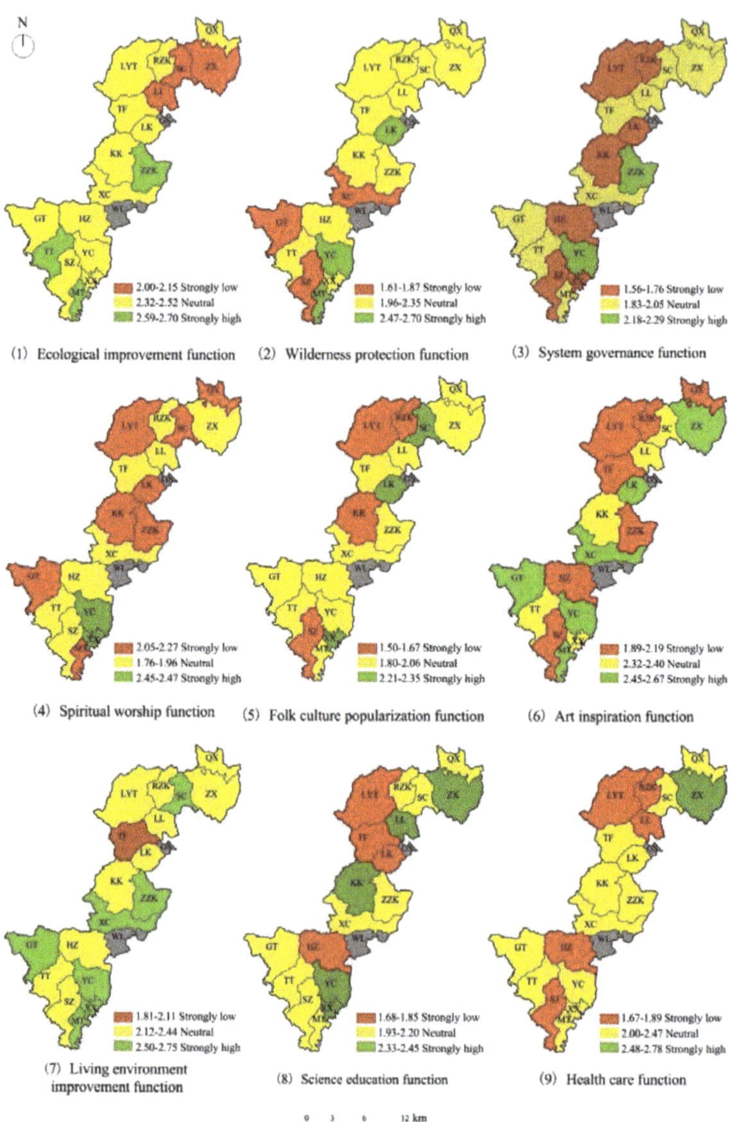

Figure 4. Spatial distribution of cultural service functions cognized by different villages. Note: the code of each administrative village is given in Table 2.

3.3. Influence of Demographics on the Cognition of Cultural Ecosystem Services

As shown in Table 5, gender had a significant impact on the respondents' cognition of the system governance function. Age prominently influenced the ecological improvement, wilderness protection, system governance, folk culture popularization, art inspiration, science education, and health care functions. Education level directly bore on the cognition of the ecological improvement, wilderness protection, system governance, spiritual worship, folk culture popularization, science education, and health care functions. Occupation had a prominent effect on the cognition of the ecological improvement, folk culture popularization, art inspiration, science education, and health care functions. The annual mean income exerted a major impact on the cognition of the ecological improvement, system governance, and science education functions.

Table 5. Influence of demographics on the cognition of cultural ecosystem services.

Demographics		Natural Experience Functions						Humanistic Concern Functions								Social Service Functions					
		Ecological Improvement		Wilderness Protection		System Governance		Spiritual Worship		Folk Culture Popularization		Art Inspiration		Living Environment Improvement		Science Education		Health Care	
		Mean	F	Mean	F	Mean	F	Mean	F	Mean	F	Mean	F	Mean	F	Mean	F	Mean	F
Gender	Male	2.45	0.14	2.10	0.08	1.95	6.49 **	2.10	0.05	1.97	2.40	2.33	2.69	2.41	0.17	2.22	0.31	2.12	1.59
	Female	2.44		2.13		1.83		2.00		1.85		2.26		2.35		1.10		2.10	
Years of residence	6–10 years	2.22		2.00		1.96		1.91		2.04		2.57		2.43		2.43		2.39	
	11–20 years	2.38	1.65	2.13	0.24	2.08	0.78	2.04	1.01	1.96	0.30	2.38	1.44	2.33	1.20	2.21	1.90	2.21	1.28
	≥21 years	2.47		2.12		1.88		2.06		1.90		2.27		2.39		2.08		2.09	
Age	≤25	2.29		1.86		2.07		2.36		1.79		2.93		2.71		2.86		2.14	
	26–40	2.61		2.48		2.09		2.21		2.23		2.48		2.46		2.39		2.38	
	41–55	2.57	6.94 ***	2.12	4.32 ***	1.93	3.58 ***	2.07	1.84	1.93	3.83 ***	2.28	3.28 **	2.40	1.93	2.23	15.80 ***	2.21	5.24 ***
	56–70	2.34		2.03		1.83		1.97		1.84		2.20		2.34		1.94		1.97	
	≥71	2.16		2.02		1.65		1.98		1.77		2.23		2.25		1.68		1.88	
Education level	Primary school and below	2.32		2.01		1.74		1.89		1.72		2.21		2.32		1.94		1.94	
	Junior high school	2.50	4.20 ***	2.18	2.50 **	2.01	4.84 **	2.17	5.42 ***	2.03	7.70 ***	2.31	1.85	2.39	1.18	2.23	6.49 ***	2.28	5.20 ***
	Senior high school and secondary technical school	2.61		2.15		2.00		2.27		2.16		2.45		2.48		2.22		2.13	
	Higher vocational school and junior college	2.67		2.48		2.14		2.14		2.14		2.52		2.57		2.52		2.38	
	Ordinary college and above	2.33		2.33		1.67		1.67		1.67		2.33		2.33		2.33		2.33	
Occupation	Farmers	2.36		2.08		1.84		2.03		1.83		2.26		2.33		1.99		2.03	
	Individual service workers	2.58		2.22		1.90		2.08		2.02		2.25		2.41		2.20		2.21	
	Enterprise employees	2.47	3.88 ***	2.06	1.55	2.00	1.05	2.00	0.97	2.18	2.76 **	2.50	2.56 **	2.56	1.99 *	2.35	6.13 ***	2.41	2.29 **
	Migrant workers	2.74		2.22		2.11		2.11		2.11		2.33		2.56		2.48		2.26	
	Students	2.13		1.63		2.00		2.00		1.88		3.00		2.75		2.88		2.00	
	Others	2.78		2.44		2.00		2.56		2.11		2.33		2.11		2.22		2.22	
Annual mean income	≤CNY 20,000	2.35		2.11		1.78		1.99		1.81		2.24		2.31		1.91		2.10	
	CNY 30,000—50,000	2.48		2.16		2.02		2.07		1.95		2.32		2.40		2.24		2.08	
	CNY60,000—150,000	2.57	3.55 ***	2.05	0.52	1.91	3.55 ***	2.17	0.98	2.07	2.12 *	2.35	0.52	2.46	1.47	2.24	6.55 ***	2.18	0.27
	CNY 160,000—300,000	2.69		2.06		1.81		2.00		2.00		2.25		2.50		2.31		2.13	
	≥CNY310,000	2.00		1.86		1.43		1.86		1.86		2.43		2.71		2.43		2.00	

Note: ***, **, and * are significance levels of 1%, 5%, and 10%, respectively.

The demographic differences in the cognition of cultural ecosystem services are summarized below: males generally perceived cultural ecosystem services better than females. Natural experience functions were the best recognized aesthetic value among those aged 26–40; the young residents at 40 and below, plus the middle-aged and elderly aged over 41, had a relatively high cognition degree of the social service functions and humanistic concern functions. The natural experience functions and humanistic concern functions were the best recognized aesthetic values among the residents who had graduated from higher vocational schools and junior colleges, while the humanistic concern functions were the best recognized functions among the residents who had graduated from senior high schools and secondary technical schools. Farmers failed to perceive any of the cultural ecosystem services highly. The ecological improvement function was the best cognized function among the respondents earning CNY 160,000–300,000 each year. System governance was the best cognized function among those with an annual mean income of CNY 60,000–150,000. The science education function was better recognized than any other function among those earning CNY 310,000–500,000 per year.

3.4. Importance of Cultural Ecosystem Service Functions

The weights of the evaluation indices for the cultural service functions were solved by the EWM (Table 6). The cognitive weights of the natural experience functions, humanistic concern functions, and social service functions were 0.3286, 0.3503, and 0.3211, respectively.

Table 6. Index weights for cognition evaluation of cultural ecosystem services of community residents.

Criteria Layer	Weight	Index Value	Index Layer	Global Weight	Intra-Class Weight	Index Value
Natural experience (B1)	0.3286	0.6885	Ecological improvement (C1)	0.0695	0.2115	0.1697
			Wilderness protection (C2)	0.1314	0.3998	0.2777
			System governance (C3)	0.1277	0.3887	0.2412
Humanistic concern (B2)	0.3503	0.7218	Spiritual worship (C4)	0.1248	0.3562	0.2561
			Folk culture popularization (C5)	0.1355	0.3867	0.2591
			Art inspiration (C6)	0.0901	0.2572	0.2066
Social service (B3)	0.3211	0.6985	Living environment improvement (C7)	0.0778	0.2421	0.1851
			Science education (C8)	0.1133	0.3529	0.2390
			Health care (C9)	0.1300	0.4050	0.2743

By the importance of each index, the three types of cultural service functions could be ranked as humanistic concern functions > natural experience functions > social service functions, whose index values were 0.6885, 0.7218, and 0.6985, respectively. On the cognition of the community residents, the humanistic concern functions are the most important cultural ecosystem service of the QNPSPA.

In terms of the natural experience functions, wilderness protection was the most important function (cognitive weight: 0.3998), followed by system governance (cognitive weight: 0.3887); ecological improvement was the most unimportant function (cognitive weight: 0.2115).

In terms of the humanistic concern functions, folk culture popularization was the most important function (cognitive weight: 0.3867), followed by spiritual worship (cognitive weight: 0.3562); art inspiration was the least important function (cognitive weight: 0.2572).

In terms of the social service functions, the first and second most important functions were health care (cognitive weight: 0.4050) and social education (cognitive weight: 0.3529); the least important function was the living environment improvement function (cognitive weight: 0.2421).

3.5. Evaluation Results on Cultural Ecosystem Service Functions

Following the FCE procedure, the FCE matrices R_{B1}, R_{B2}, and R_{B3} can be established for the natural experience, humanistic concern, and social service functions of the QNPSPA, respectively:

$$R_{B1} = \begin{bmatrix} 0.5427 & 0.3786 & 0.0613 & 0.0175 & 0.0000 \\ 0.3654 & 0.3829 & 0.1729 & 0.0591 & 0.0197 \\ 0.1947 & 0.4989 & 0.2429 & 0.0481 & 0.0153 \end{bmatrix}$$

$$R_{B2} = \begin{bmatrix} 0.3020 & 0.4486 & 0.1904 & 0.0547 & 0.0044 \\ 0.2254 & 0.4617 & 0.0394 & 0.1160 & 0.1575 \\ 0.4289 & 0.4354 & 0.1072 & 0.0197 & 0.0088 \end{bmatrix}$$

$$R_{B3} = \begin{bmatrix} 0.4902 & 0.4114 & 0.0788 & 0.0131 & 0.0066 \\ 0.3217 & 0.4661 & 0.1357 & 0.0591 & 0.0175 \\ 0.3632 & 0.3961 & 0.1554 & 0.0788 & 0.0066 \end{bmatrix}$$

Through the compound operation of the fuzzy matrices, the primary FCE sets S_{B1}, S_{B2}, and S_{B3} were obtained for the natural experience, humanistic concern, and social service functions, respectively. On this basis, the secondary FCE matrix can be constructed for the cognition evaluation of cultural ecosystem services in the QNPSPA:

$$R_B = \begin{bmatrix} S_{B1} \\ S_{B2} \\ S_{B3} \end{bmatrix} = \begin{bmatrix} 0.3366 & 0.4271 & 0.1765 & 0.0460 & 0.0138 \\ 0.3050 & 0.4503 & 0.1106 & 0.0694 & 0.0647 \\ 0.3793 & 0.4245 & 0.1299 & 0.0559 & 0.0104 \end{bmatrix}$$

Finally, the FCE set can be derived through the compound operation of the fuzzy matrices for the community residents' cognition of cultural ecosystem services:

$$S_B = K_B * R_B = (0.3392\ 0.4344\ 0.1384\ 0.0574\ 0.0306)$$

As shown in Table 7, the community residents had a moderate to high level of cognition for the cultural ecosystem service functions in the study area (3.99), according to the comments in the score set corresponding to the maximum value of the FCE set, and the principle of maximum membership of the FCE.

Table 7. Scores of cultural ecosystem service functions of community residents.

Goal Layer	Score	Criteria	Score	Index Layer	Score
		Natural experience (B1)	4.03	Ecological improvement (C1)	4.45
				Wilderness protection (C2)	4.02
				System governance (C3)	3.80
Cultural ecosystem service functions (A)	3.99	Humanistic concern (B2)	3.86	Spiritual worship (C4)	3.99
				Folk culture popularization(C5)	3.48
				Art inspiration (C6)	4.26
		Social service (B3)	4.11	Living environment improvement (C7)	4.36
				Science education (C8)	4.00
				Health care (C9)	4.03

4. Discussion

4.1. Community Residents Have Different Cognitions of QNPSPA Cultural Ecosystem Service Functions

Considering the realization of cultural functions in national parks, this study scientifically classifies the cultural ecosystem service functions of the QNPSPA from the perspective of the benefits of the community residents. Three kinds of cultural service functions were summarized, namely, natural experience, humanistic concern, and social service. Through the scientific categorization of cultural service functions, we clarified the core research objects and their basic connotations. The understanding of cultural ecosystem services among the community residents is mainly reflected by indices like the ecological improvement, art inspiration, and living environment improvement functions. These indices are easy for the residents to understand, thanks to their intuitiveness, perceptibility, and visibility. The functions and benefits of these services can be easily perceived by people in their daily life [35]. Nevertheless, the emotional and spiritual benefits of cultural ecosystem services are usually subtle, implicit, and expressed indirectly [38]. Unlike other ecosystem services, cultural ecosystem services are difficult to be felt or seen. To understand cultural ecosystem services, one must be familiar with the ecological processes, which is no easy task for the community residents. Analysis reveals some differences between the community residents in the cognition of the cultural ecosystem service functions. These differences mainly arise from the residents' internal interests and cognition difficulty [39]. Most of the residents have lived in the QNPSPA for over 20 years. Their production and life are closely linked to the QNPSPA ecosystem; it is natural for them to emphasize the cultural service functions provided by the QNPSPA landscapes in terms of the natural eco-environment. In addition, the QNPSPA is an important ecological functional area of the Yangtze River Delta, a population stronghold and socioeconomic high ground of China. The permanent residents in this area view a good eco-environment as a high-quality resource and treat it as a significant advantage over the other areas of the Yangtze River Delta. This is consistent with the results of Ridding et al. (2018), Peng et al. (2019), and Yu (2019) [40–42].

According to the results of the structured interviews, the residents who were interviewed were generally old. Most of the young people in their families seek jobs in nearby cities, such as Hangzhou and Shanghai. The permanent residents in the community are mainly middle-aged and elderly people. The internal demand for better ecological conditions and a beautiful community environment directly affects how the respondents evaluate the functions of cultural services. The functions more in line with their needs attract more attention [43]. In this study, the community residents were interviewed deeply. The results show that, under the wild animal protection policies in the national park, the QNPSPA is overflown with wild boars, which often destroy the production spaces (e.g., farmlands and vegetable fields) of the community residents and disturb their daily life. Therefore, the cultural, ecological, and economic values of forests can promote each other and may conflict with each other. Recent studies have shown that the housing price could be improved if the houses were surrounded by forests or woods, provided that the forests or woods are not too biologically diverse and are highly accessible. Otherwise, the housing price will remain low because most residents fear wild animals [44]. With the growing need for the cultural value of the forests, the contradiction between the cultural value and ecological and economic values becomes increasingly prominent. Then, people start paying attention to the balance between the cultural, ecological, and economic values of forests [45,46].

The natural experience function is highly perceived in Suzhuang Town, but not so in Hetian Township. The wilderness protection function is highly perceived in Changhong Township, but poorly cognized in Hetian Township. The results have much to do with the natural eco-environment of the regions. Suzhuang Town, the site of the original Gutian Mountain Nature Reserve, boasts a high ecological quality. The residents of the town have a natural advantage in perceiving natural experience. By contrast, Hetian Township is densely populated and dominated by farmland. The frequent human interference drags down the perception of the residents of the natural experience function. That is why the wilderness protection function is poorly cognized in Hetian Township. The high cognition of Changhong Township dwellers is possibly due to the complex geology and landform; the unique terrain of the township leads to rich geological landscapes, including hills, valleys, rocks, cliffs, and canyons. These resources push up the perception of locals for the wilderness protection function.

The participation of community, an integral part of national parks, is crucial to the sustainable development of national parks. Community-based co-management, also known as community participatory management, community cooperative management, or community co-management, is a management mode in which local residents and the government share responsibilities and obligations. The main purpose of the mode is to achieve a win-

win between ecological protection and community sustainability. The establishment of a national park has a great impact on those who have been living in the area for a long time. It is particularly important to properly handle the interests of the community residents. Our survey reveals that Qianjiangyuan National Park Administration Bureau provided the community residents with job opportunities, such as rangers, tour conductors, and sanitary workers. The residents are therefore motivated to participate in the joint management. Nonetheless, the community residents generally have not realized their right of supervision over the construction of the national park, nor exercised their supervision power in a wide range. In fact, they have not become the master of national park management. Owing to the complex natural conditions in the study area, long-term dwellers in the region have formed the traditional cultural concept of respecting and conforming to nature, which is very conducive to the ecological protection of the national park. This concept is worthy of further examination and consideration.

4.2. Demographics Significantly Affect the Cognition of QNPSPA Cultural Ecosystem Service Functions

Cultural ecosystem services are the most direct and fastest way for the community residents to enjoy the landscape values of national parks, providing an important way to promote ecosystem management [47]. From the perspective of the landscape aesthetics benefits of the community residents, the cognition of the cultural ecosystem service functions in a national park depends heavily on gender, age, income, and other demographic features. Different cognitions will result in different ecological protection behaviors (positive or negative) for the national park [48].

Except for the years of local residence, the community residents' cognition of different cultural ecosystem service functions in the QNPSPA varied significantly with their demographic features. Among them, age significantly affects all the functions of natural experience (ecological improvement, wilderness protection, and system governance), with the p-value remaining below 0.01. Age could greatly influence service functions like health care and science education ($p < 0.01$). In semi-structured interviews, many residents expressed an unwillingness to leave their community, even if the community has some defects. This complex may be related to the age of the interviewees. According to experience, old people are more nostalgic than young people. Therefore, the construction of the QNPSPA cultural functions should focus on the age differences of the audience. For the community residents of different ages, it is important to strengthen the management of landscape forests, especially the plant landscape (a semi-natural area surrounding the community) configuration and artistic conception around the community.

Meanwhile, the community residents are mainly middle-aged and elderly people. In-depth interviews show that they are very concerned about health. From the perspective of human healthcare, it is necessary to better manage and show a tendency towards "green shower" forests in the QNPSPA. Medial research has proved the health functions of natural factors like forest volatile matters and negative oxygen ions. The forest environment plays a major role in stress relief, immunity boosting, and anxiety mitigation [49]. This study also found that the community residents of different ages vary significantly in their cognition of the health care function. Drawing on field surveys, we suggested developing the forest health and wellness industry, relying on the rich forest resources and convenient traffic of the QNPSPA. It is advised to properly plan a forest health and wellness base in the south of the QNPSPA and cultivate "green shower" forests by planting antioxidant tree species, such as camphor, camphor, metasequoia, Chinese yew, Sakura, and tea.

In addition, occupation significantly affects all the functions of social service (living environment improvement, science education, health care). The significance of the science education function was $p < 0.01$. Moreover, education level significantly affects all the functions of natural experience (ecological improvement, wilderness protection, and system governance), with the p-value remaining below 0.05. Under the premise of protecting natural resources, the audience of different occupations and education levels are recommended to develop a batch of natural education bases to meet the needs of multiple age groups, based on existing rural schools, community parks, and outdoor activity spaces. In addition, the famous and ancient trees in the surroundings should be utilized to build new forms of science education spaces, e.g., forest classrooms and outdoor blackboard walls.

4.3. Different Cultural Ecosystem Service Functions Differ in Importance and Evaluation Score

When it comes to the importance of the cognition of the cultural ecosystem service functions, the humanistic concern functions were regarded as the most important, followed by the natural experience functions; the social service functions were considered the least important. In terms of specific functional indices, the community residents cognize different cultural service functions. The most important functions in their eyes include folk culture popularization, wilderness protection, and health care. However, the cognition degrees of these functions show that the QNPSPA has not fully mined the traditional aesthetic and cultural values. Deeper research and practice are wanted, combined with different regional functions and the distribution of cultural resources. This corresponds with Xiao's (2018) evaluation of the suitability of the QNPSPA for recreational use [50].

The interviews suggest that the community residents generally have high aesthetic requirements and cultural needs for the QNPSPA landscapes, as evidenced by the general preference for plant landscapes with rich colors and cultural connotations. In China, many ethnic groups and regions have forest culture traditions like holy mountains, divine trees, and feng-shui forests. These culture traditions crystallize the history of various ethnic groups and regions, nurture their survival, development, and growth, and play a vital role in the protection of forests and ecosystems. In India, sacred forests and temple forests symbolize the oldest forms of forest protection. With the overall degradation of forests across the country, these forests and vegetation have been well protected due to religious reasons [51]. Nonetheless, the excessive use of forest cultural services will negatively affect forest resources. This issue has piqued the interest of many scholars [52–54]. The relevant studies concentrate on the effects of tourism, outdoor sports, and other forest activities. The International Union for Conservation of Nature (IUCN) (2016) pointed out that, as outdoor leisure activities gain popularity, the impact of outdoor activities on wild animals invites more attention from those engaged in animal protection. Overall, the QNPSPA boasts a long history and profound cultural deposits. Many villages have preserved a considerable number of feng-shui forests and famous ancient trees, most of which are more than 300 years old. This reflects the value of traditional Chinese culture, and indirectly promotes the ecological protection of the QNPSPA.

More and more scholars have confirmed that various outdoor activities will affect ecosystem stability [55–58]. In addition, intensive recreational activities will cause many ecological problems [59]. Thus, people should not only pay attention to cultural services in national parks (or protected areas), but also look for ways to sustainably utilize cultural service functions, and to maintain their impact on other services. The community structure formed by ancient trees, famous woods, rare or unique tree species, and local tree species presents an ecological landscape in the region. This landscape defines the general features of the region, and may grow into the center of a specific history [60]. In the study area, ancient camphor trees are often considered to have the ability to drive away evil spirits and bless the healthy growth of infants. Thanks to this concept, many ancient camphor trees survive urban construction. Nowadays, these camphor trees retain a certain spiritual connection with residents and become a part of the local cultural landscape that attracts tourists.

According to the FCE of the QNPSPA cultural ecosystem service functions, the community residents, as major beneficiaries of the SPA policy, feel that the most effective functions are the social service functions, while the worst performing functions are the humanistic concern functions. These results directly reflect the interests of the respondents. Studies have shown that professional knowledge is necessary to evaluate cultural service functions [61]. That is why the traditional landscape aesthetics evaluations, e.g., the visual management system (VMS) of the US Forest Service, mostly adopt the expert paradigm (one of the four major factions of American landscape aesthetic evaluation). For the above reason, this study specially investigates a group of experts who are not core stakeholders and unifies the research methodology, such that the evaluation results can be easily compared with the survey results on the community residents. To ensure the representativeness, the study mainly surveys the experts who used to research in the QNPSPA. For example, some experts are from the National Ecological Positioning Station of Qianjiangyuan Forest Ecosystem, Zhejiang A&F University, and East China Normal University. A total of 71 effective responses were collected. Through FCE, it was learned that the composite score given by the experts to the cultural service functions of the QNPSPA was 3.92, slightly lower than the score rated by residents.

Note that the score of the humanistic concern functions (3.16) was far lower than that of the natural experience functions (4.18) and social service functions (4.30). The situation echoes with the findings of Yu Fei (2019), who studied the forest culture value of Tianmu Mountain (in the same province as the QNPSPA), evaluated by a group of experts [42]. Hence, the cultural functions with rich humanistic connotations are not easily perceived by people. To a certain extent, humanistic concern functions reflect higher spiritual needs than cultural services, a mirror of social and physical attributes. It takes a long time to construct the cultural cognition of humanistic concern functions, as stated by Han et al. [62].

Culture, a product of the interaction between human activities and the natural environment, exerts an influence over the environment and human society. Cultural activities and cultural identity can improve the toughness of rural communities against external shocks [63,64]. Zhang et al. discovered that, among agricultural cultural heritages, culture maintains the stability of traditional landscapes through its attraction and resistance [65]. According to Maslow's Hierarchy of Needs, the higher the composite index of the subject cognitive level, the more difficult it is for such subjects to realize their needs. From this perspective, the community residents have a high demand for the QNPSPA cultural ecosystem services. This means the QNPSPA should step up its efforts in cultural construction.

4.4. Limitations and Future Outlook

Cultural ecosystem services are a cross-disciplinary topic. The indices of such services should be more accurate and complete. Some studies have demonstrated the action of cultural ecosystem services on human well-being, but the action is not clearly quantified. Additionally, the existing studies mostly focus on a global or national scale, failing to tackle specific national parks. What is worse, the evaluation indices are very limited [35,66]. In future, it is important to establish index systems suitable for the cultural ecosystem services of national parks by integrating multiple disciplines and to step up the research on the relationship between cultural services and other service functions and human well-being, highlighting the importance of the application of cultural ecosystem services in planning and management decision making.

The research results provide a reference for improving the management of national parks and ease the growing contradiction between people and land. Drawing on the above conclusions, the authors suggest that regional features should be highlighted in the landscape plans of villages, in the light of the culture of specific villages, and the differences between towns/townships in cultural service functions, in addition to the protection of local ecological resources. For example, Qixi Town could expand the wild alpine azalea into a plant landscape spanning thousands of mus. Referring to the architectural features of residential houses, folk culture tourism villages like Liyangtian could plant fruit trees before and behind houses, creating profound local flavors. The QNPSPA could optimize the tree species configuration in key spaces and strengthen landscape creation in cultural venues like religious sites, red education sites, and cultural public activity spaces.

5. Conclusions

Taking the QNPSPA as the study area, this paper clarifies the main types of cultural ecosystem service functions in the national park and scientifically evaluates the importance of each function, as well as the overall level of these functions, from the angle of the community residents' functional benefits from cultural services. The main conclusions are as follows:

(1) The community residents value the benefits brought by the QNPSPA the most in terms of the ecological improvement function (2.44), and the situation is consistent across the four towns/townships. By contrast, the community residents attach the least importance in terms of its benefits to the system governance function (1.89), but the situation varies between towns/townships. Specifically, Hetian Township had the lowest cognition of the wilderness protection function (1.80), while Qixi Town had the lowest cognition of the folk culture popularization function (1.82);

(2) Except for the years of local residence, the community residents' cognition of the QNPSPA cultural ecosystem service functions may vary significantly. Among them, age and education level significantly affect all the functions of natural experience, while occupation significantly affects all the functions of social service;

(3) Concerning the importance of functional indices, the importance scores of the natural experience functions, humanistic concern functions, and social service functions are 0.3286, 0.3503, and 0.3211, respectively. Concerning the cognition of the cultural ecosystem service functions, the community residents rated the cultural ecosystem service functions in the QNPSPA as 3.99. By the principle of maximum membership, the community residents had a moderate to high level of cognition for the cultural ecosystem service functions. The different types of functions can be sorted by effectiveness as: the social service functions (4.11) > natural experience functions (4.03) > humanistic concern functions (3.86).

Author Contributions: Conceptualization, P.W. and Y.H. (Youjun He); methodology, P.W.; software, P.W.; validation, Y.H. (Yating He), N.L., and Y.H. (Youjun He); formal analysis, P.W.; investigation, P.W.; writing—original draft preparation, P.W.; writing—review and editing, P.W.; visualization, P.W.; supervision, Y.H. (Youjun He). All authors have read and agreed to the published version of the manuscript.

Funding: This research was funded by the National Natural Science Foundation of China (grant number 52008389, 31901297), the Young Talents Project of Central Public Welfare Research Institute Fund (CAFYBB2017QC006), and the Key Project of National Forestry and Grassland Administration (500102-5105).

Data Availability Statement: Not Applicable.

Acknowledgments: Thanks to Zhiyong Li for his careful guidance. Thanks to the staff of Zhejiang Provincial Forestry Bureau and Qianjiangyuan National Park Administration for their help during the questionnaire survey.

Conflicts of Interest: The authors declare no conflict of interest.

Appendix A. Questionnaire of Community Residents on Cultural Ecosystem Service Functions in the Qianjiangyuan National Park System Pilot Area

Dear Sir/Madam, Hello!

We are researchers from the Chinese Academy of Forestry, and this survey will only be used for cultural ecosystem services research. Please feel free to fill in. Thank you for your cooperation.

Questionnaire number: _________________; Village: _________________;

Functional area: _________________; Geographic coordinates:_________________

I. Basic survey of community residents

(1) Gender:

□ Male □ Female

(2) Age:_________________

(3) Your Education level:

□ Primary school and below □ Junior high school

□ Senior high school and secondary technical school

☐ Higher vocational school and junior college ☐ Ordinary college and above
(4) Your occupation type:
☐ Farmers ☐ Individual service workers ☐ Enterprise employees
☐ Migrant workers ☐ Students ☐ Others
(5) Your annual income is:
☐ ≤CNY 20,000 ☐ CNY 30,000-CNY 50,000 ☐ CNY 60,000-CNY 150,000
☐ CNY 160,000-CNY 300,000 ☐ ≥CNY 310,000
(6) How many years have you lived here:
☐ 5 years and below ☐ 6-10 years ☐ 11-20 years ☐ 21 years and above
II. Cognition of cultural ecosystem service functions
According to your daily living experience in the System Pilot Area, please score and evaluate your cognition of the following cultural ecosystem service functions.

Table A1. Cognition of cultural ecosystem service functions.

Cultural Ecosystem Service Functions	Cognitive Situation	Function Evaluation
Wilderness protection function	☐ strongly agreee ☐ agree ☐ strongly disagree	☐ very high ☐ relatively high ☐ general ☐ relatively low ☐ very low
System governance function	☐ strongly agreee ☐ agree ☐ strongly disagree	☐ very high ☐ relatively high ☐ general ☐ relatively low ☐ very low
Spiritual worship function	☐ strongly agreee ☐ agree ☐ strongly disagree	☐ very high ☐ relatively high ☐ general ☐ relatively low ☐ very low
Folk culture popularization function	☐ strongly agreee ☐ agree ☐ strongly disagree	☐ very high ☐ relatively high ☐ general ☐ relatively low ☐ very low
Art inspiration function	☐ strongly agreee ☐ agree ☐ strongly disagree	☐ very high ☐ relatively high ☐ general ☐ relatively low ☐ very low
Living environment improvement function	☐ strongly agreee ☐ agree ☐ strongly disagree	☐ very high ☐ relatively high ☐ general ☐ relatively low ☐ very low
Science education function	☐ strongly agreee ☐ agree ☐ strongly disagree	☐ very high ☐ relatively high ☐ general ☐ relatively low ☐ very low
Health care function	☐ strongly agreee ☐ agree ☐ strongly disagree	☐ very high ☐ relatively high ☐ general ☐ relatively low ☐ very low
Wilderness protection function	☐ strongly agreee ☐ agree ☐ strongly disagree	☐ very high ☐ relatively high ☐ general ☐ relatively low ☐ very low

Note: the interpretation of cultural ecosystem services was explained by the investigators to the community residents.

Appendix B. Representative Landscape of Cultural Ecosystem Services

Figure A1. Schematic representation of representative landscapes for cultural ecosystem services.

References

1. Rotherham, I.D. Bio-cultural heritage and biodiversity: Emerging paradigms in conservation and planning, indicators for large-scale assessment of cultural ecosystem services. *Ecosyst. Serv.* **2015**, *21*, 258–269.
2. Han, Z. What are the differences between Chinese and Western core values. *Seek. Truth Facts* **2014**, *2*, 50–51. (In Chinese)
3. Selin, S.; Chevez, D. Developing a collaborative model for environmental planning and management. *Environ. Manag.* **1995**, *19*, 189–195. [CrossRef]
4. Agrawal, A.; Gibson, C.C. Enchantment and disenchantment: The role of community in natural resource conservation. *World Dev.* **1999**, *27*, 629–649. [CrossRef]
5. Ferretti-Gallon, K.; Griggs, E.; Shrestha, A.; Wang, G.Y. National parks best practices: Lessons from a century's worth of national parks management. *Int. J. Geoheritage Parks* **2021**, *95*, 335–346. [CrossRef]
6. Cao, S. From "Indian Wilderness" to "Uninhabited Wilderness": The Transformation of White American Wilderness Concepts and the Expulsion of Native Americans from the Yellowstone National Park Area. *Ludong Univ. J. (Philos. Soc. Sci. Ed.)* **2019**, *36*, 26–31. (In Chinese)
7. Mannigel, E. Integrating parks and people: How does participation work in protected area management? *Soc. Nat. Resour.* **2008**, *21*, 498–511. [CrossRef]
8. Liu, J.; Dietz, T.; Carpenter, S.R.; Alberti, M.; Folke, C.; Moran, E.; Pell, A.N.; Deadman, P.; Kratz, T.K.; Lubchenco, J.; et al. Complexity of coupled human and natural systems. *Science* **2007**, *317*, 1513–1516. [CrossRef]
9. Ostrom, E.A. General framework for systems. *Science* **2009**, *325*, 419–422. [CrossRef]
10. Hiwasaki, L. Toward sustainable management of national parks in Japan: Securing local community and stakeholder participation. *Environ. Manag.* **2005**, *35*, 753–764. [CrossRef]

11. Shoji, Y.; Kim, H.; Kubo, T.; Kubo, T.; Tsuge, T.; Aikoh, T.; Kuriyama, K. Understanding preferences for pricing policies in Japan's national parks using the best–worst scaling method. *J. Nat. Conserv.* **2021**, *4*, 125954. [CrossRef]
12. Zhang, Z.W.; Yang, R. Public participation in national Park management planning. *Chin. Landsc. Archit.* **2015**, *31*, 23–27. (In Chinese)
13. Tuler, S.; Webler, T. Public participation: Relevance and application in the national park service. *Park Sci.* **2000**, *20*, 24–26.
14. Lynch, H.J.; Hodge, S.; Albert, C.; Dunham, M. The Greater Yellowstone Ecosystem: Challenges for regional ecosystem management. *Environ. Manag.* **2008**, *41*, 820–833. [CrossRef] [PubMed]
15. Zhang, J.Y.; Zhang, Y.J. On public participation in the construction of national parks. *Biodivers. Sci.* **2017**, *25*, 80–87. (In Chinese) [CrossRef]
16. Zhang, S.; Wang, M.; Wang, Z. Community involvement in India's national park tiger conservation experience and enlightenment. *J. Beijing For. Univ. (Soc. Sci. Ed.)* **2021**, *20*, 101–107. (In Chinese) [CrossRef]
17. Singh, H.; Husain, T.; Agnihotri, P.; Pande, P.; Khatoon, S. An Ethnobotanical study of medicinal plants used in sacred groves of Kumaon Himalaya, Uttarakhand, India. *J. Ethnopharmacol.* **2014**, *154*, 285–295. [CrossRef]
18. Rawat, M.; Vasistha, H.B.; Manhas, R.K.; Negi, M. Sacred forest of Kunjapuri Siddhapeeth, Uttarakhand, India. *Trop. Ecol.* **2011**, *52*, 219–221.
19. Pala, N.A.; Negi, A.K.; Gokhale, Y.; Aziem, S.; Vikrant, K.; Todaria, N. Carbon stock estimation for tree species of Sem Mukhem sacred forest in Garhwal Himalaya, India. *J. For. Res.* **2013**, *24*, 457–460. [CrossRef]
20. Lu, Y.; Tang, H. Research progress of ecosystem cultural Services: A visual analysis based on CiteSpace. *J. Beijing Norm. Univ. (Nat. Sci. Ed.)* **2021**, *57*, 524–532. (In Chinese)
21. Plieninger, T.; Dijks, S.; Oteros-Rozas, E.; Bieling, C. Assessing, mapping, and quantifying cultural ecosystem services at community level. *Land Use Policy* **2013**, *33*, 118–129. [CrossRef]
22. Brown, G. Mapping Spatial Attributes in Survey Research for Natural Resource Management: Methods and Applications. *Soc. Nat. Resour.* **2004**, *18*, 17–39. [CrossRef]
23. Raymond, C.M.; Bryan, B.A.; Macdonald, D.H.; Cast, A.; Strathearn, S.; Grandgirard, A.; Kalivas, T. Mapping community values for natural capital and ecosystem services. *Ecol. Econ.* **2009**, *68*, 1301–1315. [CrossRef]
24. Zhang, H.F.; Chen, J.C.; Shi, J.Z.; Wang, W.; Huang, L.; Ye, Q.; Ruan, X.F. Effects of spatial relationship of nature reserve on distribution of giant panda in Sichuan Area of Giant Panda National Park. *Acta Ecol. Sin.* **2020**, *40*, 2347–2359. (In Chinese)
25. Xiao, R.Q.; Zhao, X.D.; He, Y.J.; Yan, Y.Q.; Ye, B.; Xu, D.Y.; Zou, W.T. Study on asset pricing mechanism of national Park ecological resources. *For. Econ.* **2019**, *41*, 3–9. (In Chinese)
26. Wei, Y.; He, S.Y.; Lei, G.C.; Su, Y. Enlightenment of conservation easement to unified management of National parks in China: Based on American experience. *J. Beijing For. Univ. (Soc. Sci.)* **2019**, *18*, 70–79. (In Chinese)
27. Daily, G.C. *Nature's Services Societal Dependence on Natural Ecosystem*; Island Press: Washington D.C., USA, 1997.
28. Sutherland, W.J.; Freckleton, R.P.; Godfray, H.C. Identification of 100 fundamental ecological questions. *J. Ecol.* **2013**, *101*, 58–67. [CrossRef]
29. Carpenter, S.R.; Mooney, H.A.; Agard, J. Science for managing ecosystem services: Beyond the millennium ecosystem assessment. *Proc. Natl. Acad. Sci. USA* **2009**, *106*, 1305–1312. [CrossRef]
30. Wang, Z.F.; Peng, Y.Y.; Xu, C.Y. Progress and trend of practical application of ecosystem service tradeoff research. *Acta Sci. Nat. Univ. Pekin.* **2019**, *55*, 773–781. (In Chinese)
31. Zhang, H.; Ouyang, Z.; Zheng, H. Spatial scale characteristics of ecosystem services. *Chin. J. Ecol.* **2007**, *9*, 1432–1437.
32. Hatan, S.; Fleischer, A.; Tchetchik, A. Economic valuation of cultural ecosystem services: The case of landscape aesthetics in the agritourism market. *Ecol. Econ.* **2021**, *184*, 107005. [CrossRef]
33. Booth, P.N.; Law, S.A.; Ma, J.; Buonogurio, J.; Boyd, J.; Turnley, J.G. Modeling aesthetics to support an ecosystem services approach for natural resource management decision making. *Integr. Environ. Assess. Manag.* **2017**, *13*, 926–938. [CrossRef] [PubMed]
34. Costanza, R.; Arge, A.; Groot, R.D. The value of the world's ecosystem services and natural capital. *Nature* **1997**, *387*, 253–260. [CrossRef]
35. Chan, K.M.A.; Guerry, A.D.; Patricia, B. Where are Cultural and Social in Ecosystem Services? A Framework for Constructive Engagement. *BioScience* **2012**, *62*, 744–756. [CrossRef]
36. Russell, R.; Guerry, A.; Balvanera, P. Humans and Nature: How Knowing and Experiencing Nature Affect Well-Being. *Annu. Rev. Environ. Resour.* **2013**, *38*, 473–502. [CrossRef]
37. Zhao, Z.; Liu, Y.L.; Wen, Y.L. Discussion on evaluation system of socialized service function of urban forest: Based on the perspective of citizen benefit. *For. Resour. Manag.* **2019**, *4*, 1–9. (In Chinese)
38. Anthony, A.; Atwood, J.; August, P.; Byron, C.; Cobb, S.; Foster, C.; Fry, C.; Gold, A.; Hagos, K.; Heffner, L.; et al. Coastal Lagoons and Climate Change: Ecological and Social Ramifications in U.S. Atlantic and Gulf Coast Ecosystems. *Ecol. Soc.* **2009**, *14*, 8. [CrossRef]
39. Dou, Y.; Yu, X.; Bakker, M.M.; de Groot, R.; Carsjens, G.J.; Duan, H.L.; Huang, C. Analysis of the relationship between cross-cultural perceptions of landscapes and cultural ecosystem services in Genheyuan region, Northeast China. *Ecosyst. Serv.* **2020**, *43*, 101112. [CrossRef]
40. Ridding, L.E.; Redhead, J.W.; Oliver, T.H.; Schmucki, R.; Mcginlay, J.; Graves, A.R.; Morris, J.; King, H.; Bullck, J.M. The importance of landscape characteristics for the delivery of cultural ecosystem services. *J. Environ. Manag.* **2018**, *206*, 1145–1154. [CrossRef]

41. Peng, W.T.; Liu, W.Q.; Cai, W.B.; Wang, X.; Huang, Z.; Wu, C.Z. Evaluation of cultural service value of urban protected area ecosystem based on participatory mapping: A case study of Shanghai Gongqing Forest Park. *Chin. J. Appl. Ecol.* **2019**, *30*, 1–14. (In Chinese)
42. Yu, F. Study on the Impact of Forest Landscape Pattern on Forest Cultural Value in Tianmu Mountain Based on Perception Evaluation. Doctoral Dissertation, Chinese Academy of Forestry, Beijing, China, 2019. (In Chinese).
43. Kumar, M.; Kumar, P. Valuation of the ecosystem services: A psycho-cultural perspective. *Ecol. Econ.* **2008**, *64*, 808–819. [CrossRef]
44. Tuffery, L. The recreational services value of the nearby periurban forest versus the regional forest environment. *J. For. Econ.* **2017**, *28*, 33–41. [CrossRef]
45. Eggersa, J.; Lindhagenb, A.; Linda, T. Balancing landscape-level forest management between recreation and wood production. *Urban For. Urban Green.* **2018**, *33*, 1–11. [CrossRef]
46. Ham, C.; Champ, P.A.; Loomis, J.B.; Reich, R.M. Accounting for heterogeneity of public lands in hedonic property models. *Land Econ.* **2012**, *88*, 444–454. [CrossRef]
47. Gobster, P.H.; Nassauer, J.I.; Daniel, T.C.; Fry, G. The shared landscape: What does aesthetics have to do with ecology? *Landsc. Ecol.* **2007**, *22*, 959–972. [CrossRef]
48. LeRoy, C.J.; Fischer, D.G.; Lubarsky, S. How do aesthetics effect our ecology? *J. Ecol. Anthropol.* **2006**, *10*, 61–65.
49. Zhang, Z.Y.; Wang, P.; Gao, Y.; Ye, B. Current Development Status of Forest Therapy in China. *Healthcare* **2020**, *8*, 61. [CrossRef]
50. Xiao, L.L. Study on Suitability Evaluation and Management of Recreation Use in National Parks: A Case Study of Qianjiangyuan National Park Pilot Area. Doctoral Dissertation, University of Chinese Academy of Sciences, Beijing, China, 2018. (In Chinese).
51. Du, Y.H. Study on Forest Health and Traditional Forest Culture in Napan River Reserve. Doctoral Dissertation, Minzu University of China, Beijing, China, 2015. (In Chinese).
52. Head, L.; Muir, P. Nativeness, invasiveness and nation in Australian plants. *Geogr. Rev.* **2004**, *94*, 199–217. [CrossRef]
53. Yang, G.F. Value Analysis of ancient and famous trees in famous historical and cultural cities—A case study of Lijiang Ancient City. *Guangdong Agric. Sci.* **2011**, *19*, 63–65. (In Chinese)
54. Greider, T.; Garkovich, L. Landscapes-The social construction of nature and the environment. *Rural Sociol.* **1994**, *59*, 1–24. [CrossRef]
55. Larson, C.L.; Reed, S.E.; Merenlender, A.M.; Crooks, K.R. Effects of recreation on animals revealed as widespread through a global systematic review. *PLoS ONE* **2016**, *11*, e0167259. [CrossRef] [PubMed]
56. Arlettaz, R.; Nusslé, S.; Baltic, M.; Vogel, P.; Palme, R.; Jenni-Eiermann, S.; Patthey, P.; Genoud, S. Disturbance of wildlife by outdoor winter recreation: Allostatic stress response and altered activity-energy budgets. *J. Appl. Ecol.* **2015**, *25*, 1197–1212. [CrossRef] [PubMed]
57. Wolf, I.D.; Hagenloh, G.; Croft, D.B. Vegetation moderates impacts of tourism usage on bird communities along roads and hiking trails. *J. Environ. Manag.* **2013**, *129*, 224–234. [CrossRef] [PubMed]
58. Coppesa, J.; Nopp-Mayrb, U.; Grünschachner-Berge, V. Habitat suitability modulates the response of wildlife to human recreation. *Biol. Conserv.* **2018**, *227*, 56–64. [CrossRef]
59. Hammitt, W.E.; Cole, D.N.; Monz, C.A. *Wildland Recreation: Ecology and Management*; John Wiley & Sons: Hoboken, NJ, USA, 2015.
60. Stephenson, J. The Cultural Values Model: An integrated approach to values in landscapes. *Landsc. Urban Plan.* **2008**, *84*, 127–139. [CrossRef]
61. Wang, B.Z.; Wang, B.M.; He, P. Theory and method of aesthetic evaluation of landscape resources. *Chin. J. Appl. Ecol.* **2006**, *17*, 1733–1739. (In Chinese)
62. Han, L.; Shi, L.; Yang, F.; Xiang, X.; Gao, L. Method for the evaluation of residents' perceptions of their community based on landsenses ecology. *J. Clean. Prod.* **2020**, *281*, 124048.
63. Sarah, S. Enhancing the analysis of rural community resilience: Evidence from community land ownership. *J. Rural Stud.* **2013**, *31*, 36–46.
64. Beel, D.C.; Wallace, G.; Webster, H.; Nguyen, H.; Tait, E.; Macleod, M.; Mellish, C. Cultural resilience: The production of rural community heritage, digital archives and the role of volunteers. *J. Rural Stud.* **2015**, *10*, 459–468. [CrossRef]
65. Zhang, Y.X.; Min, Q.W.; Zhang, C.Q.; He, L.L.; Zhang, S.; Yang, L.; Tian, M.; Xiong, Y. Traditional culture as an important power for aintaining agricultural landscapes in cultural heritage sites: A case study of the Hani terraces. *J. Cult. Herit.* **2017**, *25*, 171–179. [CrossRef]
66. Hernandez-Morcillo, M.; Plieninger, T.; Bieling, C. An empirical review of cultural ecosystem service indicators. *Ecol. Indic.* **2013**, *29*, 434–444. [CrossRef]

 *land*

Article

Examining Social Equity in the Co-Management of Terrestrial Protected Areas: Perceived Fairness of Local Communities in Giant Panda National Park, China

Qiujin Chen [1], Yuqi Zhang [2], Yin Zhang [1,*] and Mingliang Kong [1]

1   School of Architecture and Urban Planning, Chongqing University, Chongqing 400030, China
2   Department of Cultural Geography, Faculty of Spatial Sciences, University of Groningen, 9700 AV Groningen, The Netherlands
*   Correspondence: yinzhang@cqu.edu.cn

Abstract: Social equity is imperative both morally and instrumentally in the governance of protected areas, as neglecting this consideration can result in feelings of injustice and thus jeopardize conservation objectives. Despite the progressive attention paid to conservation equity, few have linked it with co-management arrangements, especially in the context of terrestrial protected areas. This study assesses the fairness perceptions in China's Giant Panda National Park from recognitional, procedural, and distributional dimensions, to further disclose their correlations with individuals' characteristics and participation in co-management activities. The regression analysis shows that all co-management types (instruction, consultation, agreement, and cooperation) are significantly linked with certain directions of perceived social equity. One novel finding here is that alternative types of co-management activities are influencing social equity in different ways. In addition, our research discloses the effects of education across all equity categories, and location is merely significantly related to recognitional equity. These findings suggest more inclusive and empowered co-management endeavors to strive for more equitably managed protected areas. Crucial steps to advance this include extending participative channels, co-producing better compensation plans, strengthening locals' conservation capabilities, etc. Herein, this study appeals to a greater focus on social equity issues in co-management regimes, and tailored actions should be taken to tackle specific local problems.

Keywords: protected areas; co-management; social equity; fairness perception; empowerment levels

Citation: Chen, Q.; Zhang, Y.; Zhang, Y.; Kong, M. Examining Social Equity in the Co-Management of Terrestrial Protected Areas: Perceived Fairness of Local Communities in Giant Panda National Park, China. *Land* **2022**, *11*, 1624. https://doi.org/10.3390/land11101624

Academic Editors: Rui Yang, Yue Cao, Steve Carver and Le Yu

Received: 19 August 2022
Accepted: 19 September 2022
Published: 22 September 2022

Publisher's Note: MDPI stays neutral with regard to jurisdictional claims in published maps and institutional affiliations.

1. Introduction

Protected areas are essential not only to sustain biodiversity and ecosystem services, but also to support local livelihood and well-being [1]. By no means should indigenous people and local residents be forced into victims and refugees of the global expansion of protected areas [2]. Over the last two decades, there have been concerted efforts globally to make protected areas more effectively and equitably managed, mostly for the benefits of local communities [3,4]. The slogan of "equity and benefit sharing" was put forward by the Convention on Biological Diversity's Programme on Protected Areas in 2004. Furthermore, the principle that protected areas should be "effectively and equitably managed" was highlighted by the Aichi Biodiversity Target 11 in 2010 [5], which was later strengthened by International Union for Conservation of Nature (IUCN) World Parks Congress held in 2014 [6]. The better understanding and consideration of social equity issues in protected areas are believed to deliver better conservation outcomes, as protected areas can seldomly survive without strong and firm social support from their surroundings [7–9].

The recent 5 years have witnessed a considerable increase in the number of studies focusing on the social equity aspects of protected areas [10–12]. Zafra-Calvo et al. (2017) established an indicator system to assess the equitable management of protected areas from recognitional, procedural, and distributional dimensions, and later applied this framework to evaluate their interrelations among 225 protected areas globally [13,14]. Bennett et al. (2020) expanded and enriched those indicators to capture the fairness perceptions of small-scale fishermen in marine protected areas [15]. While some authors were inclined to look at social equity issues from the perspective of distribution [16,17], others paid more attention to the procedural or recognitional dimension [18,19].

Among all those researches, very few have linked social equity with co-management of protected areas. Despite the fact that there is no commonly accepted concept for co-management, this term is most frequently comprehended as the sharing of rights and responsibilities among the governments, local resource users, and other partners (Carlsson and Berkes 2005; Borrini-Feyerabend 2007) [20,21]. In addition, the majority of current studies in this aspect are set in the context of marine protected areas or fisheries [16,22,23], not in terrestrial protected areas, the co-management of which also displays significant roles in forest, grassland, and biodiversity conservation [24,25]. Although several studies have demonstrated how demographic attributes and social-economic characteristics, such as gender, education level, and household wealth, can have impact on fairness perceptions of local communities toward the co-managed marine protected areas, none of these have considered the influence of their involved co-management types [15,16]. Due to the complexity and plurality of co-management mechanisms, local stakeholders are usually involved in different co-management types and forms, showing the variability in perceptions, attitudes, and behaviors toward protected areas, which are frequently related to social equity issues [26,27]. To understand the correlation between participative co-management activities and the fairness perceptions of grassroots is vital to achieve better social outcomes of co-management in protected areas. On the contrary, the lack of this consideration in enforcing co-management programs in protected areas can result in serious social conflicts, and consequently lead to poor conservation performance [28].

In this paper, we aim to explore how participative co-management activities can have influences on locals' fairness perceptions in a newly designated terrestrial protected area in southwestern China. Our research hypotheses are listed as follows: (1) Individuals' demographic characteristics (e.g., gender, age, residency year, education, and profession) can have influences on their fairness perceptions; (2) some household features (e.g., villages, household size, migrant workers, annul income, and income sources) are associated with individuals' perceived fairness; (3) the number and type of participative co-management activities are positively linked with villagers' fairness perceptions. In the IUCN guideline of good governance of protected areas, those co-management arrangements diversified into five types, namely, instructive, consultative, agreement, cooperation, and empowerment, based on their empowerment levels [29]. Moreover, in this study, we classify diverse co-management activities in Giant Panda National Park according to the IUCN classification. With respect to the measurement of fairness perception, we largely borrow from Zafra et al. (2017) [13] and Bennett et al. (2020) [15], while making minor adjustments according to the study site. Both quantitative and qualitative methods were adopted herein. Through this explanatory study, we seek to disclose the relations between participated co-management types and fairness perceptions of locals from the recognitional, procedural, and distributional perspectives, contributing to the empirical evidence on the social equity of co-managed terrestrial protected areas, and producing practical and theoretical insights for the co-management policy and practice in protected areas.

2. Materials and Methods

2.1. Study Site

The Giant Panda National Park (GPNP) is located in the southwest China as part of the Minshan and Qionglai Mountains, covering a total area of roughly 21,978 km^2. It was first promoted as a national park pilot in 2016 and then officially recognized as one of China's first batch of national parks in 2021 [30]. The GPNP is integrated and expanded from 73 existing protected areas, and is further divided into 4 regions after being designated as a national park. With a total area of approximately 400 km², the Tangjiahe area is situated in the northeast region of GPNP in Sichuan Province, with the protection of giant pandas and their habitats as the primary conservation objectives (Figure 1). In addition, this area has been assigned as a nature reserve since 1978, with over a 40-year history of conservation.

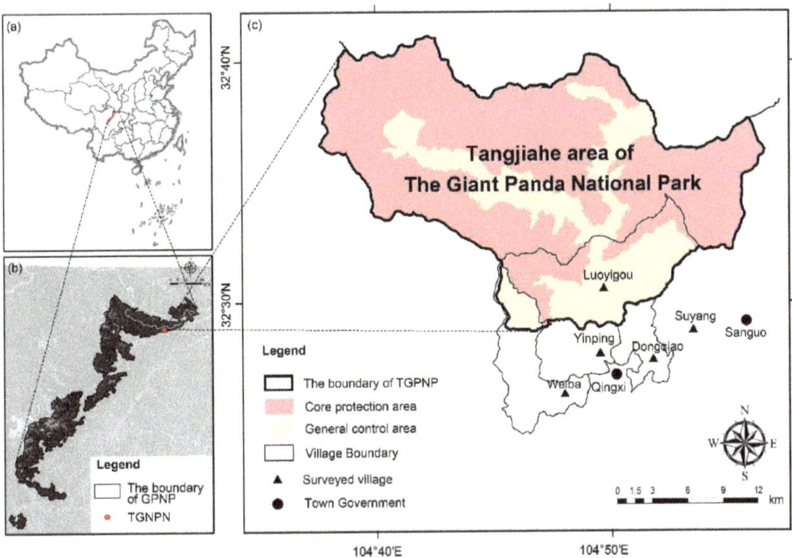

Figure 1. Map of the Tangjiahe area of the Giant Panda National Park: (**a**) Location of GPNP in China; (**b**) location of TGPNP; (**c**) location map of the TGPNP and surveyed villages. (Note: i. The figure does not include the spatial boundary of Suyang village as this information is not available for our research group. ii. The location of surveyed villages is positioned at the office of the village committee).

Tangjiahe area of GPNP (TGPNP) is selected as our study site to explore the fairness perception of community-based co-management in protected areas for the following two reasons. First, the Administration of Tangjiahe Area (ATA) has started to enforce community-based efforts (e.g., joint fire prevention and infrastructure building supports) with its surrounding communities since 1978, and has tried various co-management strategies, such as organizing co-management committees, signing co-management agreements, and arranging industrial guidance, as well as introducing foreign and domestic NGOs to develop a differentiated co-management model in surrounding villages. Those co-management arrangements appear to be super abundant and diverse until now, yet the social effects remain to be uncovered [31]. Second, as establishing community-based co-management mechanisms was put forward as one of the critical strategies in the construction of China's national park system in 2017, the GPNP positively responded to the summoning of the central government and largely facilitated community-based tourism [32]. The locals' fairness perceptions toward those co-management countermeasures are essential to be disclosed, as it might affect the conservation outcomes and performances.

There are 7 villages bordering TGPNP in Qingxi and Sanguo Town, with a population of about 9500. To identify the specific villages suitable for our in-depth survey, we consulted with officials working at the ATA, and two criteria were adopted after repeated discussion. First, there are stable and long-lasting co-management arrangements settled between villages and the ATA. Second, those co-management models need to be both representative and differentiated. By this method, villages of Yinping, Luoyigou, Weiba, Dongqiao, and Suyang were selected as a result, and the basic information was listed in Table 1. Among those villages, Luoyigou village is the only one located in the General Control Area within the boundary of the TGPNP, with ecological restoration and habitat enhancement as its main conservation objective. Due to this reason, the village has suffered from severe human-wildlife conflicts for years. Therefore, ATA has established a co-management committee since 2018 and a human-wildlife compensation program was specifically launched in 2019 in Luoyigou. In addition, as the gateway community of Tangjiahe area, Yinping village has been greatly supported financially and technically by the ATA and the Qingchuan County Government since 1997 to promote tourism development. More specifically, an agricultural cooperative was established in Suyang village to facilitate local development. Apart from the aforementioned arrangements, forest ranger programs and infrastructure construction were enforced by the ATA in all five villages, while beekeeping training was organized in four villages except for Suyang.

Table 1. Basic information of the five selected villages.

Villages	Population Size	Area	Main Industries	Key Co-Management Strategies
Luoyigou	1085	62 km^2	Tourism, agriculture, and cultivation	Co-management committee, human-wildlife conflicts compensation, tourism support, beekeeping training, infrastructure construction, and forest rangers
Yinping	1823	39.7 km^2	Tourism, agriculture, and cultivation	Co-management committee, tourism support, beekeeping training, infrastructure construction, and forest rangers
Weiba	870	66.12 km^2	Tourism, agriculture, stone production	Beekeeping training and forest rangers
Dongqiao	1160	27.92 km^2	Agriculture, cultivation, and tourism	Beekeeping training and forest rangers
Suyang	1389	22 km^2	Agriculture and cultivation	Establishing an agricultural cooperative

2.2. Survey Sampling Methods and Design

A pre-survey field was conducted in July 2017 to interview ATA staff to collect basic information about co-management arrangements of TGPNP, and both informal discussions with local people and formal interviews with ATA staff were conducted to select the most suitable criteria to assess co-management activities and the perceived fairness of locals. Questionnaires were distributed on-site from 29 June 2022 to 7 July 2022 in TGPNP. This survey took the household as the basic unit and selected one person with the most frequent contacts with the ATA, recommended by household members. Random sampling was used to select the respondents, and the specific number of respondents was determined according to the population size of the village. In this way, a total of 428 questionnaires were collected by the research team. After excluding 4 invalid questionnaires, the respondents of which came from non-survey villages, the actual valid samples reached 424, with a 99.3% effective return rate.

The questionnaire consists of four sections. The first two sections include a broad set of questions related to the demographics (e.g., gender, age, education, location, occupation) of local residents and their household characteristics (e.g., household income, household size, income sources, and residency year), as well as the co-management activities they are involved in. Those 12 types of co-management activities are inducted from a total of 15 co-management arrangements after the discussion with ATA staff (see Supplementary Materials—Table S4). Those activities are classified into four categories based on an increasing level of empowerment of local communities. While instructive co-management refers to those community-based measures where the ATA takes the lead and communities simply follow the instructions, consultative co-management means better information exchange between both sides. The responsibilities and benefits of conservation are clearly and formally divided among different stakeholders in the agreement type of co-management. Furthermore, cooperation is the co-management typology where the participants can partially be delegated in the decision-making or enforcement of conservation affairs, which is the highest empowerment level recognized in TGPNP. All those activities are assessed by Yes ("participated") or No ("not participated"), listed in Table 2.

Table 2. Co-management types and activities.

Category	Activity Number	Co-Management Activities
Instruction	A1	Energy transformation and other infrastructure building projects
	A2	Skill training and industrial support activities
	A3	Environmental educational activities
Consultation	A4	Community-based co-management meetings
	A5	Consultative meetings for planning and policy making
	A6	Easy access to co-management Information
Agreement	A7	Agreements of fire prevention and human-wildlife conflict compensation
	A8	Agreements of community-based co-management
	A9	Benefits sharing of bee farming and other cooperatives
Cooperation	A10	Fire prevention and forest patrolling work
	A11	Participation in enacting conservation rules
	A12	Accountability for some conservation affairs

The last section of the questionnaire is concerned with locals' perceptions of fairness toward TGPNP, measured through statements developed for each dimension of social equity. In this section, we borrowed from Zafra-Calvo et al. (2017) [13], Lou Lecuyer (2019) [33], Nathan J. Bennett et al. (2020) [15], and Georgina G. Gurney (2021) [16], while making minor adjustments and adding additional attributes according to our study site. For recognitional equity, an item concerning land ownership was added as land conflicts were frequently recognized by interviews. For procedural equity, we deleted the indicator of access to justice, since no conflict resolution mechanisms were found in TGPNP. From the distributional perspective, two attributions of wildlife compensation and empowerment distribution were added for the wildlife-human conflict compensation and the forest ranger programs launched in TGPNP. All those questions are measured in a 5-point Likert scale in this section, listed in Table 3.

Table 3. Selected indicators to measure social equity in GPNP.

Category	Attribute	Survey Questions
Recognition	culture	GPNP respects our local culture and traditional customs
	livelihood	GPNP imposes no negative impact on my original livelihood
	Legal and traditional rights	GPNP can sincerely respect my legal and traditional rights
	Land ownership	I declare no land ownership conflicts with GPNP
	Traditional knowledge	Traditional knowledge can be effectively involved in the management of GPNP

Table 3. *Cont.*

Category	Attribute	Survey Questions
Procedure	Decision making	I can fully express my opinion and effectively be involved in the decision-making process of GPNP
	Participation	GPNP has convenient channels and fair procedures to encourage local participation
	transparency	The information of conservation decisions and reasons for decisions are readily available
	Accountability	I understand the responsibility of ATA and know to whom to raise concerns to solve issues related to management actions
	Free, prior, and informed consent (FPIC)	When ATA issues plans and policies addressed to me, I will be informed in advance
Distribution	Conservation burdens	I fairly bear the responsibility of conservation in GPNP, compared to other local residents
	Ecological compensation	I am satisfied with the ecological compensation made by GPNP
	Wildlife conflicts compensation	I can easily get appropriate compensation from human-wildlife conflicts
	Benefits distribution	I can fairly get economic benefits from co-management, compared to other local residents
	Employment distribution	I can fairly get employment opportunities from ATA, compared to others

Qualitative methods are also used as a supplementary approach in this research. Seventeen semi-structured interviews were conducted with different stakeholders: Staff in the Community Office of the ATA, officials of Qingxi and Sanguo town governments, as well as village leaders and elites. The selection of stakeholders is based on the correlation to co-management, such as people with rights, with official information, and prestigious local people, as well as considerations of the equilibrium of gender and age. The purpose of these interviews is to identify the equitable issues of TGPNP and select the most suitable criteria to assess co-management activities and fairness perceptions. In addition, secondary data (e.g., research reports, government reports and plans, and statistics) were collected and analyzed to understand the contexts.

2.3. Data Analysis

All data analysis was completed in SPSS 26.0 (IBM Corp., Armonk, NY, USA). First, the reliability analysis was performed in this study using Cronbach's alpha index. In this study, the overall Cronbach's alpha coefficient was 0.831, which is above the eligible index of 0.7, indicating that the obtained survey results had good internal reliability. The content validity was also assessed here. The figures for all items were all significantly correlated at the 0.01 level, signifying positive outcomes in content validity.

First, we calculated the score for each equity dimension (recognitional, procedural, and distributional equity) by the average score of five indicators in this category, and then built the score for combined equity by the mean score of all 15 indicators. Second, we tested for univariate associations (one-way ANOVA and Spearman correlation analysis) between recognitional, procedural, distributional, and combined equity scores and the demographic characteristic and participative factors. While one-way ANOVA was used for categorical variables (e.g., gender, occupation, and villages), Spearman correlation analysis was utilized for ordinal variables, such as education level, household size, and income, as well as the number of participative co-management arrangements. Finally, linear regression analysis was adopted here to develop regression models for each composite social equity score using variables (e.g., age, education, annual household income) significantly correlated to equity perception, to further disclose their intertwined relations.

2.4. Sample Description

Our sample consisted of 424 residents who lived within or surrounding TGPNP (Table 4), with 46.0% male and 54.0% female. The majority of respondents were in older age brackets, with 72.6% (n = 308) older than 50. Their education levels were generally low, since most respondents (65.1%) had only completed primary or junior school and even 22.4% had never attended any school. In addition, the vast majority of respondents lived here for more than 20 years (88.3%) and made a living by farming (76.4%).

Table 4. Description of respondents involved in this survey.

Survey Item	Category	Frequency (n = 424)	Percentage (%)
Gender	Male	195	46.0
	Female	229	54.0
Age	Under 40	56	13.2
	41–50	60	14.2
	51–60	135	31.8
	61–70	94	22.2
	Over 70	79	18.6
Education	No school	324	22.4
	Primary school	23	41.5
	Junior school	36	23.6
	High school	41	8.5
	Undergraduate and above	95	4.0
Residency years	Under 10	176	4.2
	10–20	100	7.5
	Over 20	36	88.3
Professional	Farmers	17	76.4
	Employees	104	5.4
	Merchants	139	8.5
	Other	73	9.7
Villages	Luoyigou	60	24.5
	Yinping	48	32.8
	Weiba	18	17.2
	Dongqiao	32	14.2
	Suyang	374	11.3
Household size	1–3	115	27.1
	4–6	261	61.6
	7–9	41	9.7
	>10	7	1.7
Household migrant workers	0	144	34.0
	1	125	29.5
	2	99	23.3
	>3	56	13.2
Annual household income (RMB)	Less than 10,000	153	36.1
	10,001–30,000	131	30.9
	30,001–60,000	78	18.4
	60,001–100,000	37	8.7
	More than 10,001	25	5.9
Household source of income	Farming	195	46.0
	Tourism	71	16.7
	Forestry	18	4.2
	Local employment	106	25.0
	Nonlocal employment	136	32.1
	Other	68	16.0

The average household size of those respondents was five people, and most of the households had none or only one migrant worker, accounting for 34% and 29.6%, respectively. Most of the surveyed households had a relatively low annual household income, with 36.4% earning less than RMB 10,000 and 30.9% earning between RMB 10,001 and 30,000 per year. Despite the fact that the main household income sources were farming and non-local employment, there were also households that made a living by local employment, tourism, and forestry, the percentages of which were 25.0%, 16.7%, and 4.2%, respectively.

3. Results

3.1. Fairness Perceptions toward TGPNP

The descriptive analysis showed that perceptions of recognitional equity were more positive (Mean = 3.59), compared to those of procedural equity and distributional equity (Mean = 2.60 and 2.81, respectively). As shown in Figure 2, indicators related to recognitional equity were heavily skewed toward positive judgements, indicating that recognitional equity was most likely to be perceived as fair. Significantly, most respondents (67%, 70.1%, 56.8%, and 78.8% respectively) "strongly agreed" or "agreed" to the four indicators of culture, livelihood, legal and traditional rights, and land ownership. By contrast, all indicators related to procedural fairness were strongly skewed toward negative perceptions, especially regarding community participation, where most respondents (66.3%) felt they were not truly involved in the planning and management of TGPNP. Similarly, 47.1% of the respondents "disagreed" or "strongly disagreed" about effective decision-making. In addition, indicators of distributional equity showed dissimilar results. The data showed that 67.9% of respondents believed they were equally responsible for forest fire prevention, while merely one-fifth of respondents "agreed" or "strongly agreed" with the appropriate amount of ecological compensation and wildlife conflict compensation. Moreover, perceptions of distribution of benefits and employment were balanced between positive and negative, since a considerable proportion of respondents (41.7% and 33.5%, respectively) did not have access to the relevant information.

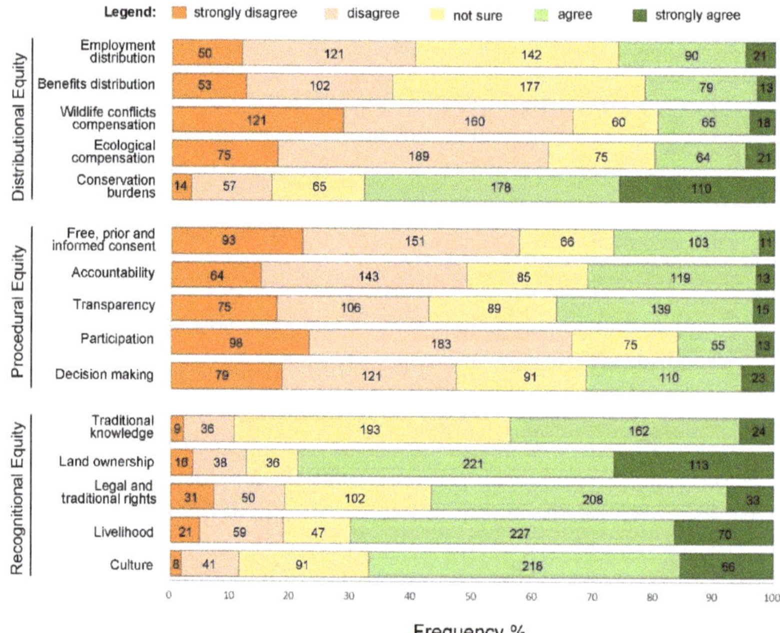

Figure 2. Stacked bar charts showing frequency distributions for all individual social equity indicators. The numbers within the bars indicated the number of respondents (further details are provided in Supplementary Materials—Table S7).

3.2. Participative Co-Management Activities

Descriptive analysis showed varied participation rates among different co-management activities (see Figure 3 and Supplementary Materials—Table S5). The most frequently participated co-management activity was energy renovation arrangements (A1 = 74.5%), followed by environmental education activities (A3 = 56.4%). The percentages for the remaining 10 co-management activities were all below 40%, with co-management agreements and conservation accountability ranking the lowest two (A8 = 9.2%, A12 = 5.4%). Furthermore, we contrasted the numbers of participants across four empowering levels of co-management, with the instruction type being the largest, followed by those of consultation and agreement, and finally, the cooperation type. It was clear that the number of participants tended to decline with the increase in co-management empowering levels. Statistics also showed that the majority of respondents (n = 273, 64,4%) were involved in less than three co-management events. Notably, 4.5% (n = 19) of respondents had none of this experience (see Figure 4 and Supplementary Materials—Table S6).

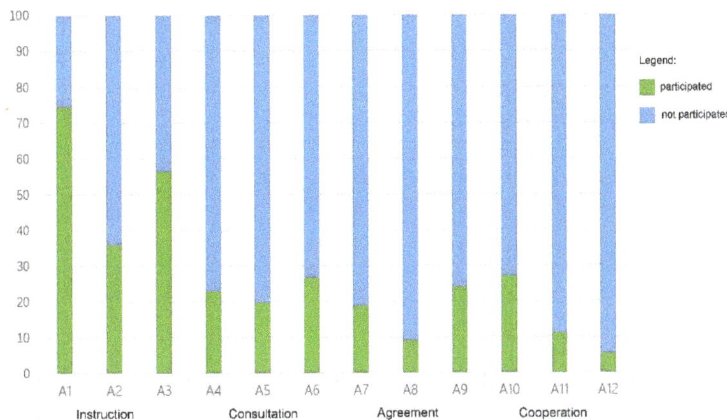

Figure 3. Participation frequency in diversified co-management activities. The horizontal coordinates represent different co-management activities, and the vertical coordinates represent the percentages of participants (n = 424).

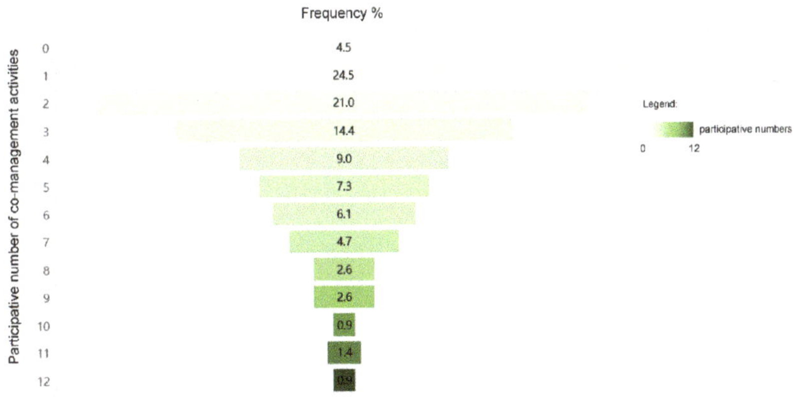

Figure 4. Frequency of the number of co-management activities in which respondents participated. The numbers on the bar chart represent the percentage of respondents for a certain participated number (n = 424).

3.3. Analysis of Correlation

Results of the Spearman correlation analysis showed that all types of co-management were significantly correlated with each dimension of social equity, listed in Table 5. When it came to the number of participated co-management activities, the results were similar. Those findings indicated that the more local residents were involved in co-management arrangements, the more likely they would have positive feelings for recognitional, procedural, distributional, and composite fairness.

Table 5. Summary of results from a univariate model of the relationship between predictors (types and numbers of co-management activities) and social equity perceptions. (Note: The data in the table showed correlation coefficients. Significance levels: * = $p < 0.05$, ** = $p < 0.01$).

Category	Recognitional Equity	Procedural Equity	Distributional Equity	Combined Social Equity
Instruction	0.190 **	0.278 **	0.321 **	0.333 **
Consultation	0.177 **	0.389 **	0.325 **	0.399 **
Agreement	0.102 *	0.216 **	0.279 **	0.250 **
Cooperation	0.169 **	0.374 **	0.392 **	0.411 **
Number of participated co-management activities	0.198 **	0.373 **	0.400 **	0.418 **

With respect to the socio-demographic features, one-way ANOVA and Spearman correlation analysis were adopted accordingly. Spearman correlation analysis revealed that education and annual household income were significantly and positively correlated with all directions of fairness perceptions, while age was negatively related to all. Results of the one-way ANOVA test showed that villages were merely significantly related to recognitional equity. Additionally, two household sources of income, tourism, and forestry, were recognized as significantly correlated factors for certain dimensions of fairness perceptions (Table 6).

Table 6. Summary of results from a univariate model of the relationship between predictors and social equity perceptions. (Note: The symbols + or − indicate the direction of the relationship between fixed factors and ordinal levels. +: Positive correlation, −: Negative correlation. Significance levels: / = Not significant, * = $p < 0.05$, ** = $p < 0.01$).

Category		Analysis Method	Recognitional Equity	Procedural Equity	Distributional Equity	Combined Social Equity
Gender		One-way ANOVA	/	/	/	/
Age		Spearman	− *	− **	− **	− **
Occupation		One-way ANOVA	/	/	/	/
Education		Spearman	+ **	+ **	+ **	+ **
Villages		One-way ANOVA	**	/	/	/
Residency years		Spearman	/	/	/	/
Household size		Spearman	+ **	/	/	/
Household migrant workers		Spearman	/	/	/	/
Annual household income		Spearman	+ **	+ *	+ **	+ **
Household source of income	Farming	One-way ANOVA	/	/	/	/
	Tourism	One-way ANOVA	/	/	**	**
	Forestry	One-way ANOVA	/	**	**	**
	Local employment	One-way ANOVA	/	/	/	/
	Non-local employment	One-way ANOVA	/	/	/	/

3.4. Regression Equation

We conducted the linear regression analysis to assess how social demographics and participative factors could have influence across all directions of perceived fairness (see Supplementary Materials—Model 1–4). The first regression model (Adjusted R^2 = 0.107, F = 6.653, *p*: <0.0001) clearly showed the impact of education, village, household size, and consultation on recognitional equity. Among them, household size has the largest effect, followed by education and consultation, with village being the smallest one. The second model (Adjusted R^2 = 0.272, F = 20.756, *p*: <0.0001) showed the influence from education, consultation, and cooperation on perceived procedural equity. In this model, cooperation has the largest effect on procedural equity, while education was the smallest. The third model (Adjusted R^2 = 0.27, F = 18.423, *p*: <0.0001) disclosed the causal relationship between the four types of co-management and perceptions of distributional equity, among which the cooperation continued to have the largest effect.

In addition, a linear regression analysis was conducted to assess the effects of those indicators on perceived combined equity, marked as model four (Adjusted R^2 = 0.321, F = 23.26, *p*: <0.0001). The significant impact was detected from variables of education, construction, consultation, and cooperation, among which the cooperation type of co-management has the largest impact. Above all, the aforementioned four models all passed collinearity diagnosis, serial correlation diagnosis, and residual normality test, thus partially reflecting the causal relationship between the relevant variables and to some extent assisting us in better understanding their influences on fairness perceptions.

4. Discussion

4.1. Relationships between Perceived Fairness and Socio-Demographic Characteristics

Our findings disclose that locals' fairness perceptions are significantly associated with several socio-demographic characteristics, including education, location, and household size. Among them, the level of education is the most widely related factor, concerning not only recognitional (Beta = 0.164, *p* = 0.004) and procedural equity (Beta = 0.165, *p* = 0.002), but also combined equity (Beta = 0.167, *p* = 0.001). This suggests that higher education levels of local residents generally lead to better exposure to CBCM information and participative opportunities, and this can consequently link to a better understanding and acceptance of conservation justice. This result aligns with researches conducted by Nathan J. Bennett et al. (2020) [15], Lou Lecuyer et al. (2019) [33], and Aires Afonso Mbanze et al. (2021) [34], demonstrating the impact of formal education on perceived fairness of conservation. By contrast, Georgia G. Gurney et al. (2021) [16] discloses the association between formal education and perceived distributional equity, which is not significant in this research.

Moreover, our results reveal that village is significantly related to their perceived recognitional fairness (Beta = 0.118, *p* = 0.014). This can be clearly illustrated by the fact that recognitional fairness perceptions of residents living inside TGPNP are the lowest (Luoyigou village, mean = 3.36), while those from the gateway community are the highest (Yinping village, mean = 3.71). This phenomenon is not complex to comprehend. For one thing, residents of Luoyigou village are more likely to develop negative judgements toward the recognitional indicators of livelihood, legal and traditional rights, as well as land ownership, as they suffer from more strict land use restrictions and more intense human-wildlife conflict, compared to villagers living outside the boundary of TGPNP. For another, the gateway community, Yinping village, has long been supported financially and technologically to develop eco-tourism by the county government and ATA, therefore locals' feelings of recognitional justice are more likely to be positive-going. Similarly, O. Digun-Aweto et al. (2018) [35] found that communities living close to the national park showed more negative attitudes toward conservation, while communities living far away from the national park were not severely impacted by wildlife-caused crop losses and consequently developed more positive perceptions. Apart from this, another factor affecting perceived recognitional equity is household size, which was discovered in our study. Similar results are noted by Ding Ya (2019) [36] and Liu Yucheng et al. (2018) [37], stating that respondents with larger household size are more likely to be satisfied with ecological compensation and the implementation of programs.

Other demographic features (e.g., gender, annual household income, and age), although they do not pass our regression analysis, have been discussed heatedly in other literature. Georgina G. Gurney et al. (2021) [16] discovered that men are more likely to develop fairness perceptions than women toward merit and equality principles. In addition, Carolina T. Freitas et al. (2020) [22] similarly believed that co-management of fisheries could promote gender equity. However, this phenomenon was not found in TGPNP, where organized co-management activities imposed no apparent gender restrictions on participants. Moreover, a significant positive correlation was detected between annual household income and perceived fairness in our study. This result is consistent with the findings of Nathan J. Bennett et al. (2020) [15], who argued that people with higher relative wealth had more earnings and therefore would have a positive perception of distributional equity. A study by Zhu Ting et al. (2012) [38] further indicated that participation in co-management programs had a significant positive impact on household income. However, Georgina G. Gurney et al. (2021) [16] argued that stakeholders with more material assets are more likely to perceive the distribution of benefits as unfair. Finally, our research also verified the findings by Nathan J. Bennett et al. (2020) [15], in which increasing age is associated with worsening perceptions of recognition, distributive, and integrated equity. This is possibly due to the fact that some co-management activities in TGPNP (e.g., forest rangers and rural tourism skills training) set age restrictions for participants, which lead to the fact that elder villagers with fewer participative opportunities were less likely to develop fairness perceptions.

4.2. Participative Co-Management Activities and Their Associated Fairness Perceptions

All participated types of co-management are positively associated with certain dimensions of fairness perceptions. First, the instructive type of co-management, where the government is completely taking control, is significantly associated with perceived distributional equity (Beta = 0.100, p = 0.041), as well as combined social equity (Beta = 0.108, p = 0.021). This is due to the fact that most of these co-management activities which are dominated by ATA help in enhancing local livelihood, such as energy transformation, industrial support, and technology improvement [31]. Those economically supportive activities, in return, are exchanged for conservation obedience of local residents, the enforcement of which can improve local perceptions toward distributional fairness [20].

Second, the consultative type of co-management is recognized as the most widely correlated factor, which is positively associated with all equity dimensions (recognition, Beta = 0.130, p = 0.026; procedure, Beta = 0.283, p = 0.000; distribution, Beta = 0.127, p = 0.016; and combined equity, Beta = 0.243, p = 0.000). This highlights the crucial function of information exchange between communities and ATA staff, if timely and sufficient, it can greatly enhance the identity recognition, participative channels, and opportunities, and promote more equitable benefits sharing of communities. This finding expands the discovery by Catherine Gross (2007) [39] in which access to adequate information is important for procedural fairness, and further detects its effects on recognitional, distributional, and composite justice. By contrast, the agreement co-management type is merely correlated with distributional equity (Beta = 0.104, p = 0.037). This is due to the fact that most of the agreements already signed focus on dealing with economic losses or the redistribution of benefits, such as the agreements of human-wildlife conflict compensation and Chinese beekeeping benefit-sharing. However, only a small proportion of local households have reached agreements with ATA, with the percentages for A7, A8, and A9 as 18.9%, 9.2%, and 24.1%, respectively. This can well explain why participated co-management agreements have no direct influence on combined equity, as the participation scope is not wide enough to exert a comprehensive impact.

Finally, involvement in cooperative co-management activities, such as forest patrolling and enacting conservation rules, has a strong effect on residents' feelings of justice, especially in the procedural, distributional, and combined dimension (Beta = 0.257, 0.300, and 0.295, respectively). It is not difficult to understand its dramatic effect since this type of co-management highly empowers locals. An interesting phenomenon here is that most of the respondents who fairly collaborated with the ATA are local elites, such as village directors, cadres, and rangers, etc. Those elites are extensively exposed to, sufficiently involved in, and fully responsible for those collaborative activities, in order that they are more likely to perceive fair procedures than the ordinary residents, and consequently bestowed with more equitably distributed benefits. This result is consistent with the finding by Haiyun Chen et al. (2012) [40] that members of village councils and co-management committees involved in more projects can enjoy more equitable treatment as a result.

4.3. Recommendations for TGPNP

Our results disclose that the relatively low empowering levels and limited numbers of participative activities in co-management, can consequently lower the fairness perceptions toward TGPNP. Most of the respondents in our survey have merely been involved in less than two types of co-management activities, and more specifically, at the lowest instructive empowering level. This dilemma has not been appropriately solved despite the fact that various co-management interventions lasted more than four decades in TGPNP, partially due to the lack of conservation capacity among locals [31]. Another reason for this phenomenon is the scarcity of participative channels, as an interviewee complained: "If the ATA asks me to give suggestions or get involved in conservation affairs, I am very willing to do; but the situation is that they would never ask me". By analyzing the negatively perceived indicators, including the livelihood and land ownership for recognitional equity, participation, FPIC, and decision-making for procedural equity, as well as ecological compensation and human-wildlife conflicts for distributional equity, we can further detect some potential issues faced with TGPNA. It is self-evident that the land grabbing and limitations on traditional livelihoods are common issues facing worldwide protected areas [41–43], and insufficient industrial support, untransparent procedures, inappropriate compensation, and other managerial shortcomings may hinder locals to develop fairer judgements toward the TGPNA.

Based on the aforementioned issues, we believe that facilitating local participation in diversified co-management arrangements can effectively promote more equitably managed protected areas. In this direction, we suggest that the ATA strengthen the publicity of co-management to locals, particularly for the elder and low-income groups, setting more channels for participation, and simultaneously, enhancing conservation awareness and capability of locals. Those countermeasures can improve the participative rates of locals and gradually enhance their empowering levels, after years of attempts and endeavors. Apart from this, the impacts of conservation initiatives on land use and traditional livelihoods need to be addressed urgently [44,45]. For this, we recommend the ATA to facilitate alternative livelihood, such as eco-agriculture and eco-tourism, and provide more job opportunities for locals to get involved in conservation, especially for residents living inside the TGPNP. Moreover, it is imperative to better inform and involve locals in the co-management meetings, planning consultations, and capability-building workshops, in order to set up with fairer participative procedures. Furthermore, we suggest strengthening wildlife monitoring and developing a more equal and reasonable compensation plan to relieve the human-wildlife conflicts and strive for better distributional equity [46].

4.4. Future Research

One novel contribution of this research is that it discloses the effects of participative co-management activities on perceptions of recognitional, procedural, distributions, and combined equity. Nevertheless, some limitations remain. First, as the co-management of GPNP is mostly conducted at the instructive level, the research findings can be different in the contexts of more empowered co-managed regime. Moreover, the survey sampling can spread to a broader range of age groups, especially for the younger generations, since more than half of the current respondents are in their fifties or sixties. Furthermore, additional in-depth interviews can be conducted with local people to capture their deeper understandings toward equity, since this concept can have differentiated meanings to different groups.

Our research finds that the location of villages is a critical element in impacting perceptions of procedural equity. In-depth, we speculate that other spatially related factors (e.g., the accessibility of the residence, the distance from the main road, and entrance of the protected area) might also have essential influence on recognitional and possibly other fairness perceptions. Therefore, we recommend that future researches should focus on this direction to explore the correlation between spatial factors and locals' perceptions of fairness. Moreover, with informal interviews, we realize that multiple stakeholder groups show different perceptions toward various co-management projects. Therefore, we suggest that in-depth interviews and participating observations should be adopted in future studies to compare and contrast fairness perceptions among different stakeholder groups [23].

Furthermore, there is a need to assess the correlation among fairness perceptions, satisfaction degrees, and conservation attitudes, in which a non-linear and complex statistic model might be required. Locals' perception of fairness may affect their satisfaction with and conservation attitudes toward protected areas, and to understand this relationship, it is conducive to achieve the conservation success [47–49].

Finally, both effectiveness and equity are essential, yet different and interdependent concepts in the conservation of protected areas (Woodley et al. 2012 [50]; Schreckenberg et al. 2016 [3]). Some scholars believe that the effectiveness of protection is often achieved without perfect social equity (Klein et al. 2015 [7]; Dawson et al. 2017 [51]), indicating that they may not be simply positively correlated. Therefore, pursuing extreme equity in protected areas is encouraged, but it is more worthwhile to explore the extent of equity that can achieve maximum efficiency. Community-based co-management, in the context of protected areas, serves as a crucial means to balance both effective and equitable management (Persha and Andersson 2014 [52]) [20]. Therefore, a greater focus on analyzing the relationship and trade-off between social equity and conservation effectiveness of co-managed protected areas can produce thought-provoking findings, and better conduct management effectiveness evaluation with consideration of social equity.

5. Conclusions

For the broader well-being of the local people and stakeholders, the equity issues in marine and terrestrial protected areas are receiving increasing attention globally. However, despite the co-management approach being widely promoted worldwide for the better governance of protected areas, little attention has been paid to its effects on fairness perceptions. This paper builds on the considerable work in social equity issues and the empowering levels of co-management to further explore the correlation between participated co-management arrangements and perceived social equity of locals, from recognitional, procedural, and distributional dimensions. The main conclusions are summarized below: (1) There is a distinct variability in fairness perceptions toward TGPNP, with the recognitional equity as positive, procedural, and distributional negative, and the combined equity as neutral. (2) The participated co-management activities and reflected empowering levels of locals are rather limited, with most of the respondents remaining at the instructive level. The number of participants declines with the increase in empowering levels. (3) Participation in diversified co-management activities is revealed to be influential on locals' fairness perceptions. While the consultative type of co-management is recognized as the most widely correlated factor, the cooperation is found to have the strongest impact. By contrast, the impact scope of instruction and agreement types of co-management are pretty narrow, mostly in the distributional dimension. (4) With regards to the demographic features, education is found to be positively related to all equity perceptions, while village is significantly merely for recognitional equity. These findings indicate that the more locals are involved in co-management activities, the fairer they are likely to perceive the protected areas. This points to the need for more empowered and widely involved co-management plans to improve social equity judgement in TGPNP. Furthermore, regarding specific affairs (e.g., ecological compensation and human-wildlife conflicts) or communities of different locations (e.g., communities inside or outside protected areas), tailored countermeasures should be taken for better consideration of social equity issues.

This paper highlights the critical importance of exploring social equity in the co-management arrangements of nationally designated protected areas. To promote the achievement of the fairness goals and conservation goals for a broader population, we encourage the global conservation community to conduct more discussions that combine social equity and co-management issues, which can consequently produce more co-management plans, principles or instructions with equity consideration.

Supplementary Materials: The following supporting information can be downloaded at: https://www.mdpi.com/article/10.3390/land11101624/s1. Table S1: Survey questions related to the demographics and characteristics of local residents; Table S2: Survey questions related to participated types of co-management local residents; Table S3: Survey questions and responses related to recognitional, procedural, and distributional equity; Table S4: Descriptive summary of survey sample including demographics and characteristics of local residents; Table S5: Descriptive summary of survey sample including the number and percentage of local residents that participated in co-management types and activities; Table S6: Descriptive summary of survey sample including the quantity of co-management participation of local residents; Table S7: Descriptive summary of responses to all individual perception indicators; Table S8: Correlation analysis between demographic characteristics and perceived fairness; Table S9: Correlation analysis of co-management participation type and fairness perception; Model S1: A regression model of recognitional equity and relevant independent variables; Model S2: A regression model of procedural equity and relevant independent variables; Model S3: A regression model of distributional equity and relevant independent variables; Model S4: A regression model of combined social equity and relevant independent variables.

Author Contributions: Conceptualization, Y.Z. (Yin Zhang) and Y.Z. (Yuqi Zhang); methodology, Q.C. and Y.Z. (Yin Zhang); software, Q.C.; formal analysis, Q.C.; investigation, Q.C., Y.Z. (Yuqi Zhang) and Y.Z. (Yin Zhang); data curation, Q.C.; writing—original draft preparation, Q.C., Y.Z. (Yuqi Zhang) and Y.Z. (Yin Zhang); writing—review and editing, Q.C., Y.Z. (Yuqi Zhang), Y.Z. (Yin Zhang) and M.K.; visualization, Q.C.; supervision, Y.Z. (Yin Zhang); project administration, Y.Z. (Yin Zhang); funding acquisition, Y.Z. (Yin Zhang). All authors have read and agreed to the published version of the manuscript.

Funding: This research was funded by the National Natural Science Youth Program, grant number 52108040, and the China Postdoctoral Science Foundation, grant number 2021M700574.

Data Availability Statement: Not applicable.

Acknowledgments: The authors would like to thank all the anonymous reviewers and editors who contributed their time and knowledge to this study. The authors also thank Peng L., Liang Y.L., Wang L.C., Li Q.Y., Shen B.S., Chen J.X. and Zuo C.L. who were involved in the tough field survey in torrid summer. Thanks to Zhang Z.Q. and Li K. for their precious support for our survey. And thanks to all staffs of ATA who provided us with documentation and materials, or helped with our interviews and questionare distribution. We also own special thanks to Dan Brockington, an interview with whom initially inspired the corresponding author to develop the preliminary idea of this research. We authors would like to thank Huang J.L. sincerely for his strong statistical support and guidance of our study.

Conflicts of Interest: The authors declare no conflict of interest.

References

1. Cardinale, B.J.; Duffy, J.E.; Gonzalez, A.; Hooper, D.U.; Perrings, C.; Venail, P.; Narwani, A.; Mace, G.M.; Tilman, D.; Wardle, D.A.; et al. Biodiversity loss and its impact on humanity. *Nature* **2012**, *486*, 59–67. [CrossRef]
2. Brockington, D. Community conservation, inequality and injustice: Myths of power in protected area management. *Conserv. Soc.* **2004**, *2*, 411–432.
3. Schreckenberg, K.; Franks, P.; Martin, A.; Lang, B. Unpacking equity for protected area conservation. *Parks* **2016**, *22*, 11–26. [CrossRef]
4. Zafra-Calvo, N.; Geldmann, J. Protected areas to deliver biodiversity need management effectiveness and equity. *Glob. Ecol. Conserv.* **2020**, *22*, e01026. [CrossRef]
5. The Strategic Plan for Biodiversity 2011–2020 and the Aichi Biodiversity Targets. Available online: https://www.cbd.int/doc/strategic-plan/2011-2020/Aichi-Targets-EN.pdf (accessed on 18 August 2022).
6. The Promise of Sydney: Innovative Approaches for Change. Available online: https://www.worldparkscongress.org/about/promise_of_sydney (accessed on 18 August 2022).
7. Klein, C.; McKinnon, M.C.; Wright, B.T.; Possingham, H.P.; Halpern, B.S. Social equity and the probability of success of biodiversity conservation. *Glob. Environ. Chang.* **2015**, *35*, 299–306. [CrossRef]
8. Cetas, E.R.; Yasué, M. A systematic review of motivational values and conservation success in and around protected areas. *Conserv. Biol.* **2017**, *31*, 203–212. [CrossRef]
9. Oldekop, J.A.; Holmes, G.; Harris, W.E.; Evans, K.L. A global assessment of the social and conservation outcomes of protected areas. *Conserv. Biol.* **2016**, *30*, 133–141. [CrossRef]
10. Gill, D.A.; Cheng, S.H.; Glew, L.; Aigner, E.; Bennett, N.J.; Mascia, M.B. Social synergies, tradeoffs, and equity in marine conservation impacts. *Annu. Rev. Environ. Resour.* **2019**, *44*, 347–372. [CrossRef]
11. Campbell, L.M.; Gray, N.J. Area expansion versus effective and equitable management in international marine protected areas goals and targets. *Mar. Policy* **2019**, *100*, 192–199. [CrossRef]
12. Moreaux, C.; Zafra-Calvo, N.; Vansteelant, N.G.; Wicander, S.; Burgess, N.D. Can existing assessment tools be used to track equity in protected area management under Aichi Target 11. *Biol. Conserv.* **2018**, *224*, 242–247. [CrossRef]
13. Zafra-Calvo, N.; Pascual, U.; Brockington, D.; Coolsaet, B.; Cortes-Vazquez, J.A.; Gross-Camp, N.; Palomo, I.; Burgess, N.D. Towards an indicator system to assess equitable management in protected areas. *Biol. Conserv.* **2017**, *211*, 134–141.
14. Zafra-Calvo, N.; Garmendia, E.; Pascual, U.; Palomo, I.; Gross-Camp, N.; Brockington, D.; Cortes-Vazquez, J.A.; Coolsaet, B.; Burgess, N.D. Progress toward equitably managed protected areas in Aichi target 11: A global survey. *BioScience* **2019**, *69*, 191–197. [CrossRef]
15. Bennett, N.J.; Calò, A.; Franco, A.D.; Niccolini, F.; Marzo, D.; Domina, I.; Dimitriadis, C.; Sobrado, F.; Santoni, M.C.; Charbonnel, E.; et al. Social equity and marine protected areas: Perceptions of small-scale fishermen in the Mediterranean Sea. *Biol. Conserv.* **2020**, *244*, 108531. [CrossRef]
16. Gurney, G.G.; Mangubhai, S.; Fox, M.; Kim, M.K.; Agrawal, A. Equity in environmental governance: Perceived fairness of distributional justice principles in marine co-management. *Environ. Sci. Policy* **2021**, *124*, 23–32. [CrossRef]
17. Wang, W.; Liu, J.; Innes, J.L. Conservation equity for local communities in the process of tourism development in protected areas: A study of Jiuzhaigou Biosphere Reserve, China. *World Dev.* **2019**, *124*, 104637. [CrossRef]
18. Friedman, R.S.; Rhodes, J.R.; Dean, A.J.; Law, E.A.; Santika, T.; Budiharta, S.; Hutabarat, J.A.; Indrawan, T.P.; Kusworo, A.; Meijaard, E.; et al. Analyzing procedural equity in government-led community-based forest management. *Ecol. Soc.* **2020**, *25*, 16. [CrossRef]
19. Massarella, K.; Sallu, S.M.; Ensor, J.E. Reproducing injustice: Why recognition matters in conservation project evaluation. *Glob. Environ. Chang.* **2020**, *65*, 102181. [CrossRef]
20. Carlsson, L.; Berkes, F. Co-management: Concepts and methodological implications. *J. Environ. Manag.* **2005**, *75*, 65–76. [CrossRef]
21. Borrini-Feyerabend, G.; Jaireth, H.; Farvar, M.T.; Pimbert, M.; Renard, Y.; Kothari, A.; Warren, P.; Murphree, M.; Pattemore, V.; Ramrez, R. *Sharing Power: Learning-by-Doing in Co-Management of Natural Resources throughout the World*; Earthscan: Oxford, UK, 2007.
22. Freitas, C.T.; Espírito-Santo, H.M.V.; Campos-Silva, J.V.; Peres, C.A.; Lopes, P.F.M. Resource co-management as a step towards gender equity in fisheries. *Ecol. Econ.* **2020**, *176*, 106709. [CrossRef]
23. d'Armengol, L.; Castillo, M.P.; Ruiz-Mallén, I.; Corbera, E. A systematic review of co-managed small-scale fisheries: Social diversity and adaptive management improve outcomes. *Glob. Environ. Chang.* **2018**, *52*, 212–225.
24. Ullah, S.M.A.; Tani, M.; Tsuchiya, J.; Rahman, M.A.; Moriyama, M. Impact of protected areas and co-management on forest cover: A case study from Teknaf Wildlife Sanctuary, Bangladesh. *Land Use Policy* **2022**, *113*, 105932. [CrossRef]
25. Islam, K.; Nath, T.K.; Jashimuddin, M.; Rahman, M.F. Forest dependency, co-management and improvement of peoples' livelihood capital: Evidence from Chunati Wildlife Sanctuary, Bangladesh. *Environ. Dev.* **2019**, *32*, 100456. [CrossRef]
26. Islam, K.N.; Jashimuddin, M.; Hasan, K.J.; Khan, M.I.; Kamruzzaman, M.; Nath, T.K. Stakeholders' perception on conservation outcomes of forest protected area co-management in Bangladesh. *J. Sustain. For.* **2021**, *AHEAD-OF-PRINT*, 1–17. [CrossRef]
27. Franco-Meléndez, M.; Tam, J.; van Putten, I.; Cubillos, L.A. Integrating human and ecological dimensions: The importance of stakeholders' perceptions and participation on the performance of fisheries co-management in Chile. *PLoS ONE* **2021**, *16*, e0254727. [CrossRef]

28. Phromma, I.; Pagdee, A.; Popradit, A.; Ishida, A.; Uttaranakorn, S. Protected area co-management and land use conflicts adjacent to Phu Kao–Phu Phan Kham National Park, Thailand. *J. Sustain. For.* **2019**, *38*, 486–507. [CrossRef]

29. Governance of Protected Areas: From Understanding to Action. Available online: https://www.iucn.org/content/governance-protected-areas-understanding-action (accessed on 18 August 2022).

30. Li, S.; Feng, J.; Li, B.V.; Lü, Z. The Giant Panda National Park: Experiences and lessons learned from the pilot. *Biodivers. Sci.* **2021**, *29*, 307–311. [CrossRef]

31. Fu, Z.P.; Shen, L.M.; Yang, Y.B.; Chen, W.; Liu, L.; Yu, B.; Liu, H.; He, W.H. Study of Status and Development of Community Management for Giant Panda Sanctuary. *Sichuan J. Zool.* **2015**, *34*, 468–473.

32. Giant Panda Park Community Co-Construction and Co-Management Plan Released and New Eco-Experience Activities to Be Carried Out in Xiling in the Future. Available online: https://baijiahao.baidu.com/s?id=1735853566322799672&wfr=spider&for=pc (accessed on 18 August 2022).

33. Lecuyer, L.; Calme, S.; Blanchet, F.G.; Schmook, B.; White, R.M. Factors affecting feelings of justice in biodiversity conflicts: Toward fairer jaguar management in Calakmul, Mexico. *Biol. Conserv.* **2019**, *237*, 133–144. [CrossRef]

34. Mbanze, A.A.; da Silva, C.V.; Ribeiro, N.S.; Santos, J.L. Participation in illegal harvesting of natural resources and the perceived costs and benefits of living within a protected area. *Ecol. Econ.* **2020**, *179*, 106825. [CrossRef]

35. Digun-Aweto, O.; Fawole, O.P.; Saayman, M. The effect of distance on community participation in ecotourism and conservation at Okomu National Park Nigeria. *Geojourna* **2020**, *84*, 1337–1351. [CrossRef]

36. Ding, Y. Farmers' Satisfaction with the Implementation of Precise Poverty Alleviation Projects and Influencing Factors—Take Geyi Village in Taijiang County as an Example. Master's Thesis, Guizhou University, Guizhou, China, 2019.

37. Liu, Y.C.; Zhang, X.L. A Study on the Herdsmen's Choice of Grassland Ecological Compensation Mode. *Ecol. Econ.* **2018**, *34*, 197–201.

38. Zhu, T.; Ganesh, P.S.; Chen, H.Y.; David, M. A survey-based evaluation of community-based co-management of forest resources: A case study of Baishuijiang National Natural Reserve in China. *Environ. Dev. Sustain.* **2012**, *14*, 197–220.

39. Gross, C. Community perspectives of wind energy in Australia: The application of a justice and community fairness framework to increase social. *Energy Policy* **2007**, *35*, 2727–2736. [CrossRef]

40. Chen, H.Y.; Shivakoti, G.; Zhu, T. Livelihood Sustainability and Community Based Co-Management of Forest Resources in China: Changes and Improvement. *Environ. Manag.* **2012**, *49*, 219–228. [CrossRef]

41. West, P.; Igoe, J.; Brockington, D. Parks and peoples: The social impact of protected areas. *Annu. Rev. Anthropol.* **2006**, *35*, 251–277. [CrossRef]

42. Ward, C.; Stringer, L.C.; Holmes, G. Protected area co-management and perceived livelihood impacts. *J. Environ. Manag.* **2018**, *228*, 1–12. [CrossRef]

43. Ayivor, J.S.; Nyametso, J.K.; Ayivor, S. Protected area governance and its influence on local perceptions, attitudes and collaboration. *Land* **2020**, *9*, 310. [CrossRef]

44. Bennett, N.J.; Dearden, P. Why local people do not support conservation: Community perceptions of marine protected area livelihood impacts, governance and management in Thailand. *Mar. Policy* **2014**, *44*, 107–116. [CrossRef]

45. Abukari, H.; Mwalyosi, R.B. Local communities' perceptions about the impact of protected areas on livelihoods and community development. *Glob. Ecol. Conserv.* **2020**, *22*, e00909. [CrossRef]

46. Xu, J.; Wei, J.; Liu, W. Escalating human–wildlife conflict in the Wolong Nature Reserve, China: A dynamic and paradoxical process. *Ecol. Evol.* **2019**, *9*, 7273–7283. [CrossRef]

47. Nsengimana, V.; Weihler, S.; Kaplin, B.A. Perceptions of local people on the use of Nyabarongo River wetland and its conservation in Rwanda. *Soc. Nat. Resour.* **2017**, *30*, 3–15. [CrossRef]

48. Mogomotsi, P.K.; Mogomotsi, G.E.J.; Dipogiso, K.; Phonchi-Tshekiso, N.D.; Stone, L.S.; Badimo, D. An analysis of communities' attitudes toward wildlife and implications for wildlife sustainability. *Trop. Conserv. Sci.* **2020**, *13*. [CrossRef]

49. Sagoe, A.A.; Aheto, D.W.; Okyere, I.; Adade, R.; Odoi, J. Community participation in assessment of fisheries related ecosystem services towards the establishment of marine protected area in the Greater Cape Three Points area in Ghana. *Mar. Policy* **2021**, *124*, 104336. [CrossRef]

50. Woodley, S.; Bertzky, B.; Crawhall, N.; Dudley, N.; Londoño, J.M.; MacKinnon, K.; Redford, K.; Sandwith, T. Meeting Aichi Target 11: What does success look like for protected area systems. *Parks* **2012**, *18*, 23–36.

51. Dawson, N.; Martin, A.; Danielsen, F. Assessing equity in protected area governance: Approaches to promote just and effective conservation. *Conserv. Lett.* **2018**, *11*, e12388. [CrossRef]

52. Persha, L.; Andersson, K. Elite capture risk and mitigation in decentralized forest governance regimes. *Glob. Environ. Chang.* **2014**, *24*, 265–276. [CrossRef]

Article

Local Residents' Social-Ecological Adaptability of the Qilian Mountain National Park Pilot, Northwestern China

Jing Li [1], Guoqiang Ma [1,*], Jinghua Feng [1], Liying Guo [1] and Yinzhou Huang [1,2]

[1] College of Earth and Environmental Sciences, Lanzhou University, Lanzhou 730000, China; jingli2021@lzu.edu.cn (J.L.); fengjh19@lzu.edu.cn (J.F.); guoly2020@lzu.edu.cn (L.G.); yzhhuang@lzu.edu.cn (Y.H.)

[2] Key Laboratory of Western China's Environmental Systems, Lanzhou University, Lanzhou 730000, China

* Correspondence: mgq@lzu.edu.cn

Abstract: Protected areas are critical for biodiversity conservation and ecosystem services. In the last few years, there has been growing recognition of the role of indigenous peoples and local communities in the management of government designated protected areas, and thus their perceptions and adaptability were paid much attention. Drawing on a survey of 487 residents in the Qilian Mountain National Park Pilot of Northwestern China, this study used the adaptive analysis framework to study the adaptability of local residents. The main contribution of this paper is to select a typical social-ecological system to study the adaptability of local residents, and using Elinor Ostrom's Social-Ecological System framework to analyze the adaptability mechanism. The results show that different types of residents had different adaptability to environmental change. People whose income mainly depends on work salary with a small part of herding have the highest level of adaptability, while people whose income mostly comes from farming with a small part of herding have the lowest level. This result is related to people's living location, as people living in the core zone and buffer zone of the reserve mainly earned from grazing, and people living in the experimental zone and peripheral zone earned mainly from outside work. Moreover, people living in the core zone and buffer zone are mostly elders and ethnic groups, while people in the experimental zone and buffer zone are Han people. To improve management effectiveness and to avoid conflict between local residents and managers, this paper suggests that more attention should be paid to these who have lived for a long time in the core zone and buffer zone. They are the most vulnerable groups and show low adaptability in almost all domains. For the long run, education quality should be improved to decrease the population in the reserve.

Keywords: adaptability; residents; perception; Qilian Mountain National Park Pilot; social-ecological system

Citation: Li, J.; Ma, G.; Feng, J.; Guo, L.; Huang, Y. Local Residents' Social-Ecological Adaptability of the Qilian Mountain National Park Pilot, Northwestern China. *Land* **2022**, *11*, 742. https://doi.org/10.3390/land11050742

Academic Editors: Le Yu, Rui Yang, Yue Cao and Steve Carver

Received: 21 March 2022
Accepted: 7 May 2022
Published: 17 May 2022

Publisher's Note: MDPI stays neutral with regard to jurisdictional claims in published maps and institutional affiliations.

1. Introduction

Protected areas (PAs) are one of the most important conservation tools for protecting biodiversity and ecosystem services [1–4]. To date, PA coverage has reached over 15% of the global land area [5,6]. Despite this extensive coverage, it is widely acknowledged that PAs are being increasingly influenced by global forces of economic development and socio-political change [7–9]. Therefore, there is a need to better understand the complex interactions between humans and protected areas [10–13].

Adaptability is a notion that was originally used in ecology to emphasize that species can change their own state and procreate species to adapt to changing environments [14–16]. Then, the application of adaptability expanded from biophysics to sociology, like how a social system adjusts its own behavior to the natural environment [17]. It was later also applied to the fields of climate change and natural disasters [18]. In these fields, adaptability was adjusted by natural ecosystems or human systems in response to actual or expected climate change and natural disasters, emphasizing risk recognition, adjustment and management [19,20]. Around the year 2000, adaptability was widely applied as an important attribute of social-ecological systems (SESs) [21–24]. In the field of SESs, adaptability referred to the capacity of actors to adjust their behaviors in response to external uncertainties and disturbances [25]. Adaptability was often associated with resilience [21–23,26,27]. It was a capacity of actors in the SES to influence resilience, and essentially to manage it [22]. At present, scholars take "SES" as the main research object, and it was an important trend of sustainable development and global change to study the adaptability to external disturbance and the adaptability mechanism. The research scale of SES adaptability mainly focused on national, regional and community levels [28–30]. At the community level, there was no lack of studies integrating livelihood capital from the Sustainable Livelihood Framework into the index system of adaptability evaluation [31,32]. However, the five dimensions of the Sustainable Livelihood Framework (physical, nature, social, financial and human capital) mainly represent the society, economy, and the ecology, which are the three pillars of SESs. Most studies did not explore the comprehensive impact of external policies and internal psychology and culture on livelihood system, and also separated the interaction between subsystems, although there were many studies that have introduced and modified the adaptive analysis framework proposed by Smit et al. (namely, adaptation to what, who or what adapts, and how does adaptation occur?) [25,33]. However, there were few studies that comprehensively constructed an analysis framework of SESs, sustainable livelihood, residents' behavior and adaptability. Therefore, it is necessary to construct an analysis framework of residents' adaptability for SESs.

The Qilian Mountain Nature Reserve, which is located in Gansu Province, western China, recently became a focus of attention due to its ineffective management. To improve the management, the reserve was then designated to be a pilot national park, and its name thereby was changed into Qilian Mountain National Park Pilot (QMNPP). However, the notion of a national park is relatively new in China. Considering that there are still many people living in the reserve and it is impossible to move all of them out, a better understanding of residents' perceptions and adaptations will benefit synergetic development of nature conservation and human welfare for the newly established national park. This study constructed an analysis framework of residents' adaptability in the QMNPP, and comprehensively evaluated the adaptability of farmers from different types and regional locations. Then, the impact factors and adaptability mechanism were also analyzed. Finally, we put forward suggestions to improve the adaptability of residents and enhance management effectiveness. The innovations of this article include the following: (1) Combining the Sustainable Livelihood Framework with the existing adaptability analysis framework to construct an analysis framework of residents' adaptability in the QMNPP, which improved the adaptability index system to some extent and provided reference for the adaptability study of residents in other protected areas; and (2) Ostrom's Social-Ecological System Framework (SESF) was used to analyze the adaptability mechanism, systematically analyze the causes of residents' adaptive behavior and the interaction within the system, and deepen the analysis of impact factors of adaptability.

2. Study Area

The QMNPP is part of Qilian Mountain range on the border of Qinghai and Gansu provinces, northwest China (Figure 1). It is a natural germplasm bank of alpine creatures and an important ecological corridor. It is protected for Picea crassifolia, Cypress chinensis, cranes and other organisms. However, snow leopard, a national first-class key protected animal, has been frequently captured by camera recently. The QMNPP is also designated as a national key water conservation forest area, a national natural forest protection project area, a national key ecological public welfare forest, and so on. The landscape covers glaciers, forests, grasslands, deserts, etc., and is a priority area for biodiversity conservation in China. The snow and glaciers on the Qilian Mountain provide precious water to more than 5 million people in the Hexi Corridor, which is located at the northern foot of the Qilian Mountain and characterized by its arid climate. Therefore, the Qilian Mountain is also called the Mother Mountain of the Hexi Corridor, making its protection much more meaningful.

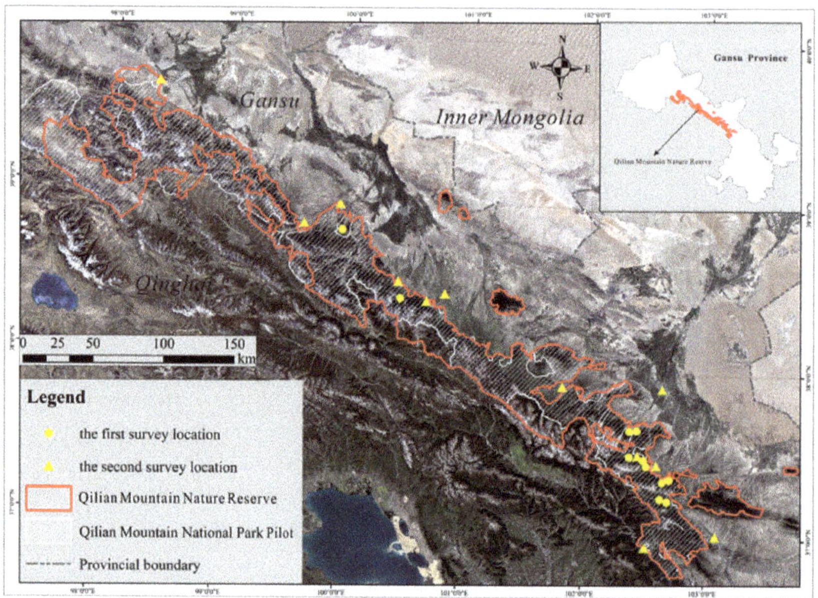

Figure 1. Study area.

People living inside the Qilian Mountain area have a very long history. However, intensive resource use began in the 1960s, when logging became an important industry for the area. Then, mining and hydropower infrastructures followed, and the number of livestock has rapidly increased in the 1980s and 1990s. As a result, serious grassland degradation was detected in the late 1990s. Therefore, restoration of grassland for this area has been paid much attention from the early 2000s. In 2015, a public warning was given to local officers by the State Forestry Administration and the Ministry of the Environment of China, and human activities, such as illegal mining, unauthorized construction of hydropower facilities, excessive waste discharge and polluting emissions by local factories, are main issues that existed in the reserve. However, things changed little in the next two years. Then, the General Office of the Central Committee of the Communist Party of China and the General Office of the State Council gave a briefing on the destruction of the ecological environment in the reserve in July 2017. Nevertheless, the operation of mining and hydropower stations have not been stopped until 2018. In particular, tourism has greatly developed from the early 2000s.

To sum up, the QMNPP has rich natural resources and a long history of human activities, and is a complex adaptive system composed of ecological subsystems and social subsystems [34,35]. In addition to the typical natural ecosystem, its environmental problems are also typical. Furthermore, human utilization of natural resources and strong dependence on resources are typical. In 2017, it was identified as one of the pilot projects of national parks, and its experience can be replicated and promoted as an example. Based on this, we selected the QMNPP as a typical social-ecological system, and there is a need to clarify the adaptability of local residents.

3. Data and Methods

3.1. Data Source

In-depth interviews and questionnaires were conducted to collect data for this paper. The survey was conducted in September 2018 and October 2020. Considering the vast territory and intra-regional diversities of the QMNPP, and residents live in a scattered distribution, 10 protection stations and 2 towns were chosen as our survey destinations (Figure 1). Questionnaires were distributed randomly by government workers and protection station managers to the local residents. At the same time, we verified the credibility of the questionnaire results through in-depth interviews. A total of 513 questionnaires were sent out; questionnaires with incomplete information and inconsistent answers were deleted, and 487 valid questionnaires were recovered. Cronbach's α was 0.749 (>0.7), indicating that the data availability is good. Although the number of questionnaires was relatively small, it was found to be well-representative after comparison with the statistical data. Among respondents, the sample Tibetan population accounted for 27.93% of the total population, 11.91% of the total population had high school education, 24.85% had junior middle school education, and 30.39% had primary school education, which was approximately the same as the local statistical yearbook (in which the Tibetan population accounted for 26.27% of the total population and the Han was 42.5%, 10.22% of the total population had high school education, 22.65% had junior middle school education, and 36.79% had primary school education).

The survey content included three sections: (i) Social-demographic characteristics of respondents (i.e., age, gender, educational degree, location of the functional zone, household income, health, labor force); (ii) respondents' knowledge, satisfaction and implementation of legal policies in the QMNPP (the legal policies including "Returning the grain plots to forestry and grass", "Fodder–livestock balance system", "Eco-migration", "Eco-compensation", etc.); (iii) residents' perception of ecological, economic and social aspects in the QMNPP (i.e., the attitudes towards natural environment, economic source, economic income, infrastructure and ethnic culture). Questionnaire indicators were assigned by a five-point Likert scale.

3.2. Conceptual Analysis Model

The adaptive analysis framework constructed by Smit et al. (1999) was adopted in this paper [25]. It is a commonly used conceptual framework for adaptive analysis [33]. In the framework, the following aspects were considered (Figure 2): Adaptation to what? Who/What adapts? How does adaption occur? It proposes a framework to promote consistency and rigor in the use of concepts and terms for adaptability. This framework provides a structure for improving the science of adaptability and its adaptability to disturbance [25,26]. In addition, it also gave us the logic that we can study adaptation scientifically. Through this analytical framework, it is beneficial to further clarify the three core elements: adaptation to what, who or what adapts, and how does adaptation occur. This paper focuses on the local residents' adaptability, which is relatively simple in structure and complexity compared with regional systems, but requires more detailed and in-depth analysis. This framework can help us solve this problem. Therefore, this paper uses Smit et al.'s adaptive analysis framework to explore residents' adaptability to the changing environment.

In this paper, "Adaptation to what?" is the disturbance of the changing environment. Environmental changes will have risk and opportunity disturbance in the regional SES. "Who/What adapts?" refers to residents' adaptability to the changing environment in the QMNPP, including the adaptability of residents to policies, economy, ecology, society, culture and psychology. "How does adaption occur?" is the behavior response of residents, which is mainly in cognition, adaptive state, impact factors and mechanism of adaptability.

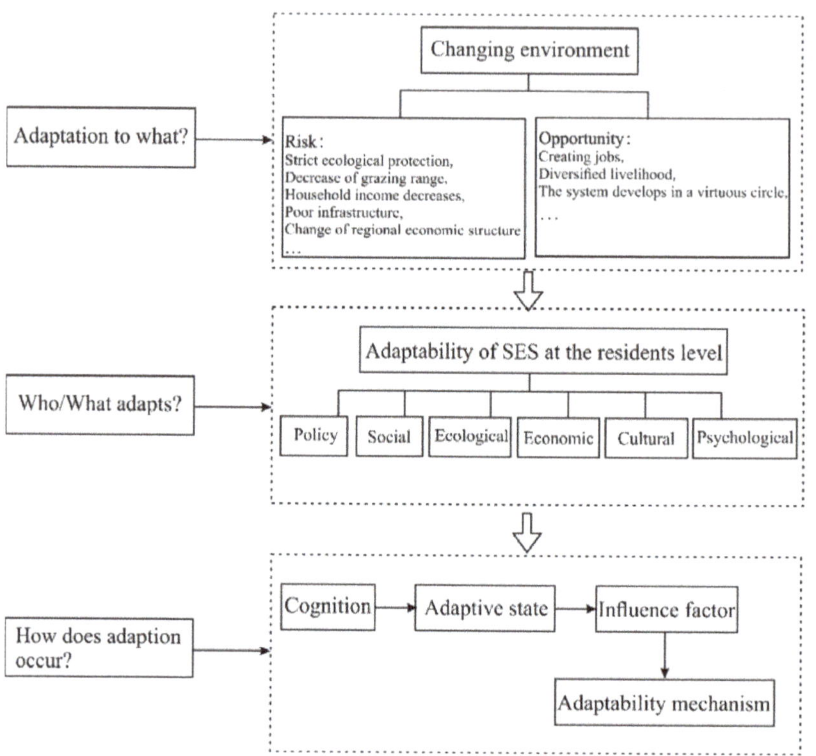

Figure 2. Analysis framework of residents' adaptability in the QMNPP (adopted from Smit et al. (1999) [25]).

3.3. Index System

Research of adaptability analysis methods usually adopts the Sustainable Livelihoods Framework [36,37], which examines residents' ability or capital to improve their quality of life and adaptability when they face natural disasters (such as drought, tsunamis, landslides), market competition, and system changes such as uncertain and changing environments. However, only using the Sustainable Livelihoods Framework to measure the adaptability of residents cannot reflect the integrity of SESs. The Sustainable Livelihood Framework [38–41] and Wu et al.'s (2018) [42] research were referred to build the indicator system. The analysis framework of residents' adaptability in this paper combines the Sustainable Livelihood Framework with the existing indicator system for adaptability evaluation, and adds the policy, culture and psychological dimensions as the "Who/What adapts?" part. Then, these indicators are classified into six domains, namely policy adaptability, social adaptability, ecological adaptability, economic adaptability, cultural adaptability and psychological adaptability. According to the characteristics of the QMNPP and availability of data, the index system was improved. For example, physical, social and financial capital of livelihood capital were integrated into the economic adaptability domain, and human and social capital were brought into the social adaptability domain. Furthermore, on the basis of the index system in Wu et al., the infrastructure was moved to the social domain, and indicators such as education, physical health and labor force were added. There were also many other such improvements.

Next, the Analytic Hierarchy Process (AHP) and the Expert Scoring method were used to weight the six domains. Eight experts including local managers, herdsmen who live in the park and scholars in ecology, sociology and geography were invited to score the indicators. The detailed steps were as follows: First, the eight experts were asked to score the six domains without communication, using the 1–5 scale method. Second, the scores of the eight experts were averaged to obtain the judgment matrix. Third, the matrix was analyzed by AHP, with weight calculation and a consistency test. The consistency of the weights was tested and the $CR = 0.065$ (<0.1), showing that the judgment matrix meets the consistency test. Finally, six domains of weights (W_i) were obtained (Table 1).

The index of measure layer was calculated via the entropy method. The result of the entropy method is objective. The detailed steps were as follows:

First, standard processing of data. All variables were normalized to a scale of 0–1, so they could be combined and compared.

Second, the specific gravity (f_{ij}) after dimensionless treatment was calculated. The formula is:

$$f_{ij} = \frac{x'_{ij}}{\sum_{i=1}^{m} x'_{ij}} \tag{1}$$

In the formula, x'_{ij} represents the normalized value of the jth term of the ith domain.

Third, we calculated the entropy value (e_j) and avail value (d_j) of the jth term, and the formula is as follows:

$$e_j = -\frac{1}{\ln m} \sum_{i=1}^{m} f_{ij} \ln f_{ij} \tag{2}$$

$$d_j = 1 - e_j \tag{3}$$

Finally, we calculated the weight of item j's index (w_{ij}), and the calculation formula is as follows:

$$w_{ij} = \frac{d_j}{\sum_{j=1}^{n} d_j} \tag{4}$$

The final results are shown in Table 1.

Table 1. Adaptability index values of residents in the QMNPP. The policies included the following: "Returning the grain plots to forestry and grass", "Fodder–livestock balance system", "Eco-migration", "Eco-compensation", etc. B3 = Labor force ÷ total number of family. B4 = Number of high school or above ÷ total number of family. D3 = 5 ≥ 70,000; 4 = 50,000–70,000; 3 = 30,000–50,000; 2 = 10,000–30,000; 1 ≤ 10,000. Natural assets owned by households include the number of cattle and sheep, grassland area and so on. D5 = Productive consumption ÷ total annual consumption of a family.

Domain (i)	W_i	Measure (x_{ij})	w_{ij}
Policy adaptability	0.25552	A1 Knowledge	0.3636
		A2 Satisfaction	0.3158
		A3 Implementation	0.3206
Social adaptability	0.17083	B1 Social network	0.0270
		B2 Infrastructure	0.0496
		B3 Proportion of household labor force	0.1749
		B4 Education	0.2543
		B5 Physical health	0.4941
Ecological adaptability	0.10227	C1 knowledge of social-ecological system	0.5281
		C2 Ecological awareness	0.4719

Table 1. *Cont.*

Domain (*i*)	W_i	Measure (x_{ij})	w_{ij}
Economic adaptability	0.24628	D1 Satisfaction of income	0.2425
		D2 Livelihood diversity	0.2941
		D3 household income (¥)	0.2695
		D4 Natural assets	0.0733
		D5 Proportion of consumption	0.1206
Cultural adaptability	0.09522	E1 Ethnic costume	0.1898
		E2 Diet custom	0.1130
		E3 Ethnic languages	0.2781
		E4 Ethnic music and dance	0.2743
		E5 Traditional festival	0.1448
Psychological adaptability	0.12987	F1 Acceptance of external culture	0.3941
		F2 Family resilience	0.4366
		F3 Acceptance of change	0.1692

3.4. Adaptability Assessment Model

The Comprehensive Evaluation of the Residents' Adaptability Index (*RAI*) in the QMNPP was measured through the linear weighed method. The model is as follows:

$$RAI = \sum_{j=1}^{6} W_i F_{ti} \tag{5}$$

$$F_{ti} = \sum_{i=1}^{m} w_{ij} x'_{ij} \tag{6}$$

In the formula, W_i is the weight of the *i*th domain layer, and *RAI* is the comprehensive evaluation of adaptability index of measure *j* under domain *i*. *RAI* was then divided into four grades [43,44]; they are extremely low adaptability ($0.00 \leq RAI \leq 0.25$), low adaptability ($0.26 \leq RAI \leq 0.50$), high adaptability ($0.51 \leq RAI \leq 0.75$) and extremely high adaptability ($0.76 \leq RAI \leq 1.00$).

4. Results

4.1. Demographic Sample Analysis

Among the 487 residents in this survey, males were the majority, accounting for 67.76% (Table 2), indicating that men dominate in the families in the survey area. There were more residents over 40 years old, of which 6.78% are over 65 years old. Except for the Han residents, there were more Tibetan residents, followed by the Yugur. Families' main source of income was grazing, supplemented by other income methods (such as planting crops, wage income obtained from ecological protection work in national parks, etc.). The annual income was mostly between 30,000 and 50,000 ¥, which is basically in line with the income characteristics of residents in pastoral areas. Residents mostly lived in the experimental zone and peripheral zone, although the grassland of the residents was still in the core zone and buffer zone, so there were still grazing activities in the core zone and buffer zone, which meets the needs of this survey.

Table 2. Demographic sample.

Survey Item	Type	Frequency (Sample = 487)	Percentage (%)
Gender	Male	330	67.76
	Female	157	32.24
Age	15–30	63	12.94
	31–40	149	30.60
	41–64	242	49.69
	Over 65	33	6.78
Nation	Han	242	49.69
	Tibetan	136	27.93
	Yugur	82	16.84
	Du	16	3.29
	Hui	9	1.85
	Mongolian	2	0.41
Annual household income (¥)	≤10,000	63	12.94
	10,000–30,000	181	37.17
	30,000–50,000	120	24.64
	50,000–70,000	77	15.81
	≥70,000	46	9.44
Source of income	Grazing	-	32.79
	Planting crops	-	9.70
	Self-employed income	-	5.82
	Wage income	-	31.77
	Government subsidies	-	9.46
	Other	-	10.47
Functional zone in protected areas	Core zone	70	14.37
	Buffer zone	109	22.38
	Experimental zone	191	39.22
	Peripheral zone	117	24.02

4.2. A General Analysis

4.2.1. Who/What Adapts?

"What adapts?" in this paper indicates the adaptability of local residents to policy, economy, ecology, society, culture and psychology. To categorize households, we classified the respondents on the basis of their income into nine types. They are H, H&F, H&W, F&H, F, F&W, W&H, W&F and W. Here H means herding, F means farming and W means having a job outside the QMNPP. If a respondent is categorized into type H, F or W, it means that he/she only has herding, farming or working as a livelihood, whereas types H&F, H&W, F&H, F&W, W&H and W&H mean that the income of the respondent depends on two sources. Taking H&F as an example, the respondent's livelihood source mainly comes from herding and a small part of farming, and the other types are alike.

Survey data (Table 3) presented that, among 487 respondents, type H&W (25.5%), W&H (18.1%) and W&F (15.6%) composed most of the respondents in the QMNPP, while type F was the lowest, meaning that very few people in the QMNPP took farming alone as their livelihoods. This result is consistent with the physical environment of the QMNPP, where the elevation is relatively high and is suitable for herding rather than farming.

Table 3. Number of livelihood type surveyed in the QMNPP.

Livelihood Type	H	H&F	H&W	F&H	F	F&W	W&H	W&F	W	Total
Number	23	18	149	3	1	9	97	84	103	487
Ratio (%)	7.6	8.2	25.5	1	0.2	2.7	18.1	15.6	21.1	100

4.2.2. How Does Adaption Occur?

(1) Adaptability analysis

The analysis of variance was used to analyze the RAI. Significant differences of RAI ($p < 0.05$) were detected, and the result is W&H > H&W > W > W&F > H > F&W > H&F > F&H (Figure 3: left).

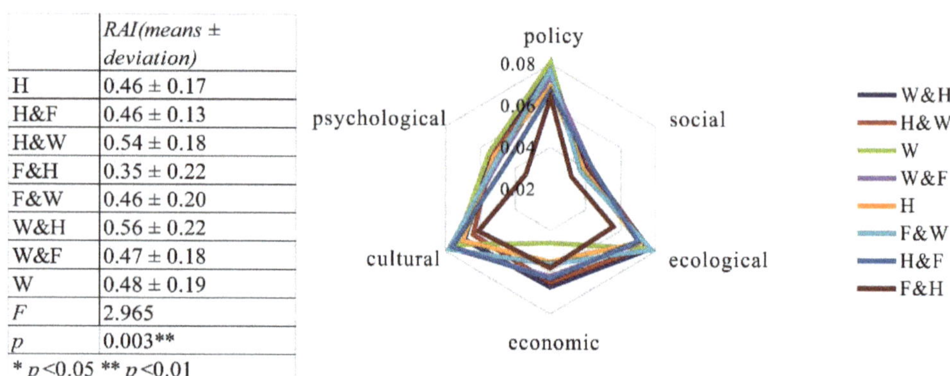

Figure 3. Result of analysis of variance and the adaptability of different livelihood types in different domains.

It can be concluded that local resident adapted policy change the most (Figure 3). However, different types of residents adapted it differently ($p = 0.002$ **, ** $p < 0.01$). Type W had the highest ability to adapt, followed by W&H and F&W, while type F&H was the lowest. Indeed, over the past twenty years, polices implemented, including "Returning the grain plots to forestry and grass", "Fodder–livestock balance system", "Eco-migration", etc., mostly aimed to limit herding or cultivating in the QMNPP. Therefore, people whose income depended on herding or farming would be affected the most, while type W people, as their income comes from outside work, have been little affected. However, it should be noted that, though these policies have negative impacts on local residents' income, they are always made up by compensation. The existing ecological compensation is mainly reflected in the following aspects: eco-migration, forest ecological benefits, water-saving projects, returning the grain plots to forestry and grass, biodiversity protection, nature reserve protection, etc. [45,46]. Therefore, it is not surprising that, compared with other domains, policy adaptability is the highest.

The p value of ecological adaptability was 0.069 ($p > 0.05$), indicating that nearly all respondents had same feelings regarding ecological change. This indicates high awareness of local residents to ecological protection. To understand this, there is a need to take a look at QMNPP's grassland degradation over the past decades. In the early 2000s, grassland degradation was a very serious issue for the area due to overgrazing, and it had seriously affected local residents' livelihood. On the one hand, policies were carried out to limit the number of livestock. On the other hand, livestock health was affected because of the degraded grassland. After years of restoration and along with increasing of income, local residents obtained a better understanding of the relationship between grassland and the number of livestock. Therefore, they were willing to take part in ecological protection.

Referring to economic adaptability, the p value was slightly higher than that of social adaptability and psychological adaptability. The economic adaptability of residents' livelihood types was significant at the 0.01 level ($F = 22.254$, $p = 0.000$ **), indicating that different types of residents had very different economical adaptability. The results are: type W&H had the highest adaptability (0.067), followed by H&W, and type W had the lowest (0.046). Generally, nomads in the QMNPP have the highest income. Therefore, the residents whose income mainly comes from grazing and also have family members working outside showed the highest adaptability. However, for type W, most of them worked as forest rangers, grassland rangers, protection station managers, and in mass prevention and mass treatment, for which salaries are very low, and thus they showed very low economic adaptability.

With reference to cultural adaptability, great discrepancies ($F = 2.650$, $p = 0.008$) are noticeable. Types F&W and W&F had relatively higher p values. This is because people in these types are mainly Han people whose culture is much more adaptable than ethnic groups, including the Tibetan, the Yugur, etc. Because the Han culture is more resilient than ethnic groups, it usually shows a strong ability to withstand external disturbance. Compared with the Han people, the most obvious difference is the language. Ethnic groups have been influenced by their own languages since childhood, and their cultural values and behavior are deeply rooted. Moreover, their native language will make it difficult for them to contact the new cultural environments. In addition, the religious traditions and customs of some ethnic groups are more conservative and strict than those of the Han people, which also lead to some restrictions on adaptive behavior. Furthermore, due to the differences in ideology, economic level and educational resources, the education of ethnic groups is weaker than Han people, which leads to the relatively strong learning ability of the Han people. This was confirmed in the in-depth interviews.

Nevertheless, residents showed very low adaptability in the social ($p = 0.045$) and psychological domains ($p = 0.055$). This is consistent with our field survey. During the survey, many respondents complained of the poor infrastructure because there are restrictions on building new infrastructure. Moreover, people living inside the reserve mostly are old or little-educated, and they need social care more than others. They have been accustomed to everyday life in the reserve, and thus the adaptability to external culture or environmental change is low.

(2)   Adaptability analysis in different regions

As different functional zones (i.e., core zone, buffer zone, experimental zone and peripheral zone) in protected areas of China are managed differently, the households were classified into four groups in accordance with their living locations. After 2017, all residents in the core zone have moved into the experimental zone and peripheral zone. In this paper, the residents in the core zone refer to those whose pasture is located in the core zone. It can be seen in Figure 4 that, though with a little difference, all zones' residents' adaptability levels show similar result. However, their adaptability to the six domains differs (Figure 5).

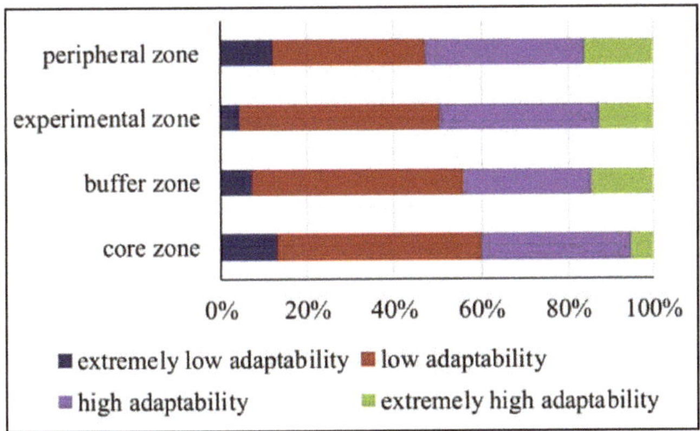

Figure 4. Adaptability of residents in different functional zones.

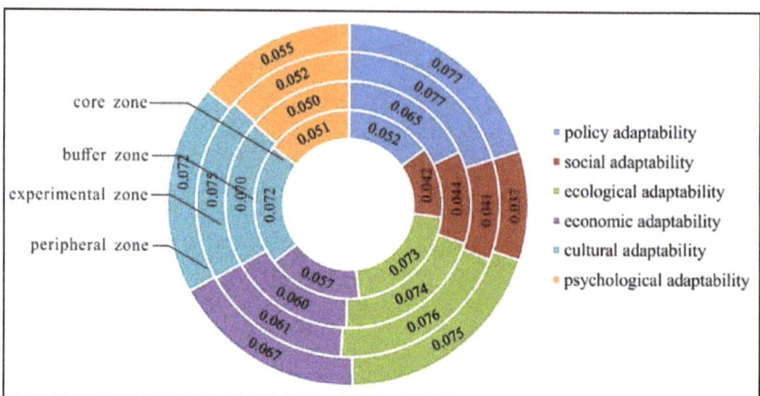

Figure 5. The adaptability of different regions in different dimensions.

As far as policy adaptability are concerned, the result presented that policy change affected the people living in the core zone the most, followed by the buffer zone, experimental zone and peripheral zone. This is easy to explain because the core zone and buffer zone of the QMNPP are located at relatively high elevations that are only fit for herding. People living in these areas are accustomed to herding and have no other job skills. Their ability to accept new things is low and they have difficulty changing their ways of livelihood. People from the experimental and peripheral zones, however, as their livelihood changed little under the new policy, showed high policy adaptability.

On the contrary, people's economic adaptability showed an opposite sequence, where peripheral zone > experimental zone > buffer zone > core zone. As a matter of fact, livelihood in the QMNPP is highly related to the elevation. People's livelihoods in the core zone and buffer zone highly depends on grazing animals, while people' livelihoods in the other two zones are much more diverse.

The *RAI* of residents of the other four domains, including social adaptability, ecological adaptability, cultural adaptability and psychological adaptability, presented little difference among the four functional zones. Only the *RAI* of residents in the periphery zone for psychological adaptability was slightly higher than that of other regions.

4.3. Impact Factors

To avoid collinearity among indicators, stepwise regression analysis was adopted to study the relationship between indicators and RAI. In the stepwise regression analysis, all indicators were considered independent and RAI-dependent. The regression equation model is as follows:

$$RAI = -1.531 + 0.502A1 + 0.566A2 + 0.223B1 + 0.180B3 + 0.175B4 + 0.465B5 + 0.341C1$$
$$+ 0.532C2 + 0.107D1 + 0.296D2 + 0.249D3 + 0.337D4 + 0.293E1 + 0.535E3$$
$$+ 0.281E5 + 0.546F1 + 0.473F2$$

The results ($R^2 = 0.849$, $F = 155.625$, $p = 0.000 < 0.05$) showed that the model was effective. It presented that the *RAI* in the QMNPP was affected by 17 factors (Table 4). To level them, these factors were divided into three grades: high influencing factors ($\beta \geq 0.200$), medium influencing factors ($0.100 < \beta < 0.200$) and low influencing factors ($\beta \leq 0.100$).

Table 4. Regression analysis results of influencing factors on residents' adaptability in the QMNPP.

Factors	Constant	Unstandardized Coefficients		Standardized Coefficients	t	p	VIF
		B	Standard Error	β			
		−1.531	0.095	-	−16.153	0.000 **	-
Policy knowledge	A1	0.502	0.090	0.139	5.563	0.000 **	1.935
Policy satisfaction	A2	0.566	0.098	0.154	5.796	0.000 **	2.213
Social network	B1	0.223	0.090	0.052	2.490	0.013 *	1.347
Household labor	B3	0.180	0.048	0.071	3.735	0.000 **	1.138
Education level	B4	0.175	0.057	0.057	3.061	0.002 **	1.090
Physical health	B5	0.465	0.043	0.204	10.825	0.000 **	1.105
Knowledge of SES	C1	0.341	0.096	0.099	3.557	0.000 **	2.410
Ecological awareness	C2	0.532	0.102	0.133	5.232	0.000 **	2.017
Satisfaction of income	D1	0.107	0.054	0.041	1.982	0.048 *	1.317
Livelihood diversity	D2	0.296	0.041	0.138	7.300	0.000 **	1.113
Household income	D3	0.249	0.060	0.079	4.118	0.000 **	1.133
Natural assets	D4	0.337	0.072	0.088	4.683	0.000 **	1.095
Ethnic costume	E1	0.293	0.068	0.110	4.311	0.000 **	2.029
Ethnic languages	E3	0.535	0.064	0.218	8.309	0.000 **	2.139
Traditional festival	E5	0.281	0.077	0.099	3.646	0.000 **	2.297
Acceptance of external culture	F1	0.546	0.072	0.189	7.585	0.000 **	1.932
Family resilience	F2	0.473	0.066	0.144	7.181	0.000 **	1.256

Dependent variable = *RAI*; D-W = 2.065; * $p < 0.05$ ** $p < 0.01$; $R^2 = 0.849$; $F(17,469) = 155.625$; $p = 0.000$.

It can be concluded that two factors (i.e., B5 and E3, namely physical health and ethnic languages, $\beta = 0.204$, $\beta = 0.218$, respectively) are high influencing factors (Table 4). From an individual's perspective, physical health reflects the vitality and sustainability of the social system. It determines people's development ability. People in poor health always show little adaptability to outside disturbance. In terms of ethnic language, people living in the reserve are mainly ethnic minorities. Most of them can only speak local languages, and thus there is an obstacle for them to adapt to the outside world.

A1, A2, C2, D2, E1, F1 and F2 are medium influencing factors. It is policy that determines living standards in the reserve. It further affects social, ecological, economic and psychological adaptability. Therefore, the higher the policy knowledge and satisfaction, the more stable the SES is. The indicator ecological awareness (C2, $\beta = 0.133$) reflects residents' willingness, attitude and behavior regarding ecological protection. Livelihood diversity (D2) directly affects household income, reflecting that the higher the livelihood diversity of residents, the better they can adapt to changing SES. The indicator acceptance of external culture (F1) is related to the adaptation of one's own ethnic culture and the diversity of livelihood. In general, the more ways of livelihood residents have, the more changes they are exposed to, and the higher their response to changes. Perception of household resilience (F2) can reflect residents' self-confidence in the face of environmental change, and directly affect their enthusiasm and initiative.

Other indicators are low influencing factors. Among these factors, social network (B1) is an important condition for the social system to be active. In the reserve, some residents live far from each other. Due to poor infrastructure, their social network is thus very poor. The structure and quality of family members are the basis of family adaptability. Household labor force (B3) and level of education (B4) directly affect the livelihood of residents, further affect the family income, and finally affect the development of the social system. Income satisfaction (D1) and annual household income (D3) affect residents' consumption elasticity and quality of life.

4.4. Adaptability Mechanism

Residents' behavior is not only affected by their own conditions, but also affected by the background environment of their region. However, the impacts of external environmental changes on residents are difficult to be measured quantitatively. The social-ecological system framework (SESF) proposed by Ostrom (2009) provides an ideal tool to analyze the adaptation mechanism [47]. SESF includes all of the resources involved in the interaction process of human society and ecosystem as well as the social, economic, political and ecological settings. There are four subsystems in the framework (Figure 6): Resource System (RS), Resource Units (RU), Governance System (GS) and Actors (A). The four subsystems interact and produce outcomes under the background of the social system and ecosystem, and emphasize actors' utilization behavior of resource units from the resource system. The interaction can effectively explain the adaptation mechanism of residents.

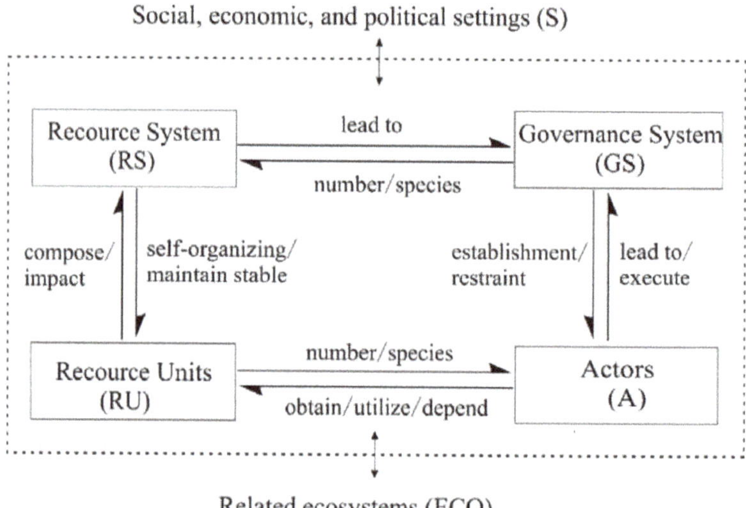

Figure 6. SES adaptability mechanism of residents (Ostrom, 2009, [47]).

The RUs in the national park are rich and diverse, including forest, grassland, wildlife, mineral resources, etc. The interaction and transformation of RUs can ensure the stability of the ecosystem through positive and negative feedback. The actors in this paper were local residents. In the early 1950s, logging was still permitted in the QMNPP. As a matter of fact, it once was treated as a major industry for local governments, which resulted in 50% loss of forest area. In the early 1980s, logging was forbidden. And from the year of 2000, a reforestation project was conducted under the Natural Forest Protection Project Policy. Then, forest area increased. In terms of grassland, here is a long history that nomads grazed in this region. As China's civil war ended in the end of the 1940s, the number of livestock increased along with a rapid human population increase, leading to severe grassland degradation. Upon this background, the GS played an important role by enacting policies to limit the number of livestock and prohibit grazing. Moreover, grassland was demarcated into plots for households rather than the previous situation in which people could graze anywhere as seasons changed.

For residents, most of these adaptive behaviors were passive and influenced by policy. When asked about the prohibition of grazing, one resident said, "We don't agree with the grazing prohibition completely. In the core zone, since grazing has been banned, grass has grown very thick, which is prone to fire in the winter. But being constrained by policy, we have no choice but to stop grazing." After grazing was forbidden, the ecological environment has improved, but the main source of income in the QMNPP was limited; "in order to maintain the normal operation of life, we have to choose to work in other places, or dig up the *Cordyceps sinensis*, or other ways to support the family." The livelihood has changed from only herding or farming to combined occupations. This change in behavior was spontaneously adopted by residents. After implementing the grazing prohibition, the GS found that the local social and economic development was restricted. Therefore, the GS alleviated social problems by providing more employment opportunities, such as forest and grassland rangers, and encouraging the development of franchising, etc. Moreover, on the premise of not destroying the stability of the grassland resources system, residents were allowed to graze properly to maintain the sustainable development of the social system. In addition, residents also adjust the ecological health of grassland system through rotational grazing and rest grazing, so as to achieve a win–win situation of ecological benefits and economic benefits.

From the analysis above, It can be concluded that a resource system maintains the balance of the resource units through self-organization. Resource units' interaction will affect the stability of the resource system in turn. The governance system plays a role in regulating the residents' behavior. Residents obtain resource units in the resource system to meet their livelihood needs, and are dependent on the natural resources. Residents have both positive and negative behaviors. Positive behaviors, such as returning farmland to forests and grasslands, will increase the amount of grassland resources, while negative behaviors such as overgrazing will reduce the amount. The negative impact will lead to the establishment of rules and policies for a governance system (environmental protection policies, such as a fodder–livestock balance system, eco-migration, etc.). The behavior of residents will be restricted by these policies. Moreover, there is a continuous and complex interaction between social systems and ecological systems. It is thus clear that the adaptive behavior of residents is caused by legal policies and survival needs. The legal policies are the basis of residents' adaptive behavior, and the survival needs are internal motivation.

5. Conclusions and Implications

As China's policy on protected areas will undoubtedly get stricter in the future, people's perception and adaptation should be considered; this is similar to the conclusions of Jia et al. (2022) [48]. For years, the reserve was managed just like other remote areas that are not reserves in China, except that logging and hunting were forbidden. Policy change could bring much adaption issues for local residents. The findings of this article are consistent with the research conclusions drawn by Yin et al. (2020) that the policy of ecological restoration is the external thrust of farmers' adaptive behavior choice [36]. This study showed that the comprehensive evaluation of residents' adaptability index in the QMNPP is at a low level. Residents of different livelihood sources and different regions had different adaptability levels. The high adaptability groups are mainly formed by the combination of high policy adaptability, ecological adaptability, economic adaptability, cultural adaptability and stable ecosystems. These groups are mainly residents living in the experimental zone and peripheral zone. However, residents' social, economic and psychological adaptability were low.

To improve the adaptability of residents and enhance management effectiveness, it is possible to suggest improving education quality in the reserve as more and more young people are going out to search for higher salaries and population is decreasing in the reserve. For the long run, people with higher education would not like to stay in the reserve any more, which will further decrease the population in it. Then, land rights should be gradually changed as people move out. Land located at the core zone and buffer zone should be purchased by the government from local residents who already work outside the reserve. However, the government should pay more attention to these who have lived for many generations in the core zone and buffer zone. Most of these people are elders and ethnic groups, who showed very low adaptability in nearly all domains. While most of them have been moved out, this paper suggest that infrastructures could be built at their new home locations. Moreover, pasture land could be set around their new homes to maintain their life style, though income from the pasture land would be very low compared to their previous pasture land.

The analysis framework of residents' adaptability in the QMNPP constructed in this study was intended to provide a tool for adaptability analysis in protected areas. This framework emphasizes the selection of adaptability indicators, which is more comprehensive than the existing indicator system, mainly taking into account various indicators of SESs, including policy, cultural, economic, ecological, social and psychological domains. In addition, this framework focuses more on the whole process of adaptability, deepens the study of the adaptability of residents from the perspective of system integration, and also provides a theoretical analysis framework for the research of residents' adaptability to global environmental change. Moreover, residents are the main actors of national parks or natural reserves. Adaptability is an important basis for the sustainable development of protected areas, and thus the results of this study can be used for reference for the community management of other protected areas. The index system needs to be verified and improved from the scientific and practical points of view. Additionally, only eight experts were included; however, we chose experts in different fields to score, which can reflect some problems to a certain extent. In future studies, we will increase the number of experts to make sure the data become more convincing. This study reflects the adaptability of residents in the QMNPP, but it is only one subsystem of the SESF (only an actor subsystem). Limited by data collection and processing methods, the current adaptability mechanism focuses on qualitative discussion. In further research, we will construct the SESF of Qilian Mountain, conduct further data collection, and use a combination of qualitative and quantitative methods to analyze the interaction between the four core subsystems of Qilian Mountain.

Author Contributions: Conceptualization, J.L. and Y.H.; methodology, J.L.; software, J.L.; validation, J.L., Y.H. and G.M.; formal analysis, J.L.; investigation, J.L., L.G., J.F. and Y.H.; resources, Y.H.; data curation, J.L.; writing—original draft preparation, J.L.; writing—review and editing, Y.H., G.M., and J.L.; visualization, J.L.; supervision, Y.H.; project administration, G.M. and Y.H.; funding acquisition, G.M. and Y.H. All authors have read and agreed to the published version of the manuscript.

Funding: This research was funded by the National Key Research and Development Program of China (Grant Number: 2019YFC0507405), the Fundamental Research Funds for the Central Universities (Grant Number: lzujbky-2020-4), and the Natural Science Foundation of Gansu province of China (Grant Number: 20JR10RA629).

Institutional Review Board Statement: Not applicable.

Informed Consent Statement: Not applicable.

Data Availability Statement: The data that support the findings of this study are available from the corresponding author upon reasonable request.

Acknowledgments: The authors sincerely thank the local managers in Zhangye and Wuwei who helped during the survey, and the anonymous reviewers for their valuable comments and suggestions to improve the quality of this paper.

Conflicts of Interest: The authors declare no conflict of interest.

References

1. Bennett, N.J.; Dearden, P. Why local people do not support conservation: Community perceptions of marine protected area livelihood impacts, governance and management in Thailand. *Mar. Policy* **2014**, *44*, 107–116. [CrossRef]
2. Ferraro, P.J.; Hanauer, M.M.; Sims, K.R.E. Conditions associated with protected area success in conservation and poverty reduction. *Proc. Natl. Acad. Sci. USA* **2011**, *108*, 13913–13918. [CrossRef] [PubMed]
3. Jones, N.; McGinlay, J.; Dimitrakopoulos, P.G. Improving social impact assessment of protected areas: A review of the literature and directions for future research. *Environ. Impact Assess. Rev.* **2017**, *64*, 1–7. [CrossRef]
4. Huang, Y.Z.; Fu, J.; Wang, W.R.; Li, J. Development of China's nature reserves over the past 60 years: An overview. *Land Use Policy* **2019**, *80*, 224–232. [CrossRef]

5. Visconti, P.; Butchart, S.H.M.; Brooks, T.M.; Langhammer, P.F.; Marnewick, D.; Vergara, S.; Yanosky, A.; Watson, J.E.M. Protected area targets post-2020. *Science* **2019**, *364*, 239. [CrossRef]

6. Bingham, H.C.; Bignoli, D.J.; Lewis, E.; MacSharry, B.; BurgessN, D.; Visconti, P.; Kingston, N. Sixty years of tracking conservation progress using the world database on protected areas. *Nat. Ecol. Evol.* **2019**, *3*, 737–743. [CrossRef]

7. Jones, P.J.S.; Qiu, W.; De Santo, E.M. Governing marine protected areas: Social-ecological resilience through institutional diversity. *Mar. Policy* **2013**, *41*, 5–13. [CrossRef]

8. Posner, S.M.; McKenzie, E.; Ricketts, T.H. Policy impacts of ecosystem services knowledge. *Proc. Natl. Acad. Sci. USA* **2016**, *113*, 1760–1765. [CrossRef]

9. Zhang, Y.; Xiao, X.; Cao, R.; Zheng, C.; Guo, Y.; Gong, W.; We, Z. How important is community participation to eco-environmental conservation in protected areas? From the perspective of predicting locals' pro-environmental behaviours. *Sci. Total Environ.* **2020**, *739*, 139889. [CrossRef]

10. Booth, J.E.; Gaston, K.J.; Armsworth, P.R. Public understanding of protected area designation. *Biol. Conserv.* **2009**, *142*, 3196–3200. [CrossRef]

11. Bennett, N.J.; Roth, R.; Klain, S.C.; Chan, K.; Christie, P.; Clark, D.A.; Cullman, G.; Curran, D.; Durbin, T.J.; Epstein, G.; et al. Conservation social science: Understanding and integrating human dimensions to improve conservation. *Biol. Conserv.* **2017**, *205*, 93–108. [CrossRef]

12. Garnett, S.T.; Burgess, N.D.; Fa, J.E.; Fernández-Llamazares, Á.; Molnár, Z.; Robinson, C.J.; Watson, J.E.M.; Zander, K.K.; Austin, B.; Brondizio, E.S.; et al. A spatial overview of the global importance of Indigenous lands for conservation. *Nat. Sustain.* **2018**, *1*, 369–374. [CrossRef]

13. Rodríguez-Rodríguez, D.; Larrubia, R.; Sinoga, J.D. Are protected areas good for the human species? Effects of protected areas on rural depopulation in Spain. *Sci. Total Environ.* **2021**, *763*, 144399. [CrossRef] [PubMed]

14. Inoue, K.; Takei, Y. Diverse adaptability in *Oryzias* species to high environmental salinity. *Zool. Sci.* **2002**, *19*, 727–734. [CrossRef]

15. Jung, K.; Kalko, E.K.V. Adaptability and vulnerability of high flying Neotropical aerial insectivorous bats to urbanization. *Divers. Distrib.* **2011**, *17*, 262–274. [CrossRef]

16. Worland, A.J. The influence of flowering time genes on environmental adaptability in European wheats. *Euphytica* **1996**, *89*, 49–57. [CrossRef]

17. May, C.K. Coastal community resilience and power in the United States: A comparative analysis of adaptability in North Carolina and Louisiana. *Environ. Manag.* **2021**, *68*, 100–116. [CrossRef]

18. Sultana, H.; Ali, N.; Iqbal, M.M.; Khan, A.M. Vulnerability and adaptability of wheat production in different climatic zones of Pakistan under climate change scenarios. *Clim. Chang.* **2009**, *94*, 123–142. [CrossRef]

19. Grothmann, T.; Patt, A. Adaptive capacity and human cognition: The process of individual adaptation to climate change. *Glob. Environ. Change* **2005**, *15*, 199–213. [CrossRef]

20. Mitter, H.; Larcher, M.; Schonhart, M.; Stottinger, M.; Schmid, E. Exploring farmers' climate change perceptions and adaptation intentions: Empirical evidence from Austria. *Environ. Manag.* **2019**, *63*, 804–821. [CrossRef]

21. Folke, C.; Carpenter, S.R.; Walker, B.; Scheffer, M.; Chapin, T.; Rockstrom, J. Resilience thinking: Integrating resilience, adaptability and transformability. *Ecol. Soc.* **2010**, *15*, 9. [CrossRef]

22. Walker, B.; Holling, C.S.; Carpenter, S.R.; Kinzig, A. Resilience, adaptability and transformability in social-ecological systems. *Ecol. Soc.* **2004**, *9*, 9. [CrossRef]

23. Chelleri, L.; Minucci, G.; Skrimizea, E. Does community resilience decrease social-ecological vulnerability? Adaptation pathways trade-off in the Bolivian Altiplano. *Reg. Environ. Chang.* **2016**, *16*, 2229–2241. [CrossRef]

24. Cottrell, S.; Mattor, K.M.; Morris, J.L.; Fettig, C.J.; McGrady, P.; Maguire, D.; James, P.M.A.; Clear, J.; Wurtzebach, Z.; Wei, Y.; et al. Adaptive capacity in social-ecological systems: A framework for addressing bark beetle disturbances in natural resource management. *Sustain. Sci.* **2020**, *15*, 555–567. [CrossRef]

25. Smit, B.; Burton, L.; Klein, R.J.T.; Street, R. The science of adaptation: A framework for assessment. *Mitig. Adapt. Strateg. Global Chang.* **1999**, *4*, 199–213. [CrossRef]

26. Smit, B.; Wandel, J. Adaptation, adaptive capacity and vulnerability. *Glob. Environ. Chang.* **2006**, *16*, 282–292. [CrossRef]

27. Stotten, R.; Ambrosi, L.; Tasser, E.; Leitinger, G. Social-ecological resilience in remote mountain communities: Toward a novel framework for an interdisciplinary investigation. *Ecol. Soc.* **2021**, *26*, 19. [CrossRef]

28. Burnham, M.; Ma, Z. Climate change adaptation: Factors influencing Chinese smallholder farmers' perceived self-efficacy and adaptation intent. *Reg. Environ. Chang.* **2017**, *17*, 171–186. [CrossRef]

29. Pandey, V.P.; Babel, M.S.; Shrestha, S.; Kazama, F. A framework to assess adaptive capacity of the water resources system in Nepalese river basins. *Ecol. Indic.* **2011**, *11*, 480–488. [CrossRef]

30. Lioubimtseva, E.; Henebry, G.M. Climate and environmental change in arid Central Asia: Impacts, vulnerability, and adaptations. *J. Arid. Environ.* **2009**, *73*, 963–977. [CrossRef]

31. Hoque, S.F.; Quinn, C.; Sallu, S. Differential livelihood adaptation to social-ecological change in coastal Bangladesh. *Reg. Environ. Chang.* **2018**, *18*, 451–463. [CrossRef]

32. Singh, R.K.; Zander, K.K.; Kumar, S.; Singh, A.; Sheoran, P.; Kumar, A.; Hussain, S.M.; Riba, T.; Rallen, O.; Lego, Y.J.; et al. Perceptions of climate variability and livelihood adaptations relating to gender and wealth among the Adi community of the Eastern Indian Himalayas. *Appl. Geogr.* **2017**, *86*, 41–52. [CrossRef]

33. Fussel, H.M. Adaptation planning for climate change: Concepts, assessment approaches, and key lessons. *Sustain. Sci.* **2007**, *2*, 265–275. [CrossRef]

34. Cumming, G.S.; Allen, C.R.; Ban, N.C.; Biggs, D.; Biggs, H.C.; Cumming, D.H.M.; De Vos, A.; Epstein, G.; Etienne, M.; Maciejewski, K.; et al. Understanding protected area resilience: A multi-scale, social-ecological approach. *Ecol. Appl.* **2015**, *25*, 299–319. [CrossRef] [PubMed]

35. Cumming, G.S. Theoretical frameworks for the analysis of social-ecological systems. In *Social-Ecological Systems in Transition*; Springer: Tokyo, Japan, 2014; pp. 3–24.

36. Yin, S.; Chen, J.; Yang, X.J. Adaptive behavior of farming household and influential mechanism in the background of social-ecological system reconstruction. *Hum. Geogr.* **2020**, *35*, 112–121.

37. Zhang, Q.; Zhao, X.Y.; Tang, H.P. Vulnerability of communities to climate change: Application of the livelihood vulnerability index to an environmentally sensitive region of China. *Clim. Dev.* **2019**, *11*, 525–542. [CrossRef]

38. Ahmed, N.; Allison, E.H.; Muir, J.F. Using the sustainable livelihoods framework to identify constraints and opportunities to the development of freshwater prawn farming in southwest Bangladesh. *J. World Aquacult. Soc.* **2008**, *39*, 598–611. [CrossRef]

39. Kelman, I.; Mather, T.A. Living with volcanoes: The sustainable livelihoods approach for volcano-related opportunities. *J. Volcanol. Geotherm. Res.* **2008**, *172*, 189–198. [CrossRef]

40. Nikolakis, W.; Grafton, R.Q. Putting Indigenous water rights to work: The sustainable livelihoods framework as a lens for remote development. *Community Dev.* **2015**, *46*, 149–163. [CrossRef]

41. You, H.Y.; Zhang, X.L. Sustainable livelihoods and rural sustainability in China: Ecologically secure, economically efficient or socially equitable? *Resour. Conserv. Recycl.* **2017**, *120*, 1–13. [CrossRef]

42. Wu, J.L.; Zhou, C.S.; Xie, W.H. The influencing factors and evaluation of farmer's adaptability towards rural tourism in traditional village: Based on the survey of 6 villages in Xiangxi Prefecture, Hunan. *Sci. Geogr. Sin.* **2018**, *38*, 755–763.

43. Guo, X.L.; Zhou, L.H.; Chen, Y.; Gu, M.H.; Zhao, M.M. Analysis on the adaptability of farmers to ecological and environmental changes and the driving factors in the Inner Mongolia Autonomous region. *Acta Ecol. Sin.* **2018**, *38*, 7629–7637.

44. Zhou, S.X.; Tian, Y.P.; Liu, L.F. Adaptability of agricultural ecosystems in the hilly areas in Southern China: A case study in Hengyang Basin. *Acta Ecol. Sin.* **2015**, *35*, 1991–2002.

45. Wang, Y.; Zhou, L.H.; Yang, G.J.; Guo, R.; Xia, C.Z.; Liu, Y. Performance and obstacle tracking to natural forest resource protection project: A rangers' case of Qilian Mountain, China. *Int. J. Environ. Res. Public Health* **2020**, *17*, 19. [CrossRef]

46. Dong, X.; Wang, L.; Cao, L.; Yang, Z.B.; Ma, J.F. Farmers' participation in ecological compensation willingness and behavior in Qilian Mountain area. *J. Arid Land Resour. Environ.* **2020**, *34*, 74–79.

47. Ostrom, E. A general framework for analyzing sustainability of social-Ecological systems. *Science* **2009**, *325*, 419–422. [CrossRef]

48. Jia, L.Q.; Wei, J.Q.; Wang, Z.B. The intention of community participation in the Qilian Mountain National Park Policy Pilot. *Land* **2022**, *11*, 170. [CrossRef]

 *land*

Article

Understanding Residents' Perceptions of the Ecosystem to Improve Park–People Relationships in Wuyishan National Park, China

Siyuan He [1],* and Yang Su [2]

[1]   Institute of Geographic Sciences and Natural Resources Research, Chinese Academy of Sciences, Beijing 100101, China

[2]   Management World Journal Press, Development Research Center of the State Council, Beijing 100026, China; suyang1@drc.gov.cn

*   Correspondence: hesy@ignsrr.ac.cn

Abstract: A healthy park–people relationship depends essentially on the fair and sustainable maintenance of rural livelihood. When a protected area is designated, rural people may face restrictions on access to land and resource use. In Wuyishan of China, we analyzed the role of traditional tea cultivation during consistent protected area management to find ways to maintain the stability of this social-ecological system in the new national park era. Based on the social-ecological system meaning perception, we used an intensive social survey to investigate residents' perception of the ecosystem in terms of tea cultivation and its interaction with conservation policies. Results showed that tea cultivation brought major household income and was associated with multiple cultural services. Protected area management affected land use, and conservation outcomes were more obvious to farmers than economic and social ones. We argue that the multi-functionality of the forest-tea system has the potential to benefit both the local people and the public through conservation-compatible activities at three levels: to regulate biophysical elements in the land plot, to link production and market at the mountain level, and to secure tenure and encourage community participation at the landscape level. This knowledge co-production approach revealed that to avoid a negative park–people relationship, traditional knowledge and people's right to benefit must be respected.

Keywords: national park; social-ecological system; ecosystem services; tea cultivation; protected area management

Citation: He, S.; Su, Y. Understanding Residents' Perceptions of the Ecosystem to Improve Park–People Relationships in Wuyishan National Park, China. *Land* **2022**, *11*, 532. https://doi.org/10.3390/land11040532

Academic Editors: Rui Yang, Yue Cao, Steve Carver and Le Yu

Received: 8 March 2022
Accepted: 2 April 2022
Published: 6 April 2022

Publisher's Note: MDPI stays neutral with regard to jurisdictional claims in published maps and institutional affiliations.

1. Introduction

In the management of a national park (NP) and other protected areas (PAs), a healthy park–people relationship depends essentially on the fair and sustainable maintenance of rural livelihood [1]. Research abounds in the park–people relation that neglect of local culture and limit to access to resources significantly affect the local community's satisfaction with park management and the conservation outcomes [2–5].

A healthy park–people relationship is especially important to current protected area (PA) management in China, where a new National Park system is under construction starting from pilots. China's national parks are similar to the National Park in the International Union for Conservation of Nature (IUCN)'s PA category [6] but the system aims to reform the previously multi-headed management of various types of PAs by optimizing spatial planning and unifying management agencies. Since 2015, 10 piloting parks have integrated separate PAs for ecological integrity while forming a unified management unit. The integration and unification bring in the new institution that affects both the geographical location of communities relative to national parks, and the resource use of local people [7]. Thus, the local community's livelihood is now a focus of NP management in China because of the reflection on the somewhat fortress approach to conservation [7]. Studies of many nature reserves, which are the main area-based conservation measures since 1956, provided evidence that the perception of cost-benefit of rural livelihoods strongly affected local people's acceptance of PAs, hence the management outcomes. Alternative livelihoods, job opportunities, eco-compensations, infrastructures, and social welfare were typical benefits rural people pursued [8–11], while restrictions on resources, energy, and productive activities, as well as human-wildlife conflicts, were frequently mentioned as major costs [12–14]. In addition, respect for local culture [10], the equity of benefits and compensation [15], and the degree of community participation [9] significantly affected the PA–people relationship. Therefore, conservation decision-making is oriented to benefit rural livelihoods through NP designation and management [16].

Rural livelihood is realized through the appreciation of multiple ecosystem services directly, or indirectly to benefit the local communities [2,17,18]. From agricultural systems adjacent to the protected area, local communities could provide ecosystem services beyond basic products to balance multiple benefits of stakeholders if they had a good relationship with the PA [19]. However, the disproportionate costs of livelihood and benefits from conservation may raise the counter effect of resource exploitation for livelihood and lead to the degradation of ecosystems and their capability of providing various ecosystem services [20,21]. Therefore, a healthy park–people relationship matters to the stability of the social-ecological system. Considering the NP piloting and the ongoing designation of official national parks, it is critical to regulate local people's activities regarding conservation goals without depriving them of their reasonable demand for ecosystem services, so that the desired park–people relationship can be built to maintain the resilience of the social-ecological system (SES) [22,23].

Perceptions and attitudes are an important approach to assessing the performance of conservation practices because they can affect people's conservation behaviors [24]. However, to assess the park–people relationship, most studies from the perspective of perceptions and attitudes of local stakeholders focused on revealing the multi-faceted costs and benefits of established PAs which have been operating for years, very few were concerned with the perceptions of a PA in designation, which could differ from the perception of a long-established PA because people had very limited experience of the new PA [25]. Furthermore, park–people relationships were seldom explored from the perspective of ecosystem services although trade-offs of certain goods and services among multiple stakeholders can define the park–people relationship as the main sources of costs and benefits to local people. This perspective on an NP in construction provides a solid theoretical base to broaden the scope of perceptions from focusing on the piecemeal cost-benefit factors associated with NP management to concerning the holistic human-nature interactions in the SES. All the various costs and benefits perceptions ultimately stem from rural individuals' perceptions of their territory and of its conservation history [26]. These perceptions can foresee their activities and provide a starting point for designing new rules to secure rural livelihoods in the NP era. Producing this knowledge with local people is also meaningful when stakeholder participation is officially promoted for NP management.

Therefore, the basic assumption is that local communities' perception of the meaning of their biophysical environment allows understanding and interpreting the human behaviors, because, during long-term interaction with nature, local communities have the ability to identify dynamic changes and multiple driving factors. Accordingly, this study aims to explore the meanings of tea cultivation to the rural people, how they interact with conservation, and how to adapt tea cultivation to conservation.

The paper focuses on three aspects of residents' perception of the ecosystem: (1) the role of tea as a major income source, (2) the role of tea beyond economic significance, and (3) previous protected area management and its impacts on the tea cultivation. By understanding the meaning of the human–nature interactions through tea cultivation, the paper provides a multi-scale NP management approach for a healthy park–people relationship. The study complements studies of the park–people relationship from a prescient perspective and may serve as an example to facilitate community participation in PA establishment in developing countries.

2. Materials and Methods

2.1. Study Area: Wuyishan National Park as a Social-Ecological System

The Wuyishan National Park (WNP) is located in Southeast China as a part of the Wuyi Mountains (Figure 1). This NP is integrated from mainly three PAs: the Wuyishan National Nature Reserve (NNR) to the west, the Nine-Bend Stream Ecological Protection Area (NEPA) in the center, and the Wuyishan National Scenic Area (NSA) to the east, with a total area of 982.59 km^2 after spatial optimization in its piloting period when the research was conducted.

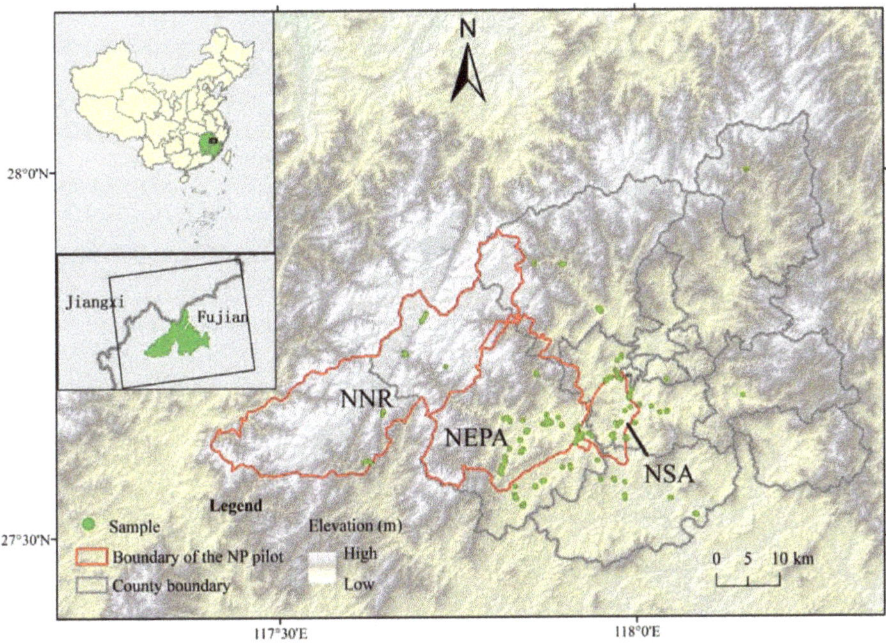

Figure 1. The location and composition of the Wuyishan national park pilot initiated in 2015.

Archaeological remains suggested that people settled in the Wuyi Mountains as early as 4000 years ago. Wuyi Mountains not only have preserved the abundant humid subtropical forest, provided suitable habitats to endangered species such as *Liriodendron chinense* and *Halesia macgregorii*, but the biophysical and geological conditions also have nurtured tea bushes dating back as least to the Tang Dynasty (618–907 AD). People's productive interaction with nature has transformed the landscape and created rich cultural landscapes such as the forest-tea system. This is a typical SES in which people adapted to the natural conditions with their traditional wisdom to keep a delicate balance with other non-human life forms [27]. Until now, this rural landscape still generates multiple ecosystem services that benefit not only locals but also domestic and international beneficiaries. Therefore, the role of humans in the past and present cannot be ignored in the study of the structure and function of the contemporary forest. In the 16th century, farmers were able to build terraces for tea cultivation with a system of dykes and drains [28]. There was also a synergy between tea bushes and natural forest [29], but the recent expansion of tea bushes and the intensification of land use can lead to forest degradation.

There are about 3000 inhabitants inside the NP and another 20,000 settled within 2 km of the park boundary. Most of the tea farmers live along the upstream zone of the Nine-Bend Stream. Rural people keep transforming the landscape mainly through forest use. In the past three decades, rural households and individuals have responded to institutional change actively, especially to reform of collective forestry rights and the management of multiple PAs during the past 40 years, in a good or bad way. Under the forestry rights reform, the forest land was treated as "a bundle of rights" when the transaction and operation rights were under the control of individual households with a clarification of resource boundaries in the collective land tenure. This reform aims at stabilizing land tenure and improving forestry efficiency, however, the flexibility without a full understanding of ecosystem multifunction has led to monoculture plantation and forest degradation [30].

Meanwhile, a series of actions were also taken to protect the biodiversity and landscape in the form of the national nature reserve (1979) and the national scenic area (1982), which are generally prioritized for preservation and tourism, respectively. Wuyishan further entered the list of the UNESCO World Heritage Sites as a mixed site because of its cultural value, natural beauty, and biodiversity value in 1999. More land-use policies were issued in the new millennium to regulate human–forestland relationships alongside the designation of PAs, such as the ban on commercial logging in 2008, and the prohibition against the expansion of tea orchards in 2011. In addition, more forests were designated ecological forests. In 2015, the NP pilot was launched and the spatial integration leads to the inclusion of more rural landscapes adjacent to the boundary of the pilot, making the role of people in the social-ecological system more prominent when new rules and regulations occur.

2.2. Conceptual Framework

Local PA–people relationships are largely defined by perceptions and attitudes of communities toward PAs. With the growing recognition of community participation in PA management and the spreading of socially inclusive conservation approaches, the governance of PAs should recognize the role of conservation culture, knowledge, and agreement of local communities [31]. Thus, PA governance is now oriented in an adaptive way that enables NPs to adapt to spatial and temporal changes in social-ecological systems and establish and maintain desired park–people relations for ecological, social, and economic outcomes [32,33].

By understanding the national park as an SES, the literature suggests the provision of ecosystem services is seldom solely natural, but part of an SES in which resource users interact with the environment to shape both the ecosystem and their culture [34,35]. According to the SES meaning perception theory, good governance of NPs secures multiple ecosystem services required by competitive stakeholders by converging their different perceptions of the same ecosystem to the largest extent [36].

This theory pointed out that, people allocated meaning to many aspects of the ecosystem, which will lead to a perception of the ecosystem service or material as a benefit (positive meaning) or a perception that benefits are reduced (negative meaning). This perception will significantly affect their activities. For example, if local people perceive forests as a commodity, they may practice timber harvesting, but if perceived as natural beauty by others, recreational activities may be preferred [36]. Therefore, balancing multiple benefits of stakeholders will eventually have impacts on the provision of ecosystem services to many stakeholders in general, and local people's livelihood in specific. Furthermore, the perception of ecosystem services is highly context-dependent; any change in the biophysical or socio-economic conditions may lead to a change in resource users' behaviors because people modify their behavior based on their knowledge and expectation concerning future changes [37]. This is obvious when new conservation policies and practices are applied in rural areas, where local people face uncertainty in livelihood and may take action to secure their benefit, thus affecting park–people relations.

Therefore, upon this perspective of perception of SES, this study analyzes the fundamental perceptions of the forest ecosystem in which the local tea farmers dwell, concerning both the meanings and the changing context. By using the SES meaning perception theory, our study demonstrates how the local residents' perception of SES contributes to a positive park–people relationship. We argue that the multi-functionality of the forest-tea system has the potential to benefit both the local people and the public by exploring its cultural meaning. The negative park–people relationship may persist if traditional knowledge and the right to benefit are still neglected. In this paper, we show how local tea farmers perceived the ecosystem during the ever-changing types of PAs and their management, and how this knowledge co-production helps build a good park–people relationship in a new NP.

2.3. Survey and Data Analysis

The research uses an intensive semi-structured interview with a sample of tea farmers living in and adjacent to the national park (Figure 1). The interview was widely used as an interactive method to gain information on specific conservation issues and understand the knowledge, values, beliefs, or decision-making process of stakeholders [38]. The semi-structured interview is more flexible for researchers to ask additional questions besides standard questions for complex issues in the studies of conservation science-policy interfaces [39]. It is also important for Wuyishan where local people are less represented in conservation decision-making. The interview was organized in different sections focusing on the following topics:

(1) General data of respondents, including personal characteristics such as age, gender, householder status, education, length of residency, household size, and production characteristics such as tea cultivation experience, labor conditions, tea plot size, tea plot umbers, distance to tea orchards, annual household income.
(2) Tea farmers' perception of the critical factors affecting the production of tea, in terms of land tenure, market competition, and natural conditions.
(3) Tea farmers' preference and assessment of the importance of ecosystem services associated with the forest and how they value tea cultivation during the construction of an NP.
(4) Tea farmers' perception of the efficiency of protected area management in terms of ecological, economic, and public welfare outcomes, considering potential land use conflict between community livelihood and ecological protection targets.

The questionnaire contained open-ended questions for topics 2, 3, and 4 to form a major part of a semi-structured interview and close-ended questions for topics 1 to 4 as a structured social survey. The respondents could answer "yes", "no", or "I don't know" to the closed questions; they could choose items all fitting their conditions from some multiple-choice questions; they could explain or provide examples to the open questions in detail. The average duration of each interview was about 40 min.

The interviews were conducted face-to-face by a team of trained volunteers from 18 to 31 July 2016. In total, 221 tea farmers participated in the study, with ages ranging between 21 and 75. This sample was a subset of a larger sample of local residents and a stratified random sampling technique was used and explained in detail in He et al., 2018 [39]. Most respondents answered all the questions. Only a few questions were left unanswered by very few people because they forgot certain numbers, did not like to comment, or had no idea of certain information. This does not affect qualitative analysis due to information saturation. In the quantitative analysis, numbers of valid data (e.g., $n = 218$) were provided to show how many people did not respond to certain questions.

Qualitative data collected from interview questions were analyzed by using the key information, and following a grounded theory approach by using the open coding and axial coding to categorize and describe tea farmers' perception of the meaning of ecosystem services and tea cultivation under PAs. Open coding is the process of decomposing, comparing, conceptualizing, and categorizing textual material and then recombining and manipulating the codes in new ways [40]. During the open coding process, the raw data from the open-ended questions were labeled to form concepts that reflect the multiple meanings of tea cultivation and the relationship with PAs. Similar concepts were further combined into categories that scaled up scattered concepts to cover the major research questions that the research aim to answer, including basically economic meanings, social-cultural meanings, park–people synergies and conflicts, protected area management outcomes, etc., laying the foundation for axial coding and provide information for the Results section. The main purpose of axial coding is to discover and establish relationships between concepts to characterize the linkages between different categories [40]. With multiple concepts and categories, the relationship between multiple meanings of tea cultivation and the current park–people interactions was built from a perspective of ecosystem service trade-off. This basically provided information for the Discussion section.

Quantitative data from the survey were entered and analyzed using Statistical Package for the Social Sciences (Version 21) and the significance value is 0.05 if not specifically mentioned. The data were analyzed in terms of descriptive statistics for general data and perceptions of the management of tea and the PAs. The non-parametric correlation was used to reveal the relationship among those general data that reflect farmers' productive behaviors, and categorical regression was used to detect the impact of variables concerning natural and human assets on the household income based on previous research [39]. For example, high tea leaf yield and more labor in a family can bring higher income [41] (Section 3.1).

For the assessment of the importance of the ecosystem services, each respondent was provided with a list of 15 ecosystem services with illustrations to assist in understanding [39]. They were asked to select and ranked five ecosystem services from the list. Ecosystem services with ranks from one to five were given a score from six to two, respectively, and those not selected were given a score of one. An average weighted score of each specific ecosystem service was calculated according to all the respondents using the equation: $\sum(S_i \times f_i)$, where S_i was the given score of a specific ecosystem service by each respondent and f_i was the frequency of respondents making this choice, i was the six selecting results. $i = 1$ to 5 when the ecosystem service was ranked from the first to the fifth, and $i = 6$ when it was not selected.

A multiple correspondence analysis (MCA) was used to explore the synergy and trade-off of the social preference for ecosystem services among tea farmers. MCA is a descriptive method that reveals patterning in a complex dataset and is widely used in studies where a large amount of qualitative data is collected [42]. Each of the 15 ecosystem services was a variable with two categories of selecting this ecosystem service and not selecting it based on the ranking procedure, making it a total dimension of 15 (30 minus 15). The calculated total inertia was 1 (the maximum number of MCA dimensions ($n = 15$) divided by the number of variables ($n = 15$)). A solution was explored with two MCA dimensions: the first accounting for 12.3% (0.123/1) of the variance and the second for 11.2% (0.112/1), yielding a total variance of 23.4% (0.234/1). Discrimination measures and a joint plot of category points were obtained. In the plot, the coordinates of each category ((non-) selection of an ecosystem service) on each dimension were displayed to determine synergy and trade-off patterns of ecosystem services as perceived by tea farmers. The distance from an object to the origin is the reflection of the variation from the "average" pattern (the most frequent category for each variable). Thus, ecosystem services that were perceived almost unanimously as important or not lie near the origin, and vice versa.

2.4. Sample Description

The average age of the 221 respondents was 49, and 62% of them were between the age of 40 to 59 as a major labor force. Males and females represent 84% and 16% of the sample, respectively. Furthermore, 71% of the sample consisted of householders and the ratios of males and females were 97% and 3%, respectively (Table 1). Most of the respondents (47%) held a secondary school degree while 30% had finished primary education at best.

Table 1. Description of the tea farmers involved in the social survey.

Factor	Category	(%)	Factor	Category	(%)
Gender	F	16		<5	10
	M	84	Tea cultivation experience (years)	6–10	12
	18–24	1		11–20	26
	25–39	17		21–30	31
Age				>30	21
	40–59	62		<30	13
	>60	20	Labor proportion (%)	30–40	15
Householder	Yes (F, M)	71 (3, 97)		40–50	12
	No (F, M)	29 (49, 51)		>50	60
	Primary and under	30		1	10
Education	Junior	47	Land Plots	2–5	46
	Senior	17		5–10	36
	College and above	6		>10	8
	<30	4		<5	6
Length of local residency (years)	30–40	14		5–10	11
	40–50	37	Land area (mu)	10–20	23
	50–60	25		20–40	38
	>60	20		40–60	6
	1–3	17		>60	17
	4–6	65		<1	3
Household size	7–9	13	Walking distance (km)	1–5	69
	>10	5		5–10	19
				>10	9
				<5	9
			Annual household income (10,000 RMB)	5–10	17
				10–50	52
				50–100	11
				>100	12

The average family size was five people. Most of the respondents (83%) had a family size of at least four people. The average ratio of the labor force in a family was 55%, and the ratio was more than 50% for 60% of the respondents. The median length of engaging with tea cultivation in the household was 20 years ($n = 218$) with a range of one year to 60 years. The median length of local residency was 47 years ($n = 217$) with a range of five to 75 years. Most respondents (52%) claimed annual household income as between 100,000 and 500,000 yuan (about 16,000 to 80,000 USD) ($n = 218$).

For the ownership of tea orchards, the median number of land plots for a household was four plots, ranging from one plot to 60 plots ($n = 202$). The median number of the total area of tea plots of each household was about 20 mu (1.33 ha) ranging from one to 400 mu (0.067 ha to 26.67 ha) ($n = 216$). The longest walking distance from home to attend to the tea bushes was 20 km and the median distance was 2.8 km. Most of the respondents (72%) had tea plots within a walking distance of 5 km.

3. Results

3.1. The Importance of Tea Cultivation as Economic Benefits

Tea was essential for livelihood. Considering the entire sample, households who had a longer residency time also had a longer engagement with tea cultivation ($p < 0.01$). In addition, households that owned more plots tended to have a larger total area of land ($p < 0.01$, $n = 202$). Furthermore, households who had more plots and a larger area of tea orchard traveled longer to their land ($p < 0.05$, $n = 199$; $p < 0.05$, $n = 213$, respectively). Families with a larger scale and higher ratio of workforce tended to own a larger area of land ($p < 0.05$, $n = 217$; $p < 0.05$, $n = 217$, respectively).

For most respondents (97%), tea was mainly for sale on the market for income. Here, 46.6% of the respondents reported that they focused on the national market and 33.5% on the local market. Tea farmers sold raw tea leaf, coarse tea, or refined tea with a certain proportion according to market conditions and their capacity. According to the respondents, one unit of refined tea was produced from two units of coarse tea dried from 10 units of raw tea leaf in the Wuyishan area. The market value of coarse tea and processed tea varied a lot due partly to the geographic location of tea orchards. The unit yield of raw tea leaves ranged between 100 to 750 kg/mu. Raw tea leaf was priced between 6 and 20 yuan/kg (0.96 to 3.2 USD/kg) and refined tea between 60 and 600 yuan/kg (9.6 to 96 USD/kg).

As tea cultivation was claimed, the most important income source (90% above), the categorical regression was used to reveal how the level of income depends on the multiple socio-economic factors specific to the tea farmers (Table 2). It was found that the annual household income level has been significantly affected by the total area of tea orchards, family size, the percentage of the workforce, the number of tea plots, and the distance of the farthest land plot, all indicating a positive relation. Therefore, the income was basically affected by land and labor.

Table 2. Impact on the household income from the analysis of categorical regression ($R^2 = 0.611$, $n = 194$).

Dependent	Independent	Beta
	Residency time	−0.076
	Time of engagement with tea	0.068
	Family size	0.157 [b]
Annual Household Income	The ratio of the workforce	0.159 [b]
	Number of plots	0.19 [b]
	Land area	0.553 [b]
	Longest walking distance	0.129 [a]
	Education	0.100

[a] $p \leq 0.05$; [b] $p \leq 0.01$.

Respondents' perceptions of tea cultivation have revealed more details of their income dynamics and critical impacting factors besides those social-economic features.

They perceived income change differently (Figure 2a). Of the respondents, 44% perceived an increase in net income since their engagement with tea plantations, but 38% claimed a continuous market fluctuation. Some tea farmers who were engaged with tea cultivation for more than 30 years had identified several critical timing in the fluctuation of market value. They described a general increasing trend over the last three decades and ascribed it to the confirmation, registration, and issuance of certificates on the right to the contracted management of forested land; while a recent (ca. 2015) decreasing trend was attributed to the increasing cost of labor by tea farmers.

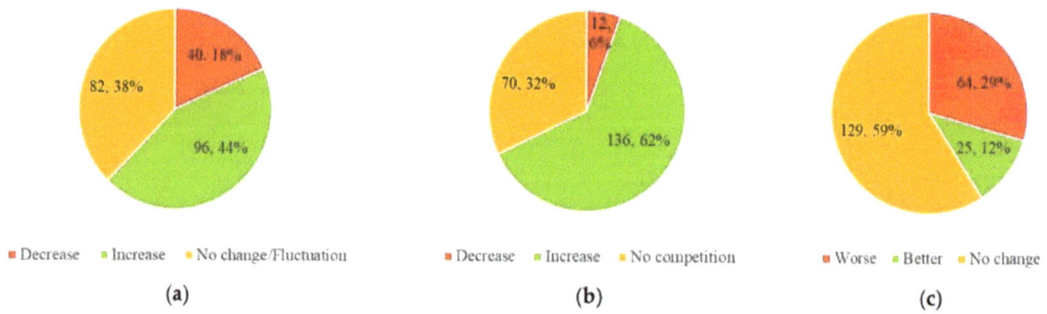

Figure 2. Tea farmers' perception of income (**a**), market competition (**b**), and environmental conditions (**c**) concerning tea cultivation and production. Illustrated by the number of respondents and the percentage.

For the intensity of market competition, more than 60% of respondents felt increasing pressure, compared to 6% who thought the opposite (Figure 2b). Interestingly, 32% of respondents reported no pressure as most of them said, "We had no feeling of competition at all because we only focused on what we can achieve". They claimed to have a stable or even fixed source of customers and their land, that is, *shan chang* (literally "the mountain") in specific geographical locations secured the tea quality. For those who had an experience of intensified competition, they ascribed it to several causes, such as unfair competition with fake commodities, farmers shifting from rice to tea planting, forcing the price down by buyers, no brand or green certificate for small-scale farmers, etc.

For the environmental conditions (Figure 2c), 59% of the respondents did not think there was a significant change regarding tea cultivation, especially regarding soil and weather conditions; but not many thought that the climate was getting any better (12%) either. Those who felt a change, especially a negative one, attributed it to climate change and human disturbance. They claimed to have experienced a higher frequency of heavy rain, drought, and spring frost, earlier warming, and more snowing days, all leading to the decrease in tea leaf yield. However, they also expressed satisfaction with the improvement of soil and water conditions due to human intervention such as weed control, fertilization, forestation, and water conservation. Furthermore, respondents mentioned that important environmental conditions for tea cultivation, including rock, soil, topography, and forest, cannot be separated but form an integrated system, the *shan chang*, which was suitable for tea bushes to gain sunshine and water.

Ownership of *shan chang* was very stable as thought by 96% of the respondents. Some pointed out that there was no way to own new land through land clearance and the only way to expand tea cultivation was to rent others' land (which was not in the same production collective) or to get subcontracted land (which was in the same production collective).

3.2. The Social-Cultural Benefit Associated with Tea Cultivation

The assessment of the importance of typical ecosystem services in the Wuyishan area by tea farmers indicated meanings of tea cultivation beyond economic importance. For all the listed ecosystem services (Figure 3), tea as a product was perceived by 95% of respondents as an important one that should rank in the top five, followed by fresh water, which was chosen by 70% of the respondents. The few who did not rank tea cultivation among the top five important ones mostly perceive eco-tourism, air purification, and local culture as more important. Eco-tourism was the most chosen cultural service as 60% of respondents thought it important, followed by the local culture which was chosen by more than half of the respondents. For regulating services, the most chosen one was air purification (41%). The scores of each ecosystem service also indicated that tea farmers definitely thought the provisioning of tea was the most important ecosystem service to them (5.3), followed by fresh water (3.2), eco-tourism (2.5), and local culture (2.3).

The MCA revealed the relationship between different ecosystem services in terms of tea farmers' perception (Figure 4). The first and second dimensions presented are, respectively, eigenvalue, 1.838 and 1.676; inertia, 0.123 and 0.112; and Cronbach's alpha, 0.658 and 0.638, which were slightly lower than the generally accepted lower limit of 0.70; however, a smaller value is acceptable in exploratory research [43]. The locations of choosing tea were very close to the origin of the coordinates, indicating that respondents had an almost unanimous assessment of the importance of tea cultivation. By contrast, locations of ES decisions far from the origin of the coordinates indicated not a unanimous perception of importance among respondents, such as the NTFPs, rice, research, and environmental education.

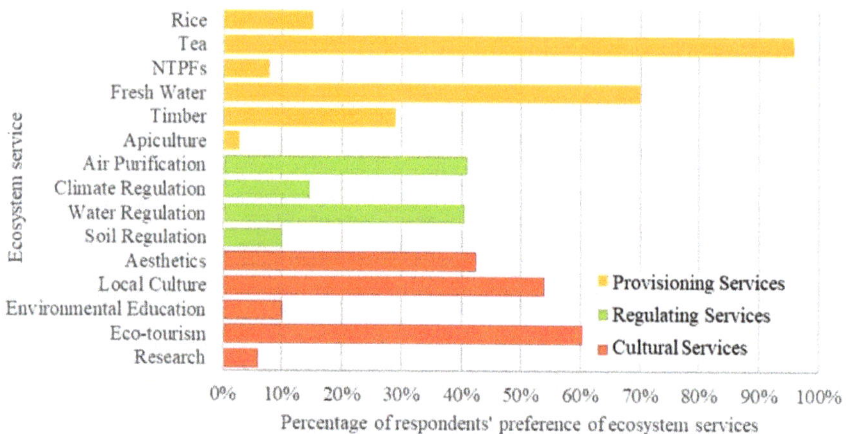

Figure 3. Tea farmers' preference of ecosystem services concerning their overall importance to life (NTFP: non-timber forest products).

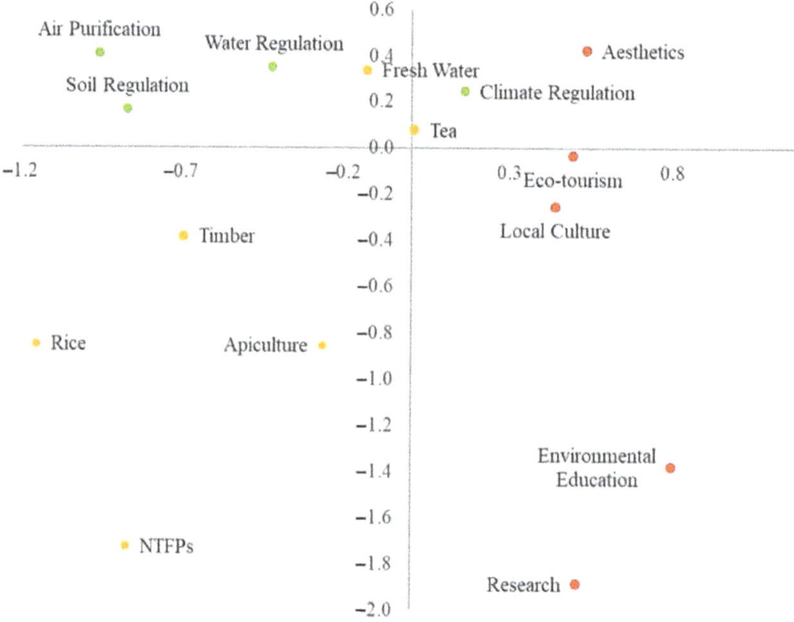

Figure 4. The MCA biplot of the preference of ES among tea farmers. Red: cultural services; yellow: provisioning services; green: regulating services.

The first axis revealed a trade-off between decisions of cultural services and other services except for tea cultivation or climate regulation, which indicates synergies between cultural services and the other two. The second axis revealed a trade-off between regulating services and other services except for fresh water, tea, and aesthetics to show synergies. Therefore, respondents who perceived tea cultivation as important or not also tend to perceive cultural services and regulating services as important or not.

Following the preference for ecosystem services beyond tea as a product, respondents identified many socio-cultural meanings of tea in the answers to the open-end question of how tea farmers value their tea orchard under the construction of an NP. Three aspects were identified after coding all the expressions (Table 3). First, engaging with tea cultivation brought individuals with physical and mental health; second, it led to the social stability of the community; third, it facilitated inherit of cultural heritage. These aspects were all confirmed as taking effects all the time, although some traditional knowledge was gradually lost. It was especially obvious that when asked about the concrete expressions or records regarding traditions associated with tea cultivation and processing, most respondents acknowledged that ceremonies were no longer practiced and folk songs and sayings were not commonly mentioned in daily life.

Table 3. Socio-cultural benefits expressed by the respondents.

Social-Cultural Meaning	Sample Expression
Physical and mental health	My view was broadened through communication during the tea sale. The natural environment secured high-quality tea which satisfied me. Regular working in the field has improved my physical condition. Drinking tea was good for people's health.
Social stability	Courtesy was practiced during tea processing and ceremonies. Engaging with tea reduced time spent on gambling and drinking. Tea processing can absorb idle labor.
Cultural inheritance	The fame of Wuyishan was promoted. We can learn from historical experience. Tea culture can be promoted. New blending and flavor of tea can be invented.

Nevertheless, they had still provided some information on the tea culture. Abundant folk songs and sayings were describing the origin of tea, the timing for attending to tea bushes, the experience of tea production, the technology of tea planting and processing, and the value of tea. They agreed that inheritance and communication of relevant knowledge were still possible. When asked about tea cultivation and processing techniques, the 221 respondents provided 285 answers, of which 35% were "through communication with neighbors" and 29% were "passed on for generations", compared to 13% of "government technology popularization" and 23% of "other sources". The respondents also mentioned the mix of practicing religion with the production and enjoyment of tea. Finally, they confirmed that some traditions, such as the ritual of the initiation of tea picking, have been gradually resumed.

3.3. Perceptions of Protected Area Management and Expectations of Future Management

Respondents were aware of the existence of the PAs and the impact of their management on tea cultivation. 83.3% of the respondents were aware of the existence of the national nature reserve and the scenic area, and the rest were not sure about the exact name (n = 221). Concerning the awareness of the geographic location of their land (Figure 5), only 15 respondents said they were not sure of the exact location, and other respondents all confirmed that they had tea plots inside of the PAs (107, 48%) or not (99, 45%). For those who had land located inside of the PAs, 61 (57%) perceived no effect of conservation management while 46 (43%) pointed out different forms of control that they thought of as disturbances to their tea cultivation. These claimed disturbances were listed in Table 4. Generally, there were two types of control; the first was a complete banning of certain land use or production way, and the second was some specific development control. These identified as prohibition and restraints were all official policies other than collective actions as informal customs. A third disturbance was also mentioned as a side-effect or accidental injury to tea cultivation during the implementation of PA management policies, such as mistakenly removal of tea bushes, contamination of tea leaves by spraying insecticide on pine forests, and lack of control on tourists who affect tea bushes.

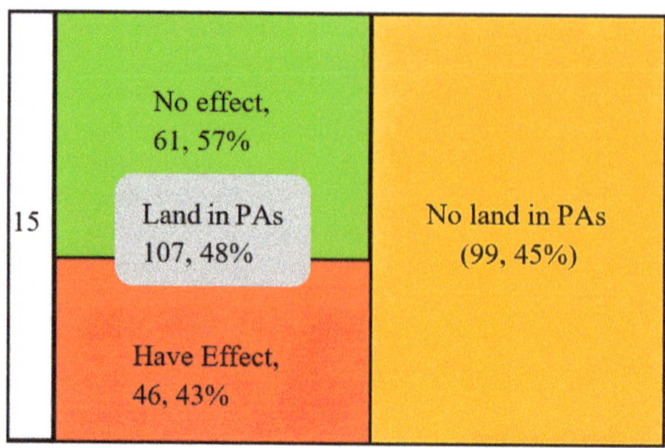

Figure 5. Perception of the impact of protected areas on tea orchard.

Table 4. Identified disturbances to livelihood activities within the protected areas.

Tea Relevant	Prohibition	Restraints
No	Harvesting Chinese fir; Collecting firewood;	Tourism; Collecting herbs;
Yes	Clearing forest; Pruning tree (to avoid shading tea bushes); Ketu (literally "guest soil"), replacing soils under tea bushes with new soils from nearby; Modernizing roads to tea terraces;	Fertilizer amount; Tea bush trimming; The flow of tea buyers; The flow of motor vehicles; The scale of the tea processing factory; Choice of varieties of tea bushes;

Respondents also hold diverse perceptions of protected area management concerning their ecological, economic, and social outcomes. Ecological outcomes were explained as direct protection results concerning elements of the ecosystem and itself. Economic outcomes were income, job position, commercial opportunities, etc., which can bring monetary benefit during PA management. Social outcomes were broader public welfare such as improvement in infrastructure and education with the existence of PA. In general, tea farmers were most happy with the ecological outcomes of conservation but the least with the realization of public welfare. For ecological outcomes, 68% of the respondents provided a positive reply (Figure 6a), while 45% of them claimed no enjoyment of any public welfare provided by the PAs (Figure 6c). Benefiting from the commercial operation of the PAs seems the most difficult to judge as the numbers of respondents holding negative, positive, and neutral attitudes were almost the same (Figure 6b).

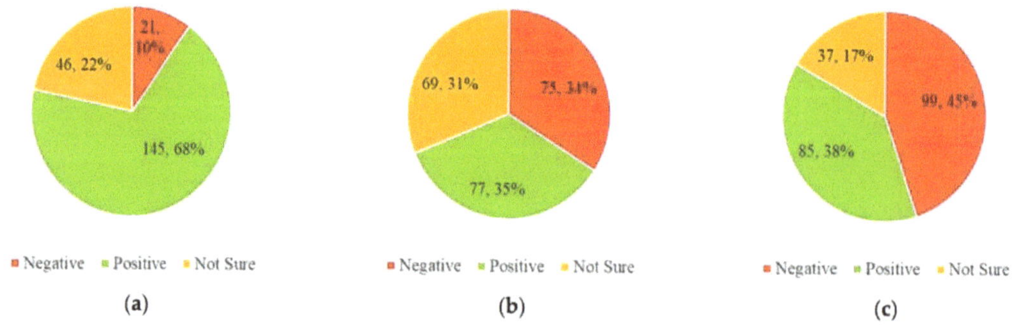

Figure 6. Tea Farmers' perception of conservation effectiveness: ecological outcomes (**a**), economic benefit (**b**), and public welfare (**c**).

Concerning the current establishment of the NP, respondents expressed their concern if tea plots would be returned to the forest. About 35% of the respondents held the attitude that there was no room for negotiation of any compensation fee as the tea orchard was the lifeline, and it was ridiculous to even think about land acquisition for other use. About 10% thought that giving up tea plantation was negotiable only if the compensation could satisfy them, the conditions included compensating according to the market value, the quality of land, and through land replacement, and the general expectation was that the living standard must not be lower than the current one. However, they acknowledged that negotiating conditions would be difficult based on their experience. The rest all preferred monetary compensation alone, 16% of the respondents asked for annual compensation, and another 39% proposed one-off compensation. However, there was a wide range of expected payments due to the productivity of the land. For the annual compensation, the expected value had a range between 7500 and 1,500,000 yuan/ha (1200 to 240,000 USD/ha); and for the one-off payment, that was between 3000 and 9,000,000 yuan/ha (480 to 1,440,000 USD/ha).

4. Discussion

4.1. Sustaining Traditions in a Protected Area under a Modern Market Economy

This study revealed how the tea farmers perceived the role of tea cultivation. One key finding is the dilemma between traditional farming and the market economy. Tea cultivation is still a traditional way of farming the forest as land and labor-intensive. Tea farmers have limited but enough access to the market by setting up a relatively fixed distribution channel in an acquaintance society. Therefore, they are eager to exploit more forested areas when engaging deeper with the market competition, but they are subject to even stricter land management rules. This dilemma is not uncommon globally [44,45], and significantly prominent in developing countries [46–48]. This study thus provides a way out of this dilemma to help conservation as well as rural development.

As in many agroforestry systems, tea cultivation in Wuyishan is not separated from the forest ecosystem, but together they formed coupled social-ecological systems generating different ecosystem services that benefit human well-being and development [49]. The provision of multiple ecosystem services is based on the agriculture of multi-functionality, which was widely supported top-down such as in the European Union (EU) [50,51] but is not well-studied in China [52]. To shed some light on this, this research showed that tea farmers think highly of local culture and eco-tourism, and tea cultivation is potentially clustered with cultural services. They all indicate that tea farmers may have the demand for commercialization of the traditional culture to fulfill other functions of tea cultivation beyond producing leaf. Thus, appreciation of cultural services can become a way of generating income besides tea production, and could possibly reduce the cases of illegal expansion of tea bushes to the forest. Concerning an even broader scope of social-cultural benefit, our study found that tea farmers cherish tea cultivation as a way of improving the well-being of a person, facilitating social stability, and sustaining a living culture. These functions of tea cultivation comply with the objective of a healthy park–people relationship and they were not unique to tea cultivation but many traditional practices in rural areas.

Therefore, farmers in rural areas can and will maintain traditional activities under PA management, if they understand the traditional culture and ecological protection concept can valorize many elements during goods production. This balance between farmers' interest and public welfare can be realized through the integration of multi-functionality and sustainability [53]: through the provision of multiple goods and services from the same social-ecological system, there is both an added-value over land expansion (ecological valorization) and diversity services over single products (cultural valorization) to resolve the dilemma of rural development and conservation.

4.2. Benefit-Sharing in the Protected Area

Equity and sustainability are important goals in natural resource management [54,55]. They also matter to the stability of a social-ecological system because resource users could perceive the benefit-sharing mechanism and react accordingly [36]. It is not surprising that tea farmers thought PAs have affected their benefit mainly because some traditions in tea cultivation were not respected, such as pruning and Ketu (Table 4). However, some of the disturbances are not true disturbances regarding "tradition", such as the prohibition of "modernizing roads to tea terraces" and the restraint of the "flow of motor vehicles". They are identified as "disturbances" usually because they affect income generation activities. Nevertheless, results from the perception of current conservation effectiveness show that PAs did not quite benefit tea farmers either through bringing income or providing more public welfare, although, at the same time, the biophysical conditions were improving. This indicates that a trade-off between maintaining the ecological functions and securing livelihood still exists.

In the newly designated NP, this issue is amplified as more working land is now within the boundary of the NP. To ensure the resilience of the system, there are three aspects of this social-ecological system worth further discussion based on the results. First, the stability and consistency of the land tenure system are important, as farmers cannot afford to lose land or affiliated products. Additionally, sustaining traditions in a market-oriented economy can benefit from treating the land property right as a bundle of rights. This means to constrain tea farmers' use rights but respect the right to benefit, as is usually practiced in conservation easements [56–58]. Second, the co-existence of forest and tea orchards has brought to tea farmers, the user system, an impression that a healthy environment is good for both people and tea bushes. The well-preserved natural conditions can transform a common product of tea into a famous brand that brings added value [59,60]. Third, the resource system is not solely tea bushes, but the integrated forest ecosystem, that is, *shan chang*. This means unnecessary human disturbance to the forest from tea cultivation should be reduced.

Therefore, the case of tea cultivation indicates that constraints to land use do not necessarily lead to instability of the system if users can conduct conservation-compatible activities that have limited disturbance to the natural environment [7]. This idea is not new and has been practiced in some areas, such as restrictions on the owner's use of land in a conservation easement. The difficulty is that this way of benefiting from conservation can be equitable and sustainable sometimes only in a long run, so some initiating stimulus and patient negotiation are necessary [61–63].

4.3. Making Community Livelihood Compatible with Conservation Goals under a National Park Concept

The national park idea promoted in the Chinese context strengthens the strict protection of large-scale ecosystems and their processes while respecting human activities conducted in harmony with nature, especially those practiced by local residents for hundreds of years. It was originated through reflection on the efficiency of fortress conservation and the need to secure multiple ecosystem services [64]. Under this idea, conservation can provide opportunities for benefit sharing through sustaining traditions if added values are realized through conserved nature instead of exploitation of forests and/or adding chemicals for quantity [16].

From tea farmers' perception of the role of tea and the relation between tea cultivation and PAs, we feel that management should be implemented on three scales to help sustain tea cultivation under conservation goals (Figure 7). This management may apply to other agroforestry systems in mountainous areas as well. It is highlighted that an efficient solution to a healthy park–people relationship based on a fair distribution of ecosystem services should not be looked at the park scale alone, but instead, on plot, mountain, and landscape scales. This enables divergent strategies at different scales and provides potentially more scale-specific and also flexible options to integrate parks and people in fair ways.

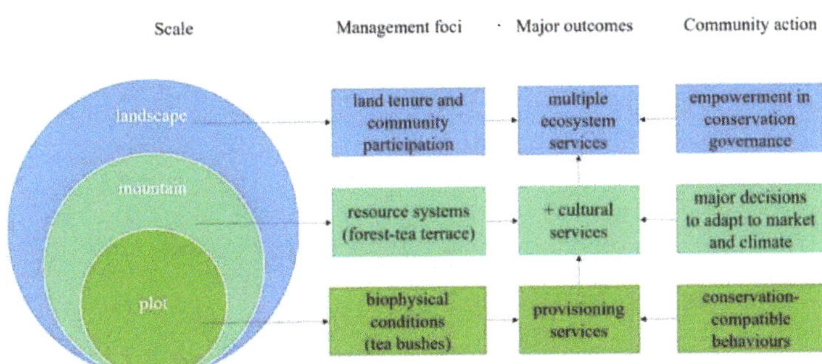

Figure 7. A scaling management strategy for adapting agroforestry systems to conservation in mountainous areas.

First, at the scale of the plot, attention should be given to species and biophysical elements for tea bushes, such as tea breeding, soil, and water conservation along the mountain slope. This is because these basic inputs sustain the growth of tea bushes and ensure the basic provisioning service. Conservation-compatible behaviors also start at this scale to avoid unfavorable activities such as killing trees.

Second, at the mountain scale, attention should be given to how the users manage the resource system. Major management decisions are made at this scale to link old wisdom with new technology to sustain the basic structure of the forest-tea terrace. Farmers are sensitive to land location and interactions with PAs mainly at this scale, and they are seeking ways to adapt to climate change and fluctuating markets. They also tend to combine provisioning services with cultural services to enlarge income sources.

Third, at the landscape scale, attention should be given to the land tenure system and the community's participation in conservation to varying degrees, such as conservation easement, payment for ecosystem services, conservation steward program, etc. Homogenization of the landscape resulting from the expansion of tea cultivation will be disastrous to the forest and is a violation of conservation goals.

5. Conclusions

National parks in China are very different from those in North America because it is difficult to find a large area of the wilderness without human activities perhaps except on the inner Tibetan Plateau. Finding ways out of the common dilemma of improving livelihood under conservation restrictions leads us to conduct this research when the newly proposed national park system provides opportunities to reflect on protected area management and learn from global experience. Wuyishan is a typical area where human activities have lasted very long with the remnant of the forest of high ecological values. This research found that conservation through setting up PAs has impacted local tea farmers' understanding of conservation regarding their demand for income. It also found that maintaining tea cultivation in harmony with the forest needs to find ways to add value to tea. In theory, this research proved that the SES meaning perception theory can reveal the potential synergies between local people and other stakeholders; in practice, the knowledge co-produced through local perception is reliable to form incentives for farmers to comply with conservation rules to secure the stability of the social-ecological system.

As revealed in this research, tea farmers are seeking equitable and sustainable benefit sharing in the PAs. The NP has the potential to secure the livelihood of tea cultivation and to promote cultural values which the tea farmers think highly of. Therefore, it is possible to maintain the stability of the social-ecological system if multiple ecosystem services can be provided, and their provision is facilitated in the management of multiple levels: the plot level where controlling and monitoring of biophysical elements are critical; the mountain level where production and market are critical, and the landscape level where land tenure and community participation are critical.

The findings are encouraging for many cultural landscapes around the world which face a similar challenge in nature conservation activities. Understanding the potential of multiple ecosystem service provision through farmers' perception will be helpful in PA designation and other ecological policy design and implementation. This three-level management may also help guide compatible production behaviors for conservation targets while securing farmers' income in populated PAs. Further research is also needed to find critical factors that could turn the potential of provision of multiple ecosystem services into real provision and income to create real multifunctional agriculture embedded and connected to PAs.

Author Contributions: Conceptualization, S.H. and Y.S.; methodology, S.H.; formal analysis, S.H.; investigation, S.H.; writing—original draft preparation, S.H.; writing—review and editing, S.H.; project administration, S.H. and Y.S.; funding acquisition, S.H. and Y.S. All authors have read and agreed to the published version of the manuscript.

Funding: This research was funded by the National Natural Science Foundation of China, grant number 42001194, and Luc Hoffmann Institute Fellowship Programme.

Informed Consent Statement: Informed consent was obtained from all subjects involved in the study.

Data Availability Statement: Not applicable.

Acknowledgments: The authors would like to thank all the anonymous interviewees who contributed their time and knowledge to this study. The authors are also grateful to the research interns from several domestic universities and local NGOs for their involvement in the household survey and preliminary analysis.

Conflicts of Interest: The authors declare no conflict of interest.

References

1. Dewu, S.; Røskaft, E. Community attitudes towards protected areas: Insights from Ghana. *Oryx* **2018**, *52*, 489–496. [CrossRef]
2. Abukari, H.; Mwalyosi, R.B. Comparing conservation attitudes of park-adjacent communities: The case of mole national park in Ghana and Tarangire national park in Tanzania. *Trop. Conserv. Sci.* **2018**, *11*, 1940082918802757. [CrossRef]
3. Allendorf, T.D. Residents' attitudes toward three protected areas in southwestern Nepal. *Biodivers. Conserv.* **2007**, *16*, 2087. [CrossRef]
4. Ite, U.E. Community perceptions of the Cross River national park, Nigeria. *Environ. Conserv.* **1996**, *23*, 351–357. [CrossRef]
5. Hough, J.L. Obstacles to effective management of conflicts between national parks and surrounding human communities in developing countries. *Environ. Conserv.* **1988**, *15*, 129–136. [CrossRef]
6. Dudley, N. (Ed.) *Guidelines for Applying Protected Area Management Categories*; IUCN: Gland, Switzerland, 2008; pp. 16–17.
7. He, S.; Gallagher, L.; Su, Y.; Wang, L.; Cheng, H. Taking an ecosystem services approach for a new national park system in China. *Resour. Conserv. Recycl.* **2018**, *137*, 136–144. [CrossRef]
8. Yan, S.; Li, Z.; Zhou, Z. Survey on attitude of community residents toward Jiugongshan Nature Reserve and countermeasures for reconciliation. *For. Inventory Plan.* **2007**, *32*, 162–167. (In Chinese)
9. Yang, J.; Zhao, Q.; Yang, X.; Ti An, L.H.; He, X.J. A survey on local people's perceptions and attitude orientation around Taibaishan Nature Reserve. *For. Inventory Plan.* **2007**, *32*, 41–45. (In Chinese)
10. Wu, W.; Liu, Q.; Liu, Z.; Zihan, L.; Libin, T.; Qiang, L. Determinants of farmer households' attitudes towards the construction of nature reserve in their neighborhood. *J. Zhejiang Agric. For. Univ.* **2014**, *31*, 97–104. (In Chinese)
11. Ma, B.; Shen, J.; Ding, H.; Wen, Y. Farmer protection attitudes and behavior based on protection perception perspective for protected areas. *Resour. Sci.* **2016**, *38*, 2137–2146. (In Chinese)
12. Li, X.; Wen, J. Study on peripheral community resident's dependence and attitude to nature reserves. *Cent. South For. Inventory Plan.* **2008**, *27*, 45–49. (In Chinese)
13. Duan, W.; Wen, Y.; Wang, C. Influence factors of farmers' attitude toward environment and Crested Ibis conservation in Crested Ibis National Nature Reserve. *Wetl. Sci.* **2013**, *11*, 90–99. (In Chinese)
14. Han, F.; Wang, C.; Zhao, Z. The combined influence of nature reserves from the perspective of farmer households in Shannxi. *Wetl. Sci.* **2015**, *37*, 102–111. (In Chinese)
15. Chang, L.; Watanabe, T. The mutual relationship between protected areas and their local residents: The case of Qinling Zhongnanshan UNESCO Global Geopark, China. *Environments* **2019**, *6*, 49. [CrossRef]

16. He, S.; Gallagher, L.; Min, Q. Examining Linkages among Livelihood Strategies, Ecosystem Services, and Social Well-Being to Improve National Park Management. *Land* **2021**, *10*, 823. [CrossRef]

17. Sirivongs, K.; Tsuchiya, T. Relationship between local residents' perceptions, attitudes and participation towards national protected areas: A case study of Phou Khao Khouay National Protected Area, central Lao PDR. *For. Policy Econ.* **2019**, *21*, 92–100. [CrossRef]

18. Wang, Q.; Yamamoto, H. Local Residents' Perception, Attitude and Participation Regarding Nature Reserves of China: Case Study of Beijing Area. *J. For. Plan.* **2009**, *14*, 67–77. [CrossRef]

19. Allendorf, T.D.; Aung, M.; Songer, M. Using residents' perceptions to improve park–people relationships in Chatthin Wildlife Sanctuary, Myanmar. *J. Environ. Manag.* **2012**, *99*, 36–43. [CrossRef]

20. Chen, X.; Lupi, F.; Liu, J. Accounting for ecosystem services in compensating for the costs of effective conservation in protected areas. *Biol. Conserv.* **2017**, *215*, 233–240. [CrossRef]

21. Zhang, L.; Luo, Z.; Mallon, D.; Li, C.; Jiang, Z. Biodiversity conservation status in China's growing protected areas. *Biol. Conserv.* **2017**, *210*, 89–100. [CrossRef]

22. Carpenter, S.; Walker, B.; Anderies, J.M.; Abel, N. From metaphor to measurement: Resilience of what to what? *Ecosystems* **2001**, *4*, 765–781. [CrossRef]

23. Holling, C.S. Resilience and stability of ecological systems. *Annu. Rev. Ecol. Evol. Syst.* **1973**, *4*, 1–23. [CrossRef]

24. Holmes, C.M. The influence of protected area outreach on conservation attitudes and resource use patterns: A case study from western tanzania. *Oryx* **2003**, *37*, 305–315. [CrossRef]

25. Nastran, M. Why does nobody ask us? Impacts on local perception of a protected area in designation, Slovenia. *Land Use Policy* **2015**, *46*, 38–49. [CrossRef]

26. Bennett, N.J. Using perceptions as evidence to improve conservation and environmental management. *Conserv. Biol.* **2016**, *30*, 582–592. [CrossRef]

27. Halliday, A.; Glaser, M. A management perspective on social ecological systems: A generic system model and its application to a case study from Peru. *Hum. Ecol. Rev.* **2011**, *18*, 1–18.

28. Bunker, N. *An Empire on the Edge: How Britain Came to Fight America*; Knopf: New York, NY, USA, 2014.

29. Zou, X. *Wuyi Zheng Shan Xiao Zhong Hong Cha (The Wuyi Lapsang Souchong)*; Zhong Guo Nong Ye Chu Ban She: Beijing, China, 2006. (In Chinese)

30. Cai, J. *Collective Forest Tenure Reform under the Perspective of Sustainable Social-Ecological Systems: The Empirical Study of Fujian Province*; Zhong Guo She Hui Ke Xue Chu Ban She: Beijing, China, 2012. (In Chinese)

31. Borrini-Feyerabend, G. Governance of protected areas—Innovation in the air. *Policy Matters* **2003**, *12*, 92–101.

32. Campese, J.; Sunderland, T.; Greiber, T.; Oviedo, G. *Rights-Based Approaches: Exploring Issues and Opportunities for Conservation*; CIFOR and IUCN: Bogor, Indonesia, 2009.

33. Timkoa, J.A.; Satterfield, T. Seeking social equity in national parks: Experiments with evaluation in Canada and South Africa. *Conserv. Soc.* **2008**, *6*, 238–254. [CrossRef]

34. Castonguay, A.C.; Benjamin, B.; Müller, F.; Horgan, F.G.; Settele, J. Resilience and adaptability of rice terrace social-ecological systems: A case study of a local community's perception in Banaue, Philippines. *Ecol. Soc.* **2016**, *21*, 14. [CrossRef]

35. Fischer, A.; Eastwood, A. Coproduction of ecosystem services as human–nature interactions—An analytical framework. *Land Use Policy* **2016**, *52*, 41–50. [CrossRef]

36. van Wyk, E.; Breen, C.; Freimund, W. Meanings and robustness: Propositions for enhancing benefit sharing in social-ecological systems. *Int. J. Commons* **2014**, *8*, 576–594. [CrossRef]

37. Young, O.R.; Berkhout, F.; Gallopin, G.C.; Janssen, M.A.; Ostrom, E.; Van der Leeuw, S. The globalization of socio-ecological systems: An agenda for scientific research. *Glob. Environ. Chang.* **2006**, *16*, 304–316. [CrossRef]

38. Young, J.C.; Rose, D.C.; Mumby, H.S.; Benitez-Capistros, F.; Derrick, C.J.; Finch, T.; Garcia, C.; Home, C.; Marwaha, E.; Morgans, C.; et al. A methodological guide to using and reporting on interviews in conservation science research. *Methods Ecol. Evol.* **2018**, *9*, 10–19. [CrossRef]

39. He, S.; Gallagher, L.; Su, Y.; Wang, L.; Cheng, H. Identification and assessment of ecosystem services for protected area planning: A case in rural communities of Wuyishan national park. *Ecosyst. Serv.* **2018**, *31*, 169–180. [CrossRef]

40. Holton, J.A. The coding process and its challenges. In *The SAGE Handbook of Grounded Theory*; SAGE: Los Angeles, CA, USA, 2007; pp. 265–289.

41. Zhang, X.; Liu, X.; Zhang, B.; Xie, Y. Will the Establishment of Nature Reserves Inevitably Lead to Low Household Income-An Empirical Study on Household Income within and nearby the Fujian Wuyishan National Nature Reserve. *Sci. Silvae Sin.* **2020**, *56*, 165–178. (In Chinese)

42. Greenacre, M.; Blasius, J. *Multiple Correspondence Analysis and Related Methods*; Chapman & Hall/CRC: London, UK, 2006.

43. Johnson, R.A.; Wichern, D.W. *Applied Multivariate Correspondence Analysis*, 6th ed.; Prentice-Hall: Upper Saddle River, NJ, USA, 2007.

44. Belsky, J.M. Community forestry engagement with market forces: A comparative perspective from Bhutan and Montana. *For. Policy Econ.* **2015**, *58*, 29–36. [CrossRef]

45. Brown, S.; Shrestha, B. Market-driven land-use dynamics in the middle mountains of Nepal. *J. Environ. Manag.* **2000**, *59*, 217–225. [CrossRef]

46. Li, Z.; Kang, A.; Gu, J.; Xue, Y.; Ren, Y.; Zhu, Z.; Liu, P.; Ma, J.; Jiang, G. Effects of human disturbance on vegetation, prey and Amur tigers in Hunchun Nature Reserve, China. *Ecol. Model.* **2017**, *353*, 28–36. [CrossRef]
47. Yu, P.; Liu, H.; Chen, S. Influences of human disturbances on vegetation of Songshan National Level Nature Reserve. *Sci. Silvae Sin.* **2002**, *38*, 162–166. (In Chinese)
48. Zeng, H.; Sui, D.Z.; Wu, X.B. Human disturbances on landscapes in protected areas: A case study of the Wolong Nature Reserve. *Ecol. Res.* **2005**, *20*, 487–496. [CrossRef]
49. Parrott, L.; Meyer, W.S. Future landscapes: Managing within complexity. *Front. Ecol. Environ.* **2012**, *10*, 382–389. [CrossRef]
50. Brunstad, R.J.; Gaasland, I.; Vårdal, E. Multifunctionality of agriculture: An inquiry into the complementarity between landscape preservation and food security. *Eur. Rev. Agric. Econ.* **2005**, *32*, 469–488. [CrossRef]
51. Kohler, F.; Thierry, C.; Marchand, G. Multifunctional agriculture and farmers' attitudes: Two case studies in rural France. *Hum. Ecol.* **2014**, *42*, 929–949. [CrossRef]
52. Song, B.; Robinson, G.M. Multifunctional agriculture: Policies and implementation in China. *Geogr. Compass* **2020**, *14*, e12538. [CrossRef]
53. Hediger, W.; Knickel, K. Multifunctionality and sustainability of agriculture and rural areas: A welfare economics perspective. *J. Environ. Policy Plan.* **2009**, *11*, 291–313. [CrossRef]
54. Brock, W.A.; Carpenter, S.R. Panaceas and Diversification of Environmental Policy. *Proc. Natl. Acad. Sci. USA* **2007**, *104*, 15206–15211. [CrossRef] [PubMed]
55. Smith, P.D.; McDonough, M.H. Beyond Public Participation: Fairness in Natural Resource Decision Making. *Soc. Nat. Resour.* **2001**, *14*, 239–249. [CrossRef]
56. Güneş, Y. Conservation Easement and Common Law Easement as a Nature Conservation Tool within the Context of Property Rights and the Economic Trade Off. *Ann. Fac. Droit D'istanbul.* **2007**, *39*, 279.
57. Rissman, A.R. Rethinking property rights: Comparative analysis of conservation easements for wildlife conservation. *Environ. Conserv.* **2013**, *40*, 222–230. [CrossRef]
58. Watson, R.; Fitzgerald, K.H.; Gitahi, N. Expanding options for habitat conservation outside protected areas in Kenya: The use of environmental easements. *Afr. Wildl. Found. Tech. Pap.* **2010**, *2*, 7–29.
59. Egelyng, H. Informed Markets as Policy Instrument for Environmental Governance of Buffer Zones around Protected Areas: A global context and European cases. In Proceedings of the ECPR Joint Sessions, Salamanca, Spain, 10–15 April 2014.
60. Owusu, A. Alleviating Rural Poverty in Ghana through Marketing of Tourism Sites and Protected Areas. *Eur. J. Bus. Manag.* **2013**, *5*, 238–245.
61. Berkes, F. Community-based conservation in a globalized world. *Proc. Natl. Acad. Sci. USA* **2007**, *104*, 15188–15193. [CrossRef] [PubMed]
62. Ingram, J.C.; Wilkie, D.; Clements, T.; McNab, R.B.; Nelson, F.; Baur, E.H.; Sachedina, H.T.; Peterson, D.D.; Foley, C.A. Evidence of Payments for Ecosystem Services as a mechanism for supporting biodiversity conservation and rural livelihoods. *Ecosyst. Serv.* **2014**, *7*, 10–21. [CrossRef]
63. Jamal, T.; Stronza, A. Collaboration theory and tourism practice in protected areas: Stakeholders, structuring and sustainability. *J. Sustain. Tour.* **2009**, *17*, 169–189. [CrossRef]
64. Xu, W.; Xiao, Y.; Zhang, J.; Yang, W.U.; Zhang, L.U.; Hull, V.; Wang, Z.; Zheng, H.; Liu, J.; Polasky, S.; et al. Strengthening protected areas for biodiversity and ecosystem services in China. *Proc. Natl. Acad. Sci. USA* **2017**, *114*, 1601–1606. [CrossRef]

 *land*

Article

The Intention of Community Participation in the Qilian Mountain National Park Policy Pilot

Liqi Jia [1], Junqing Wei [1] and Zibin Wang [2,*]

[1] School of Design Art, Lanzhou University of Technology, Lanzhou 730050, China; liqijia@lut.edu.cn (L.J.); JunqingWei@lut.edu.cn (J.W.)
[2] School of Educational Science and Technology, Northwest Minzu University, Lanzhou 730030, China
* Correspondence: wang-zibin@xbmu.edu.cn

Abstract: As a management strategy, community participation is to implement the coordinated development of communities and protected areas. In recent years, the development of China's national parks has faced many challenges related to human and environmental constraints. Community participation plays an essential role in solving such issues. As one of the critical indicators to test community participation, community residents' willingness to participate significantly impacts community participation in constructing national parks. As such, this study was conducted using the extended model of the theory of planned behavior (TPB) and the structural equation model. Taking the Tianzhu county and Sunan Yugu county as examples, and based on 230 valid questionnaires, we investigated the impacts of the Qilian Mountain National Park System Pilot Area on community residents' willingness to participate and provided relevant suggestions for amendments. The results indicated that, for the Qilian Mountain National Park System Pilot Area, behavioral attitude, subjective norms, and perceptual behavior control positively impacted the participation intention of community residents. At the same time, the variables mentioned above positively impacted the implementation of the participation intention of community residents. Specifically, the order of impacts is as follows: perceptual behavior control (path coefficient = 0.89) > participation behavior attitude (path coefficient = 0.68) > related impact system (path coefficient = 0.41) > subjective norms (path coefficient = 0.38). According to the results, we put forward three suggestions: (1) providing relevant instructions and guidance on various methods to ensure that the pilot policies on the construction of national parks can form a positive relationship with the participation intentions of the community residents; (2) making full use of the function of perceptual behavior control, so the subjective initiative of community residents can be maximized, thereby enhancing the willingness of community residents to participate in constructing national parks; and (3) strengthening the impacts of subjective norms, enhancing the soft culture of national park communities' participation, reshaping the community cultural landscapes with the goal of constructing national parks, and establishing community residents' sense of honor as the builders of national parks.

Keywords: Qilian Mountain National Park; community participation; TPB extended model; balloon dessert; structural equation model

Citation: Jia, L.; Wei, J.; Wang, Z. The Intention of Community Participation in the Qilian Mountain National Park Policy Pilot. *Land* **2022**, *11*, 170. https://doi.org/10.3390/land11020170

Academic Editors: Rui Yang, Yue Cao, Steve Carver and Le Yu

Received: 17 December 2021
Accepted: 17 January 2022
Published: 21 January 2022

Publisher's Note: MDPI stays neutral with regard to jurisdictional claims in published maps and institutional affiliations.

1. Introduction

The mechanism of community participation in national parks originated in the United States in the 1960s and 1970s [1] as an autonomous management policy that includes community residents. After a long period of evolution [2,3], it has become one of the most crucial components of the management measures of protected areas. The core idea of this concept is to encourage the residents of national park communities to participate in the construction of national parks to varying degrees as participants and beneficiaries of the national park system [4]. The majority of international research on community participation has focused on the macro level [5]. Most of this research was carried out through semi-structured interviews and analyzed individual cases or multiple cases [6]. The major factors that affect the willingness of a community's participation include management system issues, implementation progress of relevant policies, ecological compensation, and related conflicting interests [7,8]. At present, China's national parks have adopted the same policy guidance [9]; meanwhile, it reviews and compares the community participation policies implemented by the protected areas and explores the different issues encountered by the community participation policies in developing China's nature reserves [10]. In the current research related to the willingness to participate in national parks, the primary evaluation is based on measuring the perceived value of community residents [11]. As major events affecting the community, National Park System Pilot Areas have strengthened local residents' sense of place and identity [12,13], becoming one of the core elements that affect the willingness of the participation of community residents. At the same time, some scholars have claimed that the mechanism of national park community participation relies primarily on cognition, attitude, and participation [14]. Mensah believes that the impact of the perceived financial benefits of tourism has a significant influence on community engagement [15]. However, other scholars hold different attitudes toward relevant conclusions based on different research areas. These researchers claim that community engagement in tourist development is strongly dependent on gatekeepers' attitudes and communities' economic backgrounds. Ensuring community participation is more difficult in settings where economically vulnerable communities and manipulative gatekeepers are present. As a result, sustainable land and resource use practices are hindered, resulting in irreparable damage to environmentally sensitive areas [16].

Community participation is an essential part of the development of national parks. The willingness to participate is a prerequisite for the effective implementation of the mechanism of community participation; it is mostly affected by the behavioral attitudes and perceived value of community residents. Therefore, this study was conducted using the theory of planned behavior (TPB), which Ajzen proposed in 1977 [17]. This model is used to explain the behavioral attitude and behavioral intention of the research object, and a large number of scholars have used this model to study problems related to national parks. Miller used the TPB model theory to investigate and analyze the phenomenon of human–animal conflict among tourists in Yellowstone National Park in the United States and put forward practical suggestions for tourists and managers to prevent such incidents [18]. Goh et al. (2017) studied tourists' intentions to go off-trail in the Blue Mountains National Park (BMNP) in Australia and revealed that pro-environmental attitudes effectively predict general environmental worldviews [19]. Using the theory of planned behavior, Reigner et al. (2009) analyzed the relationships among visitors' attitudes, subjective norms, and perceived control over pool exploration, intentions to explore, and actual actions at pools [20]. This theoretical model is mainly used to explain the behaviors, attitudes, and intentions of the research objects. Nevertheless, the adaptability of some research subjects can hinder analysis when the model is applied to research in various fields. Behavioral intentions are often restricted by objective conditions. Ajzen also recognized the existence of such an issue and therefore pointed out that corresponding corrections or extensions are needed when utilizing the TPB theoretical model [1] in order to better adapt to research subjects when applied to different disciplines [21]. In the current research, Wenbin Zhang introduced ecological compensation mechanisms as an extension of the TPB model and analyzed the willingness for ecological protection and behavioral intentions of the residents in environmentally protected areas [22]. Han tested the established TPB model and explained consumers' behavioral intentions with regard to choosing eco-friendly hotels by using the structural equation model [23]. Yuangang Zhang introduced local theory into the TPB extended model as a research variable to analyze the impact of local emotions on tourists' traveling behaviors [24]. Using the theory of planned behavior (TPB), Wang et al. (2019) investigated the effect of the EB of a tourist spot on the ERB using a structural equation model (SEM) multi-group study (MGA) [25]. For hikers visiting a national park in Taiwan, Wang et al. (2020) investigated a behavioral model employing the latent variables of personality, environmental concern, attitudes toward activities, and environmentally responsible behavior [26]. This study will also introduce extended variables when using the theory of planned behavior (TPB): the impact of the pilot implementation of Qilian Mountain National Park on the willingness to participate of community residents (the following is the relevant impact system).

Based on the theory of planned behavior (TPB), this paper constructs a theoretical model in five aspects, including pilot implementation of national park policies, residents' behavioral attitudes, subjective norms, perceived behavioral control, and the willingness of residents' community participation. Taking the Gansu area of Qilian Mountain National Park as an example, this study focused on using a structural equation model to investigate the mechanism of the influence of policy pilots on community residents' willingness to participate in the construction of national parks. The innovations of this article include the following: (1) taking the perspective of collective choice as a prerequisite, exploring the changes in the degree of recognition of community participation through residents' subjective willingness in the circumstances of the implementation of major policies; (2) using the National Park System Pilot Area as the only influential factor among independent variables to explore whether community participation is affected, deepening the understanding of the path of influence based on the theory of planned behavior (TPB); and (3) when investigating the impact of intermediate variables (behavioral attitudes, subjective norms, perceptual behavior control) on the willingness for community participation through national park policy pilots, using differentiated analysis of the path coefficients of three intermediate variables, to a certain extent, clarifying the core influential conditions and deficiencies regarding community residents' willingness to participate and providing targeted solutions for future community participation in the construction of national parks.

2. Materials and Methods

This work aimed to investigate the various impacts on the willingness of indigenous communities to participate with regard to the Qilian Mountain National Park Pilot Area. We first made assumptions regarding the corresponding research based on the research framework. The data results were then screened using questionnaire surveys, the Delphi method, and semi-structured interviews. Finally, we analyzed the preset results and drew reasonable conclusions through the use of structural equation modeling (SEM).

2.1. The Construction of the Theoretical Model

The theory of planned behavior model was used to explain the subjective willingness of the research participants. It contains three specific stages, including the perceptional stage, attitude cognition, and behavioral intention. Existing research has analyzed residents' perceptions of the relevant measures implemented in the management regions after using the National Park System Pilot Area [27]. The stage of attitude cognition is mainly due to the subjective judgment made by the research participants after being affected by external influences [28]. Behavioral intention is primarily the subject of the dependent variables. Existing research has focused mainly on investigating the willingness to ensure environmental protection and tourism intention [24]. The following is the research framework (based on the theory of planned behavior) used in this article (see Figure 1) [2].

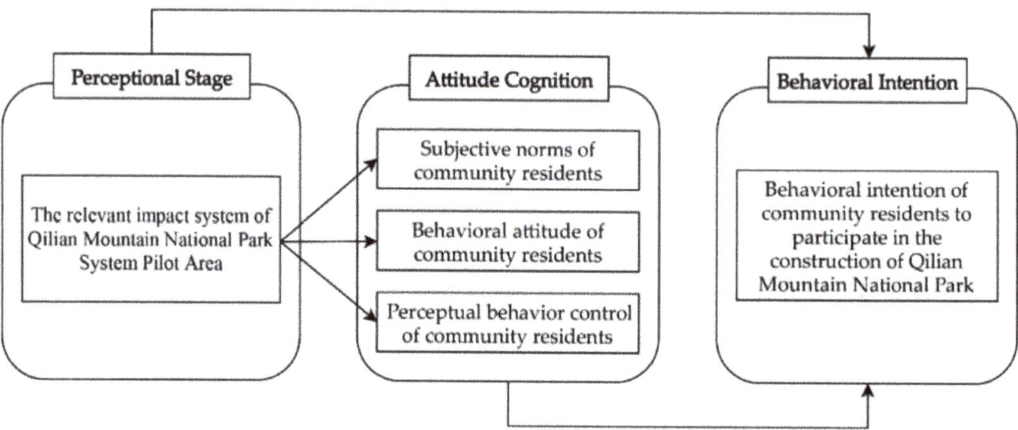

Figure 1. Extended model.

2.2. Theoretical Hypothesis

The perception element is the Qilian Mountain National Park System Pilot. As the core component of the extended proposition, it will increase the popularity of the Qilian Mountain community to a certain degree; meanwhile, the community will also become a crucial supporting point regarding implementing the functionality of the national park. Against the background of the implementation of the pilot policy, community residents' intention to participate will be affected by their subjective awareness and the external objective conditions [23]. Therefore, the hypotheses related to the relevant impact of the Qilian Mountain National Park System Pilot Area are as follows:

- H1: the relevant impact system of the Qilian Mountain National Park System Pilot Area has a positive and significant influence on the behavioral intentions to participate of its community residents.

  - H_{1-1}: the relevant impact system of the Qilian Mountain National Park System Pilot Area has a positive and significant influence on the behavioral attitudes toward participation of its community residents.
  - H_{1-2}: the relevant impact system of the Qilian Mountain National Park System Pilot Area has a positive and significant influence on the subjective norms of its community residents.
  - H_{1-3}: the relevant impact system of the Qilian Mountain National Park System Pilot Area has a positive and significant influence on the perceived behaviors of its community residents.

Attitudes and behaviors are related [23]; therefore, attitude is defined as a certain cognitive tendency and is restricted by an individual's perception and preference [24]. This article is based on the relevance of the Qilian National Park System Pilot Area and the related impact system on community residents. When community residents recognize that they benefit from the National Park System Pilot Area, their behavioral attitudes will be more positive, which has a positive impact regarding the willingness of the community to participate, and vice versa. Thus, the relevant hypothesis on behavioral attitude includes

- H2: the behavioral attitude of the participation of community residents in the Qilian Mountain National Park System Pilot Area has a positive and significant influence on the community's intention of participation.

Subjective norms can be defined as the external impetus given by others when an individual completes or performs a particular task or behavior. This impetus can be a positive expectation or negative pressure. In addition, subjective norms reflect an individual's desire to receive relevant support and approval from the public when performing a certain behavior [27,28]. Therefore, the relevant hypothesis regarding the subjective norms is:

- H3: the subjective norms of community residents in the Qilian Mountain National Park System Pilot Area have a positive and significant influence on the behavioral intentions of the community's participation.

Based on the idea of perceptual behavior control, individuals are restricted by both external and internal factors during the process of conducting some kind of behavior. In most situations, individuals' judgment about something is often restricted by their intellectual level, recognition, and external factors, rather than being based on a sense of objectivity and rationality. Therefore, the hypothesis related to perceptual behavior control is:

- H4: the perceptual behavior control of community residents in the Qilian Mountain National Park System Pilot Area has a positive and significant influence on the behavioral intentions of the community's participation.

2.3. Research Methods

Based on the existing hypothesis and present theoretical model, we collected relevant data using a questionnaire survey and analyzed the obtained data comprehensively and linearly using the structural equation model (SEM). Reasonable amendments were then made to the relevant hypothesis, finally obtaining a convincing model for the National Park System Pilot Area.

The study contains two sections. The first section details the survey design and is divided into three stages: proposition of the hypothesis, expert consultation, and prior observation. As the core part among the three stages in the survey design, the proposition of the hypothesis divided the present hypothesis into three categories. The first category is related to participants' cognition. We utilized existing research to establish a hypothesis regarding the impact of major events on the perception of community residents. The second category concerns participants' attitudes. We referred to the propositional research conducted by Yuangang Zhang [24] and Wenbin Zhang [22] to design the research hypothesis on three dimensions, including the subjective norms, the behavioral attitudes, and the perceptual behavior control of community residents. The third category involves investigating participants' behavioral intentions. We proposed the hypothesis based on relevant studies conducted by Qunming Zheng [29] and collected data using the Likert scale (1 = strongly agree, 2 = agree, 3 = slightly agree, 4 = disagree, 5 = strongly disagree).

The second section includes data analysis. Specifically, the authors conducted a component matrix rotation and tested the reliability and validity of the obtained data. After delimiting the component intervals, the structural equation model (SEM) was used to perform further tests. Because the dependent variable in this study is unobservable, and the independent variable affects multiple intermediate variables simultaneously, compared to other testing models, the structural equation model (SEM) testing method is linear and can predict the relationships of multiple interrelated variables simultaneously. In addition, it allows researchers to cope with unobservable variables in parts of models, and it helps to explain the measurement errors in the overall estimation process. The authors later referred to the study conducted by Hair et al. (1998) [30] and found that it is more reasonable to require the ratio of items to the sample number of the model to be 1/10 to 1/15 [31].

3. Data Collection and Analysis

Based on the previously detailed hypotheses and model construction, this study collected data from two gateway communities of Qilian National Park. Descriptive analysis of related variables and normal distribution tests was used to analyze the collected data.

3.1. Description of the Research Area

The research areas in this study includes Sunan county and Tianzhu county in Qilian Mountain National Park (see Figure 2). Both counties are important gateway communities of the Qilian Mountain National Park. Specifically, Sunan county is located in the middle of the Hexi Corridor and the north of the Qilian Mountain, with a total area of 23,800 km^2 (2014). With a relatively large proportion of indigenous people in the community, Sunan county has a total population of 37,579. In addition, the Sunan Yugu nationality is a unique ethnic minority group in Gansu Province. Tianzhu county is located at the eastern end of the Hexi Corridor, specifically on the northeastern edge of the Qinghai-Tibet Plateau. It is known as the "gateway" to the Hexi Corridor and borders Sunan in the northwest. Before implementing the National Park System Pilot Area, these two counties' development and industrial structures were mainly resource-oriented, including tourism, animal husbandry, and plantations. In addition, the production methods of these two counties were relatively backward, and the multi-ethnic settlements were the main form of distribution. The implementation of the National Park System Pilot Area has had a major impact on these two areas; thus, they are appropriate for the research.

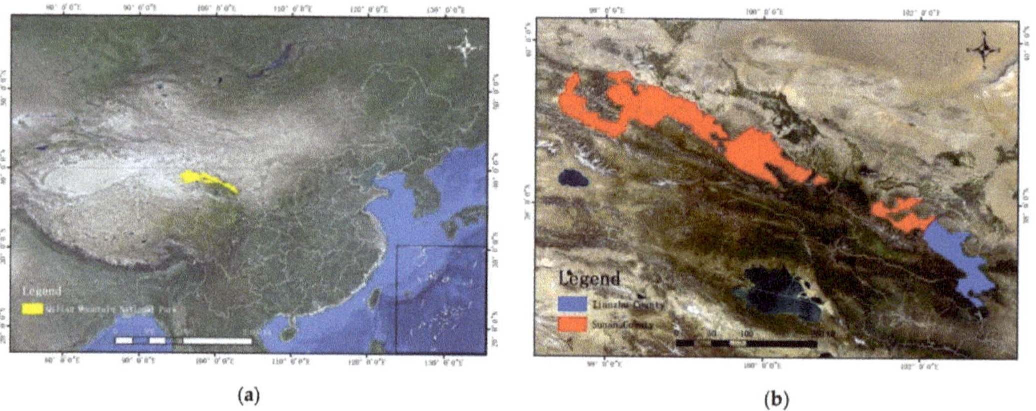

(a)   (b)

Figure 2. The location of Qilian Mountain National Park, Tianzhu county and Sunan county: (**a**) the location of Qilian Mountain National Park in China; (**b**) the location of Tianzhu county and Sunan County in Qilian Mountain National Park.

3.2. Demographic Sample Analysis

The distribution of the questionnaires comprised two methods: field distribution and online collection. The research team went to Sunan county and Tianzhu county to distribute questionnaires on site from 12 to 15 April 2020 and 1 to 3 June 2020. The online questionnaire was conducted in the form of a mobile phone app.

The survey took the family as the basic unit and selected one person from each household as the survey sample. After informing the respondents of their relevant survey obligations and obtaining permission, a one-to-one household survey was conducted. The sample selection mainly relied on the theoretical sampling method [32]. The specific number of people was determined according to the community population and industry data. The community committee recommended the industry representative family. To ensure the uniform spatial distribution of the samples and to include all representative industries, we classified the survey based on investigating the distance between household samples of different representative industries and the core areas of the national park, as well as the distance between the residence of different samples and the core area of the national park.

A total of 160 questionnaires were collected during the first distribution. After excluding invalid questionnaires, the actual valid samples totaled 147, with a 91.9% return rate. In addition, a second round of questionnaire distribution was carried out through an online collection method. Researchers distributed 90 questionnaires, and 83 valid samples were collected. The return rate was 92.2%.

Most respondents were between 26 and 45 years old, accounting for 47.8%; the educational background of respondents was mostly secondary school and below, accounting for 80.5%; and the monthly income of the majority of respondents was below 6000 RMB, accounting for 79.5%. In addition, most residents of the community were born locally, accounting for 64.3% of respondents. Moreover, the research areas of this study include minority autonomous regions, with agriculture and animal husbandry as the pillar industries, accounting for 40% of all sources of income (see Table 1).

Table 1. Demographic sample.

Survey Item	Type	Frequency (Sample = 230)	Percentage (%)
Gender	Male	120	52.2
	Female	110	47.8
Age	Under 16	4	1.7
	16–25	57	24.8
	26–45	110	47.8
	45–65	47	20.4
	Over 65	12	5.3
Education	Primary school	16	7
	Junior school	66	28.7
	High school	92	40
	College	30	13
	Undergraduate	13	5.7
	Postgraduate	2	0.8
	Other	11	4.8
Average Salary per Month	Less than 1000 (RMB)	7	3
	1001–3000 (RMB)	60	26.1
	3001–6000 (RMB)	116	50.4
	6001–9000 (RMB)	41	17.8
	Over 9001 (RMB)	6	2.7
Attribute of Residents	Native Settlers	148	64.3
	Migrants (Non-Native Settlers)	82	35.7
Source of Income	Tourism	70	30.4
	Animal Husbandry and Plantation	95	41.3
	Other	65	28.3
Distance from the Core Region	Less than 5 km	37	16.1
	5–10 km	55	23.9
	10–15 km	86	37.4
	15–20 km	52	22.6

3.3. Descriptive Analysis

Integrating the effective questionnaires (see Tables 2 and 3), we used SPSS 23.0 to perform a descriptive analysis of the distribution patterns of the mean, standard deviation, variance, skewness, and kurtosis of the survey sample items. The questionnaire in this study contained 19 items. We divided the 19 items into five potential variables according to the dimensions, including related impact system, behavior attitude, subjective norms, perceptual behavior control, and behavior intention. The items were investigated in the form of a Likert-5 scale, where the numbers 1, 2, 3, 4, and 5, respectively, represented strongly agree, agree, slightly agree, disagree, and strongly disagree. Regarding the results, the standard deviation of each item was greater than 0.60; the Likert-5 scale thus corresponded with the research expectations. The skewness value was between -0.974 and -0.298, the kurtosis value was between -0.531 and -0.091, the absolute value of the skewness was less than 3, the absolute value of the kurtosis was less than 10, and the sample data were normally distributed.

Table 2. Research aspects and measurement items.

Research	Measurement	Hypothesis	References
Related Impact System Scale of Qilian Mountain National Park System Pilot Area	Stage of Cognition	RIS1: I have a basic understanding of the related information of the Qilian Mountain National Park System Pilot Area and the communities' functions regarding the construction of the pilot area. RIS2: I understand the policy mechanism implemented in the Qilian Mountain National Park System Pilot Area. RIS3: The government actively promoted relevant knowledge to community residents in constructing the Qilian Mountain National Park System Pilot Area. RIS4: The government provided policy guidance and technical support in constructing the Qilian Mountain National Park System Pilot Area.	Zhang et al. (2017) [22] Zhou et al. (2017) [27]
Behavioral Attitude Scale of Qilian Mountain National Park Residents' Participation	Attitude Cognition	BA1: The construction of the Qilian Mountain National Park is inseparable from the participation of the community, which is the core element of the development of the national park. BA2: The Qilian Mountain National Park implements a community co-management mechanism, which is also the future trend regarding the development of national park communities. BA3: The Qilian Mountain National Park System Pilot Area can generate revenue based on eco-tourism and I can profit from it. BA4: The Qilian Mountain National Park System Pilot Area can increase awareness of my community and increase individuals' sense of pride.	Han et al. (2010) [23] Zhang et al. (2017) [24]
Subjective Norms Scale of Qilian Mountain National Park Residents	Attitude Cognition	SN1: The Qilian Mountain National Park Administration believes that community residents' awareness of participation in the construction of national parks should be raised at this stage. SN2: Schools and relevant education departments believe that community residents' awareness of participation in the construction of national parks should be raised at this stage. SN3: My friends and family members believe that community residents' awareness of participation in the construction of national parks should be raised at this stage	Zhou et al. (2014) [28]
Perceptual Behavior Control Scale of Qilian Mountain National Park Residents		PBC1: I have a basic understanding of the process of the community participation and related policies for the Qilian Mountain National Park System Pilot Area. PBC2: I can take on relevant responsibilities as a community resident after implementation of the Qilian Mountain National Park System Pilot Area. PBC3: I have an optimistic attitude towards the intentions of community residents' participation after implementation of the Qilian Mountain National Park System Pilot Area. PBC4: I have a supportive attitude towards community residents' active participation in constructing the Qilian Mountain National Park System Pilot Area.	Wang et al. (2020) [31]
Behavioral Intention Scale of Qilian Mountain National Park Residents' Participation	Behavioral Intention	BI1: As a community resident, I am willing to actively participate in constructing the Qilian Mountain National Park. BI2: I will actively cooperate with the National Park Administration to fulfill various requirements for community construction. BI3: I will encourage people around me to participate in the project actively and ask them to learn relevant information. BI4: I will actively participate in the volunteer activities needed in the construction of the national park.	Zheng et al. (2014) [29]

Table 3. Normal distribution data test.

Abbreviations of Measurement Hypothesis	Mean	Standard Deviation	Variance	Skewness		Kurtosis	
	Statistics	Statistics	Statistics	Statistics	Standard Error	Statistics	Standard Error
RIS_1	3.6478	1.14166	1.303	−0.644	0.16	−0.13	0.32
RIS_2	3.5565	1.12682	1.303	−0.595	0.16	−0.2	0.32
RIS_3	3.4652	1.19902	1.438	−0.638	0.16	−0.261	0.32
RIS_4	3.8565	1.18996	1.416	−0.974	0.16	0.133	0.32
BA_1	3.513	1.21033	1.465	−0.56	0.16	−0.531	0.32
BA_2	3.2217	1.12473	1.265	−0.298	0.16	−0.451	0.32
BA_3	3.4348	1.16821	1.365	−0.571	0.16	−0.227	0.32
BA_4	3.5	1.16255	1.352	−0.538	0.16	−0.28	0.32
SN_1	3.7087	1.20675	1.456	−0.717	0.16	−0.316	0.32
SN_2	3.6348	1.15468	1.333	−0.625	0.16	−0.294	0.32
SN_3	3.613	1.20146	1.443	−0.696	0.16	−0.177	0.32
PBC_1	3.5783	1.1599	1.345	−0.7	0.16	−0.091	0.32
PBC_2	3.5304	1.15082	1.324	−0.621	0.16	−0.203	0.32
PBC_3	3.8957	1.0645	1.133	−0.929	0.16	0.377	0.32
PBC_4	3.6565	1.14812	1.318	−0.642	0.16	−0.344	0.32
BI_1	2.4652	1.29699	1.682	0.647	0.16	−0.663	0.32
BI_2	2.5565	1.3262	1.682	0.37	0.16	−1.076	0.32
BI_3	2.2696	1.25953	1.586	0.868	0.16	−0.304	0.32
BI_4	3.2826	1.4059	1.977	−0.38	0.16	−1.089	0.32

4. Results

After analyzing the variables using descriptive analysis, we utilized Cronbach's alpha to test the reliability and validity of the measurement indicators of the questionnaire. In addition, the maximum likelihood estimation was used to analyze the structural model after the reliability and validity met the basic research requirements.

4.1. Reliability Analysis

We performed the reliability analysis using SPSS 23.0 to test 19 observable variables in 230 returned questionnaires. In most situations, the ideal Cronbach's alpha coefficient of a scale should be above 0.6. This study's overall Cronbach's alpha coefficient was above 0.8, which means that the obtained survey results had good internal consistency (see Table 4).

Table 4. Reliability analysis.

Item	Cronbach's Alpha	Standardized Cronbach's Alpha	Number of Items
Overall Scale	0.846	0.858	19
Related Impact System Scale	0.871	0.871	4
Behavioral Attitude Scale	0.886	0.886	4
Perceptual Behavior Control Scale	0.855	0.856	3
Subjective Norms Scale	0.817	0.819	4
Behavioral Intention Scale	0.879	0.883	4

Reliability analysis is a standard method to test the validity of a survey. It measures how the sample data reflects the final research contents and goals. Therefore, the higher the value of reliability, the more the survey data reflects the authentic results of the research. In general, there are two types of reliability analysis: content validity and structure validity. Because the content of the items involved in this study was reviewed and analyzed by experts in the relevant field, we did not focus on the content validity in this article.

We conducted correlation tests on the Kaiser-Meyer-Olkin (KMO) value and Bartlett sphericity. The results (see Table 5) indicated that the KMO value was greater than 0.8; and the sphericity test value was 2519. The degree of freedom (Sig) was less than 0.001 (generally, it is reasonable to conduct factor analysis when the value of KMO is greater than 0.8 and the Sig value is less than 0.001), so it was suitable for subsequent factor analysis. In terms of structure validity, we extracted and analyzed the factors through principal component analysis. According to the existing model, we extracted and rotated five common factors using the Kaiser maximum variance method. The results in Table 6 show A1 is highly correlated with RIS1–4, A2 is highly correlated with BA1–4, A3 is highly correlated with BI1–4, A4 is correlated with SN1–3, and A5 is highly correlated with PBC1–5. Combined with the results displayed in Table 7, it was found that the cumulative contribution rate of the total variance of the load factor was 71.303%, which is greater than 60% (it is generally considered that the survey has good structural validity when the cumulative contribution rate of the total variance of the load factor is greater than 0.6), so the questionnaire used in this study had good validity.

Table 5. KMO value and Bartlett sphericity test.

Scale Type	KMO Sampling Suitability Quantity	Bartlett Sphericity Test		
		Approximate Chi Square	Degree of Freedom	Significance
Overall scale	0.93	2519.525	171	0.0000
Related Impact System Scale	0.831	442.137	6	0.0000
Behavioral Attitude Scale	0.824	509.384	6	0.0000
Conceptual Behavior Scale	0.822	597.195	6	0.0000
Subjective Norms Scale	0.85	638.718	6	0.0000
Behavioral Intention Scale	0.825	916.342	6	0.0000

Table 6. Composition matrix after rotation.

Items	Composition				
	A1	A2	A3	A4	A5
RIS2	0.769				
RIS4	0.755				
RIS3	0.722				
RIS1	0.716				
BA1		0.798			
BA2		0.773			
BA3		0.734			
BA4		0.642			
BI1			−0.874		
BI2			−0.855		
BI3			−0.732		
BI4			0.569		
SN3				0.764	
SN2				0.749	
SN1				0.745	
PBC3					0.679
PBC4					0.611
PBC2					0.6
PBC1					0.579

Table 7. Variance contribution rate of load factor.

Composition	Initial Eigenvalue			Sum of Squares of Rotating Load		
	Total	Percentage of Variance	Cumulation (%)	Total	Percentage of Variance	Cumulation (%)
A1	8.603	45.276	45.276	3.092	16.276	16.276
A2	1.947	10.249	55.525	2.965	15.605	31.88
A3	1.207	6.353	61.878	2.856	15.031	46.911
A4	0.921	4.849	66.726	2.444	12.863	59.774
A5	0.869	4.576	71.303	2.19	11.528	71.303

4.2. Structural Equation Model Analysis

Based on the pre-mentioned research method, this researchers in this study used the structural equation model (SEM) to perform a linear analysis on the established extended model of the Theory of Planned Behavior (see Figure 1) to verify whether the positive relationships between variables is reasonable.

4.2.1. Parameter Fitting Analysis

It is necessary to conduct a parameter fitting analysis when utilizing the structural equation model to do research. The leading indicators contain the chi-square value (CMIN) and the degree of freedom value (DF); a ratio of these values between 1 and 3 indicates the model has a good degree of fitting. In addition, if the value of the root mean square error of approximate (RMSEA) ranges from 0.05 to 0.08, this indicates that the result is reasonable; a value less than 0.05 means that the degree of fitting is better. For IFI, NFI, AGFI, CFI, and RFI (incremental fit index, normed fit index, adjusted goodness of fit index, comparative fit index, and relative fix index, respectively), the range is usually between 0 and 1. Specifically, the closer the index is to 1, the better the fit. If the index is greater than 0.9, it indicates a good degree of fit (see Table 8).

Table 8. Model fitting index.

CMIN/DF	RSMEA	IFI	NFI	GFI	AGFI	CFI	RFI
1.697	0.055	0.959	0.959	0.899	0.968	0.958	0.938
pass	good	pass	pass	acceptable	pass	pass	pass

4.2.2. Hypothetical Test

Commonly, the hypothesized model is tested after the fitting analysis. Generally, the relationship between the absolute value of the standardized path coefficient and the variable is positively correlated. The positive and negative values represent the relevant influence directions. At the same time, the critical ratio (CR) must be satisfied when the absolute value is larger than 1.96 and the p-value is less than 0.05. When these conditions are met, the hypothesis is supported.

In this study, we analyzed the related variables of the hypothesized structural equation model based on the maximum likelihood method. The standardized path coefficients of the analyzed variables had a high significance level, and they met the relevant requirements of the CR and p-value. After verification, it was found that the hypotheses were all true (see Table 9). The specific analysis is as follows (see Figure 3).

Table 9. Model hypothesis test.

Path	Estimate	SE	CR	*p*	Result
BA<—RIS	0.706	0.077	9.199	***	Support
SN<—RIS	0.862	0.081	10.694	***	Support
PBC<—RIS	0.674	0.078	8.629	***	Support
BI<—RIS	0.413	0.292	4.714	***	Support
BI<—BA	0.676	0.136	4.738	***	Support
BI<—SN	0.38	0.149	3.257	0.037	Support
BI<—PBC	0.892	0.332	2.388	0.017	Support

Note: *** represents strong statistical significance and the value is 0.001.

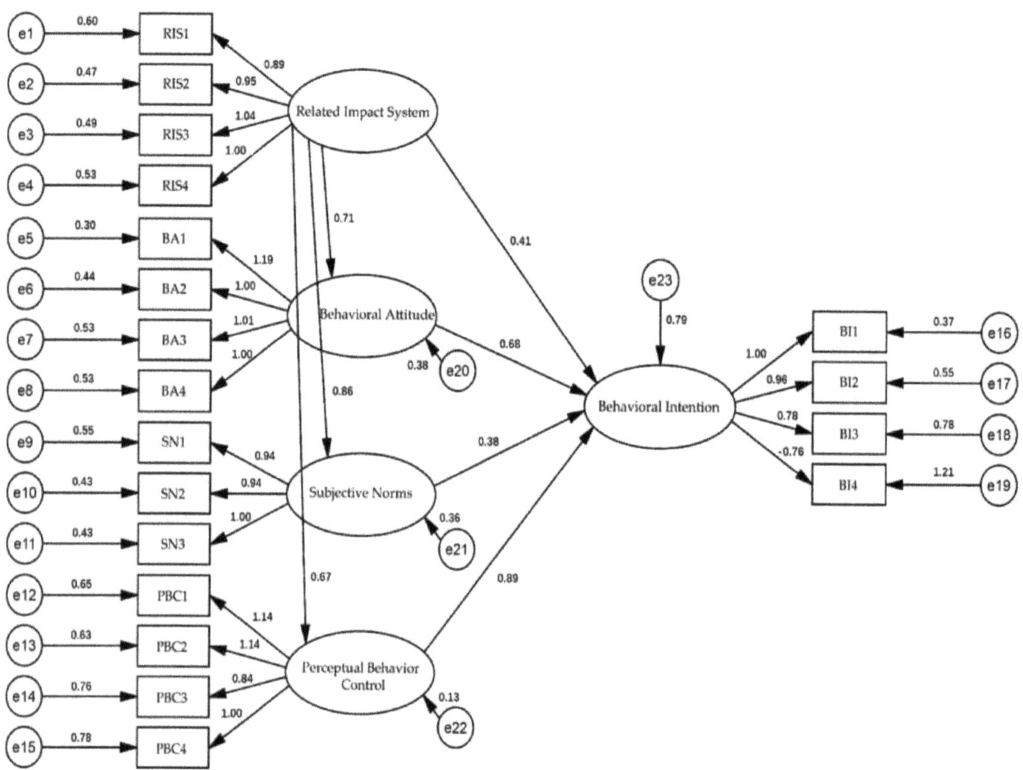

Figure 3. Structural equation model.

The critical ratio (CR) of the relevant influence system (RIS) on community residents' behavioral intention (BI), behavioral attitude (BA), subjective norms (SN), and perceptual behavior control (PBC) are 9.199, 10.694, 8.629, 4.714, respectively. These numbers are far greater than 1.96, and the *p*-values are all 0. Therefore, it can be seen that the relevant policies for the National Park System Pilot Area have tremendous potential, which influences community residents' affective commitment and community participation. From large to small, the order of these influences was subjective norms (SN), behavioral attitude (BA), perceptual behavior control (PBC), and behavioral intention (BI). Therefore, the results indicate that the National Park System Pilot Area has a significant direct impact on the behavior and ideology of the community residents and an indirect impact on community residents' behavioral intentions.

The critical value (CR) and the *p*-value of the behavioral attitude (BA) of the community residents to the behavioral intention (BI) of the community residents are 4.738 and 0, respectively. Compared with other latent variables, the behavioral attitude has a more significant positive influence on the behavioral intention. Therefore, it can be inferred that the subjective attitude of community residents is the core element that determines the willingness to participate.

The critical value (CR) and the *p*-value of the impact of community residents' subjective norms (SN) on community residents' participation intentions are 3.257 and 0.037, respectively, which correspond with the limit of the correlation coefficient of the positive path influence. The value of the standardized path coefficient is smaller among all of the latent variables, which indicates that the subjective norms of the community residents have limited influence on the participation of community residents. Therefore, it was found that relatives, friends, and community residents' subjective perceptions have no significant influence on an individuals' enthusiasm for participating in constructing the Qilian Mountain National Park.

The critical value (CR) and the *p*-value of the impact of community residents' perceptual behavior control (PBC) on community residents' participation intentions are 2.388 and 0.017, respectively, which supports positive effects. In addition, the standardized path coefficient is 0.89, which is the largest value among all the latent variables. Hence, as the latent variable preset by the model, perceptual behavior control plays a critical role in investigating community residents' participation in constructing national parks. On the other hand, the result reflects the personal will and related abilities of community residents to determine their willingness to participate in the construction of the Qilian Mountain National Park to the largest extent.

5. Discussion

The Qilian Mountain National Park System Pilot Area has a significant and positive impact on community residents' participation. It indirectly affects residents' participation intentions through intermediary variables, including behavioral attitude, subjective norms, and perceptual behavior control. The results show that the National Park System Pilot Area has a relatively low impact on the willingness of the community to participate: the related impact system (path coefficient = 0.41) < participation behavior attitude (path coefficient = 0.68) < perceptual behavior control (path coefficient = 0.89). Among the variables, "related impact path–community residents' participation" is a one-way direct effect. The influence effects of "related impact path–participation behavior attitude", "related impact path–subjective norms", and "related impact path–perceptual behavior control" are 0.71, 0.86, and 0.67, respectively.

The results indicate that perceptual behavior control is the dominant factor that affects the participation willingness of community residents in the Qilian Mountain National Park. Specifically, the order of influence is perceptual behavior control (path coefficient = 0.89) > participation behavior attitude (path coefficient = 0.68) > related impact system (path coefficient = 0.41) > subjective norms (path coefficient = 0.38).

The subject norm (path coefficient = 0.38) has the least significant impact on residents' awareness of participation. This indicates that residents in indigenous communities have a constant understanding of the significance of the existence of the Qilian Mountain National Park Pilot Area. Even the relatives and colleagues of these community residents have different perspectives on the issue regarding the community's participation, and most residents will not be affected.

6. Conclusions and Implications

Taking the Tianzhu and Sunan counties in the Qilian Mountain National Park area in Gansu province as examples, this research was conducted to investigate the impact of the Qilian Mountain National Park System Pilot Area on community residents' willingness to participate. After a comprehensive analysis, we drew the following conclusions and policy implications.

6.1. Conclusions

This section discusses this article's research conclusions and the related research conclusions drawn by previous scholars. This study analyzed the impact of major events related to the National Park System Pilot Area on the psychological perception of community residents and compared the related model to obtain the similarities and differences to related research.

With regard to the impact of significant events on community residents' perceptual intentions, this study's findings are similar to those of Zheng et al. (2014). Specifically, the National Park System Pilot Area serves as a "catalyst," indirectly deepening residents' local emotional identification and promoting individuals' willingness to build a new social network. However, different to previous studies, this research mainly focused on the impact of the three intermediate variables (behavioral attitude, subjective norms, and perceptual behavior control) on the participation intention related to the National Park System Pilot Area. That is, taking the National Park System Pilot Area as the only independent variable, and the participation intention as the only dependent variable, we investigated the community residents' perceptions and willingness to participate in major reforms (e.g., the implementation of the National Park System Pilot Area).

Most scholars have focused on environmental protection and other relevant issues for studies related to the National Park System Pilot Areas and community issues. The focus of this article is the impact on the willingness of community residents to participate. We set the National Park System Pilot Area as the only influencing factor in the preset model. They analyzed the specific relationships among each variable by influencing the intermediate variables. Based on the results, we found that the National Park System Pilot Area has a certain positive impact on community residents' willingness to participate. The findings of this article are consistent with the research conclusions drawn by Zhou et al. (2017) that related to the perceived impact of community participation. In addition, we utilized the theory of planned behavior (TPB) model to subdivide the subjective perceptual factors that affect community participation and clarify the results of different responses of the relevant factors (behavioral attitude, subjective norms, perceptual behavior control) to the dependent variable (the willingness of community residents to participate) when these factors are affected by the independent variable (the National Park System Pilot Area).

6.2. Implications

Based on the background that China's national parks are still in an exploratory stage, in order to increase the participation of community residents following the implementation of the Qilian Mountain National Park System Pilot Area, the following implications should be taken into account.

1. The National Park System Pilot Area positively increased the community residents' willingness to participate. Therefore, regarding future development, it is necessary to strengthen government guidance and popularize scientific research and instructions related to the theme of the Qilian Mountain National Park. In addition, it is critical to ensure that during the exploration period, the pilot policy can reach a critical consensus with the community in the construction of national parks and promote residents' participation intentions in relevant communities through positive relationships.

2. The National Park System Pilot Area, behavioral attitude, subjective norms, and perceptual behavior control all positively affected community residents' willingness to participate. Therefore, the government may need to consider setting up standardized and professional departments to reward and commend community participants who actively participate in constructing national parks and who demonstrate certain achievements. This method would enhance residents' sense of honor and improve the quality of community life. By "reshaping" the cultural landscape of Qilian Mountain National Park, the community will become the major component of the cultural landscape. The Qilian Mountain National Park cultural value system will be formed based on the principle of attitude priority. In addition, it is necessary to classify the

<ol start="3">
<li></li>
</ol>

study areas during the next phase of managing the national park communities and subdivide the relevant suggestions from different communities' feedback regarding the residents' perceptions of participation.

3. The impact of the construction of the National Park Pilot Area on community residents discussed in this study has a strong correlation with Wallner's three major perceptions of protected areas that affect residents (the economic situation, the history of natural protection, and the power balance between the involved stakeholders) [33]. The authors of this study believe that economic factors and the balance of interests of all parties are essential indicators for coordinating community participation. In addition, we also note that the majority of respondents were not highly educated and had limited abilities to obtain relevant information regarding the development of the national park community. Insufficient publicity of the national park concept and passive acceptance of the policy are the reasons why many respondents have a neutral or negative attitude [34,35]. Therefore, increasing the level of community participation in national parks requires not only active publicity, but also requires relevant organizations to establish effective communication mechanisms and social networks so that residents can actively interact with the National Park Management Committee on issues encountered in the construction of national parks, thereby increasing participation in constructing the national park community.

This study introduces the relevant variables of the Qilian Mountain National Park System Pilot Area into the theory of planned behavior (TPB) model. Based on field surveys and expert interviews, we hypothesized that these variables significantly impact the communities' willingness to participate, and the empirical results supported this hypothesis. At the same time, based on previous research, this study was conducted in economically underdeveloped areas dominated by ethnic minority groups in the national park communities and investigated the impacts of residents' willingness to participate.

Author Contributions: Conceptualization, L.J. and J.W.; methodology, J.W.; software, J.W.; validation, L.J., J.W. and Z.W.; formal analysis, L.J.; investigation, J.W.; resources, Z.W.; data curation, J.W.; writing—original draft preparation, J.W.; writing—review and editing, Z.W.; visualization, L.J.; supervision, L.J.; project administration, J.W.; funding acquisition, L.J. All authors have read and agreed to the published version of the manuscript.

Funding: This research was supported by the National Natural Science Foundation of China (Grant No. 51968042).

Institutional Review Board Statement: The protocol was reviewed and approved by The Lanzhou University of Technology of Science and Technology Board and the Protocol Number is LUT-SAD-51968042.

Informed Consent Statement: Not applicable.

Data Availability Statement: Some or all data, models, or code that support the findings of this study are available from the corresponding author upon reasonable request.

Conflicts of Interest: The authors declare no conflict of interest.

Notes

1. The model is extended based on the TPB theoretical model. The extension item is the relevant influence system of the Qilian Mountain National Park Pilot Area. This extension item has a particular impact on cognitive atti-tude and behavioral intention.
2. The structural equation model is a verification of the model assumptions in Figure 1. The connecting arrows represent the establishment of the influence, and the values under the arrows represent the degree of influence. Therefore, the expected assumptions in Figure 1 are all established.

References

1. Dilsaver, L.M. *America's National Park System: The Critical Documents*; Rowman & Littlefield: Washington, DC, USA, 2016.
2. Selin, S.; Chevez, D. Developing a collaborative model for environmental planning and management. *Environ. Manag.* **1995**, *19*, 189–195. [CrossRef]
3. Agrawal, A.; Gibson, C.C. Enchantment and Disenchantment: The Role of Community in Natural Resource Conservation. *World Dev.* **1999**, *27*, 629–649. [CrossRef]
4. Eneji, V.C.O.; Gubo, Q.; Okpiliya, F.I.; Aniah, E.J.; Eni, D.D.; Afangide, D. Problems of public participation in biodiversity conservation: The Nigerian scenario. *Proj. Apprais.* **2009**, *27*, 301–307. [CrossRef]
5. Ormsby, A.; Kaplin, B.A. A framework for understanding community resident perceptions of Masoala National Park, Madagascar. *Environ. Conserv.* **2005**, *32*, 156–164. [CrossRef]
6. Abukari, H.; Mwalyosi, R.B. Comparing Conservation Attitudes of Park-Adjacent Communities: The Case of Mole National Park in Ghana and Tarangire National Park in Tanzania. *Trop. Conserv. Sci.* **2018**, *11*, 1940082918802757. [CrossRef]
7. Joyner, L.; Lackey, N.Q.; Bricker, K.S. Community Engagement: An Appreciative Inquiry Case Study with Theodore Roosevelt National Park Gateway Communities. *Sustainability* **2019**, *11*, 7147. [CrossRef]
8. Ormsby, A.A. Perceptions on the Park Periphery: Resident, Staff and Natural Resource Relations at Masoala National Park, Madagascar; ProQuest, Antioch New England Graduate School. 2003. Available online: https://www.proquest.com/openview/19bcecc210949d98c4c4fb9615c66c50/1?pq-origsite=gscholar&cbl=18750&diss=y (accessed on 10 December 2021).
9. Baorong, H.; Yi, W.; Liyang, S.U.; Zhang, C.; Chen, D.; Sun, J.; He, S. Pilot programs for national park system in China: Progress, problems and recommendations. *Bull. Chin. Acad. Sci.* **2018**, *33*, 76–85.
10. Peng, H.; Zhang, J.; Lu, L.; Tang, G.; Yan, B.; Xiao, X.; Han, Y. Eco-efficiency and its determinants at a tourism destination: A case study of Huangshan National Park, China. *Tour. Manag.* **2017**, *60*, 201–211. [CrossRef]
11. Dong, Q.; Zhang, B.; Cai, X.; Morrison, A.M. Do Local Residents Support the Development of a National Park? A Study from Nanling National Park Based on Social Impact Assessment (SIA). *Land* **2021**, *10*, 1019. [CrossRef]
12. Brown, B.; Perkins, D.D.; Brown, G. Place attachment in a revitalizing neighborhood: Individual and block levels of analysis. *J. Environ. Psychol.* **2003**, *23*, 259–271. [CrossRef]
13. Milligan, M.J. Displacement and identity discontinuity: The role of nostalgia in establishing new identity categories. *Symb. Interact.* **2003**, *26*, 381–403. [CrossRef]
14. He, S.Y.; Wei, Y.; Su, Y.; Min, Q.W. A grounded theory approach to understanding the mechanism of community participation in national park establishment and management. *Acta Ecol. Sinica.* **2021**, *41*, 3021–3032.
15. Mensah, I. Effects of socio-demographic characteristics and perceived benefits of tourism on community participation in tourism in the Mesomagor Area of the Kakum National Park, Ghana. *Athens J. Tour.* **2016**, *3*, 211–230. [CrossRef]
16. Wondirad, A.; Ewnetu, B. Community participation in tourism development as a tool to foster sustainable land and resource use practices in a national park milieu. *Land Use Policy* **2019**, *88*, 104155. [CrossRef]
17. Ajzen, I.; Fishbein, M. Attitude-behavior relations: A theoretical analysis and review of empirical research. *Psychol. Bull.* **1977**, *84*, 888. [CrossRef]
18. Miller, Z.D. A Theory of Planned Behavior approach to developing belief-based communication: Day hikers and bear spray in Yellowstone National Park. *Hum. Dimens. Wildl.* **2019**, *24*, 515–529. [CrossRef]
19. Goh, E.; Ritchie, B.; Wang, J. Non-compliance in national parks: An extension of the theory of planned behaviour model with pro-environmental values. *Tour. Manag.* **2017**, *59*, 123–127. [CrossRef]
20. Reigner, N.; Lawson, S.R. Improving the efficacy of visitor education in Haleakalā National Park using the theory of planned behavior. *J. Interpret. Res.* **2009**, *14*, 21–45. [CrossRef]
21. Ajzen, I.; Driver, B.L. Application of the Theory of Planned Behavior to Leisure Choice. *J. Leis. Res.* **1992**, *24*, 207–224. [CrossRef]
22. Zhang, W.B.; Li, G.P. Ecological compensation, psychological factors, willingness and behavior of ecological protection in the Qinba ecological function area. *Resour. Sci.* **2017**, *39*, 881.
23. Han, H.; Hsu, L.T.; Sheu, C. Application of the Theory of Planned Behavior to green hotel choice: Testing the effect of environmental friendly activities. *Tour. Manag.* **2010**, *31*, 325–334. [CrossRef]
24. Zhang, Y.; Yu, X.; Cheng, J.; Chen, X.; Liu, T. Recreational behavior and intention of tourists to rural scenic spots based on TPB and TSR Models. *Geogr. Res.* **2017**, *36*, 1725–1741.
25. Wang, C.; Zhang, J.; Cao, J.; Hu, H.; Yu, P. The influence of environmental background on tourists' environmentally responsible behaviour. *J. Environ. Manag.* **2019**, *231*, 804–810. [CrossRef] [PubMed]
26. Chen, Y.S.; Lin, Y.H.; Wu, Y.J. How personality affects environmentally responsible behaviour through attitudes towards activities and environmental concern: Evidence from a national park in Taiwan. *Leis. Stud.* **2020**, *39*, 825–843. [CrossRef]
27. Zhou, R.; Zhong, L.S.; Yu, H. Community perception towards Qianjiangyuan National Park System Pilot Area administrative measures. *Resour. Sci.* **2017**, *39*, 40.
28. Zhou, L.; Li, Q.; Lin, Z. Outcome efficacy, people-destination affect, and tourists environmentally responsible behavior intention: A revised model based on the theory of planned behavior. *J. Zhejiang Univ. Humanit. Soc. Sci.* **2014**, *44*, 88–98.
29. Zheng, Q.; Xia, Z.; Lou, W.; Liu, Y. The impact of world heritage declaration on residents' sense of place: A case study of Mount Lang. *Tour. Sci.* **2014**, *28*, 54–64.

30. Hair, J.F.; Anderson, R.E.; Tatham, R.L.; Black, W.C. *Multivariate Data Analysis*, 5th ed.; Prentice Hall: Upper Saddle River, NJ, USA, 1998.
31. Wang, C.; Zhang, J.; Xiao, X.; Sun, F.; Xiao, M.; Shi, Q. Examining the dimensions and mechanisms of tourists' environmental behavior: A theory of planned behavior approach. *J. Clean. Prod.* **2020**, *273*, 123007. [CrossRef]
32. Karlsson, A. Survey sampling: Theory and method. *J. Metrika* **2008**, *67*, 241–242. [CrossRef]
33. Wallner, A.; Bauer, N.; Hunziker, M. Perceptions and evaluations of biosphere reserves by local residents in Switzerland and Ukraine. *Landsc. Urban Plan.* **2007**, *83*, 104–114. [CrossRef]
34. Dimitrakopoulos, P.G.; Jones, N.; Iosifides, T.; Florokapi, I.; Lasda, O.; Paliouras, F.; Evangelinos, K.I. Local attitudes on protected areas: Evidence from three Natura 2000 wetland sites in Greece. *J. Environ. Manag.* **2010**, *91*, 1847–1854. [CrossRef] [PubMed]
35. Abukari, H.; Mwalyosi, R.B. Local communities' perceptions about the impact of protected areas on livelihoods and community development. *Glob. Ecol. Conserv.* **2020**, *22*, e00909. [CrossRef]

Article

Sustainability and Tourist Activities in Protected Natural Areas: The Case of Three Natural Parks of Andalusia (Spain)

María Bahamonde-Rodríguez [1], Francisco Javier García-Delgado [1,*] and Giedrė Šadeikaitė [2]

[1] Research Group Local Development Institute, University of Huelva, 21071 Huelva, Spain
[2] Department of Human Geography, University of Alicante, San Vicente del Raspeig, 03080 Alicante, Spain
* Correspondence: fcogarci@uhu.es

Abstract: As a form of protection, a nature park is often created to protect and valorise natural and cultural heritage in peripheral rural areas. However, in terms of multifunctionality, new nature parks incorporate traditional productive activities, such as recreational and tourist activities, which sometimes compromise sustainability. The research objective is to study the relationship between tourism and sustainability in the nature parks of Sierra de Aracena y Picos de Aroche, Sierra Norte de Sevilla and Sierra de Hornachuelos that make up the Dehesas de Sierra Morena Biosphere Reserve in Andalusia, Spain. Therefore, selective interviews have been carried out with the stakeholders to establish their perception of sustainable tourism and the presence of dominant discourses. The main conclusions indicate: (1) the presence of different dominant discourses on sustainability, namely the conservationist and mercantilist ones, with the prevalence of the economic dimension; (2) poor awareness and adaptation to the context of global change; and (3) the presence of competitive relationships that generate difficulties for the governance of sustainable tourism.

Keywords: protected areas; stakeholders; sustainable tourism; Sierra Morena

Citation: Bahamonde-Rodríguez, M.; García-Delgado, F.J.; Šadeikaitė, G. Sustainability and Tourist Activities in Protected Natural Areas: The Case of Three Natural Parks of Andalusia (Spain). *Land* **2022**, *11*, 2015. https://doi.org/10.3390/land11112015

Academic Editors: Rui Yang, Yue Cao, Steve Carver and Le Yu

Received: 28 September 2022
Accepted: 9 November 2022
Published: 11 November 2022

Publisher's Note: MDPI stays neutral with regard to jurisdictional claims in published maps and institutional affiliations.

1. Introduction

Since the mid-20th century, there has been a progressive disempowerment of rural communities, which downgraded them to the periphery [1,2], marginalised them "to the dominant development processes" [2] (p. 2) and exposed them to external decision-making and the continuous loss of competitiveness and employment [3,4]. As a result, these processes conditioned sustainability and generated a complex institutional context of development in these areas. Agriculture lost its monopoly in rural areas [5], and the multifunctionality of uses and diversification became an opportunity to adapt to the changing reality [3,6]. As a part of adaptive strategies [6], a novel [7] non-productive socioeconomic activities were incorporated. These activities included leisure and recreation (tourism, restoration), conservation and maintenance of biodiversity and valorisation of natural and cultural heritage, residential development and traditional productive activities, which were reinterpreted [3,8].

In this context, the public administration understands the need to protect, conserve and safeguard natural and cultural resources, to establish Protected Nature Areas (herein PNA) to conserve biodiversity and ecosystems [9], to provide ecosystem services [10] and search for solutions to climate change [11]. Rural spaces, previously agricultural, turned into so-called "preserved spaces" [12], where environmental attractiveness and ease of access generate advantages, yet with certain limitations of use since local decision-making is subject to conservation criteria [2]. Nonetheless, such a nature conservation process is not exempt from contradictions [13] when attempting to turn into a natural environment socioecological system. It generates different perceptions, conflicting opinions, rejection and management problems resulting from the relationship between the social system and its environment [14] and the discourse between conservation and productivism. Therefore, it is necessary to integrate the local population into the establishment, decision-making and management processes of PNAs [15]. By integrating the PNA into the social and territorial environment through management instruments, these areas would evolve from the so-called "museum" to conservation broadly and compatibility with the rational use of resources [16,17].

Different categories of PNAs are created, which often overlap in the same territory, ranging from total protection (naturalisation) to flexible protection structures, in which the protection of natural and cultural heritage coexists with socioeconomic development and socioecological systems [10,15]. This is the case of nature parks (herein NtP) in Spain that are integrated into more conservationist models [16]. NtPs play an essential role in leisure activities [18] and tourist and recreational activities [19,20], positioning themselves in the tourist market until these PNAs form a pillar of the Community Agricultural Policy and the LEADER initiative [19,21]. In this way, the PNA appears as one of the large-scale tourism typologies of peripheral rural spaces [4]. Although environmental tourism is the classic motivation in PNAs, in this case, it is not only ecotourism or nature tourism but somewhat rural tourism, where nature-based products and services are added [18,21,22]. In the context of post-Fordism or "a la carte" tourism [23], rural tourism incorporates an advanced segmentation, the search for experiences and sensitivity to environmental issues as a response to changing demand [18,21,22,24]. However, it often goes from promoting the place to selling it [25], and tourism simultaneously produces and hides the contradictions of capitalism based "on creating attractions, or new sources of an accumulation from the very crises it produces" [26] (p. 529). Thus, three processes converge in the territory, namely: (1) patrimonialization through the protection and conservation of natural and cultural heritage, (2) enhancement of tourist value through the creation of new spaces [1] and (3) commodification of nature [26]. These coexisting processes generate discourses between conservation and exploitation [2,27], authenticity and trivialisation [28], abandonment of traditional activities and implementation of new ones, changes in use and simplification [29] or public service and private use [26]. Thus, in the context of increasing recreational and tourist frequentation [18,21], sustainability and sustainable tourism in rural areas and PNAs are perceived as a challenge given the complex and conflicting relationships [30] and exposure to risks due to their inherent fragility [31].

Since the 1980s, there has been a growing interest in applying sustainability to tourism [32–34]. According to the UNWTO, sustainable tourism is tourism "that takes full account of its current and future economic, social and environmental impacts, addressing the needs of visitors, the industry, the environment and host communities" [35] (p. 12). However, as Saarinen [34] and some other authors [36–38] in the scientific literature point out, this vision introduces the necessity of the industry, despite the need to establish limits to growth. On the other hand, it is difficult to apply the concept of sustainability imprecisely [39], which results in continuous failure [40]. It led to consider sustainable development and sustainable tourism diversely, flexibly [41] and indistinctly [40], leaving sustainability as "a 'wicked' or meta-policy problem that has led to new institutional arrangements and policy settings at international, national and local scales" [42] (p. 5). Ultimately, the rational use of resources that sustainability entails depends on values and ideologies [42] and, therefore, must be understood within the context of political–economic discourse aiming for sufficient and efficient tourism [43]. Consequently, despite contradictory, divergent or tangent discourses [44], sustainable tourism as a dominant paradigm in tourism development is identified [32,34,45]. Nevertheless, it fails to orient itself towards genuine sustainability of planning, management and policies, democratic empowerment, environmental conservation and social justice [45], or behavioural change towards sustainability [46], generating a hybridisation between neoliberalism and sustainable development [45].

Currently, a dominant discourse of sustainability and sustainable tourism appears to be somewhat rhetorical and more of a fashion to address the public [47], since sustainable tourism is considered an end. Another prevailing discussion considers sustainability as the need for neoliberal growth defined by the markets [48], based on introducing new definitions instead of solving issues. On the other hand, sustainable tourism development tends to focus on the product [40], and tourists who buy sustainable tourism products are still fewer [49]. The ideal green tourist does not consume less. Instead, they do so responsibly [50], which results in businesses focusing on responsibility rather than sustainability [51], since responsible tourists pay for it. This situation often masks the unsustainable activities of companies [34]. In other words, responsibility arises from tourist segmentation or the emotional relationship with nature (perception) [41]. Consequently, there is room for a critical analysis of the relationship between sustainable tourism and sustainable development [52], which does not emphasise establishing the limits of growth [43] and avoiding its impacts [34], understanding that "more does not mean better, and growth does not mean development" [37] (p. 131).

Nonetheless, the interpretation and application of the sustainability concept differ according to the type of destination [41], the natural environment, the characteristics of the community, the institutional framework and the management policies [53] that are necessary to be adapted to the context [54]. Thus, in PNA, including rural areas, sustainability and sustainable tourism are given significant importance, and a challenge of sustainability is perceived as the sine qua non-condition. Thus, tourism sustainability is, at the same time, a planning criterion for future development [38] and a primary instrument to increase the quality of life of the local population, maintain natural values and attractions and improve the quality of the tourist experience [55]. Tourism in rural areas or PNAs is often considered to be sustainable in itself [56] because it attracts a small number of visitors, does not require a wide range of "services, infrastructures and [types of] equipment" (herein SIEs) and tourists tend to be interested in the host community, its landscape and environmental attractions. Yet such a correlation lacks support in the scientific literature [27], and many intended measures have not effectively contributed to sustainability [56]. Moreover, many of such intents fell into perverse effects [37] by generating negative impacts, indicating that conservation strategies are essential to sustainable development [16,57,58]. Therefore, although sustainability is one, it is necessary to take into account the presence of its four interdependent and interconnected dimensions [38,41,59] as follows:

- The environmental dimension relates to the optimal use of natural resources, compatible with the maintenance of ecological processes and the conservation of biodiversity [38,60]. Understanding that tourism depends on conserving the resources that attract tourism is critical [61].
- The economic dimension focuses on economic growth, efficiency and optimisation of resources [38,60,62] for the satisfaction of material human needs and objectives [63], job creation and long-term competitiveness [38], while preventing economic growth from pressuring other sustainability dimensions [64]. Although there are increasing constraints for tourism policy, planning and management to consider and incorporate into sustainability issues [65], biases often occur towards the economic dimension [32].
- The sociocultural dimension emphasises respect for the material and immaterial culture of the community [2,38,63] and social capital [38], which results in the strengthening of equity, social cohesion and improvement of the quality of life [38,62] and contributes to intercultural understanding and tolerance [38]. The sociocultural dimension is valuable in addressing the problems of tourism development [58], fundamental in rural tourism based on a close personal interaction between residents and visitors, contributing to the revaluation of authenticity and identity [2,66].
- The political–institutional dimension concentrates on the political system and the distribution of power [62,67], including the development of management systems, governance and stakeholder participation [38,63,68], and a favourable context, defined by the regulatory framework and institutional structures [34] without which sustainable tourism cannot exist.

This multidimensional vision of sustainable tourism and the above-indicated dimensions prevail in the literature. They are considered a tool to define sustainability issues, highlighting that the interconnectivity between the dimensions is widespread as a holistic and long-term concept [41].

Therefore, this study aims to address the stakeholders' perception of the sustainability of tourism activities in a subregional area of Andalusia (Spain) forming three NtPs: NtP "Sierra de Aracena y Picos de Aroche", NtP "Sierra Norte de Sevilla" and NtP "Sierra de Hornachuelos", which together constitute the Dehesas de Sierra Morena Biosphere Reserve. The research, thus, raises the following questions: (1) Are tourism activities sustainable? (2) Which dimension of sustainability dominates? (3) What dominant discourses are present among the actors, and how are they manifested? (4) How do relationships between stakeholders influence sustainability management?

2. Materials and Methods

2.1. Data and Methods

This research applied the case study in the analysis and the prevailing discourse of the perception of sustainable rural tourism, the perception of sustainability by stakeholders [17,41,47,59], relationships between stakeholders and governance in rural spaces and PNAs [15,69–72].

This research attempts to analyse the awareness, understanding, commitment, attitudes and practices of those involved in or influencing the sustainable tourism planning process [4,73,74] through their opinions and perceptions [17,41,69,75] on three central themes: (1) sustainability, including its dimensions, and tourism [38,41,59]; (2) the presence of dominant discourses and the rhetoric of sustainability [44]; and (3) management problems derived from relationships between stakeholders [60,69,73]. For this purpose, semi-structured interviews (herein Int) were carried out with ten open questions (see Table 1), adapted from Renfors [41].

Table 1. Interview questions.

Code	Question	Topics
(q1)	What function do the nature park and biosphere reserve have in your destination (and others)?	(2,3)
(q2)	What is the value of the landscape in tourism?	(1) (2) (3)
(q3) [a]	How do you perceive sustainable tourism development in your destination?	(1)
(q4) [b]	Does sustainability have a substantial effect on the tourism development of your destination? Why?	(1)
(q5) [a]	What kind of conflicts related to sustainability is created between stakeholders?	(1) (2) (3)
(q6) [a,c]	Could you give a practical example of sustainable tourism development in your destination? What would you improve?	(1) (3)
(q7) [c]	What happens in the context of global change with your destination?	(1) (3)
(q8)	Are there difficulties in managing the tourist space?	(1) (2) (3)
(q9) [b]	Does tourism contribute to local development?	(1) (3)
(q10)	What consequences has COVID-19 had on the destination?	(1) (3)

[a] Questions based on Renfors [41]. [b] Questions adapted from Renfors [41]. [c] Control questions are aimed at the total or partial understanding of what the interviewees are being asked and to establish whether the answers respond to the awareness or dominant discourse. Authors' elaboration.

A non-probabilistic sampling method was used by conducting 40 interviews between April and July 2021 (Table 2). Some interviewees were directly identified: NtPs directors, local action groups (herein LAGs) managers, municipal stakeholders—including mayors or council members and municipal tourism technicians—and a private foundation, i.e., nature conservation NGO. Tourism companies and business associations were selected according to the type of services they provide, e.g., accommodation and tourist activities, and their local or foreign character [76]. Some were chosen based on good practices described by the interviewees, applying the snowball technique [41,77]. The territorial balance of the interviews was sought (relevance within each NtP, centrality/periphery) (Table 2). Given the restrictions imposed by sanitary measures due to COVID-19, the interviews were conducted via videoconference on Google Meet©. These restrictions prevented the conduct of systematic interviews with the local population.

Table 2. Conducted interviews.

NtP	Municipality [1]	Interview	Position/Type	Genre	Age Range
	Aracena	(Int01)	NtP director	M	50–59
	Aracena	(Int02)	LAG manager	F	50–59
	Cañaveral de León	(Int03)	Mayor	F	40–49
	Cumbres Mayores	(Int04)	Mayor	F	30–39
	Almonaster la Real	(Int05)	Councilor	F	30–39
	Cortegana	(Int06)	Councilor	F	40–49
	Aracena	(Int07)	Municipal technician	F	50–59
	Aroche	(Int08)	Municipal technician	F	40–49
	Arroyomolinos de León	(Int09)		M	40–49
	Jabugo	(Int10)		M	50–59
	Alájar	(Int11)	Tourism company	F	50–59
	Cortegana	(Int12)		F	30–39
	Los Marines	(Int13)		F	50–59
	Aracena	(Int14)	Business associations	F	50–59
	Santa Olalla del Cala	(Int15)	Foundation manager	M	20–29
	(Sevilla)	(Int16)	NtP director	M	40–49
	Cazalla de la Sierra	(Int17)	LAG manager	M	30–39
	Alanís	(Int18)	Mayor	F	40–49
	Cazalla de la Sierra	(Int19)	Mayor	M	50–59
	Real de la Jara	(Int20)	Councilor	F	30–39
	San Nicolás del Puerto	(Int21)	Councilor	M	50–59
	Las Navas de la Concepción	(Int22)	Municipal technician	F	20–29
	Cazalla de la Sierra	(Int23)		M	50–59
	Cazalla de la Sierra	(Int24)		M	40–49
	Constantina	(Int25)	Tourism company	F	40–49
	El Pedroso	(Int26)		M	20–29
	San Nicolás del Puerto	(Int27)		F	30–39
	Puebla de los Infantes	(Int28)	Business associations	M	50–59
	(Córdoba)(a)	(Int29)	NtP director	M	50–59
	(Obejo)(a)	(Int30)	LAG managers	M	50–59
	Posadas	(Int31)	LAG managers	M	50–59
	Villaviciosa de Córdoba	(Int32)	Councilor	M	>60
	Hornachuelos	(Int33)	Municipal technician	M	40–49
	Almodóvar del Río	(Int34)		F	30–39
	Hornachuelos	(Int35)		M	50–59
	Hornachuelos	(Int36)	Tourism company	F	30–39
	Hornachuelos	(Int37)		F	40–49
	Posadas	(Int38)		F	30–39
	Posadas	(Int39)		M	50–59
	Hornachuelos	(Int40)	Business associations	M	40–49

The NtP column spans groups: *Sierra de Aracena y Picos de Aroche* (Int01–Int15), *Sierra Norte de Sevilla* (Int16–Int28), *Sierra de Hornachuelos* (Int29–Int40).

[1] The seat does not coincide with the municipalities of the NtP.

The interviews were transcribed and coded, depending on whether it was verbalised by the interviewees (emic) or identified by the researcher a posteriori (etic) [78], to determine the underlying discourses [75].

The interviews were complemented with territorial recognition, i.e., patrimonial valuation, accessibility analysis and informal interviews with the local population on tourism and sustainability between September and November 2021. The fieldwork and informal discussions allow contrasting the opinions of the stakeholders interviewed with direct observation and the local population's views.

2.2. Case Study

Sierra Morena is a Mediterranean mid-mountain range that extends through the southwest of the Iberian Peninsula. The ecological and landscape richness led to the creation of six NtPs in the Andalusian Sierra Morena [79], composing the scope of the study of the three westernmost NtPs (Figure 1). Namely, NtP "Sierra de Aracena y Picos de Aroche" (herein SAPA), NtP "Sierra Norte de Sevilla" (herein SNS) and NtP "Sierra de Hornachuelos" (herein SH). These three NtPs are also the Special Conservation Areas and Special Protection Areas for Birds. They were declared as the UNESCO Dehesas de Sierra Morena Biosphere Reserve (herein DSMBR), and SNS was declared as the UNESCO World Geopark (herein UWGpSNS) [80].

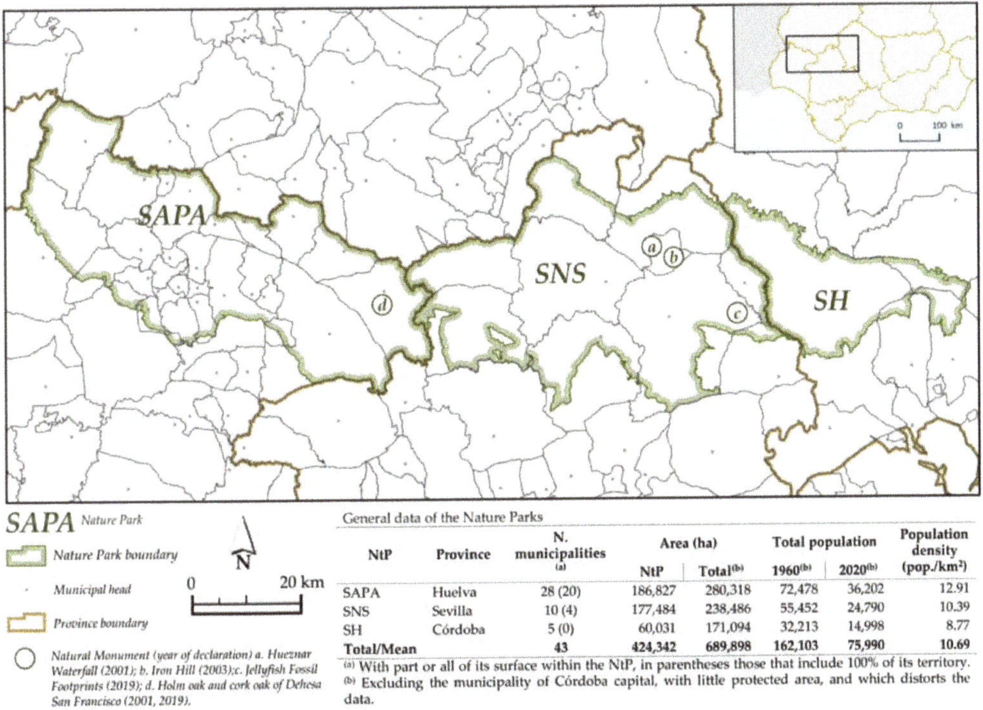

SAPA Nature Park

Nature Park boundary

Municipal head

Province boundary

Natural Monument (year of declaration) a. Huéznar Waterfall (2001); b. Iron Hill (2003);c. Jellyfish Fossil Footprints (2019); d. Holm oak and cork oak of Dehesa San Francisco (2001, 2019).

N

0   20 km

General data of the Nature Parks

NtP	Province	N. municipalities (a)	Area (ha)		Total population		Population density (pop./km²)
			NtP	Total[b]	1960[b]	2020[b]	
SAPA	Huelva	28 (20)	186,827	280,318	72,478	36,202	12.91
SNS	Sevilla	10 (4)	177,484	238,486	55,452	24,790	10.39
SH	Córdoba	5 (0)	60,031	171,094	32,213	14,998	8.77
Total/Mean		43	424,342	689,898	162,103	75,990	10.69

[a] With part or all of its surface within the NtP, in parentheses those that include 100% of its territory. [b] Excluding the municipality of Córdoba capital, with little protected area, and which distorts the data.

Figure 1. Scope of the study. Source: [80,81]. Authors' elaboration.

The specific characteristics of Sierra Morena gave rise to an agro-silvo-pastoral exploitation system that is unique in the world known as "dehesa", or "montado" in Portuguese (Figure 2), which is a cleared Mediterranean forest where forestry, livestock and the hunting vocation predominates, with an exploitation system dominated by large estates [82,83]. The dehesa has generated an exceptional landscape with high heritage values [84], yet it is subject to change processes of coverage and degradation due to abandonment or overexploitation [85]. Currently, the dehesa is facing the extreme effects of climate change and the seca (fungal disease of *Quercus ilex* and *Quercus suber*, the main species of the dehesa) [11].

Figure 2. Dehesa de San Francisco, Santa Olalla del Cala (SAPA). Dehesa of cork oaks where you can see free-range Iberian pigs.

SAPA and SNS are large NtPs with population settlements in the interior, while SH has an intermediate area and large properties predominate, lacking an internal network of settlements [20]. In addition, most of the surface of these NtPs is a private property [20].

Since the 1960s, several general and specific factors have generated the crisis and the massive rural exodus in Sierra Morena [86], which lost 52.55% of its population between 1960 and 2020 (Figure 1). Today it has an ageing population and low demographic density (10.69 inhabitants/km^2), with 18 municipalities with <10 inhabitants/km^2 [81]. Only 4 municipalities have >5000 inhabitants (2020), whereas 18 municipalities have <1000 inhabitants [81]. Traditional economic activities are linked to the dehesa [83], highlighting the Iberian pig farming in SAPA and SNS (Figure 2) and its associated industry [86,87], while in SH, hunting and forestry activities predominate [20].

SAPA and SNS have regional entities with LAGs practically identified with their territory, while SH is distributed between two LAGs (see Figure 3).

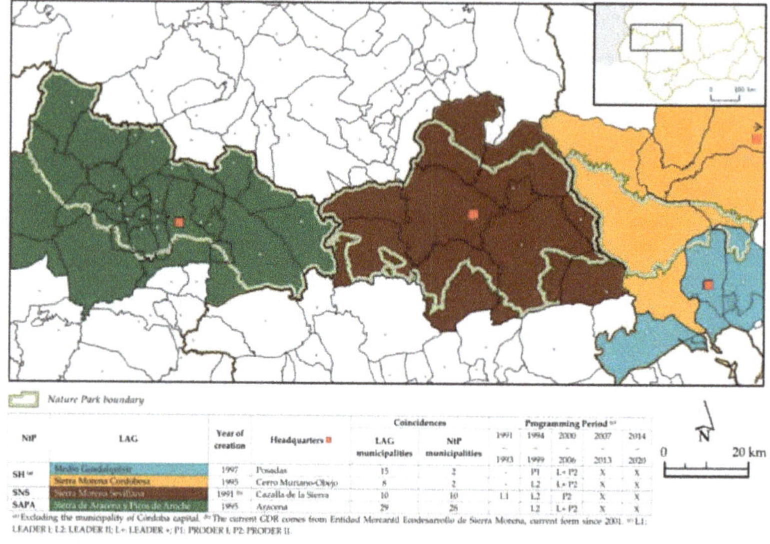

NtP	LAG	Year of creation	Headquarters	Coincidences		Programming Period [c]				
				LAG municipalities	NtP municipalities	1991 – 1993	1994 – 1999	2000 – 2006	2007 – 2013	2014 – 2020
SH [a]	Medio Guadalquivir	1997	Posadas	15	2	-	P1	L+P2	X	X
	Sierra Morena Cordobesa	1995	Cerro Muriano-Obejo	8	2	-	L2	L+P2	X	X
SNS	Sierra Morena Sevillana	1991 [b]	Cazalla de la Sierra	10	10	L1	L2	P2	X	X
SAPA	Sierra de Aracena y Picos de Aroche	1995	Aracena	29	28	-	L2	L+P2	X	X

[a] Excluding the municipality of Córdoba capital. [b] The current CDR comes from Entidad Mercantil Ecodesarrollo de Sierra Morena, current form since 2003. [c] L1: LEADER I; L2: LEADER II; L+: LEADER +; P1: PRODER I; P2: PRODER II.

Figure 3. LAGs in the scope of the study. Source: [88]. Authors' elaboration.

It is a space with a marked peripherality, poor land communications with the provincial capitals and within and between the counties. The better accessibility of the east of SAPA and south of SNS leads to developing leisure and residential functions linked to the city of Seville [83,89,90]. There was practically no tourist offer at the time of the proclamation of the NtPs [91], assuming the tourism offer organiSed itself organically [21,27]. However, the unequal activity distribution remains until the present, with a predominance of rural houses and the recent appearance of tourism activity companies [92].

3. Results

3.1. The Presence of Dominant Discourses and the Rhetoric of Sustainability

The declaration of the three NtPs protects biodiversity and guarantees territorial unanimity between NtPs directors and the Foundation. The LAGs' managers agree on the importance of protecting and conserving yet highlight the lack of territory revitalisation by the NtPs. They indicate that it is necessary to "overcome its function as a figure of conservation, as has happened in other NtPs of Andalusia" (Int30). While tourism companies and business associations highlight conservation, they only partly assign protection as a guarantor of environmental sustainability. Municipal stakeholders often perceive NtPs as a limitation imposed on the local population from outside.

For NtPs directors and LAGs managers, NtPs have been fundamental for attracting tourists, although mostly central areas benefited more due to better accessibility, tradition and tourist offers. Companies that offer outdoor activities highlight that there is still much-untapped tourism potential. Tourism companies and business associations relate NtPs (and DSMBR) with promotion and marketing opportunities, emphasising that being part of an NtP allows for ecotourism and sustainable activity for a specific tourist/visitor. Nonetheless, it is indicated that efforts to foster environmental protection are usually limited to advertising and posters, without creating a real fundamental change in the client. However, they recognise that they have become the main tourist attraction in the central municipalities over time.

Considering the above, NtPs directors and LAGs managers highlight that the declarations of the NtPs have led to the development of other protection types, including DSMBR, Special Protection Areas for Birds, Special Conservation Areas and UWGpSNS, as well as the obtaining of other certifications such as the European Charter of Sustainable Tourism and Sierra Morena Starlight Reserve (herein SMSTRE). Although the DSMBR is perceived as motivated by the NtPs directors, they affirm that its importance has not been visualized nor its potential developed because there is no management instrument (this document is currently being drafted). It is divided between three provinces, making it unfamiliar to the LAGs managers. For tourism companies, DSMBR is relatively unknown, and they even deny its existence. For municipal stakeholders, these declarations and certifications are just titles that are added to others or patrimonial protections, e.g. tourist brands and patrimonial declarations, linking their quantity directly to the inflow of visitors. Only one of the municipal stakeholders highlights that DSMBR is a recognition of the traditional way of exploiting natural resources, assuming international promotion and the receipt of public aid.

For the directors, sustainability is a context for developing NtPs as established by the regulation. The LAGs managers point out that the values of sustainability have been recognised during the pandemic, with citizen participation being crucial for achieving it, yet without undervaluing its economic costs. It is generalised among municipal stakeholders to affirm that traditional and tourist activities have always been sustainable despite the prevailing three visions, as follows:

(1)   The sustainability of traditional activities and tourism is necessary; "without sustainability there is no development" (Int03); raising awareness among companies and the local population is essential.

(2) Traditional and tourist activities are sustainable, but sustainability creates "a difficulty to compete" (Int32); urban spaces receive "water, air and recreation" (Int17) for which the NtPs "need compensation" (Int01).

(3) Sustainability is "something to sell nature" (Int33), and now "everything has to be sustainable" (Int04).

For tourism entrepreneurs, sustainability is an end. Still, they affirm that this is not the case for most tourism companies that seek sustainability because of the subsidies that can be obtained or because tourists demand it. Other tourism entrepreneurs even say that "sustainability is an invention" (Int27) because tourism activity in rural areas must necessarily be sustainable. In contrast, others claim to be learning to use sustainability "as a strength for development" (Int14).

3.2. The Pre-Eminence of One Dimension over the Others in Sustainability in Tourist Activities

The NtPs directors perceive sustainability in tourist activities because tourism is highly regulated by laws within the PNA, making it a comparative advantage for tourist satisfaction. Additionally, they tend to question the sustainability of some tourist actions, emphasising the need to exercise greater control overall. While they understand that nature tourism and ecotourism have great potential and the effects of over-frequency and overload are punctual time- and space-wise, sometimes they cause problems for the owners at harvest time. The emergency in the light of global change, with a technical and comprehensive vision, is particularly emphasised with accompanying proposals to make investments, e.g., of Next Generation EU funds, to address them. The suggested measures for sustainability are the control of access to maximum protection areas, the promotion of energy self-sufficiency in urban centers, support for active tourism and advice on diagnoses and environmental plans for companies. Implementing the Andalusian Nature Park Brand, SMSTRE and stargazing, mushroom picking, hiking and specific examples of certified companies are among the best practices. The model to follow would be that of other Andalusian NtPs.

LAGs managers agree on the increasing sustainability of tourism activities and the positive and growing influence of sustainability on the destination. However, they point out that many times the activities are sustainable for companies because of "opportunity (business) and not because of conviction" (Int02), as a way of advertising, since tourists traditionally do not choose the destination entirely only because it is sustainable. Although they point out that the trend is changing and the investments necessary to achieve sustainability are amortised thanks to the satisfaction of a "new view of the tourist" (Int31) who pays for sustainability and nature tourism and ecotourism linked to experiences, companies incorporate sustainability into their management and facilities through "personal awareness" (Int02). In the case of over-frequency and overload, LAGs managers share the views of the NtPs directors, highlighting the problems of overcrowding in urban areas and the economic impacts on farms. Likewise, while LAGs agree with NtPs directors on the global change, they indicate that "those who most notice the changes are the smallest peoples" (Int02). Although they tend to think that "it is not something imminent" (Int31) and "they only see the problems when they translate into something economical, as occurs with the drying of the oak" (Int02), they call for necessary measures to raise awareness among the population and strengthen the nature preservation legislation. They underline as measures for sustainability the promotion of energy self-sufficiency and the reduction of light pollution. Good practices focus on projects resulting from cooperation networks such as SMSTRE, the Ruta del Jabugo (herein RJ), the LongDistance Trail 48 and activities linked to the UWGpSNS. As with the NtPs, LAGs also discourse about outdoor activities, extreme sports and accommodation companies but critically assess the certifications that "tax quality" (Int17). LAGs managers do not provide role models.

Municipal stakeholders agree that tourism, including traditional activities, is sustainable. Nonetheless, there are three different opinions:

(1) Most consider sustainability inherent in traditional and tourist activities since "the environment has been preserved because traditional activities are sustainable" (Int19). They relate nature tourism and ecotourism to recreation and some complementary activities. Despite some denials, they primarily defend the need to control capacity to avoid overcrowding so that it does not damage traditional activities. Only some interviewees identify climate change as an issue, without considering it imminent, and refuse to foster legislation, expressing that "it is necessary to adapt, but we are used to it" (Int22). As measures for sustainability, these stakeholders propose betting on experiential ecological tourism and smart rural destinations, with the limitation that there are no subsidies and aid for sustainability. In contrast, others suggest sustainable investments, e.g., renewable and efficient energies, diversification of products, motorhomes, enhancement of resources, trails, paths and renovations. Sound practices are related to the development of municipal strategic plans and are exemplified by accommodation companies, agri-food companies with a tourist offer, outdoor activities and extreme sports, adventure parks and heritage rehabilitation. The model to follow as a destination refers to other municipalities of the NtP and other NtPs of Andalusia and the Basque Country.

(2) The perception of sustainability as the basis of development is limited to some municipal stakeholders who advocate that traditional and tourist activities are generally sustainable, but "not everything rural is sustainable" (Int20), making awareness necessary. They identify nature tourism and ecotourism with active and sports tourism, disconnection and personalised services and experiences. They highlight a non-massive context, especially in more peripheral municipalities, and the need to limit the tourist flow and plan. Tourist companies are held responsible for overcrowding "because they think that the more people the better instead of looking for a model of quality business" (Int03), and seasonality is very marked. They relate global change as a major issue. Collaborative projects such as SMSTRE, LongDistance Trail 48, RJ, ecovillages, municipal awareness campaigns, programs against depopulation and cultural and environmental initiatives and activities as well as accommodation companies with tourist activities (agritourism, gastro-tourism) stand out as good practices. They mention regional models with particular emphasis on cultural initiatives.

(3) Sustainability as an NtP imposition is seen by the minority (SNS, SH), which points out that tourism activity has to be developed "within a sustainable framework" (Int21). It has to be legal and certified to satisfy tourists "who seek sustainability" (Int21), and this has increased during the pandemic. They do not consider over-frequency and overload, except at specific times due to the pandemic, and do not perceive any effect of global change. The measures for sustainability are related to tourism quality, while routes oriented toward a specific segment of demand are considered an example of good practice. As a model, Navarra is mentioned.

Tourism companies and business associations agree that they act in a sustainable destination where the work of the NtPs with the business community is essential. From here, two different visions are developed:

(1) A majority group, consisting of tourism companies and business associations, attribute tourism sustainability to: (a) local companies that work for environmental and economic sustainability, while foreign companies do not carry out sustainable activities; (b) the activities that are internally monitored as sustainable versus the non-monitored unsustainable ones; (c) sustainable private business activities versus unsustainable public ones because they are unrealistic and compete with private ones. These companies agree that sustainable tourism does not exist and sustainability is not a motivation, despite some changes since the pandemic as tourists, especially youth, are progressively getting involved with sustainability and complying with the rules. Betting on nature tourism and ecotourism is done for the central values of the territory, the "silence, the place" (Int23). New types of transport such as bicycles and horseback riding are available for a tourist who does not want to go by car and is respectful of

the environment, though it costs more. Overcrowding and overload are not an issue, and conflicts are due to the lack of visitors' civility. The measures for sustainability proposed are limited to training and awareness actions, and the references to global change are few. They are reluctant to converse about good practices. However, networks such as SMSTRE, RJ, service companies such as electric bicycles, adventure parks and the creation of charging points for electric vehicles are mentioned, pointing to quality certifications as an impediment. Management models are from neighboring municipalities and NtPs, indicating companies with similar activities and providing examples such as the Pyrenees or the Spanish Ecotourism Club.

(2)   A minority group of tourism companies perceives that tourism sustainability is due to the company's efforts since many wield "the flower of sustainability, and those who have spent their entire lives working in the territory, on the other hand, do not have any recognition" (Int12). They relate nature tourism with a source of employment that provides differential value in terms of negative value. It distinguishes between ecotourists who come from abroad looking for a specific offer and sustainability and nearby travelers looking for a place for their vacations and travelling in a group. At the same time, tourist satisfaction is unrelated to sustainability. Overcrowding and overloading are considered a "cancer of the territory" (Int10) that occurs in specific attractions due to lack of action and regulation, especially in the best-connected places. Self-limitation, the non-admission of large groups and the search for under-tourism are pointed out as sustainability measures. These companies understand global change as an essential and multidimensional issue. Although it does not currently affect reserves, going so far as to point out that it is necessary to "educate ourselves and educate others" (Int13), the change of tourist activities towards sustainability and the search for new, nearby markets is needed due to the decline in international tourism in the context of global change. They primarily emphasise the individual measures for sustainability, e.g., not having a pool, eliminating chemicals, ensuring energy efficiency, creating ecotourism experiences, and realising FAM trips and environmental certifications. Good practices include the implementation of municipal 2030 agendas and programs against depopulation, promoting stargazing and bird watching and strengthening companies with specific cultural and environmental activities in the open air or extreme sports without emphasizing role models.

The Foundation voices the lack of sustainability of some traditional and tourist activities. It considers that nature tourism and ecotourism are necessary but ensures controls to avoid over-frequency or overload. The measures for sustainability include the awareness of the local population and tourists as well as raising tourist capacity controls. It focuses on climate change, especially the dry season, and notes that all activities must be made sustainable, not by prohibiting, but rather by controlling traditional practices to adapt. Activities such as bird watching and mountain biking are highlighted. The Foundation perceives itself as an example of good practice.

3.3. The Political–Institutional Dimension: The Relationships between Stakeholders and the Difficulties in the Management of the Tourist Space

The different stakeholders establish collaborative relationships with other stakeholders, among which are:

- Municipal stakeholders: collaboration with other municipalities is based on formal and informal networks.
- Municipal stakeholders and tourism companies: municipal support for companies.
- LAG managers are the generation of networks with the different LAGs and with other external local agents.

Other cooperation relationships are highlighted by only one type of stakeholder, regardless of whether it affects several stakeholders, e.g., the vertical and horizontal coordination underlined by the NtPs directors (see Table 3).

Table 3. Relations between stakeholders and the indicated causes.

Type of Relation	Interviewees	Description of the Relations	Indicated Cause (Verbalized)	Institutions and Organisations Involved [1]
Positive	(Int22) (Int40) (Int19) (Int21) (Int08) (Int07)	Tourism cooperation and complementarity between municipalities	Formation of formal and informal networks	3
	(Int15)	Cooperation with external tourism companies	Control of tourist flows (bundling)	5, 7
	(Int30) (Int31) (Int02) (Int17)	Cooperation between the LAGs and with other stakeholders internal and external to the NtPs	Existence of a network and application of the LEADER approach; outward projection	2, 3, 4, 6, 7, 10, 11
	(Int33) (Int07) (Int36) (Int26) (Int11)	Cooperation of municipalities and tourism and hospitality companies	Information; technical support; nearest administration	3, 4, 6
	(Int30)	Cooperation in development strategies with NtPs	Shared actors (NtP governing board)	1, 2
	(Int12)	Cooperation between similar or complementary companies	The joint vision of destination and trust	4
	(Int01) (Int16)	Vertical (JA-TD-NtP-SM) and horizontal (NtP-PC) coordination	Regulatory framework and organisational structure NtP; participation of the municipalities in the NtP Governing Board	1, 2, 12, 13, 14
	(Int01) (Int15)	Collaboration in nature conservation	Same conservation goals between institutions	1, 7
Negative	(Int02) (Int22) (Int03) (Int21) (Int20) (Int05) (Int28)	Competition and lack of subsidiarity between municipalities; generation of "micro-destinations"; lack of coordination between attractions; scarcity of tourist activities in municipalities	The rivalry between municipalities; different levels of development; lack of communication; the existence of municipal lobbyists; political decision-making without counting on and considering the tourism sector	2, 3, 4, 6
	(Int36) (Int39) (Int37) (Int26) (Int09) (Int24)	Conflicts of use between tourism and private property	The predominance of private property; lack of entrepreneurship; incompatibility of uses; usurpation of public space	1, 4, 8, 9
	(Int29) (Int01) (Int02) (Int17) (Int32) (Int19) (Int39) (Int27) (Int24) (Int28) (Int15)	Disagreements between municipalities, NtP and JA; management conflicts; a desire to exit NtP	Restrictive regulatory framework; different speeches, politicisation; lack of communication; lack of control of activities; technical ineffectiveness; public oversight of SIEs	1, 3, 8, 14
	(Int39) (Int37) (Int35)	Unfair municipal competition to tourism companies	Creation of SIEs with public money and private management	3, 4
	(Int39) (Int35) (Int10) (Int24)	Competition between tourism companies	Duality of local–foreign companies, main–secondary activities; lack of business culture; non-business activities; lack of originality	4, 5
	(Int01) (Int30) (Int31) (Int02) (Int17) (Int20)	The difficulty for interterritorial cooperation; lack of a DSMBR planning instrument	3 NtPs, 3 provinces, 4 LAGs, 43 municipalities, different administrations and discourses; lack of coordination; the existence of municipal lobbyist	1, 3, 13, 14
	(Int13) (Int24) (Int28) (Int14)	Non-existence of a coordinating body for tourist activity in the NtP; absence of a destination; lack of a tourism strategy (brand, destination...)	Lack of agreement between the parties and stakeholder involvement; the rivalry between municipalities; politicisation; lack of goals; lack of coordination in the regional administration	1, 2, 3, 4, 6, 12

[1] 1. NtP; 2. LAG; 3. Municipality; 4. Local tourism company; 5. External tourism company; 6. Business association; 7. Foundation; 8. Local population; 9. Large landowners (generally urban); 10. Other local stakeholders; 11. Other local, territorial entity external to NtPs; 12. Provincial Council; 13. Territorial Delegation Regional Government; 14. Junta de Andalucía (regional ministries). Source: Interviews. Authors' elaboration.

Competency relationships are of crucial importance for all stakeholders, while the most visible are the following:

- The disagreements between the regional administration, i.e., NtPs, regional ministries and the municipal stakeholders, are seen by other stakeholders, which generate management conflicts and divergences caused by the restrictive regulations, the top-down approach, the ineffectiveness of the NtPs and the lack of communication.
- The competition between municipalities, observed by LAGs, municipalities and business associations, generates a lack of coordination, tourist micro-destinations and inequality in the distribution of public and private SIEs, caused by the rivalry between municipalities and the generation of lobbyists, centre-and-periphery relations and the lack of communication.

- The administrative limits created a lack of interterritorial cooperation, as perceived by the NtPs directors, all the LAGs managers and a municipal stakeholder.
- The lack of coordination and a common tourism strategy within the NtPs, perceived by tourism companies and business associations, causes the lack of destinations and a brand.

The tourist companies verbalise other competition relations about conflicts between the tourist activities and the owners that dominate the NtPs due to the usurpation of cattle trails and the limitations that private property supposes in the NtPs. Competition between tourism companies is based on local–foreign discourse, main–secondary activity, lack of business culture and originality and the presence of non-business activities, including unfair competition from non-industrial activities and even municipalities.

4. Discussion

4.1. The Presence of Dominant Discourses and the Rhetoric of Sustainability

Each stakeholder builds their reality, expressing their interests [93] collected in the dominant discourse and a representative framework [44].

The regional administration illustrates the preservation of the natural heritage through the NtPs directors, who are assigned the role of the so-called gatekeepers [94] in the process of patrimonialization and with a conservationist discourse where sustainability is the objective established by law [95]. A conflict is generated by the management of resources between conservation and traditional activities that manifest the "nature-society dualism" [96], demanding compensation by municipal stakeholders for the right to economic development and productivism despite limitations [97], and incorporating the idea of local heritage, which opposes the collective patrimonialization that comes from outside [98].

Tourism is an attractor of ecological services that positively interferes with appreciating of natural and cultural values [99]. In this way, internal and external pressures in the PNAs foster the economic use of resources greater than the intrinsic value of natural and cultural heritage [26] and, therefore, sell products and markets places, cultures and traditions [28]. A mercantilist discourse is formed [100] and linked to a vision of development as a union of endogenous and exogenous forces, public and private, based on endogenous resources yet projected outwards in terms of the flow of tourists, the arrival of capital and funding [72].

The conservationist discourse is assumed by conviction by proactive tourism entrepreneurs [101], as identified by LAGs and other companies alike, the Foundation based on their purpose [102] and some municipal stakeholders [103]. The LAGs discuss sustainability from a broader perspective of equality and existing challenges rather than ecological thinking [41]. They position themselves on the side of conservation, but complemented with sustainable tourism as an attractor [99].

Firstly, the productivist discourse and then the mercantilist one is accepted by business associations, most tourism companies and municipal stakeholders [103]. They redefine the sustainability concept and tend to fall into contradictions when simultaneously speaking of sustainability and the elimination of limitations or the increase in the number of tourists [43].

These discourses are not permanent and tend to change [72]. Thus, in the municipal elections of 2015 and 2019, the traditional political forces of social and Christian Democrats lost the elections in several municipalities. The new leaders changed the focus of local policies, allowing us to speak of municipalities of change, as dissenting voices, environmentalists and conservationists who positively value NtPs as a guarantee of sustainability. On the other hand, in 2019, the regional elections involved a change in the regional government with a center-right coalition that promotes a change in the regional environmental administration, favoring economic activities such as tourism, which implies a more productive discourse, as perceived by the interviewees, contrasting with the previous position that separated tourism, conservation and sustainability [104].

4.2. The Pre-Eminence of One Dimension over the Others in the Sustainability of Tourist Activities

All interviewees agree that tourism and tourism activities are sustainable [3,27]. However, the different stakeholders insist on one of the dimensions of sustainability and interpret sustainable tourism differently.

The conservationist discourse is dominated by the perception that the environmental dimension of tourist activity is necessary to care for and improve the environment in the NtPs, since sustainability cannot be renounced before and outside of tourist activity [27]. Namely, NtPs directors, LAGs managers, municipal stakeholders of change, proactive companies and the Foundation are concerned about global and climate change [105]. These concerns generate uncertainties about conservation and tourism activities [37] and make tourism companies focus on changes in activities. The landscape affects their business [106], and thus they consider maintaining long-distance visitors without damaging natural capital [33] and better managing local tourism flows [107]. To mitigate the effects, they propose policies and actions aimed at reducing sources of greenhouse gas emissions through investments and legislation [11,108], sustainable tourism activities planning [108] and carrying out awareness-raising campaigns for local and tourist populations [109]. The interviewees attribute to the tourists a motivation connected with their emotions and personal relationship with nature through experiences [21,110,111], a rediscovered relationship with the environment after COVID-19 [112] and a progressive involvement in the sustainability of specific tourism [27].

Environmental sustainability is fundamental for proactive European companies that specialise in high-added-value nature tourism and ecotourism [22]. They respond to the conviction by developing sustainable products [49] to turn sustainability into an instrument of business success [113] by focusing on the viability of the company [114] instead of performance. These interviewees consider that tourism does not generate significant environmental impacts, except those derived from the spatial–temporal concentration of demand [27], pointing out that it is necessary to control the flows by regulating the physical load [24,115]. Thus, companies can limit the offer to themselves to maintain quality by betting on non-aggressive and low-intensity tourism [75], where only the NtPs directors show concern about the use of water resources [11]. Sound practices are identified with the environment and resources conservation [116], the will to preserve heritage for the future [27] and the eco-efficiency of companies [56].

However, the specific examples that emphasise environmental and economic dimensions and, to a lesser extent, sociocultural and political–institutional, have common characteristics. These characteristics include innovation [117], generation of cooperation networks [118] and employment [119], a propensity to collaborate [71], enhancement of synergies [113], entrepreneurship [120], local sourcing [114], diversification of the product supply [121] and offering quality through environmental accreditations and certifications [116]. Neo-rural businesses and foreigners have launched many of these initiatives [13,66], although there are also innovative local initiatives [55]. Generally, the initiatives mentioned are few, reiterated and concentrated in SAPA, with more significant tourism development [20]. These interviewees are concerned about seasonality, which compromises service quality and business viability [27]. Moreover, they highlight the impact of visitors who occupy private farms or steal harvested fruits with the urban idea that everything in the countryside belongs to everyone [41]. They do not follow models and only mention other Andalusian NtPs.

The economic dimension predominates and is considered the most critical [41,63,72] by the majority of municipal stakeholders, tourism companies and business associations, insisting that sustainability is not well sold due to the scarce effort of the administration and necessary public aid for companies [56]. They see tourism as a private economic activity [52] from which people live and produce economic growth, a more important objective than sustainability [32,61], curtailed by the limitations established by the NtPs. While it is not a criticism of sustainability, they understand that the restrictions do not benefit tourism companies. Sustainability is attributed to the location, origin and activities, regardless of whether they are environmentally sustainable and compatible with protection [4,76,122]. Therefore, sustainability is not considered a necessity, but an option [123].

Moreover, for tourism companies and business associations, nature tourism and ecotourism become a business opportunity to satisfy tourists [34,49], create products that emphasise natural heritage and thus increase their profitability [23] and amortize the investment. Therefore, sustainability is a learning, rhetorical discourse [46,124], in which sustainable tourism in a collective context becomes responsible tourism in the personal sphere [34,51]. Its main interest is to sell nature or receive subsidies, to benefit from the few tourists who buy sustainable products [49] and tourists who seek domesticated nature [22]. Therefore, they do not value environmental sustainability as a tourist motivation [22] but merely an attraction for tourists without considering the impacts [55].

Nonetheless, these interviewees recognise that nature tourists have increased during the pandemic [122], and companies must take advantage of it. They tend to reason that global change is not imminent and is only appreciated when it causes economic damage [11], whereas climate change requires adaptivity [125]. With few exceptions, the majority acknowledges that over-frequency and overload are common in specific places and times [27], especially in central areas and urban centres, mainly due to hikers and the perimeter closures established during the pandemic [126]. Likewise, it is perceived that the NtPs must solve the issue since the environmental dimension is exclusively their concern. However, it is not a priority matter, and some positively value the high demand caused by COVID-19. In any case, they deny the possibility of developing restrictive regulations prevailing a short-term view of local authorities [127] and tourism companies, which are committed to increasing flows instead of improving quality and sustainability.

Almost no measures for sustainability are considered, and they identify it with the implementation of plans to promote ecotourism and segmentation through smart rural destinations [117]. These interviewees vaguely speak of initiatives, activities and projects that respond to market segmentation while criticising the accreditation and certification requirements that tax the ecological [117]. The specific initiatives mentioned respond to neo-Fordist products of the Disneyization of nature [22] and are neither innovative nor original [21,22]. On the other hand, electric vehicle recharging points or the diversification of products, e.g., e-bikes or motorhomes, are considered modernity [128] without considering sustainability, e.g., in terms of carbon footprint or derived pollution [129]. The models to follow are chosen not based on sustainability but on entrepreneurial success, indicated by brand awareness and continuous tourist inflow.

Contrary to the interviewees' opinions, the sociocultural dimension of sustainability is fundamental for residents [41,68]. LAGs managers and municipal stakeholders interrelate it with the economic dimension [41,59] by linking it with the environment. The economic dimension reflects the wish to continue living in the place, maintain and improve the quality of the residents' lives, preserve vitality and address depopulation and ageing, rather than the capacity of the community to accept negative social impacts due to saturation [27]. Additionally, municipal stakeholders tend to understand social good as the maximisation of market transactions [100]. As the territorial analysis and the literature on the field of the study indicate [12,83,89,90,130], the interviewees are resistant to mentioning social impacts and latent conflicts that depend on stakeholders [58]. Given the sociocultural value of the dehesas and the fact that a large part of the population is still linked to primary activities, the interviewers insisted on abandoning and changing the predominant traditional activities [27], focusing on the economic arguments, e.g., low profitability and the abandonment of the activity, rather than social ones, e.g., uprooting, showing that they do not relate the loss of farm labour in favour of tourism with the loss of the landscape that justifies tourist activities and experiences [119]. Only some interviewees valued agritourism-based initiatives very positively [56,131] and indicated more sustainable, conservationist discourse as examples to follow.

4.3. The Political–Institutional Dimension, the Relations between Stakeholders and the Difficulties in the Tourist Space Management

Depending on the participating stakeholders, there are three levels of relationships: public–public, private–private and public–private.

4.3.1. Public–Public Level

The NtPs directors specify cooperative relations at institutional levels, i.e., horizontal coordination between NtPs, municipalities and the Provincial Councils and external vertical coordination with the regional administration. The latter is the one that sets the management guidelines, responding to a traditional top-down model in PNAs [132].

The relationship between the NtPs and the municipalities is the most competitive, observed by all types of interviewees in the three NtPs, given the non-participatory management model [19,133,134]. It results in the NtPs directors being seen as external agents to the territory [19,104], except in SAPA, where the headquarters is in the region and the physical proximity determines this perception. Nevertheless, the NtPs directors consider themselves local stakeholders [94]. According to municipal stakeholders, the regulatory framework imposes limitations from the outside [16,19,83] and restricts the right to development without offering compensation [41]. The NtPs directors perceive the constraints of economic activities and urbanisation as the explanation for conservation, yet they also see opportunities for multifunctionality and diversification, generating an economic boost [104]. However, there are also underlying, unspoken issues to be addressed:

(a) There is the presence of two dominant discourses, i.e., mercantilism and conservationism.
(b) Local politicians understand themselves as the supporters of the local population and the productive system, as the self-assigned function [72].
(c) Concerning point b, the municipal leadership's role in appropriating heritage as a local government discourse opposes collective patrimonialization with the politicisation of nature protection that is wanted to exist in the NtP [19,83].
(d) Directors perceive the municipal stakeholders as opposing the NtPs [133].

These conflicts between local administrations and NtPs do not depend on the traditional governing party in the municipality or region. They are related to the dominant economic activities [12], the tourist centrality and the lower identification with the NtPs, showing that the patrimonialization process has not been completed. Even in the periphery, the municipal stakeholders of the change see the control of activities and conservation as a collective patrimonial function for non-productive functions that must be controlled in the context of global change. They perceived the significant natural value of the above restrictive framework [17].

The NtPs directors also perceive as an issue the so-called border effect between three NtPs in 3 provinces and 43 municipalities, caused by the institutional framework [73] that limits territorial cooperation [135]. This limitation is appreciated in border municipalities, i.e., between provinces, regions and Spain and Portugal, by preventing intermunicipal collaboration.

On the other hand, municipal stakeholders mostly positively highlight the relationships between themselves through formal networks, e.g., associations, projects or routes and informal networks of shared interests [71]. Municipal stakeholders mainly indicate competitive relationships between municipalities, the concentration of tourism initiatives, the lack of coordination in the management of attractions and their lack of originality as the drawbacks [21,22] due to the prevalence of local discourses [20]. Furthermore, the municipalities of change and the most peripheral ones communicate the presence of municipal lobbyists in supramunicipal structures, e.g., municipal associations and LAGs [136], aiming to benefit their municipalities by reproducing centre–periphery models [137].

The directors agree that there is a lack of funding, material and personal resources in the NtPs [104] due to a management system based on public budgets and subsidies [24] and not on payment for ecosystem services [138]. For some municipal stakeholders, the lack of funding and continued financing translates into increased sustainability costs [56]. On the contrary, for others, the most significant matter is not funding, but that aid and subsidies are aimed to benefit the same objectives. This opinion is recurrent in business associations, pointing out that sometimes immobilised financing is waiting for decision-making [55].

4.3.2. Private–Private Level

Interviewees view tourism companies positively based on their activities and economic, social and cultural contribution. The parties see business associations as valid interlocutors, such as tourism industry networks [70]. Tourism companies and business associations appreciate many positive and negative relationships and interrelationships at the private–private level, where the work of the LAGs stands out [139]. Nonetheless, the cooperative relationships between tourism companies stand out where cooperation is based on their activities, ideological affinity and proactivity trust and complementarity relationships [140]. Despite that, competitive relationships are also generated between companies:

(a)   Local companies' origin of the promoters or investment is attributed to sustainability [76] because they are local, thus questioning the legitimacy of external initiatives [141]. They do not consider their characteristics and connections with the community [142] nor the role of the neo-rural [66,143] in neo-endogenous tourism [144] or community-based tourism, which is especially visible in SAPA [13]. Foreign companies, however, blame local companies for their lack of originality [21,22].

(b)   The professionalisation of the activity refers to companies with tourism as their primary activity, which emphasise that those with tourism as a secondary, non-professionalised activity do not take care of sustainability. Therefore, the reason is opportunism that considers tourism an attractor [99] and the lack of business culture and training [27].

(c)   The type of activity points to non-business activities as a significant issue as such activities do not have business maintenance costs and act as unfair competition. They consider the offer of cultural and environmental activities by cultural and private associations as either unregulated or illegal [20,27,104]. In contrast, unmonitored activities and autonomous tourism are perceived as unsustainable [64].

4.3.3. Public–Private Level

The interviewees highlight the relations at the public–private level of cooperation and, especially, of competition, which relate to the lack of information, communication and participation in decision-making [134].

The LAGs and their managers are valued positively by all the interviewees as internal, legitimate, public–private institutional actors that respond with collaborative work to an institutional incapacity [145]. They collect stakeholders' interests at different levels [145] and lead, coordinate and bring together projects and actions to stimulate and promote tourism [50,104]. LAGs managers point out their cooperative relationships based on the LEADER approach and decentralisation [137] with many institutional and private stakeholders and with other municipalities and counties that go beyond the border effect [73]. They aim to establish innovative territorial networks [20], projects based on a joint development strategy and diversification promoted by other entities, such as the RJ [87]. However, LAGs are only project developers who may not consistently achieve real change, partly because of stakeholder resistance to cooperation and the lack of collective learning [146].

Municipalities in the context of neoliberal governance [34,41] do not have competencies in tourism and environmental policy. They often do not have a dedicated budget, yet they act as inhibitors or facilitators of sustainable tourism development, showing local leadership to business disinterest [147] and top-down directives [67]. However, they frequently face issues related to a new specialisation [72], observed by LAGs managers as a danger of so-called "pan-tourism" where any other activities are disregarded. Municipal stakeholders specify collaboration with tourism and restaurant companies, improving business activities with advice and support, and assume the role of intermediaries, acting to enhance the tourism sector's prospects based on SIEs. The relations between municipalities and tourism companies are also competence-based, and the tourism companies explain them as follows:

(a) The benefit of the municipalities to local companies is an obstacle for exogenous companies [41], regardless of their characteristics, especially relevant in the SH.

(b) The benefit of the municipalities to the external companies for the search for external financing, investments and capital flows in the short term [31,72] as an expression of the mercantilist discourse and the development of alliances with external capital [12] that hinders internal entrepreneurship, mainly prevailing in SAPA and SNS municipalities.

(c) There is unfair competition between municipalities and companies that promote SIEs with public investment, and the direct management of tourist attractions affecting negatively private business viability and calls for a clear definition of the municipal role for accountability [67,70].

For some tourism companies, the NtPs contribute to the conservation of biodiversity and the control of agricultural practices through good practices, technical support and the limitation of urbanisation. In contrast, for others, the NtPs suppose constraints and bureaucracy for their initiatives [104], as mentioned by the LAGs managers and recognised as a deficiency by the NtPs directors. The conflicts between traditional and tourism use generate a disconnection between tourism companies and the NtPs (SNS), as it does not necessarily control illegal or unlawful activities in the PNAs nor act diligently due to the scarcity of resources, and they are working as a so-called foreign administration to the territory [55].

Tourism companies and business associations highlight the lack of coordination between public and private initiatives and stakeholders. It is caused due to the inexistence of a coordinating body for tourism activities in the territory and a plan establishing tourism bases and objectives accepted by all [41,60,93]. This absence exists as an unfavorable institutional framework [53] due to the presence of administrations at different territorial scales and the distribution of environmental zones (PNAs). Tourism competencies are also divided between two regional ministries [64], leading to ineffectiveness [104]. In addition, the lack of stakeholder involvement [41], political issues and discrepant interests result in a lack of action coordination [91] and tourist micro-destination creation by the municipalities.

5. Conclusions

The perception of stakeholders about the sustainability of tourism activities, despite the contextual differences, is not substantially different from other spaces, with elements identified by Renfors [41].

Our study shows that stakeholders recognise that sustainability is generally the purpose of the PNA. However, tourism sustainability is compromised by focusing on one of the sustainability dimensions and not on the interconnection of dimensions. Sustainability is, for some, an option. For others, it is an opportunity and, for others, a conviction. It is a threat to consider sustainability as an option when conserving PNAs, and the fight against global change relies on it [57]. On the other hand, tourism should not be underestimated as an instrument for development in an agrosystem such as the dehesa, "which has been capable of changing and reinventing itself randomly from different socioeconomic and historical contexts" [84] (p. 134).

Stakeholders mutually recognise each other [63] and acquire attributes as a result of their dominant relationships with others [60,73] and perceive themselves differently based on their roles, discourses and influence and characteristics [73]. The opposing dominant discourses manifest their differences, although they might change over time. However, the dogmatism and pragmatism of the discourses are equally dangerous in a fragile territory, requiring a compromise between the actors. The relations between stakeholders materialise through ties of cooperation and competition, which hinder the governance and management of NtPs and tourist activities. At the same time, the rigidity of the regulatory framework prevents not only reaching agreements but also proposing them.

The NtPs were created to protect, although they are inhabited spaces where citizens do not participate. It indicates a paradox of sustainability where we protect the space and restrict its use, the local population must behave sustainably and the tourist population requires environmental training.

More than three decades after the proclamation of the NTPs, tourism has developed, while sustainability remains a matter for a few, and the heritage process has not been completed. The accumulation of protection objects does not guarantee conservation. Therefore, without adequate management and financing instruments, the DSMBR continues to be, two decades later, an opportunity for sustainable tourism development.

The main limitations of this study are that it is based on the opinions of the interviewees, so it is necessary to consider to what extent to trust them [74], and the absence of in-depth interviews and/or questionnaires to the local population. On the other hand, the snowball technique can be identified as a methodological deficiency since some responses from the interviewees were recurrent.

Based on the results obtained, new lines of research are proposed, as follows: (a) examine local development processes and the impacts generated by tourism, contrasting the perception of the actors with secondary sources; (b) establish the existing relationships between landscape and sustainable tourism in the PNT; (c) study the governance and the determining factors of the relations of cooperation or competition between the actors of the PNC from the stakeholder theory, taking into account the direct perception of the local population; (d) analyse the existing relationships between proactivity, ideology and gender in the development of (sustainable) tourist activities in the PNAs.

Author Contributions: Conceptualisation, M.B.-R., F.J.G.-D. and G.Š.; methodology, M.B.-R., F.J.G.-D. and G.Š.; validation, M.B.-R., F.J.G.-D. and G.Š.; formal analysis, M.B.-R., F.J.G.-D. and G.Š.; investigation, M.B.-R., F.J.G.-D. and G.Š.; resources, M.B.-R., F.J.G.-D. and G.Š.; data curation, M.B.-R., F.J.G.-D. and G.Š.; writing—original draft preparation, M.B.-R., F.J.G.-D. and G.Š.; writing—review and editing, M.B.-R., F.J.G.-D. and G.Š.; supervision, M.B.-R., F.J.G.-D. and G.Š.; project administration, M.B.-R., F.J.G.-D. and G.Š. All authors have read and agreed to the published version of the manuscript.

Funding: This research received no external funding.

Institutional Review Board Statement: Not applicable.

Informed Consent Statement: Not applicable.

Data Availability Statement: The interviewees were informed of the purpose of this work and authorised the use of the interviews.

Acknowledgments: The authors thank all the interviewees for cooperating in carrying out this research.

Conflicts of Interest: The authors declare no conflict of interest.

Abbreviations

DSMBR	Dehesas de Sierra Morena Biosphere Reserve
Int	Interviews
LAGs	Local Action Groups
NtP	Nature Park
PNA	Protected Nature Areas
RJ	Ruta del Jabugo
SAPA	Nature Park Sierra de Aracena y Picos de Aroche
SH	Nature Park Sierra de Hornachuelos
SIEs	Services, infrastructures and [types of] equipment
SMSTRE	Sierra Morena Starlight Reserve
SNS	Nature Park Sierra Norte de Sevilla
UWGpSNS	UNESCO World Geopark Sierra Norte de Sevilla

References

1. Vaccaro, I.; Beltran, O. Consuming space, nature and culture: Patrimonial discussions in the hyper-modern era. *Tour. Geogr.* **2007**, *9*, 254–274. [CrossRef]
2. Figueiredo, E. Imagine there's no rural: The transformation of rural spaces into places of nature conservation in Portugal. *Eur. Urban Reg. Stud.* **2008**, *15*, 159–171. [CrossRef]
3. Saxena, G.; Clark, G.; Oliver, T.; Ilbery, B. Conceptualizing integrated rural tourism. *Tour. Geogr.* **2007**, *9*, 347–370. [CrossRef]
4. Ramsey, D.; Malcolm, C.D. The importance of location and scale in rural and small town tourism product development: The case of the Canadian Fossil Discovery Centre, Manitoba, Canada. *Can. Geogr.-Geogr. Can.* **2017**, *62*, 250–265. [CrossRef]
5. Mormont, M. La Agricultura en el Espacio Rural Europeo. *Agric. Y Soc.* **1994**, *71*, 17–49.
6. Wilson, G. Multifunctional quality and rural community resilience. *Trans. Inst. Br. Geogr.* **2010**, *35*, 364–381. [CrossRef]
7. Prideaux, B. Building visitor attractions in peripheral areas-Can uniqueness overcome isolation to produce viability? *Int. J. Tour. Res.* **2002**, *4*, 379–389. [CrossRef]
8. Pinto-Correia, T.; Breman, B. Understanding marginalization in the periphery of Europe: A multidimensional process. In *Sustainable Land Management*; Brouwer, F., Van Rheenen, T., Dhillion, S.S., Elgersma, A.M., Eds.; Edward Elgar: Cheltenham, UK, 2008; pp. 11–40.
9. EUROPARC. *Memoria de Actividades. EUROPARC España. 2010*; EUROPARC España: Madrid, Spain, 2011.
10. Woodhouse, E.; Bedelian, C.; Dawson, N.; Barnes, P. Social impacts of protected areas: Exploring evidence of trade-offs and synergies. In *Ecosystem Services and Poverty Alleviation: Trade-Offs and Governance*; Schreckenberg, K., Poudyal, M., Mace, G., Eds.; Routledge: Oxon, UK, 2018; pp. 222–240.
11. Atauri-Mezquida, J.A.; Muñoz-Santos, M.; Múgica-de-la-Guerra, M. *Protected Areas in the Face of Global Change Climate Change Adaptation in Planning and Management*; EUROPARC-Spain: Madrid, Spain, 2020.
12. Marsden, T.; Murdoch, J.; Lowe, P.; Flynn, A. *Constructing the Countryside*; UCL Press: London, UK, 1993.
13. Ruiz-Ballesteros, E. Consumir y consumar naturaleza. Prácticas turísticas en la naturalización de la Sierra de Aracena. *Trab. De Antropol. E Etnol.* **2018**, *58*, 1–17.
14. Riechers, M.; Balázsi, A.; Abson, D.J.; Fischer, J. The influence of landscape change on multiple dimensions of human–nature connectedness. *Ecoc. Sol.* **2020**, *25*, 3.
15. Romagosa, F.; Eagles, P.F.J.; Buteau-Duitschaever, W. Evaluación de la gobernanza en los espacios naturales protegidos. El caso de la Columbia Británica y Ontario (Canadá). *An. Geogr. Univ. Complut.* **2012**, *32*, 133–151. [CrossRef]
16. Troitiño-Vinuesa, M.A. Espacios naturales protegidos y desarrollo rural: Una relación territorial. *Bol. Asoc. Geogr. Esp.* **1995**, *20*, 23–37.
17. Zawilińska, B. Residents' attitudes towards a national park under conditions of suburbanisation and tourism pressure: A case study of Ojcow National Park (Poland). *Europ. Countrys.* **2020**, *12*, 119–137. [CrossRef]
18. Flather, C.H.; Cordell, H.K. Outdoor recreation: Historical and anticipated trends. In *Wildlife and Recreationists. Coexistence through Management and Research*; Knight, R.L., Gutzwiller, K.J., Eds.; Island Press: Washington, DC, USA, 1995; pp. 3–16.
19. Araque-Jiménez, E.; Crespo-Guerrero, J.M. Conservation versus développement? Une nouvelle situation conflictuelle dans les parcs naturels andalous. *Collect. EDYTEM. Cah. De Géographie* **2010**, *10*, 113–124. [CrossRef]
20. Garzón-García, R.; Ramírez-López, M.L. Las áreas protegidas como territorios turísticos: Análisis crítico a partir del caso de los parques naturales de la Sierra Morena andaluza. *Cuad. Tur.* **2018**, *41*, 249–277. [CrossRef]
21. Anton-Clavé, S.; Blay-Boqué, J.; Salvat-Salvat, J. Turismo, actividades recreativas y uso público en los parques naturales. Propuesta para la conservación de los valores ambientales y el desarrollo productivo local. *Bol. Asoc. Geogr. Esp.* **2008**, *48*, 5–38.
22. Arnegger, J.; Woltering, M.; Job, H. Toward a product-based typology for nature-based tourism: A conceptual framework. *J. Sustain. Tour.* **2010**, *18*, 915–928. [CrossRef]
23. Krippendorf, J. *The Holiday Makers: Understanding the Impacts of Leisure and Travel*; Butterworth-Heinemann: London, UK, 1987.
24. Aparicio-Sánchez, M.S. El Reto del Turismo en los Espacios Naturales Protegidos Españoles: La Integración Entre Conservación, Calidad y Satisfacción. Ph.D. Thesis, Universidad Complutense de Madrid, Madrid, Spain, 2013.
25. Burgess, J. Selling Places: Enviromental Images for the Executive. *Reg. Stud.* **1982**, *16*, 1–17. [CrossRef]
26. Duffy, R. Nature-based tourism and neoliberalism: Concealing contradictions. *Tour. Geogr.* **2015**, *17*, 529–543. [CrossRef]
27. Cànoves-i-Valiente, G.; Villarino-Pérez, M.; Herrera-Jiménez, L. Políticas públicas, turismo rural y sostenibilidad: Difícil equilibrio. *Bol. Asoc. Geogr. Esp.* **2006**, *41*, 199–217.
28. MacCannell, D. *The Tourist Papers*; Routledge: London, UK, 1992.
29. Riechers, M.; Balázsi, A.; Betz, L.; Jiren, T.S.; Fischer, J. The erosion of relational values resulting from landscape simplification. *Landsc. Ecol.* **2020**, *35*, 2601–2612. [CrossRef]
30. Eagles, P.F.J.; McCool, S.F.; Haynes, C.D. *Turismo Sostenible En Áreas Protegidas. Directrices de Planificación y Gestión*; UNWTO: Madrid, Spain, 2003.
31. Lane, B. Sustainable rural tourism strategies: A tool for development and conservation. *J. Sustain. Tour.* **1994**, *2*, 102–111. [CrossRef]
32. Hall, C.M. Policy learning and policy failure in sustainable tourism governance: From first- and second-order to third-order change? *J. Sustain. Tour.* **2011**, *4–5*, 649–671. [CrossRef]

33. Hall, C.M.; Gössling, S.; Scott, D. The evolution of sustainable development and sustainable tourism. In *The Routledge Handbook of Tourism and Sustainability*; Hall, C.M., Gössling, S., Scott, D., Eds.; Routledge: London, UK, 2015; pp. 15–35.
34. Saarinen, J. Is Being Responsible Sustainable in Tourism? Connections and Critical Differences. *Sustainability* **2021**, *13*, 6599. [CrossRef]
35. UNEP. *Making Tourism More Sustainable—A Guide for Policy Makers*; United Nations Environment Programme UNEP: Paris, France, 2005.
36. Sharpley, R. *Tourism Development and the Environment: Beyond Sustainability?* Earthscan: London, UK, 2009.
37. Hall, C.M. Changing paradigms and global change: From sustainable to steady-state tourism. *Tour. Recreat. Res.* **2010**, *35*, 131–143. [CrossRef]
38. Padin, C.A. Sustainable tourism planning model: Components and relationships. *Eur. Bus. Rev.* **2012**, *24*, 510–518. [CrossRef]
39. Torres-Delgado, A.; López-Palomeque, F. Measuring sustainable tourism at the municipal level. *Ann. Touris. Res.* **2014**, *49*, 122–137. [CrossRef]
40. Sharpley, R. Tourism and sustainable development: Exploring the theoretical divide. *J. Sustain. Tour.* **2000**, *8*, 1–19. [CrossRef]
41. Renfors, S.M. Stakeholders' Perceptions of Sustainable Tourism Development in a Cold-Water Destination: The Case of the Finnish Archipelago. *Tour. Plan. Dev.* **2020**, *18*, 510–528. [CrossRef]
42. Hall, C.M.; Gössling, S.; Scott, D. Tourism and sustainability. An introduction. In *The Routledge Handbook of Tourism and Sustainability*; Hall, C.M., Gössling, S., Scott, D., Eds.; Routledge: London, UK, 2015; pp. 1–9.
43. Hall, C.M. Degrowing tourism: Décroissance, sustainable consumption and steady-state tourism. *Anatolia* **2009**, *20*, 46–61. [CrossRef]
44. Bianchi, R.V. The 'critical turn' in tourism studies: A radical critique. *Tour. Geogr.* **2009**, *11*, 484–504. [CrossRef]
45. Raco, M. Sustainable development, rolled-out neoliberalism and sustainable communities. *Antipode* **2005**, *37*, 324–347. [CrossRef]
46. Myers, G.; Macnaghten, P. Rhetorics of environmental sustainability: Commonplaces and places. *Environ. Plan. A.* **1998**, *30*, 333–353. [CrossRef]
47. Blackstock, K.L.; White, V.; McCrum, G.; Scott, A.; Hunter, C. Measuring responsibility: An appraisal of a Scottish national park's sustainable tourism indicators. *J. Sustain. Tour.* **2008**, *16*, 276–297. [CrossRef]
48. Bianchi, R.V.; de-Man, F. Tourism, inclusive growth and decent work: A political economy critique. *J. Sustain. Tour.* **2021**, *29*, 353–371. [CrossRef]
49. Font, X.; McCabe, S. Sustainability and marketing in tourism: Its contexts, paradoxes, approaches, challenges and potential. *J. Sustain. Tour.* **2017**, *25*, 869–883. [CrossRef]
50. Hughes, G. Tourism, Sustainability and Social Theory. In *A Companion to Tourism*; Lew, A., Hall, C.M., Williams, A., Eds.; Blackwell: Oxford, UK, 2004; pp. 498–509.
51. Caruana, R.; Glozer, S.; Crane, A.; McCabe, S. Tourists' accounts of responsible tourism. *Ann. Tour. Res.* **2014**, *46*, 115–129. [CrossRef]
52. Saarinen, J. Critical sustainability: Setting the limits to growth and responsibility in tourism. *Sustainability* **2014**, *6*, 1–17. [CrossRef]
53. Poudel, S.; Nyaupane, G.P.; Budruk, M. Stakeholders' perspectives of sustainable tourism development: A new approach to measuring outcomes. *J. Travel Res.* **2016**, *55*, 465–480. [CrossRef]
54. Gobattoni, F.; Pelorosso, R.; Leone, A.; Ripa, M.N. Sustainable rural development: The role of traditional activities in Central Italy. *Land Use Pol.* **2015**, *48*, 412–427. [CrossRef]
55. Adamowicz, J. Towards synergy between tourism and nature conservation. The challenge for the rural regions: The case of Drawskie Lake District, Poland. *Europ. Countrys.* **2010**, *2*, 118–131. [CrossRef]
56. Belliggiano, A.; Cejudo-García, E.; Labianca, M.; Navarro-Valverde, F.; De-Rubertis, S. The "Eco-Effectiveness" of Agritourism Dynamics in Italy and Spain: A Tool for Evaluating Regional Sustainability. *Sustainability* **2020**, *12*, 7080. [CrossRef]
57. Hall, C.M. Tourism and biodiversity: More significant than climate change? *J. Herit. Tour.* **2010**, *5*, 253–266. [CrossRef]
58. Helgadóttir, G.; Einarsdóttir, A.V.; Burns, G.L.; Gunnarsdóttir, G.Þ.; Matthíasdóttir, J.M.E. Social sustainability of tourism in Iceland: A qualitative inquiry. *Scand. J. Hosp. Tour.* **2019**, *19*, 404–421. [CrossRef]
59. Twining-Ward, L.; Butler, R.W. Implementing STD on a Small Island: Development and Use of Sustainable Tourism Development Indicators in Samoa. *J. Sustain. Tour.* **2002**, *10*, 363–387. [CrossRef]
60. Timur, S.; Getz, D. Sustainable tourism development: How do destination stakeholders perceive sustainable urban tourism? *Sustain. Dev.* **2009**, *17*, 220–232. [CrossRef]
61. Bosak, K. Tourism, development, and sustainability. In *Reframing Sustainable Tourism*; McCool, S.F., Bosak, K., Eds.; Springer: Dordrecht, The Netherlands, 2016; pp. 33–44.
62. Choi, H.C.; Sirakaya, E. Sustainability indicators for managing community tourism. *Tour. Manage.* **2006**, *27*, 1274–1289. [CrossRef]
63. Marzo-Navarro, M.; Pedraja-Iglesias, M.; Vinzón, L. Sustainability indicators of rural tourism from the perspective of the residents. *Tour Geogr.* **2015**, *17*, 586–602.
64. Blanco-Portillo, R. El turismo de naturaleza en España y su plan de impulso. *Estud. Turísticos* **2006**, *169*, 170.
65. Šadeikaitė, G. Supporting Sustainable Tourism Development through Improved Measurement: A Case Study of European Tourism Destinations. Ph.D. Thesis, Universidad de Alicante, Universidad de Alicante, Alicante, Spain, 2017.
66. Cànoves-i-Valiente, G.; Herrera-Jiménez, L.; Villarino-Pérez, M. Turismo rural en España: Paisajes y usuarios, nuevos usos y nuevas visiones. *Cuad. Tur.* **2005**, *15*, 63–76.

67. Ruhanen, L. Local government: Facilitator or inhibitor of sustainable tourism development? *J. Sustain. Tour.* **2013**, *21*, 80–98. [CrossRef]
68. Trišić, I.; Štetić, S.; Privitera, D. The importance of nature-based tourism for sustainable development-A report from the selected biosphere reserve. *J. Geogr. Inst. Jovan Cvijic* **2021**, *71*, 203–209. [CrossRef]
69. Hardy, A. Using grounded theory to explore stakeholder perceptions of tourism. *J. Tour. Cult. Chang.* **2005**, *3*, 108–133. [CrossRef]
70. Beaumont, N.; Dredge, D. Local tourism governance: A comparison of three network approaches. *J. Sustain. Tour.* **2010**, *18*, 7–28. [CrossRef]
71. Brouder, P. Creative outposts: Tourism's place in rural innovation. *Tour. Plan. Dev.* **2012**, *9*, 383–396. [CrossRef]
72. García-Delgado, F.J.; Martínez-Puche, A.; Lois-González, R.C. Heritage, Tourism and Local Development in Peripheral Rural Spaces: Mértola (Baixo Alentejo, Portugal). *Sustainability* **2020**, *12*, 9157. [CrossRef]
73. Saxena, G.; Ilbery, B. Developing integrated rural tourism: Actor practices in the English/Welsh border. *J. Rural Stud.* **2010**, *26*, 260–271. [CrossRef]
74. Butowski, L. Tourist sustainability of destination as a measure of its development. *Curr. Issues Tour.* **2017**, *22*, 1043–1061. [CrossRef]
75. Paül-Carril, V.; Agrelo-Janza, L.M.; Trillo-Santamaría, J.M. Montañas de Trevinca: ¿undertourism en Galicia y overtourism en Sanabria? In *Sostenibilidad Turística: Overtourism vs Undertourism*; Pons, G.X., Blanco-Romero, A., Navalón-García, R., Troitiño-Torralba, L., Blázquez-Salom, M., Eds.; Societat d'Història Natural de les Balears: Palma de Mallorca, Spain, 2020; pp. 445–456.
76. Dinis, I.; Simões, O.; Cruz, C.; Teodoro, A. Understanding the impact of intentions in the adoption of local development practices by rural tourism hosts in Portugal. *J. Rural Stud.* **2019**, *72*, 92–103. [CrossRef]
77. Secor, A.J. Social surveys, interviews, and focus groups. In *Research Methods in Geography. A Critical Introduction*; Gomez, B., Jones, J.P., III, Eds.; Wiley-Blackwell: Oxford, UK, 2010; pp. 194–205.
78. Cope, M. Coding Qualitative Data. In *Qualitative Research Methods in Human Geography*; Hay, I., Ed.; Oxford University Press: Oxford, UK, 2010; pp. 281–294.
79. Ley 2/1989, de 18 de Julio, por la que se aprueba el Inventario de Espacios Naturales Protegidos de Andalucía y se establecen medidas adicionales para su protección. Available online: https://www.boe.es/buscar/doc.php?id=BOE-A-1989-20636 (accessed on 27 September 2022).
80. Parques Naturales. Available online: https://www.juntadeandalucia.es/medioambiente/portal/parques-naturales?categoryVal= (accessed on 23 September 2022).
81. Cifras Oficiales de Población Resultantes de la Revisión del Padrón Municipal a 1 de Enero (2021). Available online: https://www.ine.es/dynt3/inebase/es/index.htm?padre=517&capsel=525 (accessed on 23 September 2021).
82. Márquez-Domínguez, J.A.; Jurado-Almonte, J.M.; Felicidades-García, J.; García-Delgado, F.J. Paisajes Agrarios Andaluces. In *Conocer Andalucía*; Cano-García, G., Ed.; Tartessos: Sevilla, Spain, 2001; Volume I, pp. 243–326.
83. Silva-Pérez, R.; Ojeda-Rivera, J.F. La Sierra Morena sevillana, a la sombra de la urbe y el mercado. *Ería* **2001**, *56*, 255–276.
84. Silva-Pérez, R.; Fernández-Salinas, V. Claves para el reconocimiento de la dehesa como "paisaje cultural" de Unesco. *An. Geogr. Univ. Complut.* **2015**, *35*, 121–142. [CrossRef]
85. Mulero-Mendigorri, A.; Silva Pérez, R. Paisajes de Sierra Morena: Una cuestión de miradas y de escalas. *RER* **2013**, *96*, 35–64.
86. García-Delgado, F.J. Industrias Cárnicas, Territorio y Desarrollo en Sierra Morena. Ph.D. Thesis, Universidad de Huelva, Huelva, Spain, 2003.
87. Pizarro-Gómez, A.; Šadeikaitė, G.; García-Delgado, F.J. The World of Iberian Ham and its tourism potential. *Europ. Countrys.* **2020**, *12*, 333–365. [CrossRef]
88. Grupos de Desarrollo Rural. Available online: https://www.andaluciarural.org/grupos-de-desarrollo-rural/ (accessed on 23 September 2021).
89. Mercado-Alonso, I.; Fernández-Tabales, A.; Bascarán-Estévez, M.V. Turismo rural y crecimiento inmobiliario en espacios de montaña media. El caso de la Sierra de Aracena. *Polígonos* **2012**, *23*, 181–211. [CrossRef]
90. Domínguez-Gómez, J.A.; Lennartz, T. Turismo rural y expansión urbanística en áreas de interior. Análisis socioespacial de riesgos. *Rev. Int. Sociol.* **2015**, *73*, e006. [CrossRef]
91. Fernández-Tabales, A.; Hernández-Martínez, E.; Marchena-Gómez, M.J.; Velasco-Martín, A. Estrategias turísticas en el Parque Natural Sierra de Aracena y Picos de Aroche. In *Patrimonio Histórico-Artístico de la Provincia de Huelva: Ponencias de las V Jornadas del Patrimonio de la Sierra de Huelva*; Almonaster la Real; Nature Park: Andalusia, Spain, 1993; pp. 69–82.
92. Registro de Turismo de Andalucía. Available online: https://www.juntadeandalucia.es/turismoydeporte/opencms/areas/temp/rta/buscador/index.html (accessed on 23 September 2022).
93. Hall, C.M. Rethinking collaboration and partnership: A public policy perspective. *J. Sustain. Tour.* **1999**, *7*, 274–289. [CrossRef]
94. Laird, S.; Johnston, S.; Wynberg, R.; Lisinge, E.; Lohan, D. *Biodiversity Access and Benefit–Sharing Policies for Protected Areas. An Introduction*; United Nations University Institute of Advanced Studies (UNU/IAS): Tokio, Japan, 2003.
95. Ley 42/2007, de 13 de Diciembre, del Patrimonio Natural y de la Biodiversidad. Available online: https://www.boe.es/buscar/act.php?id=BOE-A-2007-21490 (accessed on 27 September 2022).
96. West, P.; Igoe, J.; Brockington, D. Parks and people: The social impact of protected areas. *Annu. Rev. Anthropol.* **2006**, *35*, 251–277. [CrossRef]

97. García-Llorente, M.; Martín-López, B.; Iniesta-Arandia, I.; López-Santiago, C.A.; Aguilera, P.A.; Montes, C. The role of multi-functionality in social preferences toward semi-arid rural landscapes: An ecosystem service approach. *Environ. Sci. Policy.* **2012**, *19*, 136–146. [CrossRef]

98. Araque-Jiménez, E.; Moya-García, E. La política de conservación de la naturaleza y desarrollo socioeconómico en las Sierras de Cazorla, Segura y Las Villas (Jaén). *Ería* **2008**, *75*, 129–142.

99. Beilin, R.; Lindborg, R.; Stenseke, M.; Pereira, H.M.; Llausàs, A.; Slätmo, E.; Cerqueira, Y.; Navarro, L.; Rodrigues, P.; Reichelt, N.; et al. Analysing how drivers of agricultural land abandonment affect biodiversity and cultural landscapes using case studies from Scandinavia, Iberia and Oceania. *Land Use Pol.* **2014**, *36*, 60–72. [CrossRef]

100. Harvey, D. *A Brief History of Neoliberalism*; Oxford University Press: New York, NY, USA, 2005.

101. Kensbock, S.; Jennings, G. Pursuing: A grounded theory of tourism entrepreneurs' understanding and praxis of sustainable tourism. *Asia Pac. J. Tour. Res.* **2011**, *16*, 489–504. [CrossRef]

102. Iannuzzi, G.; Santos, R.; Mourato, J.M. The involvement of non-state actors in the creation and management of protected areas: Insights from the Portuguese case. *J. Environ. Plan. Manag.* **2020**, *63*, 1674–1694. [CrossRef]

103. Godfrey, K.B. Attitudes towards 'sustainable tourism' in the UK: A view from local government. *Tour. Manage.* **1998**, *19*, 213–224. [CrossRef]

104. Pulido-Fernández, J.I. Gestión turística activa y desarrollo económico en los parques naturales andaluces. Una propuesta de revisión desde el análisis del posicionamiento de sus actuales gestores. *RER* **2008**, *81*, 171–203.

105. Jamaliah, M.M.; Powell, R.B. Ecotourism resilience to climate change in Dana Biosphere Reserve, Jordan. *J. Sustain. Tour.* **2018**, *26*, 519–536. [CrossRef]

106. Gómez-Martín, M.B.; Armesto-López, X.A.; Cors-Iglesias, M. Percepción del cambio climático y respuestas locales de adaptación: El caso del turismo rural. *Cuad. Tur.* **2017**, *39*, 287–310. [CrossRef]

107. Jeuring, J.H.G.; Haartsen, T. The challenge of proximity: The (un)attractiveness of nearhome tourism destinations. *Tour. Geogr.* **2017**, *19*, 118–141. [CrossRef]

108. Múgica-de-la-Guerra, M.; Puertas-Blázquez, J. Los registros de patrimonio inmaterial, una herramienta para la conservación de las reservas de la biosfera. *PH Boletín Del Inst. Andal. Del Patrim. Histórico* **2018**, *26*, 22–23. [CrossRef]

109. Buta, N.; Holland, S.M.; Kaplanidou, K. Local communities and protected areas: The mediating role of place attachment for pro-environmental civic engagement. *J. Outdoor Recreat. Tour.* **2014**, *5*, 1–10. [CrossRef]

110. Perkins, H.E. Measuring love and care for nature. *J. Environ. Psychol.* **2010**, *30*, 455–463. [CrossRef]

111. Høyem, J. Outdoor recreation and environmentally responsible behavior. *J. Outdoor Recreat. Tour.* **2020**, *31*, 100317. [CrossRef]

112. Romagosa, F. The COVID-19 crisis: Opportunities for sustainable and proximity tourism. *Tour. Geogr.* **2020**, *22*, 690–694. [CrossRef]

113. Borrello, M.; Pascucci, S.; Cembalo, L. Three Propositions to Unify Circular Economy Research: A Review. *Sustainability* **2020**, *12*, 4069. [CrossRef]

114. Courtney, P.; Hill, G.; Roberts, D. The role of natural heritage in rural development: An analysis of economic linkages in Scotland. *J. Rural Stud.* **2006**, *22*, 469–484. [CrossRef]

115. Butler, R.W. The concept of carrying capacity for tourism destinations: Dead or merely buried? *Progr. Tour. Hosp. Res.* **1996**, *2*, 283–293. [CrossRef]

116. Middleton, V.; Hawkins, R. *Sustainable Tourism, A Marketing Perspective*; Butterworth-Heinemann: Oxford, UK, 1998.

117. Naldi, L.; Nilsson, P.; Westlund, H.; Wixe, S. What is smart rural development? *J. Rural Stud.* **2015**, *40*, 90–101. [CrossRef]

118. Tracey, P.; Clark, G. Alliances, networks and competitive strategy: Rethinking clusters of innovation. *Growth Change* **2003**, *34*, 1–16. [CrossRef]

119. Bonadonna, A.; Rostagno, A.; Beltramo, R. Improving the landscape and tourism in marginal areas: The case of land consolidation associations in the North-West of Italy. *Land* **2020**, *9*, 175. [CrossRef]

120. Pato, L. Entrepreneurship and innovation towards rural development evidence from a peripheral area in Portugal. *Europ. Countrys.* **2020**, *12*, 209–220. [CrossRef]

121. Ştefănică, M.; Butnaru, G.I. Approaches of durable development of tourism. *Rev. Tur.-Stud. Si Cercet. Tur.* **2013**, *15*, 41–47.

122. Romão, J. Nature, Tourism, Growth, Resilience and Sustainable Development. In *Mediterranean Protected Areas in the Era of Overtourism*; Mandić, A., Petrić, L., Eds.; Springer: Cham, Switzerland, 2021; pp. 297–310.

123. NCM. *Monitoring the Sustainability of Tourism in the Nordics*; Nordic Council of Ministers: Copenhagen, Denmark, 2021.

124. Young, R. Sustainability: From rhetoric to reality through markets. *J. Clean Prod.* **2006**, *14*, 1443–1447. [CrossRef]

125. Gonçalves, C.; Honrado, J.P.; Cerejeira, J.; Sousa, R.; Fernandes, P.M.; Vaz, A.S.; Alves, M.; Araújo, M.; Carvalho-Santos, C.; Fonseca, A.; et al. On the development of a regional climate change adaptation plan: Integrating model-assisted projections and stakeholders' perceptions. *Sci. Total Environ.* **2022**, *805*, 150320. [CrossRef] [PubMed]

126. Navarro-Jurado, E.; Ortega-Palomo, G.; Torres-Bernier, E. Propuestas de reflexión desde el turismo frente al COVID-19. In *Incertidumbre, Impacto y Recuperacion*; Instituto Universitario de Investigación de Inteligencia e Innovación Turística de la Universidad de Málaga: Málaga, Spain, 2020.

127. Richards, G.; Hall, D. The community: A sustainable concept in tourism development? In *Tourism and Sustainable Community Development*, 2nd ed.; Richards, G., Hall, D., Eds.; Routledge: London, UK, 2000; pp. 1–13.

128. Patala, S.; Korpivaara, I.; Jalkala, A.; Kuitunen, A.; Soppe, B. Legitimacy under institutional change: How incumbents appropriate clean rhetoric for dirty technologies. *Organ. Stud.* **2019**, *40*, 395–419. [CrossRef]
129. Smith, A.; Robbins, D.; Dickinson, J.E. Defining sustainable transport in rural tourism: Experiences from the New Forest. *J. Sustain. Tour.* **2019**, *27*, 258–275. [CrossRef]
130. Fernández-Tabales, A.; Santos-Pavón, E.L. Turismo y parques naturales en Andalucía tras veinte años desde su declaración. Análisis estadístico, tipología de parques y problemática de la situación actual. *An. Geogr. Univ. Complut.* **2010**, *30*, 29–54.
131. Hjalager, A.M. Agricultural diversification into tourism: Evidence of a European Community development programme. *Tour. Manage.* **1996**, *17*, 103–111. [CrossRef]
132. Eagles, P.F.J. Governance models for parks, recreation and tourism. In *Transforming Parks: Protected area Policy and Management in a Changing World*; Hanna, K.S., Clark, D.A., Slocombe, D.S., Eds.; Routledge: London, UK, 2008; pp. 39–61.
133. Stoll-Kleemann, S. Barriers to nature conservation in Germany: A model explaining opposition to protected areas. *J. Environ. Psychol.* **2001**, *21*, 369–385. [CrossRef]
134. Maríková, P.; Herová, I. Area protection in views of its residents. *Europ. Countrys.* **2010**, *2*, 201–213. [CrossRef]
135. Mulero-Mendigorri, A. Fronteras y territorios: La gestión de las áreas protegidas en cuestión. *Cuad. Gec.* **2018**, *57*, 61–86. [CrossRef]
136. Esparcia-Pérez, J.; Noguera-Tur, J.; Pitarch-Garrido, M.D. LEADER en España: Desarrollo rural, poder, legitimación, aprendizaje y nuevas estructuras. *D.A.G.* **2000**, *37*, 95–113.
137. Navarro-Valverde, F.A.; Woods, M.; Cejudo-García, E. The LEADER initiative has been a victim of its own success. The decline of the bottom-up approach in rural development programmes. The cases of Wales and Andalusia. *Sociol. Rural.* **2016**, *56*, 270–288. [CrossRef]
138. Kaiser, J.; Haase, D.; Krueger, T. Payments for ecosystem services: A review of definitions, the role of spatial scales, and critique. *Ecoc. Sol.* **2021**, *26*, 12. [CrossRef]
139. Cárdenas-Alonso, G.; Nieto-Masot, A. Towards rural sustainable development? Contributions of the EAFRD 2007–2013 in low demographic density territories: The case of Extremadura (SW Spain). *Sustainability* **2017**, *9*, 1173. [CrossRef]
140. Panyik, E. Rural Tourism Governance: Determinants of Policymakers' Support for Tourism Development. *Tour. Plan. Dev.* **2015**, *12*, 48–72. [CrossRef]
141. Molden, O.; Abrams, J.; Davis, E.J.; Moseley, C. Beyond localism: The micropolitics of local legitimacy in a community-based organization. *J. Rural Stud.* **2017**, *50*, 60–69. [CrossRef]
142. Medina, L.K. Ecotourism and certification: Confronting the principles and pragmatics of socially responsible tourism. *J. Sustain. Tour.* **2005**, *13*, 281–295. [CrossRef]
143. Romagosa, F.; Miró, A.; Buchaca, T.; Ventura, M. Residents' Versus Visitors' Knowledge and Valuation of Aquatic Mountain Ecosystems in the Catalan Pyrenees. *Mt. Res. Dev.* **2020**, *40*, R1–R10. [CrossRef]
144. Ray, C. Neo-Endogenous Rural Development in the EU. In *The Handbook of Rural Studies*; Cloke, P., Marsden, T., Mooney, P., Eds.; SAGE: London, UK, 2006; pp. 278–291.
145. McAreavey, R.; McDonagh, J. Sustainable rural tourism: Lessons for rural development. *Sociol. Rural.* **2011**, *51*, 175–194. [CrossRef]
146. Bramwell, B.; Sharman, A. Collaboration in local tourism policymaking. *Ann. Touris. Res.* **1999**, *26*, 392–415. [CrossRef]
147. Lordkipanidze, M.; Brezet, H.; Backman, M. The entrepreneurship factor in sustainable tourism development. *J. Clean Prod.* **2005**, *13*, 787–798. [CrossRef]

Article

Estimating the Probability of Visiting a Protected Natural Space and Its Conditioning Factors: The Case of the Monfragüe Biosphere Reserve (Spain)

Marcelino Sánchez-Rivero, Juan de la Cruz Sánchez-Domínguez and Mª Cristina Rodríguez-Rangel *

Department of Economics, Faculty of Economic Sciences and Management, University of Extremadura, Avenida de Elvas s/n, 06006 Badajoz, Spain; sanriver@unex.es (M.S.-R.); jsanchezdom@unex.es (J.d.l.C.S.-D.)
* Correspondence: mcrisrod@unex.es

Abstract: Spain is the European country with the highest percentage of protected areas (27.4% of its total surface area) and the country with the highest number of Biosphere Reserves, with 53. Extremadura, the region that we analyze in our study, has a total of 89 Special Conservation Areas and 71 Special Protection Areas, Monfragüe being one of them. In this context, the aim of this paper is to determine which factors have an influence on the decision to visit Monfragüe. We perform a regression analysis using a logit model, which shows that the only four factors that influence the decision to visit Monfragüe are gender, travelling with one's partner or family, the type of accommodation, and the importance given to nature conservation. We also analyze the structural change using the Chow test, which shows that there are no structural changes, i.e., that the probability of visiting Monfragüe in the high or low season is not significantly different. In the case of Monfragüe, ecotourism is not currently practiced en masse; only 3 out of 10 tourists practice ecotourism in Monfragüe, which is important for the sustainable management of the park because the number of tourists it receives each year is within its carrying capacity.

Keywords: natural parks; regression analysis; ecotourism; Extremadura

Citation: Sánchez-Rivero, M.; de la Cruz Sánchez-Domínguez, J.; Rodríguez-Rangel, M.C. Estimating the Probability of Visiting a Protected Natural Space and Its Conditioning Factors: The Case of the Monfragüe Biosphere Reserve (Spain). *Land* **2022**, *11*, 1032. https://doi.org/10.3390/land11071032

Academic Editors: Le Yu, Rui Yang, Yue Cao and Steve Carver

Received: 24 May 2022
Accepted: 5 July 2022
Published: 7 July 2022

Publisher's Note: MDPI stays neutral with regard to jurisdictional claims in published maps and institutional affiliations.

1. Introduction

The surface area protected in the form of Biosphere Reserves continues to increase and has now reached the figure of 53 territories, which have been awarded this distinction by the UNESCO in the year 2021 in Spain.

This rise can be explained by the opportunities for conservation of the development of the sustainable use of these natural resources [1]; as part of this use, the important role played by tourism should be emphasized.

Although it is true that tourist activities carried out in an uncontrolled manner can become a threat to the conservation of these spaces, the sustainable development of these activities is desirable both to develop the local communities and to generate income for the conservation of the protected space [2]. In effect, as the authors of [3] point out, socioeconomic development around protected spaces may help to avoid adverse effects such as checking depopulation and reducing the economic disparities suffered by rural areas.

For this reason, the sustainable management of natural spaces becomes an opportunity to create wealth and wellbeing in regions with little industrial development that see in the management of their natural legacy an opportunity to generate wealth and employment by the development of the service sector.

However, to achieve satisfactory tourist management of the natural space, it is essential to use suitable segmentation strategies. The segmentation of markets has habitually been used by marketing managers to get to know and understand differences between the potential tourists of a destination [4]. Their importance to management lies in the fact that the suitable segmentation of a market allows destinations to anticipate development trends and offer highly diversified products to meet the needs of tourists [5].

Despite the considerable benefits deriving from suitable segmentation, in destinations that develop their products around resources with a high sensitivity to unsustainable development models, few studies concentrate on providing information on the differentiated profile of the tourists visiting nature reserves [6]. Thus, the main research question we propose an answer for in our paper is which are the factors that have an impact on the intention to visit a protected natural area, taking as a case study the Monfragüe Biosphere Reserve.

To make up for this lack of research, this study has the initial objective of characterizing the demand for tourism in nature reserves; to do so, it uses the National Park and Biosphere Reserve of Monfragüe located in the province of Cáceres in Extremadura, Spain, as a case study

The chosen space provides an interesting case study owing to the intrinsic characteristics of this destination. The province of Cáceres can be described as an inland destination, which due to its low level of industrial development finds in its rich natural and cultural legacy a great opportunity for achieving economic progress. As this destination is in a growth stage, it is essential for its managers to count on information to be able to plan suitable management policies.

In addition to the clear practical implications for the management of the destination, the results obtained in this applied research aim to contribute towards the profiling of the characteristics of the tourist of nature reserves in inland territories in a growth stage. They thus help to increase the information available on the differentiated profile of this type of traveler.

To do so, in the first place, this study analyses the factors that determine the probability of practicing tourism in natural spaces at the destination under study. To achieve this objective, a logit model is used based on a survey of 4683 people carried out by the Tourist Observatory of Extremadura during a calendar year. Secondly, the above analysis is complemented with Chow's test, which allows for the confirmation of the existence or otherwise of structural change because of two factors of segmentation, whether the visit is made in the high season or low season, and whether the tourist analyzed is from the Spanish or foreign market.

To achieve these objectives, this study has the following structure: After this initial introductory section, a bibliographical revision is carried out to analyze the necessary symbiosis between tourism and protected spaces. Subsequently, we analyze the demand for ecotourism, a tourist type which can include nature reserve tourism. Section 4 allows the reader to get to know the main characteristics of the natural space used as a case study. Section 5 then describes the methodology used and subsequently the major results obtained are described. Finally, the article concludes with the discussions and conclusions generated by this research.

2. Protected Natural Spaces and Tourism: A Necessary Symbiosis

Following the definition of the International Union for Conservation of Nature (IUCN), a protected area is a "clearly defined geographically space which is clearly defined, recognized and managed by legal means or other efficient means so as to achieve long-term nature conservation, the ecosystem services, and the associated cultural values" [7]. As for nature reserves, ref. [8] (p. 1) define them as "spaces in which human activities have not altered the typical environment drastically and in which as a consequence both the biotic and abiotic elements have been preserved in good condition". Protected areas are essentially governance systems [9] with spatially defined areas with natural as well as cultural attributes and services managed by a group of players with different roles and institutional frameworks [10]. These areas are organized in accordance with a variety of natural and spatial attributes that determine the conservation objectives, the protection categories, and the human activities permitted [11].

The protection of natural areas has a century-and-a-half-long history and is of a universal nature [12]. It is worth highlighting the difference between the United States creating the first Nature Reserve in 1872 (that of Yellowstone) and after, insofar as the protection objectives of the territory established [13]. For example, the Middle Ages saw the appearance of the first spaces protected for reasons related to hunting [14] and with time spaces arose in which only the royalty and nobility could hunt [13]. However, when President Grant created the first nature reserve in the USA, a new type of protected area was created, which is characterized by it being public and having recreation goals. From that year onwards, the number of protected spaces in the world has increased constantly [13], which according to [15] can be divided into three stages: (1) Between 1872 and 1975, the growth was helped by the beginning of the development of laws and regulations for the protection of the spaces, as well as by the creation of the first institutions that were specialized in the protection of the environment, both in a national and international sphere. There was also an event that contributed greatly to the declaration of new protected spaces, the first World Congress of Natural Reserves held in Seattle in 1962, since after it being held, around 80% of the protected areas of the world were created [16]. (2) Between 1974 and 1992, both the policies for environmental conservation and the laws on the subject intensified and increased in number. In this second phase, the number of protected areas and their surface became more numerous all over the world, even if there existed some differences between countries. (3) The final stage began with the Rio de Janeiro summit in 1992, which marked the introduction of a new ideology regarding conservation, which links it with sustainability and its three pillars: society, environment, and ecology.

Regarding the current situation, in 2018, the protected territories attained 14.87% of the total surface area of the world [17] and in some areas reached a much higher percentage, such as the EU with 18%. In the case of Spain, the most recent data available show that the country has protected 36.2% of the total of its land surface area and 12.3% of its marine surface area. It is also the European country that contributes the largest surface area to the Natura 2000 Network, 27.4% of the total surface area of the country, and that with the most Biosphere Reserves with 53 [18]. In the case of Extremadura, the target area of this study, the region has a total of 89 Special Conservation Areas (Zonas Especiales de Conservación, ZEC), which occupy a total of 933,772 hectares, and 71 Special Protection Areas (Zonas de Especial Protección de Aves, ZEPA) with a total surface area of 1,102,409 hectares. The protected areas are not only strategic enclaves for the protection of biodiversity and other heritage values; they also contribute towards people's wellbeing [18]. These data are the consequence of the important Spanish protectionist culture, which is based on the approval of the Law on Nature Reserves of the year 1916. Both the protected surface area and the various forms of protection have increased considerably, and the regulatory framework has been built up because of the regulations approved in the respective Autonomous Regions in accordance with the sharing of powers between them and the central government [19].

An important factor which should be taken in account is the support of local communities for the establishing of the protected areas. In this sense, the economic and social circumstances influence the decisions of people as to whether to support the establishing of the protected areas [20]; great efforts have been made to make society aware of the ecological and sociocultural values of the protected areas and to increase the involvement in the conservation process [21]. The interaction of the tourists with the protected areas through observation or activities such as the practice of sports or education, when this is permitted, provides cultural and social benefits, in addition to increasing wellbeing and raising environmental awareness [22].

However, the increase in tourism implies a series of negative impacts at both a national and international level [23]. The literature fully describes the two-faced role of tourism in the sustenance of delicate environments, communities, and cultures [24–26]. Concern for the environment and the negative effects mass tourism can have if uncontrolled have meant that sustainable tourism is attracting a great deal of attention [27]; moreover, this new mode of tourism allows people to travel independently, safely and in comfort [28].

Protected zones are a powerful way of managing land use, for sustainable development, and for nature conservation [29], although seeking to comply with conservation objectives and the resulting arbitration of land use may cause conflict [29]. For example, this occurs when the various parties have different visions of conservation objectives and one or several parties attempt to impose their interests at the cost of the interests of the remainder [30,31]. In the next section, we analyze the literature published on the demand for ecotourism in protected spaces.

3. The Demand for Ecotourism in Protected Spaces: A Revision of the Literature

Market segmentation is a relatively new concept [32]. In 1956, Wendell Smith [33] suggested that big and differentiated markets consist of many smaller and similar segments, and that targeting them allows the businesses (or in our case, the destinations) to (1) position themselves uniquely, offering a better product to their chosen target and (2) to create a long-term competitive advantage. Moreover, communicating with a smaller part of the market leads to reduced marketing expenses. The tourism industry has completely adopted and adapted the concept, and there is not a single organization without a strategy for marketing segmentation, and the focus can be, for example, on tourists of different origins or with different patterns of vacation benefits preferences [32]. The pioneer of market segmentation was Josef Mazanec who, in 1984 [34], introduced the dominant approach in tourism marketing segmentation studies: cluster analysis. Even though cluster analysis dominates, there are also other techniques that have been adopted, like neural networks methods [35].

A condition for a good market segmentation is to make the correct choice of breakdown variables [36], which usually are socio-demographic, psychological or behavioral variables, as shown in [37] for the case of the tourist accommodation market. However, Haley [38] affirms that the most effective way to segment a market is the benefits customer (in our case, tourists) seek in each product, the advantage of this methodology being that it groups customers with similar real needs, which play a decisive role in the purchase of the product [39–41]. Benefit segmentation has been applied in tourism mainly to segment tourists by their expected benefits of destinations, attractions, or activities [42]; however, studies in the accommodation sector are scarcer [43]. Usually, they are based on data extracted from interviews with customers and experts [44], or factor analysis of surveys limited to subgroups of tourists, such as business travelers in luxury hotels [45], female travelers [46], AirBnB users [43] or spa hotels [47]. In the specific case of Spain, which is the country we study in our paper, Cordente-Rodríguez, Mondéjar-Jiménez and Villanueva-Álvaro [48] analyzed excursionists in the Serranía de Cuenca National Park and divided them into two groups: those whose only motivation is to enjoy nature and natural resources and those who have multiple motivations; not only do they want to enjoy nature but also the gastronomy, as well as visiting villages to learn about their culture and traditions. The study by Carrascosa-López, Carvache-Franco, Mondéjar Jiménez and Carvache-Franco [49], also carried out in the Serranía de Cuenca National Park and in the Albufera National Park, is similar to [48] in that it presents the known segments of nature and multiple motivations, but it also presents a third new segment, reward and escape, which is related to the dimensions of nature, rewards and having fun.

Nature-based tourism is a broad term for which subgroups have appeared [50,51], such as ecotourism, nature tourism and adventure tourism. The idea of ecotourism has been very popular in recent years and is a very frequent topic in literature. Ecotourism is close to sustainable tourism because they both should be ecologically, socially, and culturally sustainable, and minimize undesirable impacts on the environment [6]. Ecotourism is the environmentally responsible travel to unmodified natural and cultural areas that promotes environmental learning while contributing to the conservation of the environment and to economic development [52]. Nature tourism is the contemplation of fauna, flora, or landscape scenery [52], so it shares only part of the ecotourism requirements: its link with nature, its attractiveness, and the experience of the visitors in natural settings [6]. The purpose of adventure tourism is to involve participants in activities that imply a degree of perceived risk or controlled danger related to personal challenges [53]. In addition to the subdivisions explained, authors such as [54] suggested four segments using a motivation-based segmentation: ecotourism, wilderness use, adventure travel and camping, [49] proposed combined, or hybrid, terms, to reflect the overlapping that tourist products present; for example, a tourist product can contain not only elements of adventure, but also natural and/or cultural attractions. Some studies have been published about benefit segmentation in the nature tourism industry. The author of [55] identified four different segments of tourists in Belize (ecotourists, nature escapists, comfortable naturalists, and passive players), while Bricker and Kerstetter [56] created four segments of tourists who decided to participate in a nature tour in the Fiji Islands (eco-family travelers, culture buffs, ecotourists and eclectic travelers). Kerstetter, Hou and Lin [57] identified three segments of tourists in the coast of Taiwan, and labelled them experience-tourists, learning-tourists and ecotourists.

As for motivations, Holden and Sparrowhawk [58] affirm that the main motivations for ecotourists are (1) learning about nature, (2) being physically active and (3) meeting people with the same interests, while Page and Dowling [59] add that ecotourists travel to satisfy their recreational and leisure needs, as well as to gather information on specific zones. Pearce and Lee [60] explain that motivating factors for travelling include relaxation, escaping, the improvement of relationships and personal development, among others. Kruger and Saayman [61] observed that tourists travel to National Parks for six main reasons: searching for knowledge, experiencing nature, to take photographs, relaxing and escaping, experiencing the park's characteristics and nostalgia, while, for South Korea, the seven main factors, following Lee et al. [62], are linked with motivation, and are self-development, interpersonal links, reward, the construction of personal relationships, escape, ego-defensive function and the appreciation of nature. As for the Republic of Serbia, Panin and Mbrica [63] divide ecotourists into four groups: social activities, health and sports activities, nature-related motivations, and educational activities.

The motivations of rural tourists, as well as their behaviors, are very different from those that do conventional tourism (López-Sanz et al., 2021) [64]. Said motivations are, as studied by Lois et al. [65], Tirado [66], Devesa et al. [67] and Leco et al. [68], linked to nature, culture, and the environment. The cited authors proposed ten different main motivations for rural tourists, which are: contact with nature, rest and calmness, cleanliness of air and water, open-air spaces and healthy environment, gastronomy, activities related with agriculture, the discovery of new cultures, hospitality of the local population, contact with the heritage and travelling back in time while having the comfort of the present.

To complete the contextualization of our study, in the following section we give a brief description of the Monfragüe Biosphere Reserve and its main characteristics.

4. The Monfragüe Biosphere Reserve: An Emblematic Protected Space in Southwestern Europe

Extremadura is an Autonomous Region of Spain consisting of two provinces, Cáceres and Badajoz, which borders on Castilla y León to the north, Andalusia to the south, Castilla La Mancha to the east, and Portugal to the west (Figure 1).

It has a total of 41,634 km^2, which corresponds to 8.2% of the surface area of Spain [3]. The area subject of our study, the National Park of Monfragüe, can be found in the province of Cáceres; it is the largest area of Mediterranean woodland and the best preserved in the world, and it also has great biodiversity thanks to the rivers and reservoirs that irrigate it [69]. According to data published by the Ministry for Ecological Transition and the Demographic Challenge (MITECO) in the Report of the Network of National Parks for 2019, Monfragüe National Park has a surface area of 18,396 hectares, a peripheral protected area of 97,764 hectares, and an area of socioeconomic influence of 195,500.73 hectares [70]. Monfragüe has been a Special Protection Area since October 1998 [71]; this protection is recognized in the legislation of Extremadura by the Decree 232/2000 of 21 November. This recognition is highly relevant owing to the ornithological richness of Monfragüe as many tourists travel there to see the birds which nest in the park in their natural habitat, among which the black vulture stands out [72]. In 2003, Monfragüe was recognized as a Biosphere Reserve [72] and in 2007 the approval of Law 1/2007 of 2 March meant that Monfragüe was declared a National Park. Finally, Monfragüe was designated a Special Conservation Area in 2015 [73], which is reflected in the legislation of Extremadura by Decree 110/2015 of 19 May. The vegetation of the park makes it even more attractive with its holm oaks, cork oaks, and alders, among other species. As far as tourism is concerned, the latest data available of accommodation businesses and restaurants in Monfragüe, from December 2020, are shown in Table 1.

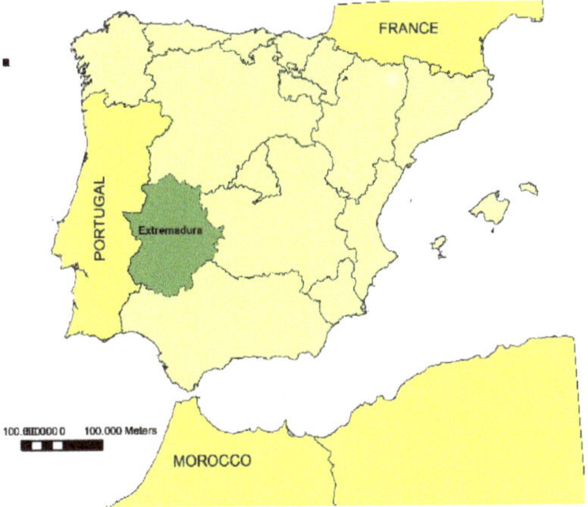

Figure 1. Location of area of study. Source: own elaboration.

Table 1. Accommodation establishments and restaurants in Monfragüe National Park.

	Total Number	**Breakdown**
Hotel-type accommodation	23	9 hotels 8 budget hotels 6 boarding houses
Non-hotel-type accommodation	60	47 rural accommodation establishments 9 tourist apartments 3 hostels 1 campsite
Restaurants and catering	94	77 restaurants 3 catering companies 14 banquet halls

Source: Sánchez-Oro et al. (2021) [74].

As for the number of travelers, in 2020, a total of 38,235 visited Monfragüe and accounted for a total of 79,571 overnight stays; they remained in the park for an average of 2.08 days. These figures represent a decrease of 49% in the number of travelers and of 42.5% in the number of overnight stays compared with 2019 [74]. To conclude, although some aspects have already been mentioned, we highlight below the main natural characteristics of Monfragüe National Park [70]: (1) Monfragüe is the largest area of preserved Mediterranean woodland in the world and is crossed by the river Tagus. The great variety of natural environments explains the wide variety of both animal and plant species in the park. (2) The landscape is characterized by being the result of human action. The dehesa woodland and pastureland system is the most outstanding example of sustainable interaction between man and the environment. (3) The birds nesting in the park include the griffon vulture, the black stork, the peregrine falcon, and the eagle owl. (4) With regard to vegetation, Monfragüe has holm and cork oak groves, heaths, and populations of maples, ashes, and alders.

5. Methodology

5.1. Estimated Probability of Tourist Visits to the Monfragüe Biosphere Reserve

The data used to estimate the previous logit model have been obtained from the surveys carried out by the Tourism Observatory of Extremadura on the Network of Tourism Offices of the region. The sample size was 4683 tourists, with no distinction being made between Spanish and foreign tourists and between the high and low season.

This questionnaire was used with the aim of finding out the profile and motivations of tourist demand in Extremadura. For this purpose, surveys were carried out randomly, establishing the minimum number of observations necessary for the sample to be representative in each territory, in each of the tourist offices in the region. The survey included a total of 12 questions distributed in different thematic blocks: socio-demographic profile of the tourist (questions 1 to 3), characteristics of the trip to the region (questions 4 to 7), activities carried out and places visited (questions 8 and 9), tourist expenditure (question 10) and degree of satisfaction with the visit (questions 11 and 12). Of all these questions, those that have been used for the present study were formulated as follows: "9. Which natural spaces have you visited, or do you plan to visit during this trip? (Monfragüe National Park as an option)"; "1. Gender"; "3. Age", "4. Who are you travelling with?"; "7. What type of lodging have you selected?". In more detail, the questionnaire used aims to find out the motivations of tourists visiting the region. As well as requesting socio-geographical data (gender, origin and age), the questionnaire includes questions related to the way of travelling (type of travel company, accommodation chosen for overnight stays, etc.), the tourist activities to be carried out, the places to be visited, daily tourist expenditure and the evaluation of the tourist services used.

To estimate the probability of visiting the Monfragüe Biosphere Reserve, a regression analysis has been used in which the dependent variable (Y_i) is a binary variable, which will have the value of 1 if the tourist has visited Monfragüe during his/her visit to the region (Extremadura) and the value 0 if he/she has not. Given the binary nature of this dependent variable, the following binary logistic regression model (or logit model) has been proposed:

$$P(Y_i = 1) = \frac{exp(z)}{1 + exp(z)}$$

with

$$z = \beta_0 + \beta_1 \, GEN_i + \beta_2 \, AG1_i + \beta_3 \, AG2_i + \beta_4 \, COMP1_i + \beta_5 \, COMP2_i + \\ \beta_6 \, H1_i + \beta_7 \, H2_i + \beta_8 \, H3_i + \beta_9 \, VAL\_ALOJ_i + \beta_{10} \, VAL\_REST_i + \beta_{11} \, VAL\_EMP_i + \\ \beta_{12} \, VAL\_NAT_i \tag{1}$$

in which $P(Y_i = 1)$ represents the probability that the tourist i visits Monfragüe, and in which the explanatory variables of the model may be grouped in three main categories:

Sociodemographic variables:

GEN: gender (1 = male; 0 = female).

AG1: age (1 = 35 years or less; 0 = others).

AG2: age (1 = between 35 and 55 years of age; 0 = others). Note: Over 55 years of age (AG1 = AG2 = 0).

Variables of trip characterization:

COMP1: type of travel (1 = as a couple or as a family; 0 = others).

COMP2: type of travel (1 = with friends or in a group; 0 = others). Note: Alone (COMP1 = COMP2 = 0).

H1: type of lodging selected for overnight stay (1 = hotel; 0 = others).

H2: type of lodging selected for overnight stay (1 = rural lodging; 0 = others).

H3: type of lodging selected for overnight stay (1 = apartment, campsite or budget hotel; 0 = others). Note: other lodgings (H1 = H2 = H3 = 0).

Variables of assessment of destination:

VAL_ALOJ: assessment on a scale of 0 to 10 points of the accommodation on offer.

VAL_REST: assessment on a scale of 0 to 10 points of the restaurants on offer.

VAL_EMP: assessment on a scale of 0 to 10 points of the tourist activity company.

VAL_NAT: assessment on a scale of 0 to 10 points of the conservation of the natural heritage.

With the inclusion of sociodemographic, trip characterization, and destination assessment variables in model (1), we aim to identify the variables that condition (or which could condition) the probability of tourist visits to be estimated.

The logit model [75–78] has been frequently used in the field of tourism research. It has, for example, been used for issues as diverse as identifying the factors determining innovation in tourism [79,80], establishing space–time relations between hotels in urban tourism destinations [81], determining the influence of High Speed Rail on the probability of returning to visit a destination [82], studying the consumption of local food in rural tourism [83], analyzing the behavior of tourists in terms of the consumption of certain products [84], analyzing the air quality of museums [85], and determining the predictive factors of tourists' loyalty to a destination [86]. This methodology is therefore widely used in the field of tourism research.

5.2. The Presnece/Absence of Structural Change in the Estimation of the Probability of Visiting Monfragüe

The test for structural change known as the Chow test [87] is habitually used with conventional regression models to determine whether on dividing a model into two subsamples there is stability in the model parameters. In a conventional regression model, this Chow test includes an F statistic in which the sum of squares of the errors of the model estimated based on the total sample (restricted model) are compared against the sum of squares of the errors of the models estimated based on each subsample (non-restricted model).

However, when the estimated regression model is a binary logistic regression model, as in this case, this Chow test is conducted as a likelihood ratio test between the restricted (pooled) logit model (model (1)) and the non-restricted logit model. The latter model defines the z function as follows:

$$\begin{aligned}
z = {} & \beta_0 + \beta_1\,GEN_i + \beta_2\,AG1_i + \beta_3\,AG2_i + \beta_4\,COMP1_i + \beta_5\,COMP2_i + \\
& \beta_6\,H1_i + \beta_7\,H2_i + \beta_8\,H3_i + \beta_9\,VAL_ALOJ_i + \beta_{10}\,VAL_REST_i + \beta_{11}\,VAL_EMP_i + \\
& \beta_{12}\,VAL_NAT_i + \beta_{13}\,D_i + \beta_{14}\,GEN_i \times D_i + \beta_{15}\,AG1_i \times D_i + \beta_{16}\,AG2_i \times D_i + \\
& \beta_{17}\,COMP1_i * D_i + \beta_{18}\,COMP2_i * D_i + \beta_{19}\,H1_i * D_i + \beta_{20}\,H2_i * D_i + \beta_{21}\,H3_i * D_i + \\
& \beta_{22}\,VAL_{ALOJ_i} \times D_i + \beta_{23}\,VAL_{REST_i} \times D_i + \beta_{24}\,VAL_{EMP_i} \times D_i + \\
& + \beta_{25}\,VAL_NAT_i \times D_i
\end{aligned} \tag{2}$$

in which the D_i variable is a control variable and assumes the value of 1 in the case of the presence of a certain characteristic and 0 if the characteristic is absent.

In our case, and given the fact that the output of the Gretl results provides the logarithm of the Log-likelihood function, the contrast which has been used is the log-likelihood ratio test between both models as shown in the following equation:

$$D = -2[log(\Lambda_1) - log(\Lambda_2)] \tag{3}$$

in which $log(\Lambda_1)$ is the logarithm of the log-likelihood function of the restricted model (model (1)) and $log(\Lambda_2)$ is the logarithm of the log-likelihood function of the non-restricted model (model (2)).

Wilks [88] demonstrates that the D statistic follows an asymptotic χ^2 distribution with $df2 - df1$ degrees of freedom, in which $df1$ and $df2$ represent the degrees of freedom of the models (1) and (2), respectively. If the p-value associated with this D statistic is lower than the level of significance, the presence of a structural change may be admitted; it would therefore be possible to conclude that significant differences exist in the adjustment of the binary logit model for the high and low seasons and for Spanish and foreign tourists.

Although this test for conventional structure change (i.e., that based on a classic regression model) has also been used quite frequently in tourism research [89–93], its use with logistic regression models and therefore its contrast through a likelihood ratio test is practically non-existent in tourism research. This study thus presents a methodological novelty in the field of tourism research.

6. Results

6.1. Estimated Probability of Tourist Visits to the Monfragüe Biosphere Reserve

The results of the model estimation (1) using the Gretl statistics package are shown in Table 2. It can be appreciated in it that gender (GEN), travel in the company of one's partner or family (COMP1), the type of accommodation chosen (H1, H2, and H3), and the assessment given to the conservation of the natural heritage are the only factors which condition the probability of visiting Monfragüe, considering a degree of statistical significance of 5% or less.

Table 2. Estimation of the binary logistic regression model (1).

Explanatory Variables	β	S.E.	Z	Wald	p-Value	Sig. [a]	$Exp\,(\beta)$
GEN	−0.140	0.065	−2.137	4.565	0.033	**	0.87
AG1	0.085	0.099	0.862	0.743	0.389		1.089
AG2	−0.005	0.08	−0.066	0.004	0.509		0.995
COMP1	0.461	0.133	3.467	12.022	0.001	***	1.585
COMP2	0.01	0.149	0.068	0.005	0.946		1.01
H1	0.23	0.079	2.913	8.488	0.004	***	1.259
H2	0.282	0.103	2.74	7.508	0.006	***	1.326
H3	0.383	0.106	3.617	13.08	0	***	1.467
VAL_ALOJ	0.022	0.034	0.638	0.407	0.524		1.022
VAL_REST	−0.002	0.036	−0.057	0.003	0.566		0.998
VAL_EMP	−0.046	0.026	−1.780	3.168	0.075	*	0.955
VAL_NAT	−0.080	0.033	−2.435	5.927	0.015	**	0.923
Constant	−0.300	0.305	−0.984	0.968	0.325		0.74

Log-likelihood: −2801.020
Schwarz criterion: 5711.912
Akaike criterion: 5628.040
Hannan-Quinn criterion: 5628.040
McFadden's R^2: 0.0141
Number of cases correctly predicted: 3298 (70.4%)
Ratio likelihood test: Chi-Square (12 df) = 79.9742 (p-value: 0.000)

* Significant at 10% level; ** significant at 5% level; *** significant at 1% level. Source: own elaboration.

In the first place in relation to gender, the negative value of coefficient β and the value of less than 1 of $exp\,(\beta)$ imply that in the case of ceteris paribus the probability that a male tourist (GEN = 1) will visit Monfragüe is lower than the probability that a female tourist (GEN = 0) will do so. In any case, the proximity to the unit of $exp\,(\beta)$ determines in this case small differences between both possibilities.

The association between the variables H1, H2, and H3 and the estimated probability of visiting Monfragüe is, however, much clearer. For these three explanatory variables, the coefficient β is positive, which determines an $exp\,(\beta)$ value exceeding 1. In this case, the types of accommodation that induce a greater predisposition to visit the Monfragüe BR are tourist apartments, campsites, and hostels ($exp\,(\beta)$ = 1.467). It therefore seems clear that the tourists visiting this protected natural space show a clear preference for these types of accommodation. To a lesser extent than those above, the tourists consulted are also more likely to visit Monfragüe if they stay in rural accommodation (casas rurales or rural hotels) (*exp* (*β*) = 1.326). Variable H1, which is associated with lodging in hotel-type establishments (in contrast to variables H2 and H3 which are clearly associated with non-hotel-type accommodation), was also statistically significant and therefore determines, under the assumption of ceteris paribus, a greater predisposition to visit the Monfragüe BR than tourists lodged in other establishments (which will not be the types mentioned above, which are those associated with higher estimated possibilities). Finally, the tourists with the lowest probabilities of making a tourist visit to Monfragüe are those lodged in other accommodation types (mainly budget hotels or boarding houses, inns, their own houses, and those of friends or relatives). Consequently, the estimation of model (1) has allowed for the identification of an empirically demonstrated association between a higher probability of visiting the Monfragüe BR and the use of non-hotel-type tourist accommodation when staying in the territory under study.

However, the most evident statistical association identified by means of the estimation of model (1) is that existing between the fact of travelling as a couple or with one's family and the probability of visiting Monfragüe. In effect, the coefficient *β* estimated from the COMP1 variable (0.461) and the clearly different value of 1 of *exp* (*β*) (1.585) determine that, ceteris paribus, the probability of making a tourist visit to Monfragüe is significantly greater among tourists who travel as a couple or with their family than among those who travel in other company. This would therefore seem to confirm that the practicing of ecotourism is of an eminently family type, at least in the protected natural space being considered in this study.

Finally, the estimated coefficient *β* of the variable VAL_NAT is negative and thus determines an *exp* (*β*) value of less than 1. This circumstance implies that if the remainder of the explanatory variables remain constant it is not the tourists who value most highly the conservation of the natural heritage of Monfragüe who are most likely to visit it. In other words, if the highest probabilities of making a visit to this protected natural space are shown by those giving the lowest score, it can be concluded that lovers of ecotourism are demanding as to the environmental protection of the natural space they visit, in such a way that the estimated values appear to recommend in this case extra effort in the conservation of the natural heritage of the Monfragüe BR.

However, the act of considering the demand for ecotourism in the Monfragüe BR in an aggregate manner, without differentiating for example between the high and low season on the one hand or between Spanish and foreign tourists on the other, may mask certain statistically significant relationships between certain explanatory variables and the probability of practicing ecotourism.

It is therefore necessary to introduce the season (high or low) and the origin (Spanish or foreign) of the tourists analyzed as control variables in the model (1) to determine if this segmentation of the ecotourism demand in accordance with the season and the origin of tourists results in differentiated behavior.

6.2. The Presence/Absence of Structural Change in the Estimation of the Probability of Visiting Monfragüe

Two control variables (D_i) are considered: one to measure the potential influence of the tourist season on the probability of visiting Monfragüe and the other to determine the effect of the tourism market of origin on this probability. In the case of the tourist season, therefore, the control variable D_i has a value of 1 for the high season (April to September) and 0 in all other cases. As for the case of the tourism market of origin, the control variable is given a value of 1 for Spanish tourists and 0 for foreign tourists.

After estimating the model (2) and taking the value of the logarithm of the log-likelihood function of this model while considering the two control variables, the results of the log-likelihood ratio test are shown in Table 3.

Table 3. Log likelihood ratio test of both control variables.

Control Variable	$log(\Lambda_1)$	$log(\Lambda_2)$	D	d.f.	p-Value
Season	−2801.020	−2794.745	12.55	13	0.4831
Tourist market	−2801.020	−2796.086	9.87	13	0.7047

Source: own elaboration.

It is therefore evident that there is no structural change in the estimated logit model either when considering the tourist season or the tourism market of origin as control variables. It may thus be concluded that the probability of visiting the Monfragüe BR is not significantly different in the high season or low season. This means that at least in the natural protected area analyzed the tourist season does not appear to have a significant influence on the demand for ecotourism, which is an advantage as a tourist destination since, unlike other types of tourism (such as sun or beach tourism, music festival tourism, or even MICE tourism, an acronym for Meetings, Incentives, Conferences and Exhibitions), ecotourism in the Monfragüe BR is a type of tourism which appears to have equal demand rates on a year-round basis.

On the other hand, and to conclude, no structural changes have been detected when considering the market of origin as a control variable. This means that the probability of visiting Monfragüe is not conditioned by the nationality of the tourist. Furthermore, it may not be necessary for tourism promotion campaigns of the Monfragüe BR to present different elements based on whether they are aimed at the Spanish or international market.

7. Analysis of the Variability of the Estimated Probabilities of a Tourist Visit in Terms of the Characteristics of the Tourist Profile

Given that only four of the explanatory variables of the model (1) are statistically significant at 5%, we give below an estimate of the probability of visiting Monfragüe based on the following reduced logistic regression model:

$$z = \beta_0 + \beta_1\,GEN_i + \beta_2\,COMP1_i + \beta_3\,H1_i + \beta_4\,H2_i + \beta_5\,H3_i + \beta_6\,VAL\_NAT_i \quad (4)$$

The relative frequency histogram of these estimated probabilities is shown in Graph 1. The mean value of these estimated probabilities is 0.2951, with a standard deviation of 0.0562. This means that ecotourism is not currently practiced on a large scale (it is not a mass tourism practice) given that it is estimated that about 3 out of 10 of the tourists visiting the territory, which is the subject of this study practice ecotourism in Monfragüe. This conclusion could help the sustainable management of the park because the number of tourists it receives every year is within its tourism carrying capacity.

The analysis of the values of the estimated probability, together with the relative frequencies corresponding to the same, which is shown in Figure 2, allows for the identification of three different levels in the possibility of getting to know and enjoying Monfragüe as part of a visit to the region:

(a) Low probability of visiting: tourists with an estimated probability of visiting Monfragüe of less than 25% (lower values than those of average probability, which are more than a typical deviation away from the same).

(b) Average probability of visiting: tourists with an estimated probability of visiting Monfragüe of between 25% and 35% (estimated values no more than one typical deviation away from average probability).

(c) High probability of visiting: tourists with an estimated probability of visiting Monfragüe of over 35% (values higher than average probability, which are more than one typical deviation from the same).

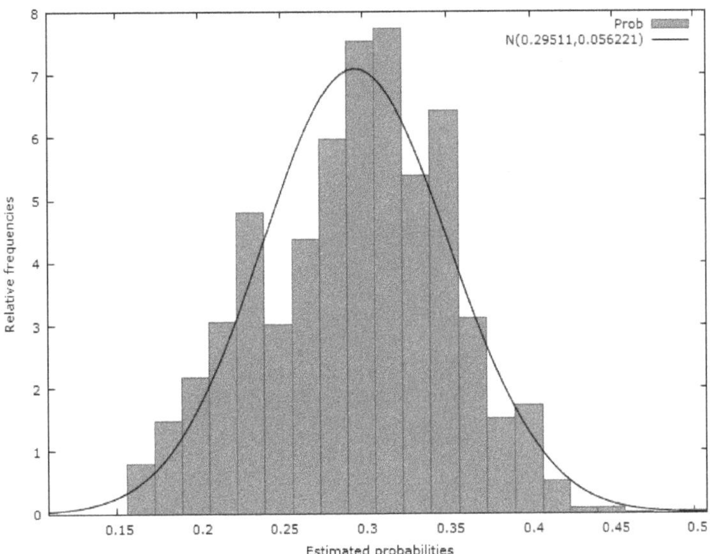

Figure 2. Frequency histogram of the estimated probabilities of visiting the Monfragüe BR. Source: own elaboration using Gretl software.

As from this segmentation of the demand based on likelihood criteria and statistical coherence, we analyze below the relationship between the explanatory variables of the estimated logit model (gender, travel type, accommodation type, and the assessment of the conservation of the natural heritage) and these three segments or levels identified.

To begin with gender (Table 4), it can be appreciated that the probability of visiting the Monfragüe BR among men (28.01%) is slightly lower than among women (30.91%). Indeed, while over a quarter of female tourists are to be found in the segment of high probability (25.9%), only 7.4% of male tourists are from this segment. In any case, to confirm empirically that this difference, although slight, is statistically significant, we have carried out test t. After confirming the hypothesis of the equality of variances with Levene's test (F = 3.328; p-value = 0.068), the high negative value of the statistic t (-18.240) and its low associated p-value (<0.0001) allow for the rejection of the hypothesis of the equality of estimated average probabilities between men and women, and therefore the confirmation that the probability of visiting the Monfragüe BR is higher among women than among men.

Table 4. Relationship between gender and the estimated probability of visiting the Monfragüe BR.

Gender	% of the Total Sample	Low Probability	Average Probability	High Probability	Mean Value of Probability
Female	51.9	15.4	58.7	25.9	0.3091
Male	48.1	29.5	63.2	7.4	0.2801

Source: own elaboration.

On the other hand, Table 5 shows the distribution of the various travel types used by the tourists consulted in each of the probability segments identified. In this case, the differences between the average probabilities are more marked than in the case of gender. Therefore, travel as a couple or with one's family are the only two travel types registering average probabilities of visiting Monfragüe of higher than 30% (32.02% and 31.90%, respectively). Indeed, compared with the almost total absence of tourists travelling alone, with friends, or in an organised group in the category of high probability of visiting, the percentage of tourists travelling as a couple or with their family and showing a probability of visiting Monfragüe of over 35% is quite high (23.1% in the first case and 24.3% in the second). The differences between these average probabilities have been ascertained by means of an ANOVA test (F = 1112.663; p-value < 0.0001) and two contrasts of independence between the lines (travel type) and the columns (probability segments) of Table 5 (Pearson's chi-square = 2310.397 with p-value < 0.0001; likelihood ratio = 2335.927 with p-value < 0.0001).

Table 5. Relationship between the type of travel and the estimated probability of visiting the Monfragüe BR.

Type of Travel	% of the Total Sample	Low Probability	Average Probability	High Probability	Mean Value of Probability
Travelling alone	8.0	65.1	34.1	0.8	0.2369
Travelling as a couple	47.1	3.7	73.2	23.1	0.3202
Travelling with friends	16.3	67.7	32.2	0.1	0.2310
Travelling with family	24.6	5.2	70.5	24.3	0.3190
Travelling with a group	4.0	75.1	24.9	0.0	0.2290

Source: own elaboration.

Similar conclusions can be obtained from analyzing Table 6, which shows the relationship between the type of accommodation used by the tourists consulted and the three segments of estimated probabilities of visiting identified. Indeed, the average estimated probabilities are within a range that exceeds 10% of the probabilities, as these are to be found between a minimum value of 24.82%, which is recorded among those who take lodging in an inn or in another type of accommodation, and of 35.41% of those who stay at a campsite. The results of the ANOVA test (F = 170.766; p-value < 0.0001) and of the contrasts of independence (Pearson's chi-square = 920.407 with p-value < 0.0001; likelihood ratio = 1009.584 with p-value < 0.0001) confirm that these differences in the probability of visiting Monfragüe depending on the type of lodging chosen by the tourists are statistically significant.

Table 6. Relationship between the type of lodging and the estimated probability of visiting the Monfragüe BR.

	% of the Total Sample	Low Probability	Average Probability	High Probability	Mean Value of Probability
4- or 5-star hotels	13.5	13.3	69.1	17.6	0.3084
1-, 2- or 3-star hotels	24.7	18.8	59.7	21.5	0.3044
Budget hotel or boarding house	7.8	40.8	58.9	0.3	0.2546
Spa	2.1	13.4	64.9	21.6	0.3085
Inn	2.8	45.7	54.3	0.0	0.2482
Casa rural	11.0	13.1	65.0	21.9	0.3179
Rural hotel	3.6	12.4	64.1	23.5	0.3213
Tourist apartment	4.9	5.2	47.0	47.8	0.3413
Campsite	4.5	2.3	44.6	53.1	0.3541
Tourist hostel	4.0	14.5	71.4	14.1	0.2924
Own home or that of friends or family	17.4	38.2	60.4	1.4	0.2578
Other lodging types	3.7	40.8	59.2	0.0	0.2483

Source: own elaboration.

Finally, Table 7 shows the distribution of the assessments given by the tourists regarding the conservation of the natural heritage of the Monfragüe BR for the three levels of probability defined. If we leave out the assessments between 0 and 4, which are completely marginal (only 0.3% of the total sample and therefore barely representative of the population being analyzed), a fairly clear association can be observed between a low score (5 or 6 points) and a high probability of visiting the Monfragüe BR (47.7% and 42.2%, respectively) and also between a very high score (9 or 10 points) and an extremely high proportion of cases with an average probability of visiting the Monfragüe BR (73.5% and 62.0%, respectively). It therefore seems clear that those who value most highly the conservation of the natural heritage of Monfragüe are not those who have the highest probability of visiting it. Indeed, and apart from the average probabilities associated with scores of between 0 and 4 points (which are not considered in this analysis owing to their very low or zero representation), the highest average probabilities of visiting Monfragüe are recorded among those giving a score of 5 of 6 points, while the lowest average probabilities occur among those who give a very high score (9 or 10 points). This apparent negative relationship between the assessment of the conservation of the natural heritage and the probability of visiting Monfragüe was found inferentially on calculating Pearson's correlation coefficient between both variables (-0.413) and the p-value associated with the hypothesis of the lack of correlation between the same (p-value < 0.0001).

Table 7. The relationship between an assessment of the conservation of the natural heritage and the estimated probability of visiting the Monfragüe BR.

Assessment	% of the Total Sample	Low Probability	Average Probability	High Probability	Mean Value of Probability
0	0.0	0.0	0.0	100.0	0.4990
1	0.0	0.0	0.0	0.0	0.0000
2	0.1	0.0	0.0	100.0	0.4174
3	0.0	0.0	0.0	100.0	0.3707
4	0.2	0.0	60.0	40.0	0.3684
5	0.9	13.7	38.6	47.7	0.3486
6	3.5	5.4	52.4	42.2	0.3438
7	12.0	12.8	55.1	32.1	0.3236
8	32.5	15.2	53.5	31.3	0.3066
9	27.0	23.7	73.5	2.8	0.2870
10	23.7	38.0	62.0	0.0	0.2633

Source: own elaboration.

8. Discussion, Limitations, and Future Lines of Research

The various types of protection of the territory are becoming an interesting resource for achieving sustainable management models for natural spaces by means of the development of ecotourism.

The responsible management of these spaces allows for the generation of income both to reinvest in the conservation of the natural resource itself and to generate employment and wealth to allow for the mitigation of adverse effects on the territory such as depopulation and the generation of territorial imbalances. It is for this reason that symbiosis between tourism and protected natural areas is clearly necessary.

To ensure that these spaces can achieve their objectives, it is necessary to obtain information on the profile of the tourist who is likely to practice this activity. Therefore, a satisfactory segmentation of the market would help the managers of destinations to design efficient strategy plans.

The starting point of this study is its objective of analyzing the probability that a tourist visiting the region under consideration, Extremadura in Spain, will practice this type of tourism based on a series of sociodemographic characteristics, the travel type, and the assessment given by the tourist of the destination by means of the designing of a logit model.

The results achieved have allowed us to discover that the gender of the traveler, the travel type, the accommodation chosen, and the level of assessment given by the tourist concerning the conservation of the heritage are variables that may be used to segment the market, as these characteristics influence the probability that a tourist visiting the region will practice tourism in a nature reserve.

The results obtained show similarities to and discrepancies with those found by previous research. For example, the greater preference of women for the practicing of this activity coincides with the results obtained in various studies concentrating on analyzing the profile of the tourist of natural spaces [6,94,95]. However, previous work shows a discrepancy regarding the age variable to segment the market. The authors of [6,95] find in their studies that there is a greater preference in middle-aged tourists for the carrying out of this activity. However, both the study carried out by [94] and the results of our research rule out age as a distinguishable variable of the profile of the tourist of natural spaces. It would be interesting to find out the reasons for this discrepancy, i.e., whether it may be due to characteristics of the destinations selected or if on the contrary age should be rejected as a variable allowing the segmentation of the tourist of protected natural spaces.

In parallel to this conclusion, we detected two variables to which little attention has traditionally been paid on segmenting the market for tourism of natural spaces in the existing literature, which are the travel type and the accommodation used; they have a strong influence on the probability of practicing this type of tourism. To be precise, it is confirmed that this type is strongly associated with family holidays and travel with a partner and that these tourists show a greater preference for staying in non-hotel-type accommodation. These characteristics must be considered by the managers of the destination with a view to designing capture and development strategies in line with the preferences of the tourists.

Finally, based on the knowledge gleaned from previous studies, which propose as a segmentation variable for tourism of natural spaces the seasonal component, the time of the year when the visit is made, high season vs. low season, and the market of origin, whether Spanish or foreign [6,94–96], the aim was to check the suitability of these characteristics for segmenting the market. To do so, a structural change test was carried out to allow for the analyzing of the influence of these factors by means of dividing the samples into two subsamples to subsequently confirm the stability in the parameters of the model proposed.

The results obtained from this analysis confirm that there are no structural changes depending on the tourist season (high or low), which demonstrates that this variable is not suitable for segmenting this market. These results contrast with those of [94], who found in their study on the nature tourism market of Norway that this variable was valid for segmenting the market. This discrepancy may be due to the limitation of the market in each of the studies, as the paper on Norway concentrates on nature tourism in a wide sense, while our research focuses more on the tourism of nature reserves. If this is so, it would be even more necessary to study the tourism of protected spaces as a market niche within nature tourism, as differential characteristics are detected, which may be interesting to consider. For its part, the market of origin of the tourist (Spanish or foreign) did not show any discrimination capacity in the practicing of this activity either.

As was indicated at the beginning of this study, the results obtained by this research help to take a closer look at knowledge of the profile of the tourist of natural spaces and in their turn constitute a valuable tool for the management of the destination analyzed.

To conclude this study, its main limitation is that of the designing of the research carried out, since as it is a case study some of the results obtained could be differentiated characteristics of the destination studied and therefore might not be comparable to the market of tourists of natural spaces. To overcome this limitation and as a future line of research it would be interesting to replicate this methodology in other similar destinations to reach conclusions that allow for the consolidation of knowledge on the niche of tourism of nature reserves.

Author Contributions: Conceptualization, M.S.-R. and M.C.R.-R.; methodology, M.S.-R.; formal analysis, M.S.-R.; investigation, J.d.l.C.S.-D.; writing—original draft preparation, M.S.-R., J.d.l.C.S.-D. and M.C.R.-R.; writing—review and editing, M.S.-R., J.d.l.C.S.-D. and M.C.R.-R.; visualization, J.d.l.C.S.-D.; supervision, M.S.-R.; project administration, M.S.-R.; funding acquisition, M.S.-R. All authors have read and agreed to the published version of the manuscript.

Funding: This paper is part of the research conducted for the project entitled "Analysis of critical factors for the tourism development of Extremadura (IB-18015)". It is funded by Junta de Extremadura (GR-21089) and the European Regional Development Fund (ERDF).

Institutional Review Board Statement: Not applicable.

Informed Consent Statement: Not applicable.

Data Availability Statement: Not applicable.

Conflicts of Interest: The authors declare no conflict of interest. The funders had no role in the design of the study; in the collection, analyses, or interpretation of data; in the writing of the manuscript or in the decision to publish the results.

References

1. Van Cuong, C.; Dart, P.; Hockings, M. Biosphere reserves: Attributes for success. *J. Environ. Manag.* **2017**, *188*, 9–17. [CrossRef]
2. Siikamäki, P.; Kangas, K.; Paasivaara, A.; Schroderus, S. Biodiversity attracts visitors to national parks. *Biodivers. Conserv.* **2015**, *24*, 2521–2534. [CrossRef]
3. Jaraíz-Cabanillas, F.J.; Mora-Aliseda, J.; Jeong, J.S.; Garrido-Velarde, J. Methodological proposal to classify and delineate natural protected areas. Study case: Region of Extremadura, Spain. *Land Use Policy* **2018**, *79*, 310–319. [CrossRef]
4. Perdue, R.R. Target Market Selection and Marketing Strategy: The Colorado Downhill Skiing Industry. *J. Travel Res.* **1996**, *34*, 39–46. [CrossRef]
5. Andriotis, K.; Agiomirgianakis, G.; Mihiotis, A. Tourist Vacation Preferences: The Case of Mass Tourists to Crete. *Tour. Anal.* **2007**, *12*, 51–63. [CrossRef]
6. Marques, C.; Reis, E.; Menezes, J. Profiling the segments of visitors to Portuguese protected areas. *J. Sust. Tour.* **2010**, *18*, 971–996. [CrossRef]
7. United Nations Environmental Programme-World Conservation Monitoring Centre, International Union for Conservation of Nature. Protected Planet: The World Database on Protected Areas (WDPA) and World Database on Other Effective Area-Based Conservation Measures (WD-OECM). Available online: https://www.protectedplanet.net/en (accessed on 26 April 2022).
8. Sánchez-Martín, J.M.; Rengifo Gallego, J.I.; Martín-Delgado, L.M. Tourist Mobility at the Destination Toward Protected Areas: The Case-Study of Extremadura. *Sustainability* **2018**, *10*, 4853. [CrossRef]
9. Siltanen, J.; Petursson, J.G.; Cook, D.; Davidsdottir, B. Diversity in Protected Area Governance and Its Implications for Management: An Institutional Analysis of Selected Parks in Iceland. *Land* **2022**, *11*, 315. [CrossRef]
10. Ostrom, E. Background on the Institutional Analysis and Development Framework. *Policy Stud. J.* **2011**, *39*, 7–27. [CrossRef]
11. Watson, J.E.M.; Dudley, N.; Segan, D.B.; Hockings, M. The performance and potential of protected areas. *Nature* **2014**, *515*, 67–73. [CrossRef]
12. Eagles, P.F.J.; McCool, S.F.; Haynes, C. *Sustainable Tourism in Protected Areas. Guidelines for Planning and Management*; International Union for Conservation of Nature: Gland, Switzerland; Cambridge, UK, 2002.
13. Martín-Delgado, L.M.; Rengifo-Gallego, J.I.; Sánchez-Martín, J.M. Hunting Tourism as a Possible Development Tool in Protected Areas of Extremadura, Spain. *Land* **2020**, *9*, 86. [CrossRef]
14. European Environmental Agency. Protected Areas in Europe an Overview. Publications Office of the European Union, Luxembourg, 2012. Available online: https://www.eea.europa.eu/publications/protected-areas-in-europe-2012/download (accessed on 26 April 2022).
15. Tolón, A.; Lastra, X. Los Espacios Naturales Protegidos. Concepto, evolución y situación actual en España. *Rev. Electr. Medio Amb.* **2008**, *5*, 1–25.
16. Possingham, H.; Wilson, K.A.; Andelman, S.J.; Vynne, C.H. Protected areas: Goals, limitations, and design. In *Principles of Conservation Biology*, 3rd ed.; Groom, M.J., Meffe, C.R., Carrol, G.K., Eds.; Sinauer Associates: Sunderland, MA, USA, 2006.
17. IUCN. Annual Report 2018. Available online: https://www.iucn.org/about/programme-work-and-reporting/annual-reports (accessed on 26 April 2022).
18. EUROPARC-España. Anuario 2020 del Estado de las Áreas Protegidas en España. Available online: https://redeuroparc.org/wp-content/uploads/2022/01/anuario2020finalweb.pdf (accessed on 26 April 2022).
19. Rengifo-Gallego, J.I.; Sánchez-Martín, J.M. La repercusión turística de la declaración de Monfragüe como Parque Nacional. *Cuad. Geogr.* **2019**, *58*, 121–140. [CrossRef]
20. Engen, S.; Fauchald, P.; Hausner, V. Stakeholders' perceptions of protected area management following a nationwide community-based conservation reform. *PLoS ONE* **2019**, *14*, e0215437. [CrossRef] [PubMed]

21. Chan, K.M.; Balvanera, P.; Benessaiah, K.; Chapman, M.; Díaz, S.; Gómez-Baggethun, E.; Turner, N. Why protect nature? Rethinking values and the environment. *Proc. Natl. Acad. Sci. USA* **2016**, *113*, 1462–1465. [CrossRef]

22. Nabout, J.C.; Tessarolo, G.; Baptista Pinheiro, G.E.; Matos Marquez, L.A.; Assis de Carvalho, R. Unraveling the paths of water as aquatic cultural services for the ecotourism in Brazilian Protected Areas. *Glob. Ecol. Conserv.* **2022**, *33*, e01958. [CrossRef]

23. Sica, E.; Sisto, R.; di Santo, N. Are Potential Tourists Willing to Pay More for Improved Accessibility? Preliminary Evidence from the Gargano National Park. *Land* **2022**, *11*, 75. [CrossRef]

24. Liu, J.C.; Sheldon, P.J.; Var, T. Resident perception of the environmental impacts of tourism. *Ann. Tour. Res.* **1987**, *14*, 17–37. [CrossRef]

25. Puczkó, L.; Rátz, T. Tourist and resident perceptions of the physical impacts of tourism at Lake Balaton, Hungary: Issues for sustainable tourism management. *J. Sustain. Tour.* **2000**, *8*, 458–478. [CrossRef]

26. Blackstock, K.L.; White, V.; McCrum, G.; Scott, A.; Hunter, C. Measuring responsibility: An appraisal of a Scottish National Park's sustainable tourism indicators. *J. Sustain. Tour.* **2008**, *16*, 276–297. [CrossRef]

27. Frey, N.; George, R. Responsible tourism management: The missing link between business owners' attitudes and behaviour in the Cape Town tourism industry. *Tour. Manag.* **2010**, *31*, 621–628. [CrossRef]

28. Setola, N.; Marzi, L.; Torricelli, M.C. Accessibility indicator for a trails network in a Nature Park as part of the environmental assessment framework. *Env. Impact. Asses.* **2018**, *69*, 1–15. [CrossRef]

29. Bishop, M.; Ólafsdóttir, R.; Árnason, Þ. Tourism, Recreation and Wilderness: Public Perceptions of Conservation and Access in the Central Highland of Iceland. *Land* **2022**, *11*, 242. [CrossRef]

30. Redpath, S.M.; Bhatia, S.; Young, J. Tilting at wildlife: Reconsidering human–wildlife conflict. *Oryx* **2014**, *49*, 222–225. [CrossRef]

31. Redpath, S.M.; Young, J.; Evely, A.; Adams, W.M.; Sutherland, W.J.; Whitehouse, A.; Amar, A.; Lambert, R.A.; Linnell, J.C.; Watt, A.; et al. Understanding and managing conservation conflicts. *Trends Ecol. Evol.* **2013**, *28*, 100–109. [CrossRef]

32. Dolnicar, S. Market segmentation analysis in tourism: A perspective paper. *Tourism Rev.* **2020**, *75*, 45–48. [CrossRef]

33. Smith, W.R. Product differentiation and market segmentation as alternative marketing strategies. *J. Mark.* **1956**, *21*, 3–8. [CrossRef]

34. Mazanec, J.A. How to detect travel market segments: A clustering approach. *J. Travel Res.* **1984**, *23*, 17–21. [CrossRef]

35. Mazanec, J.A. Classifying tourists into market segments: A neural network approach. *J. Trav. Tour. Mark.* **1992**, *1*, 39–60. [CrossRef]

36. Nessel, K.; Kościółek, S.; Wszendybyl-Skulska, E.; Kopera, S. Benefit segmentation in the tourist accommodation market based on eWOM attribute ratings. *Inf. Technol. Tour.* **2021**, *23*, 265–290. [CrossRef]

37. Rondán -Cataluña, F.J.; Rosa-Díaz, I.M. Segmenting hotel clients by pricing bariables and value for money. *Curr. Issues Tour.* **2014**, *17*, 60–71. [CrossRef]

38. Haley, R.I. Benefit segmentation: A decision-oriented research tool. *J. Mark.* **1968**, *32*, 30–35. [CrossRef]

39. Loker, L.E.; Perdue, R.R. A benefit-based segmentation of a non-resident summer travel market. *J. Travel Res.* **1992**, *31*, 30–35. [CrossRef]

40. Kotler, P.; Turner, R.E. *Marketing Management: Analysis, Planning, and Control*; Prentice-Hall: Englewood Cliffs, NJ, USA, 1993.

41. Frochot, I.; Morrison, A.M. Benefit segmentation: A review of its applications to travel and tourism research. *J. Travel Tour. Mark.* **2001**, *9*, 21–45. [CrossRef]

42. Paker, N.; Vural, C.A. Customer segmentation for marinas: Evaluating marinas as destinations. *Tour. Manag.* **2016**, *56*, 156–171. [CrossRef]

43. Guttentag, D.; Smith, S.; Potwarka, L.; Havitz, M. Why tourists choose airbnb: A motivation-based segmentation study. *J. Travel Res.* **2018**, *57*, 342–359. [CrossRef]

44. Kim, D.; Hong, S.; Park, B.J.; Kim, I. Understanding heterogeneous preferences of hotel choice attributes: Do customer segments matter? *J. Hosp. Tour. Manag.* **2020**, *45*, 330–337. [CrossRef]

45. Chung, K.Y.; Oh, S.Y.; Kim, S.S.; Han, S.Y. Three representative market segmentation methodologies for hotel guest room customers. *Tour. Manag.* **2004**, *25*, 429–441. [CrossRef]

46. Khoo-Lattimore, C.; Prayag, G. The girlfriend getaway market: Segmenting accommodation and service preferences. *Int. J. Hosp. Manag.* **2015**, *45*, 99–108. [CrossRef]

47. Ahani, A.; Nilashi, M.; Ibrahim, O.; Sanzogni, L.; Weaven, S. Market segmentation and travel choice prediction in spa hotels through TripAdvisor's online reviews. *Int. J. Hosp. Manag.* **2019**, *80*, 52–77. [CrossRef]

48. Cordente-Rodríguez, M.; Mondéjar-Jiménez, J.; Villanueva-Álvaro, J. Sustainability of nature: The power of the type of visitors. *Env. Eng. Manag. J.* **2014**, *13*, 2437–2447. Available online: http://www.eemj.icpm.tuiasi.ro/pdfs/vol13/no10/3_659_Cordente-Rodriguez_14.pdf (accessed on 12 June 2022). [CrossRef]

49. Carrascosa-López, C.; Carvache-Franco, M.; Mondéjar-Jiménez, J.; Carvache-Franco, W. Understanding motivations and segmentation in ecotourism destinations. Application to a Natural Park in Spanish Mediterranean area. *Sustainability* **2021**, *13*, 4802. [CrossRef]

50. Buckley, R. Climate change: Tourism destination dynamics. *Tour. Recreat. Res.* **2008**, *33*, 354–355. [CrossRef]

51. Fennell, D. *Ecotourism*; Routledge: London, UK, 2003.

52. Weaver, D.B.; Lawton, L.J. Overnight ecotourist market segmentation in the Gold Coast hinterland of Australia. *J. Travel Res.* **2002**, *40*, 270–280. [CrossRef]

53. Swarbrooke, J.; Horner, S. *Consumer Behaviour in Tourism*; Butterworth-Heinemann: Burlington, VT, USA, 2007.

54. Eagles, P.F.J. International Trends in Park Tourism. In Proceedings of the Europarc Federation Conference, Matrei, Austria, 3–7 October 2001.
55. Palacio, V. Identifying ecotourists in Belize through benefit segmentation: A preliminary analysis. *J. Sust. Tour.* **1997**, *5*, 234–243. [CrossRef]
56. Bricker, K.S.; Kerstetter, D.L. *Ecotourists and Ecotourism: Benefit Segmentation and Experience Evaluation*; Division of Forestry Recreation, Park and Tourism Resources Program, West Virginia University: Margantown, WV, USA, 2002.
57. Kerstetter, D.L.; Hou, J.-S.; Lin, C.-H. Profiling Taiwanese ecotourists using a behavioural approach. *Tour. Manage.* **2004**, *25*, 491–498. [CrossRef]
58. Holden, A.; Sparrowhawk, J. Understanding the motivations of ecotourists: The case of trekkers in Annapurna, Nepal. *Int. J. Tour. Res.* **2002**, *4*, 435–446. [CrossRef]
59. Page, S.J.; Dowling, R.K. Ecotourism: Themes in Tourism. In *Edinburgh Gate: Prentice Hall*; Pearson Education: Essex, UK, 2002.
60. Pearce, P.L.; Lee, U.I. Developing the travel career approach to tourist motivation. *J. Travel Res.* **2005**, *46*, 226–237. [CrossRef]
61. Kruger, M.; Saayman, M. Travel motivation of tourists to Kruger and Tsitsikamma National Parks: A comparative study. *S. Afr. J. Wildl. Res.* **2010**, *40*, 93–102. Available online: https://hdl.handle.net/10520/EJC117327 (accessed on 12 June 2022). [CrossRef]
62. Lee, S.; Lee, S.; Lee, G. Ecotourists' motivation and revisit intention: A case study of restored ecological parks in South Korea. *Asia Pac. J. Tour. Res.* **2014**, *19*, 1327–1344. [CrossRef]
63. Panin, B.; Mbrica, A. *Potentials of Ecotourism as a Rural Development Tool on the Base of Motivation Factors in Serbia*; Institute of Agricultural Economics: Sofia, Bulgaria, 2014; p. 597.
64. López-Sanz, J.M.; Penelas-Leguía, A.; Gutiérrez-Rodríguez, P.; Cuesta-Valiño, P. Sustainable development and consumer behavior in rural tourism—the importance of image and loyalty for host communities. *Sustainability* **2021**, *13*, 4763. [CrossRef]
65. Lois, R.C.; Piñeira, M.J.; Santomil, D. Imagen y oferta de alojamiento en el medio rural de Galicia. *Rev. Galega Econ.* **2009**, *18*, 1–20. Available online: https://www.redalyc.org/pdf/391/39111901004.pdf (accessed on 12 June 2022).
66. Tirado Ballesteros, J.G. La Funcionalidad turística de los espacios rurales: Conceptualización y factores de desarrollo. *Cuad. Geogr.* **2017**, *56*, 312–332. Available online: https://dialnet.unirioja.es/descarga/articulo/6280961.pdf (accessed on 12 June 2022).
67. Devesa, M.; Laguna, M.; Palacios, A. Un modelo estructural sobre la influencia de las motivaciones de ocio en la satisfacción de la visita turística. *Rev. Psicol. Trabajo Organ.* **2008**, *24*, 253–268. Available online: https://scielo.isciii.es/scielo.php?pid=S1576-5962 2008000200007&script=sci_arttext&tlng=en (accessed on 12 June 2022).
68. Leco, F.; Hernández, J.M.; Campón, A.M. Rural tourists and their attitudes and motivations towards the practice of environmental activities such as agrotourism. *J. Environ. Res.* **2013**, *7*, 255–264. [CrossRef]
69. UNESCO. Monfragüe Biosphere Reserve. 2020. Available online: https://en.unesco.org/biosphere/eu-na/monfrague (accessed on 5 April 2022).
70. Ministerio Para la Transición Ecológica y el Reto Demográfico. Memoria de la Red de Parques Nacionales 2019. 2021. Available online: https://www.miteco.gob.es/es/red-parques-nacionales/divulgacion/memoria-red-2019_tcm30-525158.pdf (accessed on 5 April 2022).
71. Natura 2000 Network. Standard Data form for Monfragüe y las Dehesas del Entorno. 2022. Available online: https://natura2000.eea.europa.eu/Natura2000/SDF.aspx?site=ES0000014 (accessed on 5 April 2022).
72. Guillén Peñafiel, R.; Cabanillas Jaraíz, J. Las TIG como recurso didáctico para el estudio de paisajes culturales. Un diseño de intervención en Monfragüe. *Rev. UNES* **2018**, *12*. Available online: https://revistaseug.ugr.es/index.php/revistaunes/article/view/12187 (accessed on 5 April 2022).
73. Natura 2000 Network. Standard Data form for Monfragüe. 2022. Available online: https://natura2000.eea.europa.eu/Natura2000/SDF.aspx?site=ES4320077 (accessed on 5 April 2022).
74. Sánchez-Oro Sánchez, M.; Nieto Masot, A.; Cárdenas Alonso, G.; Prieto Ramos, A.; Gutiérrez Gallardo, J.D.; Ríos Rodríguez, N. *Memoria Turística de Extremadura por Territorios, año 2020. Extremadura Tourism Observatory*; Universidad de Extremadura: Cáceres, Spain, 2021.
75. Peng, C.Y.; So, T.S.H. Logistic regression analysis: A premier. *Understanding Statistics* **2001**, *1*, 31–70. [CrossRef]
76. Cameron, A.; Trivedi, P. *Micro Econometrics: Methods and Applications*; Cambridge University Press: Cambridge, UK, 2005.
77. Wooldridge, J.M. *Introduction Econometrics: A Modern Approach*; Thomson South-Western: Mason, OH, USA, 2006.
78. Morley, C. Technique and theory in tourism analysis. *Tour. Econ.* **2012**, *18*, 1273–1286. [CrossRef]
79. Divisekera, S.; Nguyen, V.K. Determinants of innovation in tourism evidence from Australia. *Tour. Manag.* **2018**, *67*, 157–167. [CrossRef]
80. Nordli, A.J. Information use and working methods as drivers of innovation in tourism companies. *Scand. J. Hosp. Tour.* **2018**, *18*, 199–213. [CrossRef]
81. Li, M.; Fang, L.; Huang, X.; Goh, C. A spatial-temporal analysis of hotels in urban tourism destination. *Intl. J. Hosp. Manag.* **2015**, *45*, 34–43. [CrossRef] [PubMed]
82. Pagliara, F.; La Pietra, A.; Gómez, J.; Vasallo, J.M. igh Speed Rail and the tourism market: Evidence from the Madrid case study. *Transp. Policy* **2015**, *37*, 187–194. [CrossRef]
83. Frisvoll, S.; Forbord, M.; Blekesaune, A. An empirical investigation of tourists' consumption of local food in rural tourism. *Scand. J. Hosp. Tour.* **2016**, *16*, 76–93. [CrossRef]

84. Sabbatini, V.; Manthoulis, G.; Baourakis, G.; Drakos, P.; Angelakis, G.; Zopounidis, C. Tourists behavioral analysis on olive oil consumption: Empirical results. *Int. J. Tour. Policy* **2016**, *6*, 136–146. [CrossRef]

85. Bucur, E.; Danet, A.F.; Lehr, C.B.; Lehr, E.; Nita-Lazar, M. Binary logistic regression—Instrument for assessing museum indoor air impact on exhibits. *JAPCA J. Air Waste Manag. Assoc.* **2017**, *67*, 391–401. [CrossRef]

86. Frangos, C.C.; Karapistolis, D.; Stalidis, G.; Fragkis, C.; Sotiropoulos, I.; Manolopoulos, I. Tourist loyalty is all about prices, culture and the sun: A multinomial logistic regression of tourists visiting Athens. *Procedia-Soc. Behav. Sci.* **2015**, *175*, 32–38. [CrossRef]

87. Chow, G.C. Tests of equality between sets of coefficients in two linear regressions. *Econometrica* **1960**, *28*, 591–605. [CrossRef]

88. Wilks, S.S. The large distribution of the likelihood ratio for testing composite hypotheses. *Ann. Math. Stat.* **1938**, *9*, 60–62. [CrossRef]

89. Arfa, F.; Kaboli, S.; Yazdanfar, S.A.; Mohammadi, H. The effective factors of increasing visits of international tourists to a recognized cultural or natural heritage in UNESCO World Heritage List. *Int. J. Hum. Cult. Stud.* **2016**, *1*, 1353–1363.

90. Anggraeni, G.N. The relationship between numbers of international tourist arrivals and economic growth in the Asean-8: Panel data approach. *J. Dev. Econ.* **2017**, *2*, 40–49. [CrossRef]

91. Rodríguez-Rangel, C.; Sánchez-Rivero, M. La influencia de la presencia en redes sociales sobre el grado de ocupación de los establecimientos turísticos. *Rev. Anál. Tur.* **2016**, *21*, 1–10.

92. Holik, A. Relationship of economic growth with tourism sector. *J. Bus. Policy* **2016**, *9*, 16–33. [CrossRef]

93. Gunter, U.; Smeral, E. European outbound tourism in times of economic stagnation. *Int. J. Tour. Res.* **2016**, *19*, 269–277. [CrossRef]

94. Tkaczynski, A.; Rundle-Thiele, S.R.; Prebensen, N.K. Segmenting potential nature-based tourists based on temporal factors: The case of Norway. *J. Trav. Res.* **2015**, *54*, 251–265. [CrossRef]

95. Morais, J.; Castanho, R.A.; Pinto-Gomes, C.; Santos, P. Characteristics of Iona National Park's visitors: Planning for ecotourism and sustainable development in Angola. *Cogent Soc. Sci.* **2018**, *4*, 1490235. [CrossRef]

96. Taczanowska, K.; González, L.M.; García-Massó, X.; Zięba, A.; Brandenburg, C.; Muhar, A.; Pellicer-Chenoll, M.; Toca-Herrera, J.L. Nature-based tourism or mass tourism in nature? Segmentation of mountain protected area visitors using self-organizing maps (SOM). *Sustainability* **2019**, *11*, 1314. [CrossRef]

MDPI
St. Alban-Anlage 66
4052 Basel
Switzerland
Tel. +41 61 683 77 34
Fax +41 61 302 89 18
www.mdpi.com

Land Editorial Office
E-mail: land@mdpi.com
www.mdpi.com/journal/land